# LIFE AND LOSS AMONG THE REEDS

## A Tale of Two Pied-Billed Grebe Families

***By Andrew Abate***

***Celestial Armadillo Press***
Planet Earth

***Dedicated to the West Lagoon and South River Pied-billed grebe families whose lives I had the privilege of sharing during this crucial life phase.***

***…And to all of the people (past, present and future) who have joined the fight to save our magnificent planet and its wondrous diversity of life.***

*Celestial Armadillo Press*
Planet Earth



**Life and Loss Among the Reeds: A Tale of Two Pied-Billed Grebe Families**

**ISBN 978-0-578-62954-4**

# Table of Contents

# Preface

Most people who live in urban and suburban areas are so wrapped up in their human-centered world that they are scarcely cognizant of the other (nondomestic) life around them. When they do pay attention to or show interest in other life forms it seems this is more likely to be in situations in which the life form is being targeted for doing something that people consider to be offensive to their human-centered existence or otherwise upsets their human sensibilities. This oblivious attitude toward other life forms is one of the things that helps foster the general disconnect that exists between people in "modern" societies and the natural world—and whether or not people want to acknowledge it, despite all of the human contrivances, ultimately it is still the natural world that maintains our existence.

In respect to wildlife, urban and suburban areas are woefully impoverished. However, they are not devoid of it. While many species are driven out, excluded or locally extirpated by the destruction of their requisite habitats and subsequent replacement by artificial human ones, there are those that are able to make a living within these. Some (such as common pigeons, house sparrows and finches, European starlings, great-tailed grackles, Brewer's blackbirds, mourning doves, mallards, various seagull species, raccoons, and various squirrel and other rodent species to name a few) have even been able to thrive in these human created environs. In some areas of habitat encroachment by cities and towns some more spectacular species such as black bears and mountain lions may be seen ranging in streets and neighborhoods. Even the largest cities may harbor some surprises such as peregrine falcons in New York City and the regular occurrence of coyotes in Los Angeles. There are even some endangered species that are found within or along the edge of urban and suburban areas (for example, the San Joaquin kit foxes of Bakersfield, California, and the sea otters that inhabit the coastal waters off the heavily urbanized/suburbanized Monterey peninsula). This general situation, of course, exists not just in the United States but globally. Included are some urban and suburban areas that are known for harboring some rather unusual species for town or city settings such as the African penguins of Simon's Town, South Africa and the large population of grey-headed flying foxes in Melbourne, Australia.

The point here is that regardless of whether the species are notable or common, people living in cities and towns are still surrounded by other (nonhuman) life. If people would just bother to look around and take note they will see it in the streets, in their yards, in the trees in the neighborhood, on light posts and power lines, in vacant lots, in and around bodies of water (i.e., ponds, streams, rivers, coastlines, etc. that are in the city or town), on buildings, flying overhead, etc., etc. People don't even have to leave their homes to have some interesting wildlife viewing. For example, when I was a graduate student living in a second floor apartment in Tempe, Arizona I discovered an Anna's hummingbird nest in a tree that stood right outside my bedroom window. Thus, merely by looking out of my bedroom window every day I was able to witness the entire incubation and chick rearing periods for this female hummer (males of this species don't assist with incubation or chick rearing; as is the norm for the species two chicks were produced and both successfully fledged).

Cities and towns typically do make some acknowledgement of the human connection to the natural world and attempt to cater to this through the creation of parks (which may contain ponds or lakes) and/or small nature preserves situated within, along the outskirts of or just a short distance outside of the town (the "naturalness" and wildness of these local preserves varies greatly from place to place). In fact, it is in one such local preserve located just outside the edge of Ventura, California where the studies that are the subject of this book were conducted.

So, why the concern about city and town people being so human-centric and paying such little heed to the other organisms in their world? Quite simply, this concern is due to the current, ongoing human-caused mass extinction crisis that our species has thrust upon the planet. We are perpetrating a loss of species so devastating that, in scale, it rivals the impact of an immense comet or asteroid. Ultimately, the root of this problem lies in the pervasive, arrogant, selfish, childishly self-centered human attitude of primacy, superiority, self-importance, dominion and privileged entitlement. This attitude is borne out of a self-serving, self-aggrandizing, fallacious view and belief in the utter uniqueness of humanity; a belief that humans are a class of life separate from all other life, that humans are inherently better than and inherently unique from all other life in the universe; a belief that humans are the masters of this world, that the Earth and all of its organisms are here specifically for the benefit of humanity to use and exploit as we see fit. To foster this view humans of various religions have fabricated a mythical supreme being that conveniently created only humanity in its image and furnished the Earth and all of its organisms for its chosen special beings to utilize and reign over.

This attitude of inherent uniqueness along with its assorted baggage has remained entrenched among the majority of people. This mind-set persists despite the myth of inherent human uniqueness having been unequivocally debunked by the work done in various scientific disciplines (notably biology). It is a biological fact that our species is not inherently unique, that we are a product of the same natural processes through which every other species on this planet was produced, that we have an evolutionary connection to all of these other species. We do not have primacy here. As with every other extant species (i.e., all life forms on the planet including the flora, fauna and all other types of organisms) we are the current product of an estimated 3.8 billion years of earthly evolution. We have no special, endowed entitlement of dominion. We don't own this planet; instead we share it with millions of other species that have the same length evolutionary lineages as we do. [Estimating the total number of species present on Earth has proven to be a challenging endeavor and estimates have varied wildly. For example, some more recent estimates range from 8.7 million (this estimate doesn't include prokaryote organisms such as bacteria) to 2 billion to one trillion. At this point the consensus is that the total number is probably somewhere in the millions.]

One of the most important (if not the most important) aspects in attempting to stem and limit the damage of this human-caused mass extinction crisis is to change this attitude of ownership, dominion, and inherent superiority and uniqueness. It is a characteristic of human nature that people are more likely to take positive actions on behalf of things that they develop an appreciation for and genuinely care about. And they are more likely to care about and appreciate things that are a conspicuous part of their daily lives and experiences. Therein lays the crux of the matter as it relates to this issue. People who live in urban and suburban areas are distanced from the natural world. They are ensconced in an artificial, human created environment that contains only a sparse diversity of wildlife. Their daily lives are centered in the human world and not the natural world; they may not even take any real notice of the sparse wildlife that does exist around them. This is not the kind of situation that is conducive to causing people to take notice, care about and appreciate the natural world and its varied life. It's not a favorable environment for promoting the desperately needed change in attitude that hopefully could enlighten people and compel them to finally undertake serious measures toward halting this human-caused mass extinction crisis.

Yet, this is the situation in which the majority of humans live. A 2014 estimate by the United Nations noted that 54% of the global population was urban. It is projected that this trend toward urbanization will continue and that by 2030 two thirds of the global population will be urban. If things continue as at present then this will only serve to widen the general disconnect between humanity and the natural world. Obviously, much more effective means must be found to reconnect people to the natural world and its wondrous diversity of life; means that are capable of breaking through that pompous façade of primacy, inherent uniqueness, superiority and dominion and hammering in the reality that it is not their artificial human-created world that is maintaining their existence but the natural processes and environment of the natural world.

Is it even possible that anything can accomplish this attitude change in a large enough majority of people? Or that this would actually spur them into taking the required extensive actions within the critically short window of time that this must be accomplished? Judging by humanity's past track record the odds of this are rather poor. The usual pattern of humans in dealing with problems of this nature that they have created is that of not taking any serious, substantial remedial actions unless or until their backs are firmly against the wall. In other words, not unless or until the problem reaches the point where they are forced to do so because they themselves are being so significantly impacted by it that they can no longer refrain from taking action (the usual reasons for not taking actions in the first place are for the purpose of continued exploitation, continued reaping of profit or various other human interests; and of course, let's not forget simple apathy).

This pattern is already evident in regard to the human-caused mass extinction crisis. It is also already evident in another current dire environmental problem that our species has created: human-caused climate change (and of course, human-caused climate change is one of the numerous human-related agents that are driving the mass extinction crisis). As usual, despite forewarning our species has failed to act in time to stop it and its effects are already being manifested, notably in aspects such as higher annual temperatures, rising sea level, severe drought and flooding, alterations of important ocean and atmospheric currents, a tendency for storms to be more severe, and ocean acidification. We are thus into the period where the best that we can do is to try to limit the damage—and this, knowing that even if the required remedial actions were implemented today there would still be a lag period before these took effect.

Despite knowledge of this dire reality, in international meetings of nations to specifically address this problem humanity still refuses to take any more than what amount to token actions that fall well short of what is required. Furthermore, the nation (the United States) that has historically contributed more to the creation of human-caused climate change than any other has withdrawn from cooperating even in this less than stellar global effort. And why? For the convoluted priority of placing economic considerations and interests (particularly those of the fossil fuel industry) over the continued viability of the planet that sustains us!

Regardless of the poor odds we must continue the fight to stem this human-caused mass extinction crisis, at the very least in an attempt to save as many species as possible from this catastrophe. Conscientious, committed environmentalists must continually face up to and deal with the fact that they do not comprise the majority or most powerful and influential segment of humanity. If they did then we would not now be in the midst of a human-caused mass extinction crisis or still facing a steep uphill climb to put an end to human-caused climate change. Staunch environmentalists are greatly outnumbered and are up against an opposition (which notably includes large corporations and other economic interests and the politicians and other influential parties that they purchase) that has far superior combined wealth, power and resources. Perhaps people can best picture the situation when placed in the crass metaphorical terms of that favorite human pastime: war. Because environmentalists are up against a vastly superior foe that they can't defeat outright they are being forced to engage in guerilla warfare. In their resistance they make focused strikes here and there, attempt to gain allies and are even able to win some battles that are fought on their "turf." They must hope that with continued pressure and resistance they can somehow turn the tide; that perhaps through shear tenacity they can outwait their foe as changes over time ultimately shift the majority outlook on the situation in their favor. Unfortunately, time is not on our side. The mass extinction crisis is raging on as we speak. According to some estimates species are being lost at an average rate of 150-200 per day! For many species it is already too late to help them.

To be honest, the outlook appears grim. This is particularly so considering that the main factor underlying this crisis is the huge and still growing human population. If for example there were only 500 million people on the planet instead of the 7.6 billion and counting the problems would be more localized and humans would be taking up much less space and using much less of the Earth's limited resources. At our current population level people are literally crowding other species out of existence, using excessive amounts of the planet's resources and increasing their negative impact on the environment. The continuing population growth will only serve to exacerbate a problem that is already overwhelming and has brought about this crisis. Frankly, I don't see how this mass extinction crisis can possibly be halted under these conditions of severe human overpopulation before the losses reach staggering numbers. The many wars, famines, epidemics, etc. haven't made a dent in the continued overall human population growth and it is utter folly to even imagine that people around the world will voluntarily check their reproduction/population numbers for the sake of the other life on the planet.

Once again, despite the grim outlook environmentalists mustn't concede anything. They must not allow themselves to become so disheartened that they abandon the fight. They should refrain from viewing this matter from the gross, black and white perspective of winning or losing. Even if the "war" is ultimately lost there are still important battles that can be won, species and vital habitats that can be saved. To ward off feelings of futility and helplessness environmentalists should wrap their minds around the following thought: the situation may indeed end up bad but imagine how much worse it will be if the faction of humanity that is callously, greedily and thoughtlessly driving this crisis is allowed to do so unopposed.

As an example they need look no further than the situation regarding the Arctic National Wildlife Refuge. This large (19.3 million acres), magnificent refuge in northern Alaska is considered the crown jewel of the National Wildlife Refuge System. It comprises an immense tract of fragile arctic wilderness that is of importance to many species (for example, it is seasonally used for feeding and nesting by migratory bird species from six continents) including iconic animals such as the muskox, polar bear, grizzly, caribou and wolf. For about the past four decades the oil and gas industry and their supporters have made one attempt after another to get the refuge opened up for drilling. Yet, environmentalists and their supporters were able to thwart each of these attempts. Unfortunately, in 2017 the corporate and fossil fuel industry backed Republican controlled Congress passed their new tax bill to which they attached a provision that opened up a 1.5 million acre section of the refuge to oil and gas drilling. However, this probably isn't the final word on this issue. It's estimated that any oil production in the refuge probably couldn't commence until about a decade from this point. In the meantime environmental groups have vowed to fight against this. In addition, congressional Democrats (clearly the lesser of the evils of the two major political parties in regard to environmental issues but also corporate owned and beholden) have stated that if they win back control they will work on legislatively reversing the drilling provision and seek to bestow permanent protection to the refuge. At this point it is uncertain as to whether efforts to keep drilling out of the refuge will succeed. However, what is certain is that without the efforts by environmentalists and their supporters the refuge would already have been lost to drilling; also there is no doubt that without their continued efforts there will be no chance of keeping drilling in the refuge from being a certainty.

If environmentalists are feeling overwhelmed by the great disparity between the high average daily losses versus the comparatively meager compliment of what they have been able to protect and save they need to remind themselves of realities such as the following: the great majority of the listed threatened and endangered species in the U. S. gained their listing through pressure, petitions or legislative actions undertaken by environmentalists (advocacy groups and individuals) rather than by actions initiated by the two federal agencies that are tasked with listing imperiled species (the U. S. Fish and Wildlife Service and the National Marine Fisheries Service). One way or another this mass extinction crisis will ultimately come to an end. Regardless of how severe the outcome, more species and habitat will be saved with as opposed to without the efforts of environmentalists.

I am under no illusions that every person can be made to care about and appreciate the other life on our planet to the point where they will want to protect and preserve it. Certainly there is scant to zero chance of this regarding bottom-line people who place profit and the almighty dollar far above any concerns for the planet and its other life (and in many cases, even above human life). It is a sad reality that the ranks of the leaders, directors and owners of powerful, influential entities that have the potential to do great harm to life and the environment are rife with such greedy, uncaring, shortsighted people. Nevertheless, there is the need to recruit as many people as possible to the cause. There are various strategies and ideas about how to reach people in a manner that can make them care about and appreciate nature and the many organisms that share the Earth with us. One such approach is based on the following premise. It is a basic character of human nature that people are more likely to care about and be sympathetic to a particular cause when they have a strong emotional connection to it. For whatever reason, people tend to be more susceptible to forming an emotional connection to a cause when it is related to them on a personalized, individual level.

The two case studies presented here do just that. They do more than just present a highly detailed description of the chick rearing process within these two grebe families. Because of the focused, daily observations on the individuals of these families the studies also provide a private narrative of their lives during the respective study periods (this is especially so for the South River family). Readers can come to know each bird, to see and appreciate each as an individual life. In this knowing they may develop affection for each individual, come to care about them and, as they read through the chronologically arranged studies, feel an emotionally tinged hope that each survives and succeeds at its task of living. In these studies readers will find "noble" values and attributes that people like to claim as being exclusively human such as dedication and devotion to offspring and mate, self-sacrifice, heroism and even love (all things that people, in their arrogance and attempts to elevate themselves, like to diminish and categorize as merely "instinctive" when exhibited by any being other than a human).

I can only hope that these "personalized" studies can impart this sense of the individuality of each organism and its intrinsic value as such; that readers can extrapolate this to the many other organisms that inhabit the Earth. Perhaps this can help them see that the world is full of beings that are basically doing the same as us: just trying to have success with their offspring and get on from day to day. I hope that this intimate look into the lives of these grebes will help people realize that there is much fascinating life and drama going on around them besides that of their human-centered world. I hope that this will compel them to spend some time observing and acknowledging this other life—and that this will help them to finally realize that we have no greater claim to this planet nor a greater right to exist than they do.

# *Chapter 1*

## Introduction to the West Lagoon Family Study

For animals whose maturation process includes being parented from the time of birth (or hatching) until reaching independence, this is the most crucial and influential period of their lives. This is the time when the individual is at its most vulnerable, the period during which its very existence is dependent on the watchfulness and care of others. It is during this period that the individual must develop its ability to survive on its own. This is the formative stage of the individual's life that has an influence on all that follows. To state the primary importance of this period at its most basic level, if the individual fails to survive this initial stage of its life then, quite obviously, no other stages will follow. As a mammal who has personally gone through this stage I have acquired firsthand experience of just how vital it is.

Parenting occurs in a wide variety of species. Among this array of species parenting is quite variable in the manner in which it is conducted as well as in the various aspects and duties that it entails. For example, depending on the particular species the parenting may involve both parents (including cases in which they are assisted by other individuals) or just the mother or father (in addition, the parent-offspring relationship and the interactions between them also vary greatly among this wide array of species). As another basic example consider that, depending on the species, the parenting duties may or may not include feeding the offspring. Because of the uniquely mammalian trait of nursing offspring with milk produced by the mother, the young of all mammal species receive at least this type of feeding from the mother (and therefore, the young of all mammal species receive at least this minimal amount of parenting). However, the situation is quite different among the many bird species. In species such as the piping plover the precocial young are capable of feeding themselves from the time of hatching; thus, though the parents watch over and protect their chicks during the parenting period they don't feed them. In contrast, nest-bound altricial young such as barn swallow chicks must receive all of their food from the parents until they fledge and begin catching food on their own. Furthermore, the specific manner in which parents feed their young is also extremely variable among the many species that do so. Aside from the above, the duration of the parenting period is quite variable, ranging from periods of weeks to as long as 18 years or so as occurs with humans.

Among the young of the many species who receive parenting during their developmental period, there is also extreme variability in regard to their role in this endeavor and in what they must accomplish during this crucial stage. In the simplest of cases, this is mainly just a matter of obtaining sufficient food and surviving through this period of physical development and greater vulnerability. Other cases involve various additional levels of complexity such as: the development of particular skills, abilities and endurance; competition and/or cooperation among sibs; acquiring/accumulating certain types of experience; and various degrees of learning (this may include the learning of physical and/or social skills and obtaining necessary knowledge about the environment).

This subject has fascinated me since my childhood days when I first became interested in wildlife. What was the journey to adulthood like for the many adult animals that I see in the wild? Unfortunately, during my work as a biologist there have only been a few occasions (involving a few different species) when I had the time and opportunity to observe and study this vital life stage in the detailed manner that I prefer (this preferred method involved conducting a descriptive behavioral case study of the individual young of the litter/brood and their parent(s) as they progressed through the maturation/rearing process). When on 9/6/16 I discovered another such opportunity I couldn't resist seizing it. On this particular day I was hiking along the estuarine area at the terminal end of the Ventura River (in Ventura, California) when I happened to sight a pair of pied-billed grebes with six young chicks in one of the lagoons there. At this point in my life I had recently concluded a four year stint working on a wildlife project for an endangered species but hadn't yet found a new project to work on. Because I now had the available time, the thought quickly struck me that here was an opportunity to conduct a case study on this grebe family, a chance to observe it all the way through to the point when the chicks became independent. This then began a study period of 42 consecutive days during which I observed this grebe family. The study presented here is but a mere taste of the great variety of forms that parenting takes among the myriad species on our planet.

As a brief, general introduction to the species, pied-billed grebes are small, thick-bodied grebes. They possess a thick, stout beak. The plumage of the sexes is similar. The pied-billed grebe is basically an aquatic species. These birds are seldom seen in flight and spend essentially all of their time on the water. They prefer quiet waters such as lakes, ponds, marshes, sluggish streams and estuarine lagoons that are ringed with tall, dense, emergent growths of reeds. Their diet consists of fish, crayfish and other aquatic vertebrates and invertebrates which they dive to obtain. In fact, they are very skilled divers and are even capable of controlling their level of buoyancy at the surface (for example, an individual can remain mostly submerged with only its head above the surface). In regard to reproduction, adults form male-female pair bonds and this is the basic breeding unit. Both parents share in the tasks of nest building, incubation and chick rearing. This species has an extremely large range, the largest of any North American grebe species. They are distributed coast to coast through much of North and Central America. Within this wide range resident and migratory populations can be found.

There are several unusual or uncommon aspects associated with this study. As is obvious from the above, it was spontaneously undertaken. Therefore, there was no chance for any pre-study planning and preparations in regard to methodology or other aspects. Fortunately, the effects of this situation were fairly minor. I already knew the general method with which I preferred to conduct studies on this particular subject. Furthermore, my preferred method of conducting a completely naturalistic, descriptive behavioral study allowed me to be able to step right into the study despite not having had the chance for any pre-study preparations. Nor was it of any consequence that I had no time to review the literature on pied-billed grebe breeding and chick rearing. This is because I prefer to conduct observations of a species that I am studying for the first time (as well as follow-up studies as would be the case in 2017) with "fresh" eyes in order to avoid any potential biasing to my observations. Thus, I wouldn't have reviewed that literature anyway. (Note that prior to this study I wasn't very familiar with pied-billed grebes. My scant total knowledge of the species came from just a handful of brief, casual observations and a few basic snippets from field guides.)

On a more basic level it should be noted that detailed case studies of individual animals in the wild are fairly uncommon. This particular case study may be the only one of its kind for this species.

Another unusual aspect about this study is that I initially undertook it simply as a personal project for my own interest and curiosity. Because I was doing it merely as a "private" study for my own interest I didn't bother with some of the conventions and points of information that one always records in one's field notes when conducting a formal study. Most notably, I didn't bother to record the exact times during which I conducted my observation periods. Overall, I recorded my notes as a kind of cross between a wildlife journal and a behavioral monograph. Despite its "private" nature, out of professional habit I ended up with a very detailed study of the chick rearing process for this family which was filled with highly detailed accounts of specific behaviors and events. I realized that the study should be shared because it contained too much information to just keep it to myself; that it contained much that would be of interest to biologists (especially those who are studying this species), birders and members of the general public who are interested in wildlife (and considering the fairly recently acquired popularity of "reality" stories and shows, it is noteworthy that besides the information on the chick rearing process this study also provides a personalized story of the lives of the individual family members during this period).

This study was conducted during the southern California dry season in the period from 9/6/16 to 10/17/16. The aim of the study was to chronicle and describe in detail the chick rearing process as specifically observed in this pied-billed grebe family (in other words, it was a case study). As noted above, this was a completely naturalistic study—it was purely observational. There were no manipulations of the study subjects whatsoever and the individuals in the family were identified solely by their natural features (i.e., no banding or other artificial means of individual identification were employed). The only equipment utilized for this study was a pair of 12 x 42 binoculars, a handheld video camera to visually document some of what was observed and of course, field notebooks. This was strictly a descriptive behavioral study; no specific quantitative data were collected. Observations were conducted in an open, continuous, ad-lib manner with a general focus on aspects associated with the chick rearing process and the development of the chicks. For the reason noted above, I didn't record the times during which each of the daily observation periods were conducted. However, overall the durations of the daily observation periods ranged from approximately 1.5 to 3.5 hours (with some being a little shorter or longer) and they were conducted between the hours of 10:30 to 16:00. During observations an attempt was made to keep track of all members of the family simultaneously. This, of course, was impossible given how often members of the family were split among the parents or otherwise widely dispersed, or various individuals were out of sight within the reeds or in other areas where they couldn't be seen. Daily observations were concluded on 10/17/16. Thereafter, biweekly follow-up checks were conducted until the end of the year.

The study site was located in the estuarine area at the terminal end of the Ventura River in Ventura, California. It is described in detail in the field notes below. This area is owned by the California State Park System though jurisdictional aspects are somewhat complex and involve various other agencies also. Management, protection and restoration of this estuary area is shared, notably by California State Parks and Ventura Land Trust, a nonprofit, community-based environmental group headquartered in Ventura (in fact, during my many visits to this estuary there were a number of occasions in which I came upon Ventura Land Trust crews that were engaged in clean-up and invasive species removal work).

The study is presented here just as it was written. The chick rearing process and story thus can be followed in chronological order as it unfolded. The length of the daily observation notes varies considerably—some days were relatively routine and fairly uneventful while others offered much to be recorded.

A photo section for each family is included following their respective field notes. The photos were taken (all by me) in order to provide photographic documentation of the studies. It should be noted here that I had only a poor quality camera available to me and that my photographic expertise is lacking. These deficiencies are reflected in the aesthetic quality of the photos. However, the utility of the photos lies in their content (rather than in their aesthetic value) and they should be appreciated from this perspective. They illustrate much of what is discussed in the field notes and also allow the reader to "meet" the individual family members whose lives are being documented in these studies.

# Chapter 2

## Observations and Notes on a Pied-Billed Grebe (*Podilymbus podiceps*) Family (9/6/16-10/17/16)

## [The West Lagoon Family Study]

*[Overall, daily observation periods ranged in length from approximately 1.5 hours to 3.5 hours and were conducted between the hours of 10:30 and 16:00.]*

**9/6/16:** Discovered a pair of pied-billed grebes with six young chicks. The family is situated in a lagoon located at the terminal end of the Ventura River (in Ventura, California). At this terminus the river (which flows south) forms two side by side marshy lagoons that lie at the upper end of a wide, sandy beach [*see the diagram of the study area*]. The lagoons are separated by a fairly moderate expanse of marsh through which at least one narrow channel connects the two lagoons (aside from the latter, the two lagoons are fed by the river which passes at their north ends). The river only periodically (and for only a short period on such occasions) cuts all the way across the beach into the ocean. When it does so, it typically flows from the east side of the easternmost of the two lagoons, though a narrow flow channel does occasionally break through from the west lagoon. [Note: Periodically the river flows very strongly. When it does, it cuts right through the west lagoon straight into the ocean. At such times this area is no longer a lazy lagoon but instead is transformed into a river proper (note also that with the nearby wide connection with the ocean the water in this section becomes brackish). With the river flowing strongly straight through the former west lagoon, the east lagoon then receives less water and becomes considerably smaller. When they do happen, such periods of strong flow typically occur during the rainy season, not the dry season when the grebes nest.] The grebe family inhabits the westernmost of these lagoons. This marshy lagoon is ringed all around (except on its south end where it is bordered by the sandy beach) by dense growths of tall, thick-stemmed reeds (bulrushes and cattails). Other than the latter, the interior area of the lagoon is just open water with no emergent plant growth. A railroad bridge runs across (over) the northern end of the lagoon *[this long bridge spans the northern ends of both lagoons and continues east past the East Lagoon for a roughly similar distance]*. I lack any specific information regarding the water depth in the lagoon. However, as a general qualitative assessment it doesn't appear that the water is particularly deep in any section of the lagoon—I would guess that there are no areas in the lagoon where the bottom isn't well within the range of a diving adult grebe. *[Postscript: The above is based* *on the climate conditions and habitat features of the study site that I had observed since arriving in the area in April, 2012. During this period the region was experiencing severe drought conditions, including poor, sub-par rainy seasons. However, the rainy season that followed this study was impacted by an especially strong El Niño event. This brought a temporary, seasonal break to the drought, causing record rainfall in much of the state. As will be noted in the South River family study and some subsequent chapters the local heavy rains also affected a significant change to the west lagoon habitat.]*

The chicks are quite young and small but they are beyond the developmental stage in which young chicks climb onto the backs of their parents (though they possess the plumage typical of younger chicks). There is already very noticeable size disparity among the six chicks. Two of the chicks (and one of these in particular) are significantly larger than the others. In fact, when first sighted it appeared that they probably belonged to a different brood. This was due not only to their noticeably larger size but also because at that initial sighting they were all the way across the lagoon from where the other four chicks were (they were on the west side, initially by themselves, and the other 4 chicks were on the east side with one of the parents). However, later in the observation it became clear that they were part of the same brood when they joined with the other four chicks and the parents. Besides these two larger chicks there are two who are noticeably smaller than the rest and one of these is particularly runty. I am not familiar with this species' nesting habits but the considerable size disparity among this young brood suggests that hatching was staggered in this clutch (that is, sufficiently staggered to produce such disparity, probably on the order of at least several days).

**9/7/16:** All six of the chicks seen. Several feedings were observed and both of the parents were seen feeding the chicks. In each instance the food item that was captured and fed to the chicks was a crayfish. The parents hunted (dived and searched underwater) for the crayfish along the edges of and within the dense emergent growths of tall reeds along both the southeast and southwest sides of the lagoon. Chicks accompanied one or the other parent to the area where the parent was foraging (sometimes both of the parents foraged in the same area) and watched and waited for the parent to surface from a dive with a prey item. When a parent engaged in a foraging dive the accompanying chicks milled about in the area where the parent had dove or moved about this area in an apparent attempt to anticipate where the parent would resurface and thus perhaps gain an advantage over its sibs by being closer to the parent should it resurface with a prey item (when underwater on a foraging dive the parent typically moved at least a short distance away from its original dive point before resurfacing—i.e., it typically did not resurface in the same spot where it had dove). During this observation period the foraging parent(s) was accompanied by complements of from 4 to all 6 of the chicks. It was observed that when a parent surfaced with a crayfish that it had captured, it moved toward the chicks while holding the crayfish in its beak. Upon sighting the parent the chicks then rushed at speed to it, eager to obtain food. In the one observed case in which the crayfish was small and of a size that the chicks could easily swallow, the parent allowed the crayfish to straightaway be taken out of its beak by the chick that had out-competed its sibs for access to grab this prize (in this instance it was one of the two larger chicks who took and ate the crayfish). After grabbing the crayfish from its parent the chick quickly positioned the crayfish in its beak and swallowed it tail-first (with the tail partly doubled over on itself), opening its beak extremely wide while raising it straight up and vigorously working the crayfish down its throat, jerking its head with several consecutive short up and back thrusts as it apparently used a gravity assist to aid in working the crayfish down its throat.

In the other observed feedings the crayfish was too large to be readily swallowed by a chick. In these instances the parent did not immediately relinquish control of the crayfish to the chick(s). Instead (while being excitedly mobbed by the intensely competitive chicks who were maneuvering around the parent in an attempt to gain access to the crayfish), it maintained its control of the crayfish until it had worked it a bit and rendered it more accessible for the chicks to consume. It did this by making breaks in the exoskeleton which then made the overall crayfish "package" more pliable and opened up points of access to the meat inside it. It accomplished the latter using its stout, powerful beak, by strongly clamping down on the crayfish and breaking through the exoskeleton, then giving it a quick, vigorous shake and dropping or essentially tossing the crayfish onto the water in front of itself. The parent then picked up the crayfish again (sometimes waiting several moments before doing so as though perhaps allowing water to seep into and widen the break that had been made through the exoskeleton) and repeated this action as many times as necessary. The parent also tore off tidbits of meat from the crayfish and handed these to the chicks and this served to reduce the bulk of the carcass. [Note: The grebes generally swallow their prey whole but with its exoskeleton still intact a large crayfish is too bulky and rigid to swallow, even for an adult. Creating breaks in the exoskeleton makes the crayfish's body more pliable/re-formable and thus facilitates swallowing. In the case of feeding chicks this also creates the opportunity for a chick to peck through the openings in the exoskeleton. Depending on the situation a chick can then do one of the following: if in control of the crayfish it can pluck out bits of flesh to aid in progressively reducing the bulk of the crayfish so that it can eventually swallow it; or if the crayfish has not yet been controlled, a chick can attempt to quickly grab a tidbit of meat while competing with its sibs. The crushing bites also apparently kill or severely disable the crayfish.] When the parent had apparently worked the crayfish sufficiently in this manner it then allowed the swarming chicks to compete with themselves over access and control of it (the parent typically handing the crayfish off to the chick that had maneuvered itself into the most advantageous position for receiving it). The parent then followed and remained close to the competing chicks, overseeing the intense proceedings and retrieving the crayfish on the occasions when it was dropped and sank too far below the surface for the chicks to reach. The parent did this until one of the chicks took decisive control of the crayfish and broke from the mass. Typically the chick with the crayfish was hotly pursued by several of its sibs and fled from them, generally at a good pace, and either worked on swallowing it while fleeing its competing sibs or attempted to lose them so that it could more securely work on swallowing the crayfish (in a few cases the chick with the crayfish had moved a considerable distance before it finally swallowed its prize).

In the observed feeding which involved the largest of the crayfish it took a noticeably longer time before one of the chicks took sole, decisive control of the crayfish and eventually swallowed it. Prior to this the competing chicks (all 6 were present) intensely swarmed around the body (which was often floating on the surface during this period while the parent worked on it) and often pecked at it—it appeared that a number of the chicks were able to pluck off a tidbit from the crayfish. The largest chick finally seized control of the crayfish and swallowed it.

In two instances a parent emerged from the reeds with a large crayfish, was quickly swarmed upon by chicks but after working the crayfish the parent kept and swallowed it. These cases were obvious reminders that besides providing for the chicks the parents also need to feed themselves.

In each of the feedings that were observed today it was one or the other of the two large chicks that took control of and swallowed the crayfish that was offered by the parent. If, as this suggests, these two have dominated the feedings up to this point then it is no wonder why they are so noticeably larger than the other 4 chicks. These two chicks also appear farther along in another respect: they were the only chicks that were observed to engage in foraging dives of their own (though neither was seen to have captured a prey item; or perhaps at this stage they weren't actually foraging but were instead just working on their diving skills).

**<u>9/8/16</u>:** All six of the chicks seen. Some feedings were observed and these were conducted basically as described in the 9/7 notes.

It has already become quite obvious how vital the dense reedy habitat that rings the lagoon is for the grebes (especially the emergent portions). These tall, dense growths of reeds provide cover and concealment for the grebes and it is within and along the edges (i.e., of the emergent portions) that the chicks and parents take their rest periods. In addition, it is within this emergent reedy habitat that the parents do their foraging. In fact, up to this point in my observations of this grebe family the parents have done their foraging exclusively along the edges of and within this habitat (ditto for the two large chicks on the occasions when they have been observed engaging in foraging dives).

**<u>9/9/16</u>:** All six of the chicks seen. Several feedings were observed and these were conducted in the same general manner as has been previously observed (once again the prey item in each case was a crayfish). Also, there was another observation of a parent retaining a prey item for itself. As in the two previous such observations the parent emerged from the reeds with a large crayfish in its beak and moved toward a waiting group of chicks. The chicks then swarmed around the parent, jockeying for position among themselves for a shot at obtaining the food item. However, after working the crayfish the parent retained and swallowed it.

One of the medium-sized chicks was seen briefly flapping its wings. At this time its wings are small and poorly developed: they are covered in downy feathers and I could not as yet see any conspicuous flight feather growth (though my view was somewhat distant).

**9/10/16:** All six of the chicks seen. After having now observed this grebe family for five days, the general ranging and associative patterns of the chicks is becoming apparent. Thus far the chicks have basically kept to the southern half of the lagoon. They have ranged to the dense, reedy habitat on both the east and west sides of the lagoon. However, ultimately the chicks (and both parents as well) have consistently been seen making their way back to, gathering (in various numbers and combinations), resting and hanging out along the edges of and within the dense reeds at a particular location on the mid-western edge of the lagoon. This particular spot appears to function as a kind of home base area for the chicks. On a number of occasions there were chicks who I did not sight upon my arrival at the lagoon but eventually were seen as they emerged from within the dense, emergent reed bed in this area (the chicks having been out of view within the reeds where they were likely resting, just remaining hidden or perhaps even engaging in some foraging—thus far, only the two large chicks have been observed engaging in foraging dives; some of the other chicks have also been seen diving but their dives seemed much too brief to have been foraging dives).

During the times that I have observed this grebe family there have been periods when all six of the chicks have been together (in company either with or without a parent). However, for the majority of my observation time the chicks have been split into groups of various numbers. Often they have been seen split among the parents into two groups with each parent being accompanied by a group of the chicks. In these instances it is common for the parent/chick groups to be on opposite sides of the lagoon. There are also times when one group is accompanying a parent and another group is hanging out by itself in a different area, or when groups of chicks (none in company with a parent) are hanging out in different areas, including on opposite sides of the lagoon. When in separate groups there has been a tendency for the two large chicks to associate together and the four smaller ones to associate. The largest chick has been seen ranging by itself notably more than any of the others; this includes some independent foraging (neither it nor the other large chick has yet been seen actually capturing any prey though this certainly doesn't preclude the probability that they have done so when out of my view). This largest chick appears to be noticeably farther along than the other chicks (especially the four smaller ones) in the process of achieving independence. When the large and smaller chicks are together, the large chicks continue to decisively outcompete the smaller ones in regard to obtaining food items from the parents. This is probably the reason for the tendency of the smaller chicks to associate together when accompanying a parent—it is quite probable that they deliberately look for opportunities to do so at times when neither of the large chicks are present in the vicinity (they can thus avoid having to constantly lose out to their more dominant large sibs). Fortunately for the smaller chicks, the larger ones do become temporarily satiated after consuming one or two closely spaced feedings and for a short period thereafter cease looking to take food from the parents. This thus leaves the smaller chicks in a more favorable situation in which they are left to compete among themselves or with just one of the two larger, more dominant chicks.

As for the parents, along with what is noted above, sometimes they are together (either with or without accompanying chicks) while at other times one may be off foraging (or just sitting on the surface) independently (including in the far northern end of the lagoon) while the other is with chicks. Occasionally both parents have been seen alone (i.e., away from each other and with no accompanying chicks). In general, the parents (whether or not accompanied by chicks) spend a great deal more time apart from each other rather than in close proximity.

The chicks generally stay close to, along the edges of, and within the dense growths of the emergent reeds that line the edges of the lagoon. When not with a parent the chicks always do so, the only exception being when one or a group of chicks (singly, or together or strung out as a group of two or more) cross the open water of the lagoon to get to a parent or the other side of the lagoon. However, chicks will remain for periods in open water when with a parent. It is common for a parent to conspicuously park itself out in open water in the lagoon where it may take a short rest period, preen itself or otherwise just sit there and look about. (It appears that this behavior may be an element of territorial maintenance: lying in the open in this manner provides a very visible presence that probably serves to advise/warn other pied-billed grebes that they have claimed and occupied this territory. Such a conspicuous presence may thus also serve to intimidate other grebes from entering their territory. Note also that, besides keeping vigilant for predators and other hazards, the manner in which the parent birds look about indicates that they are actively keeping watch for possible intruders as well.) When accompanying a parent that does this, the chicks typically (but not always) stay with the parent. Sometimes, when in a group, some of the chicks stay with the parent while one or more continue on to the reed beds.

It appears that when the chicks accompany or remain with a parent they do so for two main reasons: security and feeding. Given the competition among the chicks and the manner in which the parents feed them (i.e., not by regurgitation but instead by catching a sizeable food item and handing it off, ultimately to a single chick) the chicks utilize a strategy of accompanying and following the parent while it forages (though I have observed a number of instances in which an unaccompanied parent that obviously was in "chick feeding mode" captured a crayfish while it was a considerable distance away from any of the chicks and carried it to one or a group of chicks, including all the way over from the opposite side of the lagoon). During times when a parent isn't foraging (notably when taking a rest period) it is common for a chick to solicit and attempt to coax it into foraging and catching food for it. The chicks use a standardized solicitation behavior: The chick animatedly positions itself in front of the parent's head, assumes a low profile and fully extends its neck/head out toward the parent's head and briefly contacts or brushes close to the parent's beak with its own beak. While performing this maneuver the chick paddles its feet in an exaggerated, energetic fashion that very conspicuously churns up the surface water along each side beyond the rear of its body. The chick may excitedly move from one side of the parent's head to the other in this manner, in a full or partial semi-circle, always remaining anterior to and facing the parent (apparently making sure that the parent sees the solicitation). The chick generally performs one such maneuver at a time, though after a brief pause it may repeat this action several more times. If after several such solicitations the parent does not respond and remains stationary the chick ceases its solicitations. However, as long as the parent remains stationary the chick typically will periodically renew its solicitations. If in a group, more than one of the chicks may perform solicitations though I have not seen them do so simultaneously. Such solicitations are especially likely to be seen in scenarios in which one or more chicks are with a parent that has parked itself out in open water.

**9/11/16:** All six of the chicks seen. Several more feedings were observed. These were conducted in the same general manner as all of the previous feedings that have been observed though in one of the feedings it took longer than usual before the crayfish was finally consumed by the chicks (once again, all of the prey items were crayfish). In this particular feeding all six of the chicks were in the home base area at the time, though a bit spread out. One of the parents, who was foraging by itself (unaccompanied by any chicks) in the dense reeds well south of the home base area, emerged from the reeds with a very large crayfish. While holding the crayfish in its beak it immediately began swimming straight toward the chicks in the home base area. Before reaching the home base area it was spotted by one of the smaller chicks who raced out and met it en route. When the chick reached it, the parent began working the crayfish and quickly made some sizeable breaks in its exoskeleton, notably in the tail section, as the chick eagerly looked on. With the crayfish lying on the surface in front of it, the parent tore out a decent-sized chunk of meat and handed it to the chick who immediately began working this down its throat. They were then joined by a second chick and a third one soon after. At this point the crayfish carcass was still much too large for any of the chicks to swallow. So, with the three chicks swarming and maneuvering about it, the parent continued to work the crayfish with its beak. As it did so, several of the chicks managed to grab and swallow scraps of meat that either became detached as the parent worked the crayfish or that the chick had plucked out by itself. Eventually, the second largest of the chick brood (one of the three who had joined the group around the parent) seized and took control of the crayfish. It immediately attempted to swallow the crayfish but was unable to because the carcass was still too large and bulky. While continually darting about and keeping ahead of its two closely pursuing sibs (who continued their fervent attempts to get at the crayfish) this chick continuously worked the crayfish in an attempt to reduce and reform it into a "package" that it would be able to swallow. As usual, the parent closely followed the chick that possessed the crayfish, overseeing this intense scramble and ready to retrieve the carcass should it be dropped and fall too far below the surface. This intensely competing group of chicks soon worked its way over near the edge of the dense emergent reed bed in the home base area where a fourth chick then joined the swarming mass that was vying for the crayfish (the largest of the chicks was also nearby just inside the edge of the reed bed but showed no interest in joining the fray—it is most likely that it had only recently had a significant meal and was thus temporarily satiated). The second largest of the chicks remained in control of the crayfish but as it worked the carcass there were a few occasions when a pursuing chick grabbed and swallowed a scrap from it. At one point (when it had gained momentary separation from its pursuers) this controlling chick set the carcass on the water in front of itself and managed to pluck out a sizeable chunk of meat which it quickly swallowed. It then picked up the carcass, resumed working it and was soon finally able to swallow it. From the time when it initially grabbed and assumed control of the crayfish it had taken about 2 ½ minutes before it was eventually able to swallow it.

**9/12/16:** All six of the chicks seen. The runty chick (i.e., the one that has very noticeably been the smallest of the chicks since the time when this brood was first seen) continues to lose ground to its five sibs. Its physical development is well behind that of the others and by comparison appears to have nearly stagnated. The size disparity between it and the other chicks continues to increase. Even the second of the pair of smaller chicks (the second smallest of the brood) is also now conspicuously larger than the runt (in fact, this chick has closed the gap slightly to the two medium-sized chicks). The runt is spunky and readily mixes it up with its sibs in the intense competitions to obtain food from the parents during feedings. It does manage to get some food but apparently not enough to keep pace. Overall, it loses out in these battles with its sibs, its small size allowing its larger, stronger sibs to more readily push aside, outmaneuver and dominate it.

One of the feedings that was observed today provided a prime illustration of the points noted above regarding the runt's difficult situation during competitive feedings. In this case one of the parent birds was being accompanied by only two of the chicks: the runt and one of the mid-sized chicks. When the parent emerged from the reeds with a large crayfish the two chicks excitedly swarmed around it, maneuvering amongst themselves for a shot at taking possession of the crayfish. However, the crayfish was too large for these chicks to swallow so the parent retained control and began working the prey, quickly opening up several breaks in the exoskeleton as both chicks swarmed around it. Eventually, at one of the times when the parent had momentarily laid the crayfish on the surface while working it, the mid-sized chick picked it up, made a brief, awkward attempt to swallow it but soon dropped the still too bulky carcass onto the water. The parent immediately grabbed it and began vigorously working it again with both chicks continuing to swarm around it. Shortly after (with the carcass lying on the surface in front of it) the parent plucked a large scrap of meat from the carcass and handed it off to the mid-sized chick. While that chick was busy working its scrap down its throat the parent immediately picked up the carcass and held it out to the runt. The runt wasted no time in grabbing the crayfish but held it for only about a second before losing its grip and dropping it onto the water (though it appeared that the chick managed to pluck and eat a small scrap of meat as the carcass slipped from its grasp). The carcass quickly sank but the parent ducked completely below the surface and retrieved it. It then held it out again for the runt. The runt initially grabbed and dropped the crayfish but the parent immediately picked it up and held it out for the chick a second time. This time the runt succeeded in grabbing and holding onto the carcass. It then began moving away with it (by this time the other chick had swallowed its scrap and was moving toward the runt). While moving away, the runt made a futile attempt to swallow the bulky carcass but after only a few seconds it again lost its grip and dropped the carcass onto the water. The parent quickly retrieved the carcass from the water, held it in its beak and passed it off to the closely following mid-sized chick. This chick promptly began working and attempting to swallow the still bulky carcass. As it did so it intently kept on the move ahead of the hotly pursuing runt (the parent also closely pursued the pair as it continued to oversee this feeding). The chick led the pursuing pack into the edge of the reed bed where (after having possessed the carcass for about 12 seconds) it lost its grasp and dropped it into the water. Seizing the opportunity, the closely pursuing runt then picked up the carcass and began swimming away with it, the roles having switched with the other chick now in pursuit. Keeping on the move, the runt engaged in more futile attempts to swallow the carcass. After about 15 seconds it again lost its grasp on the carcass and dropped it into the water. The still overseeing parent immediately picked up the carcass. As it do so the other chick promptly grabbed it and began moving away, continuing with its attempts to swallow the carcass, the runt (and parent) following closely behind. It continued with its futile attempts for approx. 80 seconds before another change of possession occurred—the runt again seized possession of the carcass when the other chick dropped it. In typical fashion, the runt attempted to swallow the carcass while keeping on the move ahead of its pursuing sib. By this time the largest chick (who was in the general vicinity) had caught sight of the competing sibs and was rapidly approaching them. It quickly caught them, moved into the fray and grabbed the carcass right out of the runt's beak. It then wasted no time in quickly swallowing it (though the carcass was too bulky for the smaller chicks, with its much greater size the large chick had no problem swallowing it).

Under such a competitive disadvantage, one has to wonder how much longer the runt will be able to hold on—especially considering that this situation feeds on itself in that as the size disparity increases, so does its competitive disadvantage.

At the other end of the spectrum, the largest of the chicks is noticeably more advanced physically and developmentally than its other sibs, including the other of the two larger chicks. It is also more advanced behaviorally. In particular, it spends considerably more time on its own and foraging independently. Conversely, it has been spending increasingly less time accompanying parents on foraging forays and thus is taking food from the parents increasingly less often (and therefore becoming less dependent on its parents for obtaining its food). Though I still haven't actually seen this chick catch a prey item on one of its foraging dives there seems no doubt that it does have successes. This probability is strongly supported by the facts noted above: that it is spending less time with its parents and obtaining less food from them yet its development continues at an undiminished pace (it must also be noted that I have only observed a tiny percentage of its foraging dives and probably just haven't yet been fortunate enough to witness any of its successful dives; as a side note, thus far the two largest chicks remain the only ones in this brood that I have observed engaging in foraging dives). Consider also that, given its dominant position over the other chicks and the relative ease with which it can outcompete them, if it was not having some success obtaining food on its own it would be expected that it would be spending more time (not less) accompanying parents on feeding/foraging forays in order to meet its daily food requirements. All indications are that this chick is on track to fledge and become independent sooner than any of its brood mates.

**9/13/16:** All six of the chicks seen. The runt appears to be in serious trouble. There was a very notable change in its general demeanor today: As previously noted, though the runt is commonly bested by its sibs in competitive feeding situations it has nevertheless always been spunky and readily mixed it up and fought hard with its sibs for a share of the food proffered by the parents. During the feedings observed today the runt's battling to obtain food seemed rather half-hearted, its usual "fire" no longer evident. Oftentimes it either quit the competition early or didn't even bother to join in, instead just sitting nearby as its sibs competed intently for the food (as usual, in each case the prey item that was offered was a crayfish). In general, it appeared as though the runt is just running out of steam; that it has little left to fight with. The situation does not look promising for this small chick.

On a number of occasions (including today) I have observed what appears to be a deliberate strategy employed by the runt and the other smaller chicks to place themselves in a situation that greatly improves or assures their chance of obtaining the food item from a parent during a feeding. Specifically, the apparent aim of the chick is to place itself in a situation in which it doesn't have to compete with any of its sibs for the food or in which it only has to compete with just one of the other smaller chicks (and thus significantly increases its chances of obtaining food by being with a sib that is either below or only just above it in the dominance hierarchy among the chicks—in these cases the smaller chicks are seeking to avoid having to compete with the larger, more dominant chicks, situations in which they typically lose out). In practice, this strategy consists of the chick being vigilant for opportunities in which it might be able to isolate itself with one of the parents (or with a parent and just one of the smaller chicks). Such opportunities have not appeared to be particularly numerous but do seem to occur a number times each day. Most of the observed cases have arisen out of instances in which one of the parents who was unaccompanied by any chicks at the time passed near (or briefly re-established contact with) its mate while the latter was being accompanied by a large group of chicks (3 or more) and continued on its way moving away from the group. In these cases one or two of the smaller chicks then left the larger group and tagged along with the parent who had been unaccompanied. I have also seen instances in which one or two of the smaller chicks who had been lying or resting either alone or with a group of chicks spotted an unaccompanied parent and moved to and began accompanying that parent.

**9/14/16:** Only five of the chicks were seen today. Sadly, the runt was missing. This is not surprising given how poorly the runt was doing and especially considering how serious its condition looked yesterday. It is possible that perhaps this smallest of the brood was just tucked inside the reed beds out of view during my observation period. Unfortunately, this seems unlikely. There have been days previously when one or more of the chicks initially were not seen. However, up to this point in my daily observations of this grebe family, over the course of several hours of observation all of the chicks have eventually been sighted on each of the days. It is as yet too soon to draw any definitive conclusion about the runt's status but the sad reality is that it probably has died (if so, almost certainly due to starvation though one couldn't rule out the possibility of predation). At this point I can only wait and see whether or not the runt is seen tomorrow.

With the exception of the runt who has lagged behind, the other chicks have grown considerably since they were first seen on 9/6. The largest chick has remained well on pace and is still very noticeably bigger and farther along in its development than the others. This chick has not afforded me a definitive look at its wings so I don't know how far along its flight feather development has progressed (in fact, I have not seen any of the chicks spread or flap their wings since 9/9). However, given its large size it seems quite likely that its flight feathers are fairly well developed at this point and that this chick probably isn't that far away from fledging. It continues to spend increasingly more time foraging independently but does still spend time accompanying and taking food from the parents.

Lately, the feeding competition among the chicks has begun ratcheting up to a new level. When in a group accompanying a parent the more dominant chicks (notably the two largest) have started to peck at and briefly chase subordinate chicks away from the vicinity of a foraging parent. In doing so, the more dominant chick thus is attempting to outright exclude competing sibs or create a situation in which it will be closest to and able to gain first access to the parent when it captures a prey item. Prior to this, though some pecking and brief chasing occurred, the intense competition between the chicks to gain access to the food item consisted mainly of pushing aside and attempting to outmaneuver competing sibs.

**9/15/16:** Once again, only five chicks were sighted. For a second consecutive day the runt was not seen. It now seems a virtual certainty that the runt has died, most likely due to starvation (though because its death wasn't actually observed or investigated one cannot rule out predation as a possibility). The runt fought valiantly but because of its small size it had much difficulty competing with its sibs during feedings.

One of the mid-sized chicks managed to isolate itself with one of the parents for a considerable amount of time today.

Having observed this grebe family for 10 days it is now clear that the parents' territory comprises the entire west lagoon and approximately the western half of the narrow channel that makes a connection between the two lagoons through the marshy area that separates them. In other words, the waters and bordering emergent marsh area from the south end of the lagoon (which begins at the upper or north end of the wide beach) all the way to the railroad bridge at the north end of the lagoon, plus the western half of the narrow channel through the marsh that connects the east and west lagoons. As previously noted, the terminal section of the Ventura River (which feeds both of these lagoons) lies on the immediate north side of the railroad bridge. *[Note: This is the section that I would refer to as the "south ribbon of the river" in my study of the South River family during the 2017breeding season]* This section of the river appears to be an area of overlap between three adjacent grebe territories: the pair that I am observing which occupies the west lagoon, a pair that occupies at least the northern two thirds of the east lagoon (plus the eastern half of the narrow channel between the lagoons) and a pair that occupies the segment of the Ventura River that lies immediately adjacent to (north of) this terminal section. Members of all three pairs have been observed utilizing this area, though not simultaneously (of particular note, individuals of the pairs have utilized the dense, emergent reed beds that line the north shore of this section for foraging).

Only the pair that I am observing in the west lagoon currently has dependent chicks. I have sometimes seen a pair of fledged, well-grown independent chicks in the south section of the east lagoon that very possibly may be the offspring of the east lagoon pair. Also, I have on a few occasions sighted a fledged, well-grown independent chick in the section of the river to the immediate north of this overlap area that may possibly be an offspring of this pair. From the above observations it appears that the west lagoon pair is either working on rearing a second brood for this breeding season or one that for some other reason was hatched later in the season (note that I am not familiar with the typical breeding season pattern of this species).

**9/16/16:** Five chicks seen today. For a third consecutive day the runt was not seen so it now appears certain that it has died.

The largest chick continues to spend much of its time foraging and ranging independently. The second largest chick also spends time foraging independently but not to the extent of its larger sib.

One of the mid-sized chicks was observed briefly flapping its wings today providing me with the first definitive view that I have had regarding the wing development on any of the chicks since 9/9. While its wings are still as yet poorly developed, its primary and secondary flight feathers are well on their way, having grown to perhaps one quarter or so of their ultimate length.

An interesting observation occurred today involving a parent and one of the chicks. At the time, the parent was parked out in open water in the south end of the lagoon, accompanied by the chick that, with the loss of the runt, is now the smallest of the brood (this chick had managed to isolate itself with this parent). This pair was soon joined by one of the mid-sized chicks at about the time the parent began to preen itself. Upon reaching the parent, this chick engaged in one of the beak to beak feeding solicitations that has been previously described (see notes for 9/10). The parent, however, just continued to vigorously preen itself, both chicks remaining close around the parent as it preened. Shortly after, as the parent continued its preening it plucked a feather from its breast area and appeared to deliberately hand it to the mid-sized chick. The chick readily took it, initially dropped it on the water surface but quickly picked up and swallowed it. It appeared rather obvious that the parent had deliberately plucked and "fed" this feather to the chick. On several previous occasions I have observed a chick pick up and swallow what appeared to be a feather that was lying on the surface of the water. However, in each case the chick was too distant for me to be certain that the object that was swallowed was a feather. After having witnessed the above interaction I can now feel more confident that the object that was swallowed in those cases was indeed a feather.

**9/17/16:** All five of the remaining chicks were seen. Among the feedings that were observed today, two were particularly notable in that for the first time since I began observing this grebe family I witnessed feedings that involved prey items other than crayfish. In both of these cases the prey item appeared to be a fish. The first instance involved a fish that was small enough for the chick to swallow with little problem. The parent initially worked the fish (perhaps to incapacitate it) before holding it out for the chick to take. The second case involved what appeared to be a larger fish—still fairly small but easily too large for the chick to handle. In this instance the parent continually worked the fish, pulling off chunks of it and handing these to the chick. The parent also ate pieces of this fish. In both of these cases the parent was being accompanied by only one chick.

An osprey pair that apparently is currently residing near the river well north of the lagoon has been regularly hunting for fish (either singly or with both present in the area) over both the west and east lagoons and just offshore along the beaches nearby. When hunting over the west lagoon the osprey(s) is typically quite conspicuous in flight (one was even observed catching a fish in the south end of the lagoon). However, neither the parents nor chicks have as yet displayed any noticeable fear or apprehension of these large birds (though they typically haven't happened to have been out in open water during the times when the ospreys have visited the lagoon). This includes a few occasions when one of the ospreys was perched for a period on a dead (but still rooted) tree that is located on the west side of the southern section of the lagoon near the home base area of the chicks. One is thus left to wonder if the grebes somehow know that they are not the targeted prey of these raptors?

**9/18/16:** All five chicks seen. As a general assessment, all five of the chicks appear to be progressing at this point. The largest chick remains well-advanced over its sibs and continues to be on course to achieve independence before the others. In the progression toward achieving independence the second largest chick is still intermediate between the largest chick and the other three. The two mid-sized chicks look healthy and steady on-course. The smallest chick has gained some ground toward catching up with the mid-sized chicks but overall, it is still lagging behind its sibs. It is not stagnating but its progress is proceeding at a relatively slow pace.

With the runt now gone and the largest chick spending so much time foraging and ranging independently (and the second largest chick also spending some time foraging independently or near the largest chick), the other chicks have more often than previously been in situations in which just two or even only one of them are accompanying a parent during foraging forays.

**9/19/16:** All five chicks seen. Among the feedings observed today was another one in which the prey item was a small fish (all of the other observed feedings involved crayfish). In this case the parent was being accompanied by four of the chicks (all but the largest chick) when it surfaced with the fish. The parent worked the fish only briefly before the second largest of the chicks grabbed it out of the parent's beak. The chick then circled about ahead of its pursuing sibs and was able to swallow the fish rather quickly (the fish was of a size that it could easily handle and it took the chick only about 10 seconds to swallow it). To date, this is the only feeding that I have observed which occurred in fairly open water (i.e., not right next to or within the reed beds) though it still wasn't very far out from the reed beds.

The chicks have mainly kept to the southern half of the lagoon though the largest chick has also been venturing into the northern half during its independent ranging.

It is common during the course of a day for the parents to be a considerable distance away and/or out of sight of each other (in fact, they spend little time in close company as they go about their day). However, it is clear that the parents keep track of each other's whereabouts and after these periods of separation they essentially always deliberately seek to re-establish contact. This is accomplished in a quite standardized manner. The initial settings may vary a bit but the final outcome is always similar. In the many observed cases, initially: one or both of the parents may be accompanied by one or more chicks; both of the parents may be actively moving about or one of the parents may be lying at rest or restricting its activities to a localized area while the other is actively moving about (often in such cases the former is located in the southern half of the lagoon while the latter has been foraging alone in the far northern half of the lagoon or even north of the railroad bridge); the parents are in open water where they are visible to each other or one of them is at least far enough out from the reed beds where it can be seen by the parent that makes the initial move to re-establish contact. When the decision has been made to re-establish contact, initially either by just one or both of the parents, the parent(s) begins the move toward its mate (typically, the parent(s) seeking to reconnect can be seen looking about to visually locate its mate). The parent initiating the reconnection may move nonstop to its mate or may pause or briefly stop one or more times during its approach (in these latter instances, the intent to eventually meet up still appears obvious). In cases in which one of the pair initially had not seen its mate approaching, this parent either just waits for its mate to arrive, or moves out to meet it en route (regardless of whether just one or both initiate the move to re-establish contact, one of the pair generally ends up travelling a much greater distance). Upon meeting they perform a standard greeting: they make physical contact, briefly touching breasts and or beaks as they either partially circle closely about each other or one of them continues past a short distance after touching, typically spinning about and facing its mate as it does. Sometimes the pair may remain in proximity for a period after they greet but usually the parent who travelled the long distance to reconnect (who oftentimes is unaccompanied) will continue along away from its mate (if the chicks had been in a large group accompanying the more stationary of the parents it is often at such times that a smaller chick might attempt to seize the opportunity to isolate itself with the unaccompanied parent that is moving away). In all of the reconnection episodes that I have observed the parents have always moved to each other by swimming along the surface (paddling with their feet), despite the long distances that they commonly have to travel. I have never observed either of the parents taking flight as a means of closing the distance and reconnecting (in fact, up to this point I have never observed either of the parents to take flight, period!).

While these regular interactions by the parents to re-establish contact obviously serve to aid in maintaining the pair bond, they also appear to be tied in to territorial maintenance/defense. When at a distance it is obvious that parents do visually scan for each other. However, in addition it also appears likely that they are vigilantly looking about for any other grebe that might intrude into their territory.

When initially spotting its mate at a distance it is quite probable that, though it probably perceives that this distant adult grebe is most likely its mate, it doesn't know this for certain. Lending some support for this latter point is the observation that as the approaching parent nears its mate its demeanor changes in that it tenses as it apparently gears itself up for either of two imminent actions: a greeting if it is indeed its mate, or an aggressive encounter if it is instead an intruder.

**9/20/16:** All five chicks seen. A somewhat curious feeding was observed today. At the time that this episode began all five of the chicks were gathered in proximity (though a bit scattered) among the reeds in their home base area along the mid-western edge of the lagoon (neither of the parents were with them). One of the parents was all the way across the lagoon from them, foraging among the reeds along its southeast edge. The parent emerged from these reeds with a very large crayfish in its beak and immediately began making its way across the lagoon at a fast pace, heading directly for the chicks. This thus made it obvious that the parent's intent was to carry the crayfish to the chicks and feed it to them. As the parent neared the home base area with the crayfish it was spotted by two of the chicks (the largest and the smallest chicks) who then began rapidly swimming out (from different directions) to meet the approaching parent. The largest chick easily won the race (and by a considerable margin) and thus was the first to reach the parent (not only was the largest chick closer initially, but with its larger size and more advanced development it was also a much faster swimmer). The parent apparently saw the approaching chicks and began working the crayfish about 5 seconds before the first chick arrived (as per usual, the parent laid the crayfish onto the surface while working it with its beak). Just as the largest chick reached it, the parent picked the crayfish up and held it out toward the chick. The chick made a single rather half-hearted, ineffective grab at the crayfish and it fell away back onto the surface (at this point the crayfish was still much too bulky for this large chick—and even the parent—to swallow). The parent then immediately resumed vigorously working the crayfish. A few seconds later the parent again picked up the crayfish and held it out toward the chick. This time the chick made a peck-like grab and held the crayfish only momentarily before it dropped to the surface (it appeared that the chick may have plucked a tidbit from the carcass as it slipped from its grasp). In general, this largest of the chicks just didn't appear to be very enthused about securing and eating this crayfish (perhaps it had recently eaten and was fairly satiated or maybe it was put off by the still far too bulky size of the crayfish). At this point (about 11 seconds after the largest chick had first reached the parent) the smallest chick finally arrived at the scene and was quickly joined by two other chicks. The parent continued to vigorously work the crayfish. As it did so, the four chicks initially swarmed around the parent. A few of the chicks pecked out at the carcass as it was being worked and possibly were able to pluck a scrap from it. Surprisingly, within approximately 20 seconds (with the parent still working the carcass) all but the smallest chick suddenly lost interest in the crayfish and began preening themselves (the four chicks were still fairly tightly grouped around the parent). The smallest chick continued to intently follow the parent as it worked the carcass, pecking at it several times and perhaps plucking a few scraps. However, only about 9 seconds after the other three chicks had lost interest, the parent began attempting to swallow the carcass (the carcass was still rather bulky but had now been worked to the point where the parent could handle it). About 8 seconds later the parent was able to swallow it. Just before the crayfish had disappeared down the parent's throat the fifth chick joined the group.

The final outcome of this episode was a bit surprising given that the parent had carried the crayfish to the chicks all the way over from the opposite side of the lagoon. It is thus obvious that its initial intent was to feed this crayfish to the chicks. This still appeared to be its intent when it first contacted the chicks as it clearly was working the carcass for their benefit and even held it out several times for one of the chicks to take. The explanation for what finally transpired in this episode is not clear. Did the quick loss of interest by all but one of the four chicks indicate that, in the end, this crayfish had been too large all along for all but the parent (with three of the four chicks quickly perceiving this)? After attempting to feed this large crayfish to its chicks, did this realization finally come to the parent, prompting it to eat the crayfish itself?

**9/21/16:** All five chicks seen. In regard to developmental progress the largest chick remains the most notable and advanced of the brood. It still spends some time accompanying and taking food from the parents but it has become increasingly more independent. In fact, I generally see it less often than the other chicks because it spends considerable time foraging and ranging independently. The second largest chick continues to remain more advanced than the other three chicks but notably less so than its larger sib. It also spends time foraging and ranging independently (at times in company with the largest chick) but not to the extent of the larger chick. Besides the two largest chicks (who have been observed engaging in foraging dives for some time now), some probable foraging dives have also been observed among the three smaller chicks lately (mainly the two mid-sized chicks). The two mid-sized chicks continue on a steady course and have even closed the gap a bit in relation to the second largest chick. The smallest chick has progressed but is still lagging behind the others. In regard to size disparity it doesn't appear to have gained any more ground relative to the mid-sized chicks (it is still conspicuously the smallest chick in the brood). During feedings it remains at a competitive disadvantage to its larger, more dominant sibs and (despite its spirited attempts) is routinely bested by them.

**9/22/16:** Only four of the chicks were seen today. The smallest chick was not sighted. It is too soon to know for certain if this chick has died. It's possible that it may just have remained out of my view within the reed beds. However, this is unlikely given that it has been seen on every previous observation day (it is especially unlikely that, at its current dependent stage, it would not be seen accompanying a parent on feeding forays). Hopefully, it will show up tomorrow but at this point the situation does not look promising regarding this small chick.

One of the chicks was observed briefly flapping its wings. This revealed that the development of its flight feathers is coming along well but both the primaries and secondaries are clearly still much too short for flight.

**9/23/16:** For a second consecutive day the smallest of the five surviving chicks was not seen. It thus now appears rather certain that this small chick has died. Though cause of death isn't known, starvation seems most probable (however, here again one can never rule out the possibility of predation). As with the runt, this chick was quite spirited and very willingly mixed it up with its larger sibs during feedings. Unfortunately, with is size disadvantage it regularly was bested by them. In the end, it appears that it just wasn't able to acquire enough food to keep up with the prodigious energy and nutritional demands that are placed on a young, (rapidly) growing chick.

This is now the second chick that has been lost from the originally seen brood of six (obviously, I don't know if the brood actually originally consisted of more than six chicks—I only know that there were six small chicks present when I first discovered this brood on 9/6). In both instances it was the noticeably smallest chick in the brood that was lost.

Several of the chicks were observed briefly flapping their wings. As with the chick seen doing so yesterday, their flight feathers are developing well but are still noticeably too short for flight.

As for the remaining four surviving chicks all appear to be healthy and progressing well at this point. The largest chick remains advanced in its development over the other chicks. It continues to spend considerable time foraging and ranging independently but also spends time accompanying and taking food from the parents. As has seemed apparent since the first observation of this brood, this chick appears to be on course to achieve its independence before its sibs are able to. In regard to size disparity the second largest chick has closed the gap to its larger sib. However, this perhaps is more a case of the larger chick having reached a stage of growth at which the rate of increase naturally slows as it approaches its ultimate adult size. This chick also continues to forage and range independently but not to the extent of the larger chick (and is thus noticeably more dependent on the parents). The two medium-sized chicks (now the smallest of the reduced brood) have further closed the size gap to the second largest chick but they remain clearly subordinate to it. Lately, they have also been observed engaging in some foraging dives (all of their observed dives were unsuccessful). In general, these two chicks look healthy and at this point appear to be on course toward achieving independence.

**9/24/16:** Once again (for a third consecutive day) the smallest chick was not seen. There now seems no doubt that this chick has died. All of the four surviving chicks were seen today.

Besides this grebe family there are members of other water bird and wading bird species that use this lagoon. Up to this point, none of the ones that have been seen have presented any problems for the grebes. In fact, no interactions have been observed between them and the grebes and all have basically just ignored each other while going about their business. Among the regulars are numerous American coots, several mallards, a regular sprinkling of double-crested cormorants (mainly immatures), an immature snowy egret, and several spotted sandpipers. Infrequent but fairly regular sightings have been made of a few soras and a Virginia rail among and along the edge of dense reed beds, including within the southernmost sections of reeds used by the grebes.

The American coots are easily the most common and numerous of the water birds in this west lagoon. They are by far the ones that the grebes encounter most as they go about their day (and they are the only ones that the grebes encounter regularly during their day). In fact, various coots and members of the grebe family are commonly in proximity throughout the day. This is largely because the coots mainly utilize the same sections of the lagoon as the grebes: notably, the dense reedy sections that ring the lagoon. Like the grebes, the coots feed mainly along the edges of and within the reeds (though they also do some feeding in open water—like the grebes, they are divers). However, they are not competitors in this respect because the coots feed on plant matter so there is no friction between members of the two species in this regard. Also like the grebes, the coots use the reeds for cover and concealment. The grebes often pass by or are otherwise in close proximity to coots throughout the day. Sometimes they even rest in proximity. Despite the many times that I have seen them in proximity I have yet to see any noticeable interactions between any members of the grebe family and the coots. I haven't even observed them pay any more than the most casual attention to each other. They basically just ignore each other and go about their own business.

Three mallards have been residing within the territory of the grebe parents since the time when my observations began. They have posed no problems for the grebes. Here again, the mallards and grebes pay no heed to each other and simply go about their own business.

Double crested cormorants (mainly immatures) regularly forage for fish within the lagoon and in the river section north of the railroad bridge. They do so intermittently, typically with only one or two doing so at any one time. In addition, one or several individuals sometimes stand along the edge of the lagoon or on the base of a railroad bridge pillar to rest or sun their spread wings. In the course of their activities the grebes and cormorants sometimes come into proximity. Once again, no interactions have been observed between them and they basically just ignore each other.

The immature snowy egret has come by on most days to spend some time foraging in the shallow water just off the strip of beach sand at the south end of the lagoon. (It is from this strip of beach along the south end of the lagoon that I have observed the grebe family. Because of my daily presence there, this egret has become quite habituated to me and has reached the point where it will now forage right in front me, exhibiting no apparent concern as long as I remain seated.) While foraging here the egret focuses on catching the very small (guppy-sized or slightly larger) fish that regularly utilize the shallow water here (sometimes the egret moves about quite actively while attempting to catch one of these fish). Occasionally one or more of the grebes pass within approximately 10 meters to the north of the egret (including while it is actively pursuing a fish). On these occasions, despite its large height the grebes passed by paying no noticeable attention to the egret.

Considering what goes on in the lagoon "next door" (i.e., the east lagoon), one particular type of water bird has been notably absent from the west lagoon. The east lagoon is an area of continual presence and activity for many dozens of sea gulls. In the time since I began the grebe observations numerous members of three gull species in particular (western, California and Heermann's) have been present in the east lagoon on a daily basis, constantly coming and going (with many regularly passing over the west lagoon while en route to and from the east lagoon). Yet, during this period I have not seen a single gull land in the west lagoon. Whatever the reason(s) for this, it has been a favorable circumstance for the grebe family given that gulls (notably the westerns) are potential chick predators.

Another potential chick predator common to this area is the great-blue heron (though I don't know if any of these herons have ever been known to take pied-billed grebe chicks it certainly appears to be a possibility). While I haven't seen any of these large birds in the west lagoon during the periods that I have observed this grebe family, I have seen them in this lagoon fairly regularly in the past and I have commonly seen them in the east lagoon and the marsh area between the two lagoons during this period. In addition, it is certainly possible that herons may have come into the west lagoon unseen by me during the long periods between my daily observations of the grebes.

Thus far, I have not seen any natural mammalian predators such as coyotes, foxes, raccoons, bobcats or opossums or any signs of these (footprints, scat, etc.) in the area around the lagoon (though some of these are known to range in this general area so it is quite probable that they do at least occasionally forage at this lagoon). However, I have seen feral cats around the lagoons and people commonly walk their pet dogs through the beach area at the south end of the lagoons (on several occasions in the past I have even seen pet dogs swimming in the south end of the west lagoon to retrieve a stick or ball while playing "fetch" with their caretaker; I have also regularly seen thoughtless, irresponsible caretakers allow their dogs to chase birds along the shorelines and swim out into the lagoons to chase after wading and sitting birds).

**9/25/16:** All four of the surviving chicks seen. Lately, the two mid-sized chicks (now the smallest of the brood) have been seen engaging more regularly in short foraging dives (all of the dives observed have been unsuccessful). The entire remaining brood now engages in foraging dives (the two largest chicks have been doing so since the first days that this brood has been observed).

Both parents continue to be regularly accompanied by and provide food for the chicks. It is still common for the brood to be split amongst the two parents.

**9/26/16:** All four chicks seen. One of today's feedings that involved the largest chick was particularly notable. What was so noteworthy was the large size of the food item that it was able to swallow with little difficulty. In this incident the large chick and one of the smaller ones was accompanying one of the parents at the reed beds along the edge of the lagoon in its southeastern section. The two chicks were milling about and waiting as the parent foraged amidst the emergent reeds for prey items to feed to the chicks. The large chick was busy preening itself when the parent emerged nearby with a fairly large crayfish. The large chick ceased preening and quickly made its way to the parent, easily beating the smaller chick to the prize. By the time that the large chick arrived, the parent had only been working the crayfish for about 5 seconds and thus had only just begun the process of rendering the large crayfish more palatable. The parent worked the crayfish several seconds more and then, despite its essentially unworked state, held it out for the large chick. The chick grabbed the crayfish and immediately began moving west across the lagoon (the smaller chick didn't bother to follow—in fact, it hadn't even reached the pair before the large chick headed off with the crayfish). Shortly after moving away from the parent the large chick briefly submerged completely (holding the crayfish in its beak) while continuing on its course. Upon re-surfacing and while en route, it soon began attempting to swallow the crayfish in the typical grebe manner. While holding its beak extremely wide open and pointed skyward with the crayfish oriented tail first in its mouth (the crayfish actually being partially doubled over on itself along its ventral surface), it repeatedly jerked its head up and down. In this fashion it worked the crayfish down its throat fairly quickly, doing so a mere 12 seconds after it had taken it from the parent and without having worked it any further. This was impressive and it indicated that this large chick was now capable of handling/swallowing large prey on a par with adult grebes. This largest chick of the brood continues to spend considerable time ranging and foraging independently (though it also still spends time accompanying and taking food from the parents) and it appears that its independence is not far off.

Lately, the chicks have been ranging a bit farther into the northern half of the lagoon (the largest chick has been doing so for quite some time but has predominantly kept to the southern half of the lagoon). Prior to this the chicks have mainly restricted their ranging to the southern half of the lagoon.

**9/27/16:** Three chicks definitely seen and almost certainly a fourth. During today's observation a total of only three chicks were visible at any one time though it is highly likely that during the course of the observation period all four of the chicks were seen. Being conservative, it must be acknowledged that the chicks often were a bit too distant to allow for 100% certainty regarding the identification of individual chicks. However, the degree of uncertainty was slight and I feel extremely confident that all four of the chicks were identified.

The largest chick was observed briefly stretching its wings (it was ranging alone at the time). Its flight feathers are well-developed and perhaps now even capable of flight. This raises a salient point regarding the recognition of just when these chicks actually reach the hugely important milestone of fledging (the most minimal definition of fledging is being considered here: the time when a chick is first capable of short flights). The problem here is that pied-billed grebes are rarely seen in flight (in fact, I still have never seen either of the parents take flight). Thus, it is quite possible that the chicks will be capable of flight before (and perhaps even well before) I actually observe them in flight. On the other hand, it may be that as a natural course in their developmental process, chicks will begin taking short flights as soon as they are physically capable of doing so. Regardless, whichever of the above is the actual case it will not be apparent to me because I am not familiar with this species and this is the first time that I am observing pied-billed grebe chicks go through their developmental process (as a precaution in aiding against the possibility of my observations being influenced or prejudiced I have refrained from reading any existing literature on this subject or pied-billed grebes in general nor have I read about this species in the past; my previous observations of pied-billed grebes have all merely been of a very casual nature).

Among the feedings that were observed today, one in particular stood out. In this incident one of the parents was being accompanied by a single chick (it appeared to be the largest chick) when it captured a very large crayfish. The chick eagerly swarmed around and followed the parent, waiting for the parent to make the crayfish available or to offer it up. However, the parent never relinquished control. Staying mostly on the move, the parent vigorously worked the large crayfish for a lengthy period (approximately two minutes) before finally swallowing it. From its actions, it appeared that the parent never had any intent other than keeping this crayfish for itself. Was this a kind of weaning behavior by the parent? And if so, was it specifically directed at this large chick (that is, was it specifically done for the benefit of this largest chick only)? Or was this simply a case of a parent feeding itself or, more specifically, deciding to keep for itself a crayfish that it felt was beyond what a chick could handle/swallow (as I have observed the parents do on previous occasions)?

Up to this point in my observations of this grebe family, in all but four of the observed feedings the prey item that was captured and offered by the parents was a crayfish. In three of these exceptions the prey item was a small fish and the other involved what appeared to be a large insect nymph (perhaps that of a dragonfly or damselfly?).

**9/28/16:** Same situation as on 9/27: three chicks definitely seen and almost certainly the fourth. The general circumstances were also the same (only three chicks were visible at any one time during the observation, etc.).

As noted for 9/26, the chicks have begun venturing farther into the northern half of the lagoon. This trend has been continuing and today two of the chicks were seen accompanying a parent under the railroad bridge and into the terminal river section that lies north of (and feeds into) the west and east lagoons. By doing so they thus left their parent's territory and entered into the common (overlap) area that their parents share with the two neighboring pairs of grebes—i.e., the pair that resides in the northern two thirds of the east lagoon and the pair that resides in the section of river to the immediate north of this common section of river.

An incident occurred today which leaves little doubt that the parents have definitely entered into the process of weaning their chicks toward independence. In the time prior to this incident both parents and three of the chicks had been lying in spread proximity in open water at the north end of the lagoon (just south of the railroad bridge). At about the same time, both parents began moving away in opposite directions. One parent moved north and was accompanied by two of the chicks. This parent led the two chicks under the railroad bridge and into the overlap area. The other parent moved slowly to the south, initially by itself. Shortly thereafter, the third chick set off after this parent (note that all three of the chicks were too distant for me to be able to identify individuals). When the parent stopped to preen itself (still in open water) the chick caught up, moved in close and went into its food solicitation pose. At this point the parent reacted in a manner that I hadn't previously seen from either parent. With obvious intent it moved directly at the chick and firmly rebuffed the youngster, pecking at it twice in the process and even briefly chasing it for a meter. The chick immediately fled from the parent's rebuff but remained in proximity. After clearing the parent (who again began preening), the chick circled around it but remained near (within about 4-7 m). It was clear that the chick wished to accompany the parent, specifically on a feeding foray. Toward this end the chick attempted to coax the parent into a foraging/feeding foray. Several times the chick engaged in food solicitation behavior toward the parent but (in the wake of this firm rebuff) it didn't do so in the standard manner. Instead of closing in and making beak to beak contact as per usual, the chick was obviously mindful of the rebuff and swam by a short distance in front of the parent while engaging in its food soliciting pose, clearly making sure that it passed far enough away in front of the parent to avoid another physical rebuff. The parent remained in place and continued to preen. As the apprehension generated by the rebuff began to fade, the chick gradually worked its way in close to the parent with no further rebuffs at that time.

**9/29/16:** All four chicks seen. Over the past days the chicks have been spending increasingly more time in the northern half of the lagoon (as well as in the overlap area north of the railroad bridge) though they still are most often seen in the southern half of the lagoon.

There were a number of instances today in which a parent was observed rebuffing a chick. These were generally similar in nature to the incident noted yesterday. The parent either pecked at the chick or intently moved directly at it briefly (usually pecking at the chick at least once as it did so) to drive it away from its close proximity to the parent. The majority of these rebuffs occurred while the chick was soliciting food from the parent. While the chicks did move out of close proximity to the parent after a rebuff they typically still remained in proximity and continued to accompany the parent, soon moving in close again. One of these cases was generally similar to the (non)feeding incident that was noted on 9/27, except no rebuffs were made on that occasion. The parent (accompanied by two chicks) had captured a very large crayfish and vigorously worked it for about two minutes before finally swallowing it. Here again, it appeared that the parent never had any intention other than keeping this crayfish for itself (and it didn't even bother trying to move away from the chicks). While the parent was intently working the crayfish, one of the chicks eventually realized that the parent was not going to relinquish it and swam away. However, the other chick continued to eagerly swarm around the parent in hopes that it would offer up the crayfish. Within only a second after the parent finished swallowing the crayfish, the chick moved in close and began soliciting the parent for food. The parent put a quick end to the chick's solicitation by immediately responding with two successive firm pecks that drove the chick about a meter away. The chick remained within about a one meter radius and when it moved in close about 8 seconds later the parent was again quick to rebuff it: this time the parent aggressively chased the chick for a few meters, pecking at it once as it drove the chick away. Though the chick fled it didn't go far and continued to remain in proximity to the parent, initially at a distance of about 3-4 meters.

Despite the intermittent rebuffs, the parents are still feeding the chicks. However, they are apparently making it clear to the chicks that, at this stage, they had better start securing some food on their own. And in fact, all of the chicks are now engaging in at least some independent foraging, oftentimes several doing so while in proximity to each other.

**9/30/16:** Only 3 chicks and one parent seen at any one time and these were all that could be confirmed with certainty for today (though it is probable that all four chicks and both parents were seen).

Though the chicks can still be readily distinguished from their parents by the size disparity between them (as well as by plumage), they are nevertheless fairly large now. Actually, at this point they aren't very far away from achieving adult size. As each of the chicks in turn has neared its ultimate size, its rate of growth has slowed. This has allowed the smaller chicks to pretty much catch up to their larger brood mates. As a result, the size disparity between the chicks is currently the smallest that it has ever been during the time that I have been observing this grebe family. In fact, the size disparity is now so reduced that when at a distance I can no longer distinguish between the individual chicks by way of this physical characteristic.

More parental rebuffs of chicks were observed today. Lately, these have become a common daily occurrence.

In the entry for 9/14 it was noted that the nature of the intense competition between the chicks during feeding situations involving two or more chicks had begun rising to a new level. At that point their competition to gain access to the food item being proffered by a parent (or to a parent who was foraging for such) was shifting from a strategy of mainly pushing aside and attempting to outmaneuver competing sibs to one of direct aggressive actions taken by dominant chicks toward subordinate ones aimed at driving the subordinate away from the immediate area (the actions by the dominant chick consist of pecking at and/or chasing the subordinate). This trend has continued to the point where it is now commonly observed, the competition between the chicks having intensified. A prime example of this was seen today. In this particular case two of the chicks were accompanying a parent who had captured a very large crayfish. During the period in which the parent vigorously worked the crayfish, the larger more dominant chick among these two repeatedly pecked at and/or chased the subordinate in order to drive it away from the immediate vicinity (after being driven away to a perimeter location the subordinate chick would soon work its way back in closer, only to be aggressively driven away again). Having by its actions essentially achieved sole access to the parent, the dominant chick took the crayfish when offered by the parent and assumed control of it.

<u>**10/1/16**</u>**:** All four chicks seen. More instances of intense competition during feeding situations were observed today. All of the chicks now spend at least some time foraging independently. The largest chick spends the majority of its time ranging and foraging independently but still spends some time accompanying the parents and taking some food from them. However, judging by the lesser frequency with which it takes food from the parents, the indications are that this chick is obtaining more of its food on its own than from the parents (though I still haven't been fortunate enough yet to see it or any of the other chicks capture prey on their own there seems no doubt that they do). In fact, this largest chick appears to be fairly close to achieving independence. Frustratingly, it hasn't afforded me the opportunity to get a look at its wings since 9/27. Thus, I haven't been able to assess the state of its flight feather development. In speculating on this matter, based on the large overall size of this chick it appears quite possible that its flight feathers may be fully developed at this point (and thus may even be capable of short flights).

<u>**10/2/16**</u>**:** Sightings of only three of the chicks could be confirmed, though the fourth probably was seen also. The chicks spent much of the time that they were observed today in the northern end of the lagoon near the railroad bridge. Despite the fact that all of the chicks spend at least some time ranging and foraging independently (either alone or in proximity with one or more of its brood mates), it is still common to see members of the brood split between the two parents. For example, at one point

today two of the chicks were accompanying one of the parents in the northeast section of the lagoon while one of the other chicks was accompanying the second parent in the northwest section.

Up to this point in my observations of this grebe family the parents have (with only one exception) conducted their foraging exclusively along the edges of and within the emergent dense reeds. The chicks have also followed this pattern.

**<u>10/3/16</u>:** All four of the chicks seen. The chicks were quite mobile today. They ventured the complete length of their parents' territory and beyond, from the far south end through the far north end and into the common (overlap) area.

As is now usual, the chicks spent time accompanying their parents and also ranging and foraging independently. At one point three of the chicks were foraging independently as a group in and along the dense reeds lining the north bank of the river in the overlap area (this is the farthest north that the chicks can go without entering the territories of the neighboring grebe pairs—up to this point in time the chicks have restricted their ranging solely to within their parents' territory and the overlap area).

A few of the chicks were seen briefly stretching their wings. Their flight feathers are well developed and if not already capable of short flights these two chicks are right on the verge.

**<u>10/4/16</u>:** All four of the chicks seen. Competitive interactions among the chicks were particularly prevalent today. This was largely because for a considerable amount of time during today's observations one of the parents was accompanied by three of the chicks—a setting that readily promotes such interactions, especially when feedings are involved. These interactions today mainly involved the three smaller chicks (the largest chick was mostly off by itself throughout today's observations).

These competitive interactions don't occur solely in active feeding contexts (i.e., when the parent is offering up or foraging for a prey item). They also occur at times when the parent is resting in open water while accompanied by two or more chicks. Typically in such cases the seemingly ever-hungry chicks eventually become restless and begin soliciting the parent for food, trying to coax it into engaging in a foraging/feeding foray. At such times a larger, more dominant chick may begin to assert itself over its subordinate brood mates. As one of the latter engages in a solicitation, the dominant chick may peck at and drive it away from the immediate vicinity of the parent (the chases by the dominant chick are generally short, only about 1-2 meters or less—it apparently doesn't want to move very far away from the parent and moves back in close to the parent after the brief chase). In this manner the dominant chick will sometimes control the proceedings among the chicks, repeatedly driving the subordinate chick(s) away from the parent. Even after they succeed in coaxing the parent from its rest period and the parent begins to make its way toward the dense reed beds (to forage for them) the dominant chick might continue to assert its dominance in order to maintain a position closest-in around the parent. The subordinate chicks usually are undeterred by this and follow along on the perimeter. In order to urge the parent on, the chicks commonly will continue with their solicitations while the parent is en route to the reed beds. Because they are being chased away from close proximity to the parent, the subordinate chicks modify their food solicitations accordingly. Being prohibited from coming in close they are unable to engage in the usual beak to beak touching with the parent. In being compelled to deviate from the standard display pattern, they omit that characteristic aspect of soliciting and instead just pass by a short ("safer") distance in front of the parent while assuming the typical exaggerated solicitation posture (low profile with head and neck stretched out in front at a low angle) and the typical brief burst of exaggerated paddling while in this posture. Even these modified solicitations may draw the ire of the dominant chick who sometimes will peck at and briefly chase a chick that was engaging in such.

As previously noted, despite being continually (though not always) bested and driven to the periphery by a dominant chick in group feeding situations, the subordinates nevertheless usually remain in proximity to the parent. This is hardly surprising, given that the chicks are dependent on the parents for their food. They also seem to realize that, if patient, their situation will likely improve. The dominant chick will typically move away with a prey item that it has taken from the parent as a means of trying to keep its competing sibs from attempting to take it away. By remaining around the parent in the immediate aftermath the subordinate chick(s) then has first crack at obtaining the next prey item that the parent catches and offers up, providing this occurs while the dominant chick is still busy with the prey item it had taken (obviously, this favorable scenario for the subordinate chicks won't materialize in situations in which the parent oversees the feeding all the way to the point at which the dominant chick swallows the prey item). In addition, after eating a prey item (especially a large one) or two the dominant chick commonly becomes temporarily satiated and will cease seeking food from the parent for a period. During this period the subordinate chicks are presented with a golden opportunity in which they have a chance to obtain food from the parent without the overpowering presence of that dominant chick. At their current advanced stage it is common to see the subordinate chicks engage in some foraging dives of their own in the general vicinity of a foraging parent in feeding situations in which a dominant chick is asserting itself. In these situations it appears that (perceiving that they will likely be bested by the dominant chick) the subordinate chicks are biding their time, waiting for the dominant chick to remove itself from the picture after taking a prey item from the parent, while in the meantime taking a shot at catching a prey item on their own.

All four of the chicks are now capable of handling and eventually swallowing large crayfish with little or no pre-working by the parent. An interesting example of this was observed today. It stemmed from a group feeding situation involving the three chicks noted above (the second largest chick and the two smaller ones, the second largest being dominant over the other two). What was particularly interesting here is that it was one of the infrequent cases in which a subordinate chick bested the more dominant chick and assumed control of a large crayfish offered up by the parent. [The dominant chick had claimed the prime position close to the parent while the parent was working the crayfish. When this subordinate chick tried to move in close to the parent the dominant chick immediately charged at and chased the subordinate. However, in a very slick move, while being chased away the subordinate chick deftly made a quick turn to the left and rapidly circled back to the parent and took the crayfish from it before the dominant chick could stop its subordinate sib from doing so.] The parent had only briefly worked the crayfish and it appeared to be essentially still intact and moving. The chick immediately moved away to the west with the crayfish and began working it on the go (the chick eventually ended up well out in open water after having traveled about midway across the lagoon; interestingly, neither of the other chicks pursued it, choosing instead to remain with the still foraging parent—it seemed somewhat surprising that the more dominant chick didn't pursue this subordinate chick and try to take the crayfish from it). In a manner similar to the parents, the chick vigorously worked the crayfish largely by biting down forcefully on it and giving it a quick, firm shake, essentially flinging the carcass onto the surface as it slipped from the chick's beak. The chick continually repeated this process—on several occasions the carcass sank down in the water and the chick had to dive completely below the surface in order to retrieve it. In this fashion the chick was able to make numerous breaks in the exoskeleton and gradually reduce the mass of the carcass by plucking out and swallowing bits of exposed flesh. After considerable such work the chick finally reduced and rendered the carcass flexible enough to the point where it was able to swallow it. This entire process took approximately 2 ½ minutes.

Considering how much time the largest chick is spending ranging and foraging independently, it appears that should something happen to the parents it would now be able to survive on its own.

**10/5/16:** All four chicks seen. Over the past few days (including today) several of the chicks have been observed stretching and briefly flapping their wings. Their wings appear to perhaps be well enough developed for flight though I haven't seen any of them fly. However, as previously noted, members of this grebe species are rarely seen in flight: given their basically aquatic lifestyle, they simply don't have much reason to fly during the course of a typical day. Thus, I am uncertain at this point as to whether or not the chicks are actually capable of flight (I still have yet to see either of the parents take flight).

One of the observed feedings today stood out because of the late assistance the chick that was in control of the prey item received from the parent. After taking a rather large crayfish from its parent, this chick immediately moved away into the open water in the mid-section of the lagoon where it was alone and quite distant from any of its brood mates. There, in relative seclusion the chick was able to intently work the crayfish, free of any competition and immediate worries of theft. After having worked the crayfish for a little over two minutes the chick attempted to swallow it. The chick tried for about 18 seconds to work the crayfish down its throat before finally giving up, the crayfish still being too bulky to swallow. Following this unsuccessful attempt the chick resumed working the carcass. Though some distance away, the parent saw the great trouble that the chick was having in its attempt to swallow the crayfish. As the chick continued with its efforts, the parent intently swam directly toward the chick. The parent reached the chick approx. 17 seconds after the youngster had aborted its attempt to swallow the crayfish. At that point the chick was still busy working the crayfish and when it dropped the carcass onto the surface the parent picked it up and worked it very briefly. It did so for only 4 seconds before the chick took the carcass form the parent (it was obvious that the parent's intent was to assist the chick and not to take the carcass away for itself). Shortly thereafter, the chick again attempted to swallow the crayfish. This time it succeeded, taking only about 9 seconds to work the carcass down its throat. It was not clear whether or not the parent's very brief working of the carcass had actually assisted in rendering the carcass capable of being swallowed (probably not because it seemed much too brief a period—the chick's continued working of the carcass after its unsuccessful swallowing attempt is likely what actually did the trick). What was interesting about this incident is that it demonstrated that even at this latter stage of the chicks' development (a time when they have become capable of processing and swallowing large prey items on their own) the parents are still monitoring their progress after having taken a proffered food item and are ready to offer assistance if needed. However, unlike when the chicks were younger, the parents no longer closely shadow the chick that has the prey item while monitoring it (i.e., when the chick moves a distance away after taking the prey item; they do still closely shadow the chick when it remains in the vicinity of the parent after taking the prey item).

**10/6/16:** Three of the chicks definitely seen; the fourth was possibly seen but the probable sighting was too distant to be certain.

Observations of at least a few 'parent to chick' rebuffs have become a daily occurrence. A few of the rebuffs seen today were a bit more aggressive than usual and one of them was very notably more sustained. Parental rebuffs are typically brief (though they may be repeated a number of times within a short timeframe), generally consisting of one or a few pecks and/or a short charge/chase of about a half to a few meters. This particular incident continued for approximately 25 seconds. As is usually the case with parental rebuffs, this one began when the chick moved in close and solicited the parent for food. The parent responded with an aggressive peck at the chick. The chick immediately began to flee but got only about a meter away before the parent charged, pounced on and pecked it. This time the chick dove and fled underwater, resurfacing approximately 9 meters away from the parent. However, the parent wasn't finished with its rebuke of the chick. The parent quickly spotted the chick when it surfaced and promptly and intently began moving directly toward it. In turn, the chick quickly caught sight of this and fled from the fairly distant but approaching parent. The chick initially headed toward mid-lagoon but then circled back toward the west side. Ultimately, its maneuvering was to no avail as the parent steadily closed the gap. When the parent got to within about two meters behind the fleeing chick it put on a final burst, caught and aggressively pounced on and pecked at the chick. The chick again immediately dove and fled underwater. This time the parent did not pursue the chick when it resurfaced and the incident thus was concluded. It was not clear as to why the parent persisted in its rebuff of this chick for such a long period—this was easily the longest such interaction that I have observed.

The chicks ranged extensively today, venturing from the south end of the lagoon all the way to the north end of the overlap area (all keeping within their known range). While two of the chicks were in the overlap area I finally had the chance to witness an inter-group interaction between individuals of the neighboring pairs/family. Up to this point I have only seen situations in which either an individual(s) of only one of the three neighboring pairs was in the overlap area or an individual(s) from two of the pairs was in this area but they were very widely separated and never interacted. Today one of the chicks had the misfortune of passing close by an adult from one of the neighboring pairs. When the two got close the adult wasted no time in attacking the hapless chick. It forcefully pecked at the chick and then grabbed and held the youngster by the scruff and viciously shook its beak while maintaining its hold on the back of the chick's neck. The chick was soon able to slip the adult's grasp and immediately dove and fled underwater (the attack lasted approximately 4 seconds). The adult did not pursue the chick, who remained a long, safe distance away but did not leave the overlap area at that time (the adult soon worked its way out of the overlap area). Fortunately, the chick showed no outward signs of having sustained any injuries from this encounter. [Interestingly, both of the parents still retain the black throat patch characteristic of adult breeding plumage while this is absent from the adults of the two neighboring pairs (now in winter plumage). This serves as a convenient aid in distinguishing the parents from these neighboring adults.]

**10/7/16:** Three of the chicks were definitely seen; the fourth was possibly seen but the pertinent sightings were too inadequate to confirm this and only three chicks were simultaneously visible at any one time.

Most of the feedings observed today occurred while the parent was being accompanied by only one chick (at times each of the parents simultaneously had a chick accompanying it). With the absence of a competing sib in these feedings, the chick had no reason to move far away with the prey item and remained in the general vicinity of the parent. As has always been the case in such situations, the parent shadowed the chick as it worked the prey item (a crayfish in each instance) and occasionally assisted the chick. In the situations today when a parent was being accompanied by two chicks, some of the usual competitive interactions between chicks occurred, including brief chases by the more dominant chick aimed at keeping the subordinate at a distance and thus securing the closest, most advantageous position to a foraging parent. Also as has now become usual, some parental rebuffs of chicks were seen.

**10/8/16:** All four chicks seen. The largest chick continues to spend the great majority of its time ranging and foraging independently and no longer spends much time accompanying and taking food from the parents (it is mainly due to the latter points that I see this chick considerably less often than its three sibs). In fact, at this point it seems to essentially be independent though it remains within its natal range and does still get an occasional assist from the parents. There seems little doubt that if for some reason the parents were lost, this chick would now be able to survive on its own.

The other three chicks also continue to spend time ranging and foraging independently but not near to the extent of their larger sib. They clearly are progressing well toward independence but their dependence on the parents still seems obvious. However, one of these chicks has begun spending noticeably less time accompanying the parents than the other two.

**10/9/16:** Two chicks definitely seen today and probably a third (only two chicks were simultaneously visible at any one time today and possible sightings of other individuals were too distant for positive identification).

Though all of the chicks spend time ranging and foraging independently and are closing in on achieving independence, chicks are also still commonly seen seeking out parents for feedings—and they may move a considerable distance to join up with a parent. A prime example of this was observed today. One of the chicks was initially ranging and foraging independently among the dense reeds in the southwest corner of the lagoon. It then swam out of this corner into more open water, looking about as it did. By its manner it was clear that the chick was visually searching, almost certainly looking to sight one of the parents. It soon caught sight of a parent with one of its sibs, who were across the lagoon near the reed beds along the southeast edge of the lagoon. The chick then quite purposefully set off toward the pair, crossed the lagoon and joined them.

Since the time when all of the chicks became capable of working and swallowing a large crayfish on their own, a notable change has taken place in regard to the feedings. Prior to this (especially when the chicks were still small), it was often the case that several of the chicks would obtain some food from a single proffered crayfish. This was because the feeding parent had to significantly work a large crayfish before one of the chicks was finally able to take control of and swallow it. While the parent was working the crayfish, one or more of the chicks might pluck off a scrap of flesh from the carcass (or grab a scrap that had detached) or the parent might pluck off a scrap and hand it off to a chick.

However, when the chicks became capable of working a large crayfish on their own the feeding situation changed into one in which usually only a single chick obtained food from each proffered crayfish. This is because in a group feeding situation a chick could quickly take the crayfish from the parent (since it no longer needed the parent to work the crayfish) and immediately move well away from the pack and work the crayfish on its own. Nevertheless, the odd instance has still occurred when more than one chick obtains food from a single proffered crayfish. The few observed instances of this have typically been cases in which a subordinate had control of the crayfish and had worked and eaten a portion of it before the remaining portion was taken away by a more dominant chick.

**10/10/16:** Only two of the chicks definitely seen today. The largest chick has not been seen for the past two days. It appears almost certain that it has fledged and left its natal range. Though I never observed it take flight, its wings were very well developed and looked quite capable of flight (also, since my daily observations cover only a small sample of the chicks lives it is certainly very possible that the chick has flown at times when I wasn't present to witness this). In addition, in the period leading up to the last sighting, this chick was already essentially functioning independently. Because I didn't actually observe the chick leave, the possibility that it may have died or perhaps become a victim of predation cannot be discounted. However, given the general circumstances it is most probable that the chick has fledged and left.

Over the past few days only two of the chicks (the two who were definitely seen today) have been accompanying, soliciting and taking food from the parents on a regular basis (though they also spend considerable time ranging and foraging independently). At times during today's observation each of the parents simultaneously had one of these chicks accompanying it (and thus, neither chick had a competitor that it had to contend with). It is interesting that one of these chicks is the second largest of the brood (now the largest with the former largest apparently having fledged away—though the size disparity between this and the other two chicks is no longer very noticeable). What is interesting here is that this chick began diving, ranging and foraging independently a notable amount of time before these other two chicks were observed doing so. It was thus well ahead of them in this respect. One might then expect that, like the largest chick, it would now be spending notably more time ranging and foraging independently than those two. Instead, it is spending more time accompanying the parents than either of them. Perhaps it is just employing a different strategy from that of the largest chick by taking advantage of now being the most dominant of the remaining chicks? As such, it is pretty much assured that it will win out in competitions for food items proffered by a parent. In this case, then why not spend considerable time taking advantage of the superior hunting skills of the parents (i.e., as long as the parents are still willing to conduct feedings—the parents continue to sometimes rebuff chicks but also continue to conduct feedings)?

The third chick also continues to accompany and take food from the parents but now does so notably more intermittently during the course of an observation period than the other two chicks.

**<u>10/11/16</u>:** Three chicks seen today. The fourth chick (the largest of the brood who now appears to have achieved its independence) may possibly have been sighted in the overlap area. However, the sighting was much too distant to be certain of the identity. In addition, the other three chicks could not be accounted for visually at the time of this sighting so it could not be ruled out that the chick which was sighted was one of them.

The chicks spent much of the observation period ranging and foraging independently in the northern end of the lagoon and overlap area. Here again, similar to the parents, they foraged exclusively along the edge of and within the dense emergent reed beds that ring the edges of the lagoon and the north bank of the river in the overlap area. What was particularly notable today was that I finally had an observation of one of the chicks catching a prey item (a crayfish) on its own. Though this was the first time that I actually had the chance to witness a chick catching a prey item, the chicks have undoubtedly been doing so for quite some time. Considering the lesser amount of food that the chicks have been taking from the parents since they began spending considerable time ranging and foraging independently, it seems they would otherwise not be meeting their daily nutritional needs.

One of the chicks was observed briefly flapping its wings. Its flight feathers appear to be fully developed. It is quite probable that all of the chicks are now capable of at least short flights. The fact that they haven't yet been seen taking flight may just be down to the point previously noted, that pied-billed grebes generally are seldom seen flying.

Overall, these three remaining chicks look healthy and have progressed very well. They are near adult size and their body colors are essentially similar to adult plumage. However, on the head and neck plumage they still sport the black striping on a white background that is characteristic of juveniles (I use individual differences in the pattern of the black striping to identify the individual chicks), nor have their beaks quite attained fully adult proportions and color. In general, these chicks are well advanced and appear to be just days away from becoming fully independent.

**<u>10/12/16</u>:** All three of the remaining chicks seen. They ranged widely throughout the lagoon. Their activities and behavior today were basically routine for their current stage of development. They ranged and foraged independently and also spent some time accompanying the parents, soliciting and taking food from them. At one point all three of the chicks were accompanying one of the parents. Also as is now usual, some competitive interactions among the chicks and some parent to chick rebuffs were seen.

**10/13/16:** Two of the chicks definitely seen and possibly the third. Today's observations were very significant in that both of these chicks were seen taking a short, skipping flight just above the surface of the water. This confirms what has been suspected from my views of the chicks wings over the past days (i.e., that they appeared to be well enough developed for at least short flights). With confirmation of the chicks having now fledged (i.e., by the most minimal definition of this term) and their growing ability to range and forage independently it appears certain that it is just a matter of days before they become fully independent.

Besides ranging and foraging independently, the chicks continued to spend some time accompanying and taking food from the parents. The two chicks seen today remain the only ones who continue to accompany, solicit and take food from the parents on a regular basis (the third chick has become notably more independent in this regard but as of yesterday's observations was still doing so intermittently).

**10/14/16:** All three of the remaining chicks seen. The same two chicks (and one of them in particular) continue to be the only ones who currently spend some time accompanying and taking food from the parents on a regular basis (i.e., besides ranging and foraging independently). The third chick does so occasionally but is notably more independent than the other two in this regard.

An interesting reversal of dominance in the competitive interactions between the two chicks noted above was observed today. As this intriguing incident began, the more dominant of the three remaining chicks (the one that had been the second largest in this initial brood of six) was accompanying one of the parents. This parent was just sitting stationary in open water in the mid-area of the lagoon, resting, preening and apparently simply presenting a conspicuous territorial presence. The chick remained close around it, periodically soliciting the parent in an attempt to coax it into engaging in a feeding/foraging foray. At this point the second chick (who had always previously been subordinate to this other sib) was spotted well out in the northeast section of the lagoon. It was a considerable distance away from the pair but was obviously swimming toward them. Curiously, as the approaching second chick neared within approximately 2 meters, the first chick (who was sitting close to the parent) suddenly moved several meters away to the west as though deferring in a displacement interaction. As the second chick reached the parent it immediately engaged in a food solicitation as it passed by a short distance (just out of reach) in front of the parent (this was the type of modified solicitation that lacked the typical beak to beak contact, as described in the notes for 10/4). The chick continued on a meter or so farther to the west as it concluded its solicitation. While it did so, the first chick moved back to the parent and (in the typical manner) also solicited it. These back to back solicitations from the two chicks succeeded in mobilizing the parent who then set off toward the southwest section of the lagoon. Initially, as they set off the first chick followed along next to and right behind the parent while the second chick proceeded several meters in front. When the trio had entered well into the southwest section of the lagoon (and while still in open water) the parent then slowed, nearly to a stop. As it did, the second chick (still proceeding a few meters in front of the parent) promptly circled back to and solicited the parent in an obvious attempt to coax it into continuing on to the reed beds and engage in feedings. Curiously again, as the second chick drew close to the parent while soliciting it, the first chick (who was right behind the parent) quite deliberately moved away from the parent and began swimming west toward the reed beds (once again, as though deferring to the second chick's approach). Upon seeing this chick moving away, the second chick immediately switched from concluding its solicitation to directly following/pursuing the first chick. In what was clearly a dominance interaction the first chick paddled at a fairly quick pace toward the reed beds in its attempt to get away from its intently pursuing sib. In its focused pursuit the second chick followed closely behind, maintaining an oscillating gap of from one meter to a half meter. As the pair neared the reeds the second chick put on a sudden burst of speed, charged and pounced at the first chick. The first chick immediately responded by rapidly turning north and putting on a burst of speed of its own and was thus able to pull out a short gap. The second chick continued its hot pursuit and shortly thereafter attempted a second charging burst but the first chick responded quickly with a speed burst of its own and was able to maintain a gap. With its sib still in pursuit, the first chick again changed course and headed out toward mid-lagoon and open water. Midway out in the lagoon the first chick attempted a different strategy to elude its sib: it dove and fled underwater, resurfacing approximately 9 meters to the southeast. It then began making its way toward the southwest corner of the lagoon. However, the second chick remained surprisingly persistent. It quickly caught sight of its resurfaced sib and promptly resumed its pursuit. The path of this chase took the chicks right by their still stationary parent (still in the same location where it had previously stopped) and, as though not wanting to waste any opportunity, each chick in turn very briefly solicited the parent on the go as it passed by (the solicitations had no effect and the parent remained in place). When the pair entered into the southwest corner area the first chick dove again in its attempt to elude its sib. It resurfaced a few seconds later and this time its sib significantly eased off on its pursuit. Its focused intent clearly subsided, its pursuit waned into a rather casual follow and the intensity faded away. The distance between the two thus greatly increased as the second chick just more or less kept track of its sib's general movements. The first sib soon worked its way into the dense reeds in the southwest corner where it went out of view. The second chick followed into this area shortly after and also went out of view. I was thus unable to see if anything further transpired between them (though the situation had pretty much defused by then).

The duration of this interaction was quite remarkable. It lasted for approximately 2 ½ minutes, which is far greater than any other such dominance interaction that I have observed (such dominance interactions between sibs have typically lasted only seconds). The reversal of dominance between these two chicks was quite sudden and there were no clues in my previous observations as to what may be responsible for it. And until I obtain further observations of interactions between these two I will not know if the reversal observed today is permanent or merely a temporary or situational thing. Regardless, the incredible persistence of this chick was suggestive of an individual working hard to press and consolidate a newly gained advantage.

**10/15/16:** All three of the remaining chicks seen. Today was a very significant day in the lives of these chicks. All three of them were seen taking several fairly long flights (essentially the length of the lagoon from the north end to the south end and vice versa) just above the surface of the water. The flights occurred over a short period. After the first of the chicks took flight this appeared to spur the others on as each then also engaged in flights. Given the nature and sudden appearance of these longer flights and the fact that the chicks all appear to be capable of feeding themselves, I suspect that when I check on this grebe family tomorrow I will find that the chicks are gone.

The chicks are large though not quite fully adult size. The size disparity between them and their parents is minimal but still just enough to be able to differentiate the parents from the chicks by size alone. However, when at a distance and the plumage isn't discernable (the chicks still retain their juvenile plumage) it is now difficult to tell them apart strictly by size, especially if at least one of each isn't present for a ready side-by-side size comparison.

Though all of the chicks now appear to be capable of feeding themselves they still spend some time accompanying and taking food from the parents, notably the two that have been doing so on a more regular basis (in fact, several feedings were observed today).

Unfortunately, no further competitive interactions were seen today between the two chicks noted yesterday who displayed a reversal in their dominance order. Thus, it couldn't be determined whether or not this reversal between them has held.

**10/16/16:** Both parents were seen today but none of the chicks. It thus appears that, as suspected would be the case, the longer flights observed yesterday were of significance and the chicks apparently have left their natal range and are now independent (nor were they seen in the immediate area—I also checked for them in the adjacent section of the river to the north and the east lagoon).

I will continue to check for them over the next several days in order to confirm that they have left (I cannot discount the possibility, however slight, that they were present but just tucked in amongst the reeds and hidden from view). It also needs to be determined if their move away at this time has been permanent—in other words, will they return and continue to use their natal range (intermittently or otherwise) in the immediate future. Because I didn't actually see the chicks leave the area and haven't seen them since yesterday it can't be ruled out that one or more of them were victims of predation (or otherwise have died). Of the probable conclusions, this seems highly unlikely. Rather, given the circumstances (i.e., the sudden appearance of the longer flights by all three yesterday, the fact that they appeared to be ready for independence, and the fact that all three went missing on the same day) it is almost certain that the chicks have left and are now independent. If this is indeed the case, it would be interesting to know if they left as a group and are staying together at this point.

**10/17/16:** Once again, both parents were seen but none of the chicks. The chicks were not seen in the surrounding area either (the adjacent section of the river to the north and the east lagoon). At this point it appears safe to conclude that the chicks have left their natal area and are now independent. Unfortunately, it isn't known where and how far away the chicks have moved.

Checks for the presence of the chicks in their natal area and the immediate vicinity will continue but from this point the checks will be bi-weekly rather than daily.

**11/10/16:** In the bi-weekly checks that have been conducted since daily observations were concluded on 10/17 both parents have been seen each time but none of the chicks. However, the situation was different today. For the first time since the three remaining chicks left their natal area (on 10/15 or 10/16) and became independent, one of the chicks was seen in the lagoon. The chick was initially sighted as it emerged from the reed beds in the southwest corner of the lagoon where it apparently had been foraging. At the time, both of the parents were out in the mid-section of the lagoon. One of them spotted the chick and began slowly moving toward it from its distant location. The chick caught sight of the parent while it was still distant and engaged in a rapid, sinking dive, apparently in an attempt to elude the approaching parent. [Note: In specific situations pied-billed grebes utilize a distinct method for submerging. Besides launching headfirst into a dive, they have the capacity to just sink below the surface while remaining sitting in place and they can do so quite rapidly (similarly, they have the ability to control their level of buoyancy—for example, a grebe can keep only its head at the surface while the rest of its body is submerged). This method of submergence is typically employed in circumstances in which the individual wants to remain inconspicuous or escape from a hazardous or potentially hazardous situation (including another grebe). The obvious advantages of utilizing this method in such cases is that it is quick and is a much less conspicuous way of submerging—the physical movement involved is minimal, thus making the individual more difficult to detect by means of its movement. Once submerged and hidden from view the individual can then employ either of several general strategies to elude the source of distress: it can remain hidden underwater for a period of time (i.e., within its capability) long enough for the hazard to clear the area and then fully resurface in the same location; it can travel underwater into nearby cover; or it can travel underwater for a long distance and resurface well away from (and hopefully unseen by) the source of distress.] The chick traveled out of sight underwater all the way across to the east side of the lagoon and resurfaced there, near the reed beds. The parent soon re-sighted it and altered its course toward the chick. Shortly after resurfacing, the chick spotted the still approaching parent. It again engaged in a sinking dive (in what seemed an obvious attempt to elude the parent) and was not sighted again (perhaps resurfacing somewhere hidden within the reeds?). The parent continued on to the area where the chick had dropped from sight and looked around for it. It remained in this area for several minutes searching for the chick. After failing to re-sight the chick it then moved back out and joined its mate in the mid-area of the lagoon.

This interesting incident obviously was precipitated by the presence of the chick within the parents' territory, the chick apparently having returned to its natal area to forage. Judging by the chick's elusive behavior and the keen approach and search by the parent it appears that (after having fledged and moved away) the chick(s) may no longer be welcome in the parents' territory (i.e., its natal range). It is also possible that (given the long distance from which it was sighted) the parent didn't specifically know that it was its fledged chick and was merely approaching what at that point was an unidentified grebe in its territory. Regardless, the chick's behavior indicated that it was definitely trying to elude the parent (though, in a similar vein, it can't be known if the chick knew that the approaching adult grebe was one of its parents). As a further consideration it should be noted that my bi-weekly checks here comprise only a tiny sample of the total time. It is thus quite possible that this chick and/or others from the brood may have returned to the lagoon during this vast majority of time when I wasn't present to witness it. If such is the case for this chick then this opens up the possibility that its elusive response to the parent could be based on a previous (negative or aggressive) encounter with a parent after returning here.

---------------------------------------------------------------

During my bi-weekly checks through to the end of the year one or both of the parents were sighted on each occasion but there were no further chick sightings.

**Basic summary notes:** The parents have claimed the west lagoon as their territory and it is within this and the modest overlap area north of the railroad bridge that the chicks were reared. It is not known whether this brood was their second clutch of the season or a late start on a first clutch.

The parents were quite proficient and did an admirable job rearing their brood, successfully fledging 4 of 6 chicks. Both of the chicks who were lost were notably smaller than their sibs, right from the time when this brood was first seen. They had much difficulty competing with their larger sibs during feedings (competition among the chicks during feedings in group situations was rather intense; an apparent strategy employed by chicks to reduce or avoid competitive feeding situations was to attempt to isolate themselves with one of the parents). Conversely, one of the chicks in particular was notably larger and more advanced than its brood mates right from the start of my observations and achieved its independence about 5 days or so before the other three were able to (the very notable size disparity among the chicks suggests that hatching in this clutch was staggered over a period of at least several days). Three of the chicks achieved independence and moved away from their natal range 40 days after first being sighted; the largest chick did so in about 35 days (the chicks were small when first seen on 9/6 but I don't know exactly how old they were at that point—they were already beyond the stage in which young chicks climb onto the backs of their parents but still retained the plumage of younger chicks). The chicks fledged (i.e., were capable of at least short flights) several days (and probably sooner) prior to moving away and becoming independent.

In working their way toward independence the chicks followed a general pattern of spending increasingly more time ranging and foraging independently. As part of the weaning process, parental rebuffs of chicks (first observed on 9/28) became daily occurrences though feedings by the parents nevertheless continued right on through the last day before the last of the chicks left the natal range and became independent.

Observed feedings consisted almost solely of crayfish. Only four of these involved a prey item other than crayfish: three involved a small fish and one involved what appeared to be a large insect nymph. Both of the parents fed the chicks and it was common to see the brood split between the two parents.

The habitat of dense reeds that ringed the lagoon was vital for the chicks and parents. The parents did all (with one exception) of their observed foraging along the edge of and within the dense stands of emergent reeds; the chicks also did so when they began foraging. The reeds were also important as a place in which to rest and seek cover. A particular location among the reeds on the mid-western edge of the lagoon served as a kind of home base area for the brood (perhaps it was within the reeds in this area that the nest had been located).

# West Lagoon Family Album

*A few general notes regarding the West Lagoon family photos:*

- In the following photos the parents (i.e., the West Lagoon pair) have been identified by their gender. This gender identification is something that was done after the fact—specifically it wasn't done until a little over 8 months after this study was concluded. In other words, I did not know which of the pair was the male and which the female during the period that the study was conducted. This is because the plumage of pied-billed grebes is similar for both sexes. Therefore the gender of the individual members of the pair couldn't be distinguished by this prominent physical trait or by any other physical trait that I was aware of at the time. Furthermore, no copulations (the pair's brood had already hatched out when first discovered) or other unequivocal gender specific behaviors were seen during the course of the study that could have aided in gender identification. Thus, I had no idea as to whether the more subtle physical differences that I used to distinguish the individuals of the West Lagoon pair from each other were gender related or just down to general individual variability. Fortunately, during my study of the South River pair in the following breeding season I had the opportunity to observe 5 copulations (the relative positions and behavior of the copulating individuals are gender specific so it is unequivocal as to which one is the male and which the female) and through these I discovered a way to physically distinguish males and females of the species. Of particular note here is the fact that the more subtle physical traits that I was using to distinguish the individuals of the South River pair from each other were the same general traits that I had used for distinguishing the West Lagoon pair. The specifics of this matter are discussed in the comments section of the 6/24/17 field notes for the South River family and the reader is referred to these.

- While viewing Figures 1, 3, 6, 7, 10, 12, 13, 14 and 21 the reader should take special note of the lush, dense, continuous stand of reeds that line the shore in the background. This comprises the east shore reed habitat of the West Lagoon. What is of particular interest here is that this lush continuous stand of reeds that lined the east shore would be utterly and completely obliterated (except for the section around the mouth of the inter-lagoon channel) during the ensuing rainy season. This occurred during periods of uncharacteristically heavy rains which caused the river to flow very wide and very strongly (when the river flows strongly the West Lagoon is transformed from a lazy lagoon to a proper river). The east shore comprises the outside bank of the river in this section and thus takes the brunt of its erosive force. In this case the east shore reeds were completely scoured away by the erosive force of the river. This matter is discussed in the South River family study as well as in sections of Chapters 5 and 6. A reminder here of how vital the reedy habitat is for breeding grebe pairs and their broods in this particular locality.

**Figure 1.** The six West Lagoon family chicks with the female parent. Note the considerable size disparity that exists among the brood.

**Figure 2.** The notable size disparity among the West Lagoon family chicks is strikingly illustrated in this photo of a large and small chick.

**Figure 3.** The competition between the chicks for feedings was quite intense. Here, after having just received a crayfish from a parent, a chick flees from two of its hotly pursuing sibs in an attempt to keep them from stealing the crayfish.

**Figure 4.** Three of the chicks swarm around the male parent as he administers a crayfish feeding.

**Figure 5.** Three of the chicks with the male parent.

**Figure 6.** A chick swallows a crayfish that it had obtained from a parent. The chick is being pursued by two of its sibs. The parent (third in the line from the left) also follows along in order to oversee the feeding. An American coot is visible in the background.

**Figure 7.** One of the chicks flaps its as yet poorly developed wings. Note in particular how short the growing flight feathers are at this point (the photo is from 9/23/16).

**Figure 8.** A chick works on swallowing a crayfish that it has received from the female parent. The chick is being pursued by one its sibs while the parent follows along to oversee the feeding.

**Figure 9.** One of the West Lagoon family chicks (photo taken on 9/29/16).

**Figure 10.** Two of the chicks with the male parent (located between the two chicks). Note how the juvenile markings on the side of the head and side and back of the neck of the chicks differ from the solid coloring of the adult parent.

**Figure 11.** A dominance interaction among the chicks: a more dominant chick aggressively pecks, pounces on and chases a smaller, subordinate chick away from close proximity to the trailing parent (the male), thus securing this more favorable and advantageous position for itself.

**Figure 12.** One of the chicks in amongst the reeds.

**Figure 13.** A chick works on swallowing a crayfish that it has just taken from a parent. The parent remains alongside to oversee the feeding while a second chick pursues the first to contest it for the crayfish.

**Figure 14.** One of the chicks moving within the reeds.

**Figure 15.** One of the West Lagoon family chicks.

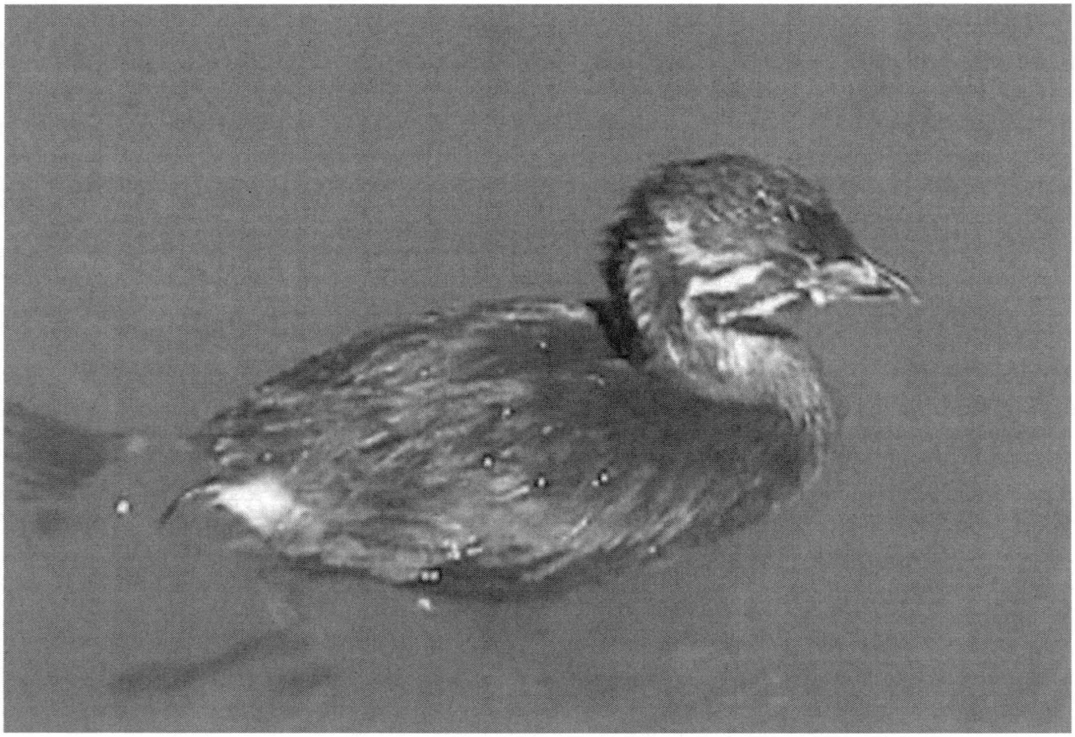

**Figure 16.** One of the West Lagoon family chicks.

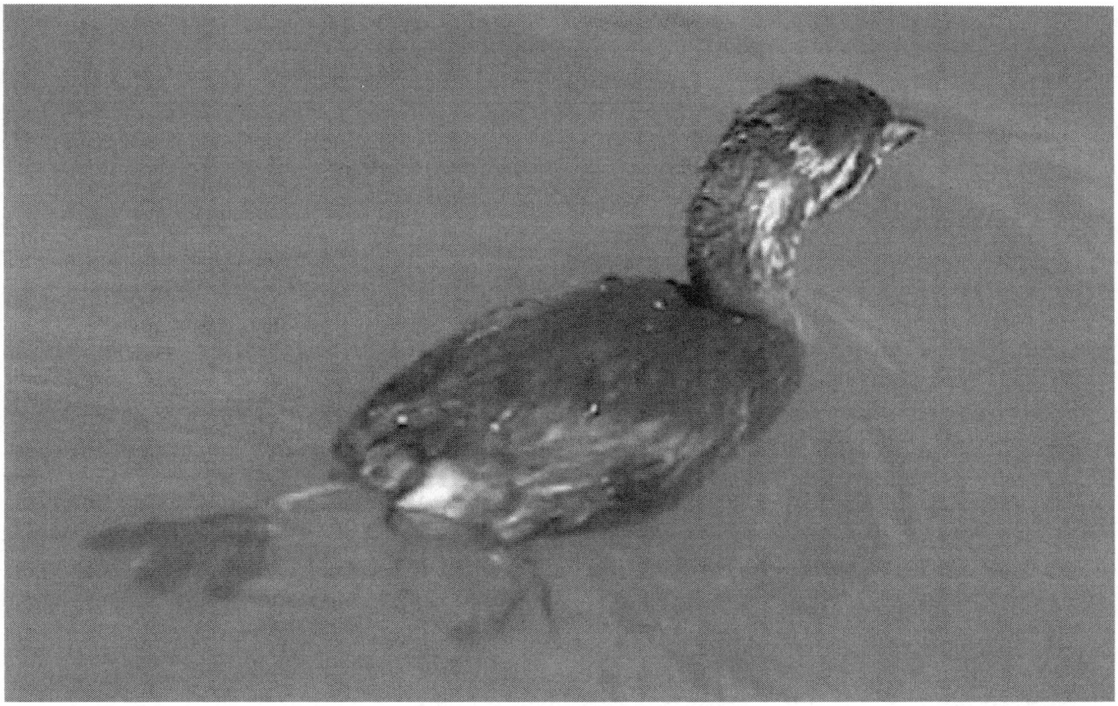

**Figure 17.** In this view of the same chick portrayed in Fig. 16 note the distinctive partially webbed, tri-lobed feet that the grebes use to good effect when paddling about on the surface (which is by far their most commonly utilized mode of locomotion).

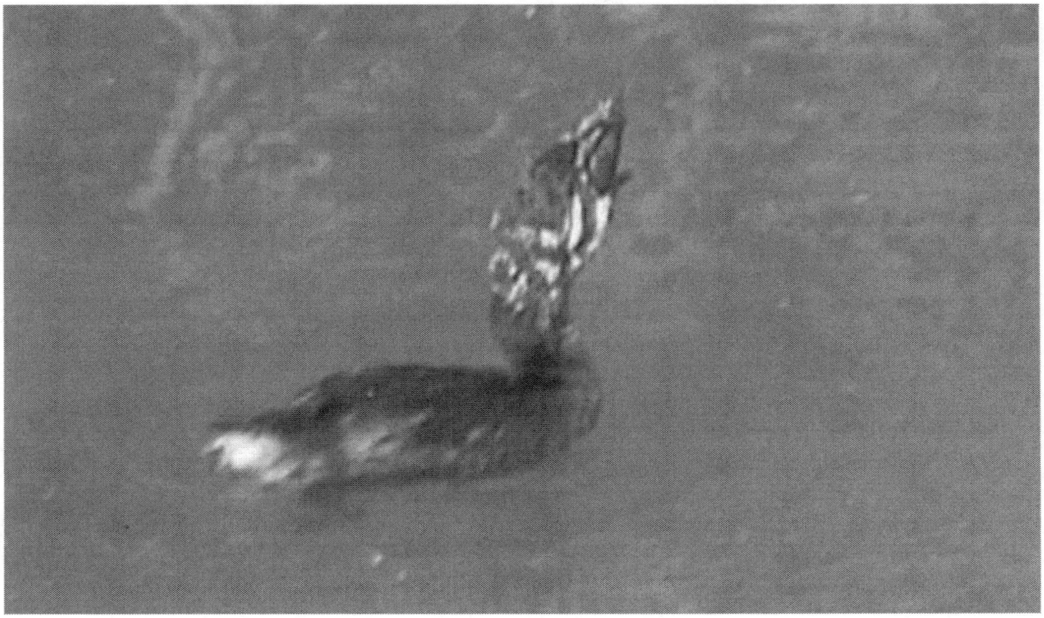

**Figure 18.** A chick swallows a crayfish that it had obtained from a parent.

**Figure 19.** A chick solicits the female parent. The chick is wary of a possible rebuff from the parent so it cautiously solicits her from a distance rather than from up close.

**Figure 20.** A chick accompanies a parent in the hopes that the parent will begin foraging and capture prey that it will then feed to the chick.

**Figure 21.** The male parent.

**Figure 22.** One of the chicks flaps its wings. At this point the flight feathers are fully grown or nearly so (this photo was taken on 10/11/16). Contrast the flight feather growth here with that seen in the photo that was taken earlier on 9/23/16 (Fig. 7).

**Figure 23.** A parent (second grebe from the right) and the three remaining chicks (at this point (10/15/16) only three chicks remained in the brood). This was the last day of the chick rearing period. By the next day all of the chicks had left the natal area and moved away. Note the stand of massive metal support beams for the railroad bridge in the background.

**Figure 24.** Now finished with their breeding/chick rearing for the season (and having successfully reared 4 of the 6 chicks of their brood to independence) the West Lagoon pair is here seen coming together for a greeting. One of the massive support pillars for the railroad bridge is visible in the background (this photo was taken on 10/17/16).

# Chapter 3

## Introduction to the South River Family Study

After conducting my study of the West Lagoon pied-billed grebe family in the 2016 breeding season I decided that I wanted to conduct a follow-up study in the 2017 season in the same general location (i.e., the estuarine area at the terminal end of the Ventura River). My plan was to conduct another case study on one of the grebe pairs in this area as the pair reared its brood. In my 2016 study I discovered that there were three mated grebe pairs who held territories in this area during that breeding season. These included the pairs whom I designated as the West Lagoon pair (my 2016 study pair), the East Lagoon pair and the South River pair. The West Lagoon pair's territory included the entire West Lagoon and the western half of the inter-lagoon channel (see the diagram of the study area). The territory of the East Lagoon pair comprised the northern two thirds to three quarters of the East Lagoon and the eastern half of the inter-lagoon channel. The South River pair's territory was located at the south end of the river immediately north of the West Lagoon (in other words, it bordered the north end of the West Lagoon pair's territory)—it consisted of the section of river extending north from the north end of the West Lagoon (marked by the railroad bridge) to about halfway from there to the Highway 101 overpass bridges.

Of these three pairs I considered only the West Lagoon and South River pairs as potential candidates for the projected case study. I ruled out the East Lagoon pair as potential study candidates because there are too many areas within their territory that I'm not able to view. My hope was that at least one of these two pairs would have a brood and thus provide me with the opportunity to conduct the study. If both had a brood then my preference would be for the West Lagoon pair. This would give me the chance to observe the chick rearing process for two different broods that were reared by the same pair. If the South River pair turned out to be the only option (as ultimately was the case) this would nevertheless also provide an interesting scenario. It would increase my sample size of observed pairs to two and allow me to compare the chick rearing process between the two pairs and families. While a paltry sample size of two isn't much better than one at least it begins to open the window on the kind of variability that exists in the chick rearing process for this species.

As with the West Lagoon family study, the aim of this one was to observe and describe the chick rearing process in detail for the chosen pair/family. I was especially hopeful of being able to observe the entire chick production/ chick rearing process this time. Recall that the West Lagoon brood had already hatched and had recently become mobile when I first discovered it. I thus never had the opportunity to observe the nest building, egg laying, incubation, hatching and early post-hatching periods. Fortunately, as things turned out I got the opportunity to observe all of these periods and, in fact, did get to observe the entire chick production/chick rearing process for the selected study family.

Over the past years this area (and California in general) has been experiencing a severe drought (the nearly unanimous consensus is that this drought is yet another product of human-caused climate change). A particularly strong El Nino event brought a temporary break to the drought during the 2016-2017 rainy season, bringing record rainfall to much of the state. These rains caused the Ventura River to flow very strongly during this period. By the time that the West Lagoon reformed in the dry season its habitat had been significantly altered by the erosive force of the river during the period when it was flowing strongly. It is probable that it is for this reason that the West Lagoon pair did not remain here and reassert their territorial claim for the 2017 breeding season (this matter is discussed in detail in the field notes and will not be repeated here). However, the South River pair did remain and they re-established themselves in their territory. Ultimately, they did nest there and were selected as the subjects for this study.

The study of the South River pair/family chick production/chick rearing period was conducted during the interval from 6/13/17 to 10/10/17 (a preliminary observation occurred on 6/9/17 during which the South River pair's nest was first discovered). The pair/family was observed on every day during this four month period. Thereafter, biweekly post-chick rearing follow-up observations were conducted until 10/30/17. Similar to the West Lagoon family study, this one was completely naturalistic. It was strictly observational with no manipulations whatsoever of the study subjects (also, only natural markings and features were used for the identification of individuals; no artificial means of individual ID were utilized). As with my 2016 study, the study site was located in the Ventura River estuarine area which is situated in the terminal section of the river. The study site mainly comprised the South River pair's territory (see the field notes below for details). Note that the pair's territory changed during the course of the study when the northern half of their territory was usurped by a neighboring pair (their territory initially did not include the West Lagoon—this was annexed to their territory as a compensatory addition after they lost the northern half).

Unlike the 2016 study which was spontaneously undertaken, I had time for planning and preparation prior to beginning this one. Most significantly, this meant that I had time to think about, design and refine the methodology that I wanted to use for the study. As noted above, this was another descriptive behavioral case study in which the selected pair/family was observed during the entire chick rearing period. The aim of the study once again was to describe in great detail the chick rearing process for this family and, if the opportunity arose, the entire chick production and nesting process. For this study I wanted to describe the chick rearing process in even greater detail; I wanted to provide a much higher resolution (and more intimate) picture of how this family progressed through this important endeavor. In contemplating this matter I felt that it could be best accomplished by shining a brighter spotlight on and more intimately following (documenting) the lives and roles of each individual in the family as well as providing a higher resolution, moment to moment look at the family's daily life. In order to accomplish this I designed a methodology in which the foundation of the daily observations was a "formal" observation period of one hour to one hour and 15 minutes in length (with a minimum length of one hour). During this formal observation period a continuous account was kept (as much as was possible) on what was occurring with the family and what each individual member was doing. This running account thus provided a much more detailed picture of the family, its structure and functioning, and the lives of each member. It brought more focus on each individual; also the mundane, routine aspects of their lives were recorded and not just those that were more noteworthy. Besides the above, these daily formal observation periods (which were daily one hour samples of the family's life) presented a look into the rhythm, patterns and flow of life for the family, the kind of happenings and events that confronted them as they went about their lives and what affect, if any, these had on the family (or just particular members) and their usual activities. (For lack of a better term these more specifically chronicled continuous, ad-lib observation periods will be referred to as "formal" observation periods as a means of differentiating them from the less formal "regular" continuous, ad-lib observations that were employed in both of the studies.)

My choice of a one hour time length for these formal observation periods was based on what I had experienced from my observations of the West Lagoon family. I found that in general, things often were fast paced. I realized from those observations that attempting to follow and continually record the activities of the family and its individual members would be difficult, intense and mentally fatiguing, and that at times things would happen simultaneously with different members of the family or in such quick succession that I wouldn't be able to keep up (also, depending on the distance and clarity of the view, identifying individuals could take some extra time and require extra-keen focus). Under the circumstances I felt that a one hour observation period would be more than enough for me to consistently handle. I also felt that because these observations were to be conducted on a daily basis a one hour period would be sufficient for my purposes. As things turned out, the one hour length worked out well, especially since there were other things that I wanted to attend to and accomplish during the time that I spent observing the family.

While I had also employed continuous, ad-lib observations in my study of the West Lagoon family the previous year, I didn't continuously chronicle what was happening from moment to moment in their lives as I did for the South River family. For that study I recorded my observations by summarizing what I saw as the observation period progressed as well as including detailed descriptions of any noteworthy behaviors, interactions and events that occurred. In comparing the two studies it is obvious how the addition of the "formal" observation periods to the South River family study enhanced its detail and completeness.

The plan for when to initiate the formal observation periods into the daily observations was dependent on the situation. Ideally, it was hoped that the study pair's breeding effort would be first discovered at a time when the pair had built (or was constructing) the nest but had not yet laid any eggs. In this scenario then the formal observation periods would begin when the first egg was laid (only "regular" daily continuous observations would be conducted prior to this) and continue through the duration of the period of daily observations (daily observations were to continue until at least several days after the chick rearing period had concluded; thereafter a period of biweekly follow-up "regular" observations would be conducted). If eggs had already been laid and incubation was in progress then the formal observation periods would be initiated when hatching began (or a few days prior if this could be determined). If hatching had already occurred and chicks were present then formal observation periods would begin immediately (i.e., from day one of the study). The study initially began under the second of the above noted scenarios: at the time that their breeding effort was first discovered the South River pair had already laid their eggs and were in the process of incubating them. However, the situation changed when disaster struck and both the clutch and nest were lost. As a result the unfortunate pair then had to restart from scratch, this then creating the first of the scenarios noted above. Thus, as things worked out the formal observation periods were initiated on 6/26, which was three days before egg laying began.

Even after the formal observation periods were initiated, the daily observations of the family consisted of much more than just these. At the conclusion of each day's formal observation period, regular continuous observations of the pair/family continued, typically for a period of about one hour to two and a half hours longer. Any noteworthy behaviors, interactions and incidents that occurred during these continued observations were duly recorded in the field notes. The utility of these continued observations extended beyond just increasing the amount of time that the family was observed. There was a somewhat restrictive aspect associated with the formal observation periods. This was rooted in the requirement of having to continually keep track of what was happening with the family as a whole; of attempting to simultaneously keep tabs on each member of the family. In the continued regular observations conducted after the formal observation period was concluded I was freed from this regimen. I was thus free to do other things such as taking some time to focus my attention on and follow just one particular individual or just one subgroup when the family was split up.

An important task that I regularly attended to during these continued observation periods was to work on chick identification. Being able to identify the individual chicks was vital for obtaining the kind of detailed observations and information that were required to fulfill the aims of this study. From the point when the young chicks first emerged from the reeds I spent time each day during these continued observation periods studying each chick's identifying markings. Recall that this was a completely naturalistic study so I had to discover and use natural markings and features that were unique to each chick in order to identify individuals. The most significant and useful of these (i.e., the ones that I mainly keyed on) were the unique (and sometimes rather subtle) individual differences in the general pattern of the plumage on the head and neck that is characteristic of chicks and juveniles of this species. The plumage on the chicks changed as they matured. Therefore, various markings that I had been keying on for each individual eventually disappeared as they matured and I had to find new markings/features to key on. In order to keep abreast of this dynamic situation and stay sharp on my chick identification I spent some time each day taking a focused look at each individual chick and studying its identifying markings/features.

Other than two exceptions, this was strictly a descriptive behavioral study. These two exceptions comprised the only things for which quantitative data were specifically collected. In regard to one of these I had the good fortune of getting to witness some copulations among the South River pair and thus was able to identify the gender of each member of the pair. Having obtained this knowledge I was curious about how equally the male and female of the pair would share incubation duty. Therefore, for each of the formal observation periods during the incubation period I kept track of how many minutes each of the pair spent incubating the clutch and how long each of their incubation shifts lasted.

I was also curious about the relative amounts of each general prey type that the parents fed to the chicks. In addition, it would be interesting to see if there was any difference in the predominant type of prey that was fed to the chicks during various stages of their development (notably, from when they were very young versus when they were older). To monitor this I recorded the type of prey that was fed in each observed feeding that occurred in each of the formal observation periods throughout the chick rearing period. Only feedings in which one or more chicks were actually seen swallowing the proffered prey item were counted (the few exceptions to this were instances in which, though not actually seen, there was absolutely no doubt that the chick ate the prey item).

Every day after the daily observations were completed (and sometimes also in between the formal observation period and the continued regular observations) I took a walk through the general areas surrounding the South River pair's territory in order to check on what was going on in these areas. I did this in order to keep abreast of the contextual situation in the general study area. The South River pair didn't exist in a vacuum. They were a part of their general area and their lives could be affected by incidents and situations (both physical and biological) around them. For this reason it is important to continually monitor the contextual situation when conducting studies such as this.

Due to the nature of the vegetation and topography surrounding the South River pair's territory the observation points from which to conduct the study were extremely limited. Each of the two main sections of the pair's territory (the Semicircle and the West Lagoon) had only a single good observation point. In fact, the section which I referred to as the Semicircle had only one realistic viewing point, period. The West Lagoon contained more than just one observation point from which the grebes could be seen. However, they were so distant from the sections of the lagoon that the family typically utilized that identifying the individual chicks from them was extremely difficult and often impossible. The reality was that there was only a single observation point that was close enough for consistent identification of the chicks—and even this was fairly distant, thus making chick ID more difficult (certainly more so than in the Semicircle) and sometimes impossible. [Of these two main areas, viewing typically was closer and much better in the Semicircle (especially regarding ease of chick identification). In this respect, it was fortunate that the Semicircle was the core area of the pair's territory (accordingly, this is where the pair nested) and the area that was used most by the family.] Unfortunately, neither of these observation points allowed for simultaneously viewing into both of the main sections. In other words, the West Lagoon couldn't be seen from the Semicircle observation point and vice versa. Nevertheless, there was an observation point from which all of the South River pair's territory was visible: this was from the west end of the railroad bridge. However, the bridge wasn't suitable as a regular observation point. The regular train traffic made it hazardous (I had to retreat from the bridge whenever a train approached it) and having to remain vigilant for approaching trains meant that I couldn't fully focus on the observations. Furthermore, if I was in the middle of an interesting observation when a train was approaching I had to abandon the observation. In addition, observations from the bridge were typically fairly distant. This greater distance meant that observations could not be as detailed and chick identification often ranged from difficult to impossible. On the other hand, the all-encompassing view made the bridge ideal for brief checks for individuals within the pair's complete territory (i.e., when members of the family were well spread out within the territory).

I restricted myself somewhat in regard to the time of day when I conducted my observations. This was basically due to the sad human situation that existed in this area. It is a disgrace that a nation as prosperous as the United States should have so many homeless people. This problem is particularly evident in Ventura and a number of these homeless people were camped out and living in the more natural habitat of my general study area. Of particular relevance here is that my most important observation point (the one in the Semicircle) happened to be in a location that was highly favored and regularly utilized by a number of the homeless people who inhabited this area. It was an aesthetic, fairly private location that was out of view of the main trail in this section. Because of its relative privacy it was used by some of the human inhabitants here as a place to sleep, to wash and bathe, to change clothing, to eat meals, as an access point for taking an occasional dip in the water, and as a peaceful place to just enjoy the day. I quickly learned that this location usually wasn't used by them during the period from approximately 10:00 to 15:00. This location in essence was their home so out of respect and common courtesy I tried to confine my observations to this midday period when they were absent from there. Occasionally one of these people was at my observation point when I arrived. In each such instance the person who was present was very cordial and readily invited me to please stay and conduct my observations. Though personally I always felt that I was imposing, I was always made to feel welcome (in fact, some of the homeless people became interested in my study and regularly asked how the grebe family was doing). [Note that the study area was closed to the general public from sunset to sunrise. Thus, even if I had wanted to I was not able to observe how the grebes spent their nights.]

# *Chapter 4*

## Observations and Notes on the Nesting and Chick Rearing Periods of a Pair of Pied-Billed Grebes (*Podilymbus podiceps*)

## [The South River Family Study]

**6/9/17:** While hiking through the area around the two lagoons at the terminal end of the Ventura River an individual from the South River pied-billed grebe pair was sighted in the end section of the river immediately north of the railroad bridge. This individual appeared to be sitting on a nest tucked inside the dense bed of emergent reeds along the north bank of a sort of side cove situated along the west side of the south end of the river, just north of the West Lagoon. My view at this time was too inadequate to be certain that this individual was actually on a nest. Nor did I have the available time on this occasion to stay and continue observing this individual in order to confirm whether it was nesting. I thus decided that I would return sometime soon to check on this pair again. [Note that on this occasion I had come to this area for the specific purpose of checking for the three pied-billed grebe pairs that I had discovered here last year (the West Lagoon pair, the East Lagoon pair and the South River pair) during my observations of the West Lagoon family. In particular, I was checking to see if the West Lagoon or South River pairs were currently in the process of nesting or rearing a brood of young chicks. My intention here is that I am hoping to have the opportunity to conduct another case study on the chick rearing process for one of these pairs as a follow-up to my study of the West Lagoon family from last year. I am especially hopeful of also having the opportunity to observe the nesting process, something which I wasn't able to do last year because the chicks had already hatched when I first discovered that family. Ideally, I would like to have the opportunity to observe the entire chick production process from nesting on through to chick independence. (The reason that I did not include the East Lagoon pair as potential study candidates is because there are too many areas within their territory that I'm not able to view.)]

**6/13/17:** The apparent nest belonging to the South River pied-billed grebe pair that was noted on 6/9 has been confirmed! One of the pair was again seen sitting on the nest. At one point this grebe got off the nest for ~ 1-1¼ minutes and sat on the water beside it. It was at this time that at least 3 and probably 4 (or possibly more) eggs were seen in the nest, indicating that incubation was in progress. This parent also was seen adding a bit of vegetation to the nest before going back onto it. When the parent went back onto the nest it spent some time repositioning and rotating the eggs before settling down onto them to resume incubation. While it incubated, the other member of the pair was seen preening itself out in open water ~ 40-50 m SE of the nest.

The nest is a floating structure with a central bowl section raised above the surface and appears to be constructed of local vegetation, mainly reed material. It is situated within a small clump of emergent reeds that lies just in front of the dense stand of tall reeds that rings the west bank of the river along its terminal (southern) section (these reeds—bulrushes and cattails—extend in a continuous band from the immediate north side of the railroad bridge at its south end—i.e., bordering the north end of the West Lagoon—to the Highway 101 overpass bridges well to the north). The reeds in this clump are only about half the size (though still fairly tall) of the main reed bed behind it. The gap between this clump and the main bed is barely noticeable, less than one third of a meter. Overall, the nest and sitting parent are well concealed and quite difficult to see.

More generally, the nest is located within a sort of side cove along the west bank of the river at the immediate north side of the railroad (RR) bridge. This "side cove" is actually just a fairly large, deeply scalloped section of the west bank here that is roughly semicircular in shape. Thus, though the river proper (which runs north to south) generally has east and west banks, this semicircle area has north and south banks and a curving west bank. The east end of this semicircle is the section that opens up along the river proper. The nest is located in the reeds at the mid-area of the north bank. The semicircle begins at the immediate north side of the west end of the RR bridge and extends west of it (adjacent to the section of the tracks on the land west of the bridge) and it is the main section of the river that lies east of the semicircle that runs under the bridge into the west lagoon. While this semicircular side cove is actually just part of the river here, for the purpose of better identifying locations in the notes I will treat it as a distinct geographic area. It will be referred to as the "semicircle" and henceforth will be denoted by the abbreviation "SC."

As a further clarification, in reality it appears that this entire southern end segment of the river is a lagoon. Because the territory of the West Lagoon pair ended at the RR bridge, I arbitrarily marked that as the northern boundary of the West Lagoon. In actuality, the wide expanse of the water at the southern (terminal) end of the Ventura River (at least, at this time) extends from the West Lagoon all the way to the area just north of the Highway 101 overpass bridges (this section, including the East Lagoon, comprise the estuarine segment of the river). Upriver from this the river is basically nothing more than a small brook most of the time. This is because there is a reservoir well upriver to the north and the flow of the river generally is controlled by human manipulations (i.e., other than any natural drainage that may occur during seasonal rains in the area south of the reservoir). The typically meager flow at this southern terminus thus usually doesn't have enough "steam" to plow through the beach to the ocean and therefore backs up in this section, forming the lagoons. This situation occasionally changes during an adequate rainy season (as happened this past season which saw record rainfall in the state) when the river runs strongly, breaking through to the ocean, and the lagoon is then transformed into a proper river (during such times this diverts much of what is typically fed into the East Lagoon when the system is backed up and this lagoon becomes much smaller during these periods). This reality should be kept in mind as I arbitrarily refer to the segment of water north of the RR bridge as the southern segment of the river rather than as a lagoon. The arbitrary divisions that I have made to the study area have been done for the purpose of better denoting the locations of the grebes in the notes.

From this point forward the plan is to conduct daily observations of the South River (SR) pair (and their imminent brood) from nesting right on through the chick rearing period until the chicks become independent. I anticipate that, should all go smoothly (which is never a guarantee in wild situations) this will encompass a period of about three months.

At least three young chicks (not yet fully mobile but hanging around in a feeding station area) were seen around the edges of and within the emergent reed beds along the east shore of the East Lagoon approx. a third of the way up (north) from the south end of the lagoon. Two feedings were observed. In both instances the parent had captured a small fish a fair distance away and carried it to the chicks who were waiting for feedings to be brought to them in the feeding station area.

**6/14/17:** The South River (SR) pair continues with the incubation of their clutch. There were several times when neither of the parents was on the nest and eggs were clearly visible. Once again, at least three eggs were definitely seen. However, my view of the nest is rather limited. Because it is tucked inside the reeds, my view of it is somewhat analogous to looking through a picket fence with varying sized gaps between the lattices. I thus couldn't be certain of just how many eggs were there and it is possible that four or more could be present.

While incubating, the parent that was initially doing so when I began my observations was rather active. It often stood and repositioned itself over the eggs and/or repositioned/rotated the eggs. Periodically it also left the nest and was observed diving in its vicinity, typically resurfacing with some vegetation that it had collected during its dive which it then added onto the floating nest. At one point, the other parent went to the nest while its mate was settled on it and both of them then spent ~ 7 minutes engaging in nest maintenance/enhancement activity. Each made dives in the vicinity of the nest collecting vegetation during the dives and adding it onto the nest. At the end of this activity one of the pair climbed onto the nest and resumed incubation (my view was too inadequate to determine if this was the same parent who had been incubating prior to this activity or if a nest exchange had occurred).

**6/15/17:** Being short on time today I could only manage a brief check of the nest (~ 15 minutes). When initially sighted, one of the pair was well-settled on the nest incubating the eggs and the other was diving in its vicinity, collecting vegetation and adding this onto the nest. After several such dives this parent left the nest area and went farther out in the river, east of the SC. It returned to the nest ~ 7-8 minutes later and an interesting nesting exchange then took place. After slowly approaching the nest this parent paused for several seconds when it reached the east side of the nest and gazed at its mate who was well-settled on the eggs, facing away to the NW. It then climbed onto the nest and carefully walked onto the back of its still incubating partner, its mate accommodating this by flattening its posture (as during a copulation) and thus making it easier for the mounting bird to maintain its position on top. It remained standing on top of its mate for ~ 6-7 seconds. While on top it rubbed its rump/cloacal area across the dorsal rear end section of its mate for several seconds, as though in a pseudo-copulation (it was clear that no copulation occurred because the bird was too far forward nor did its mate raise its rump to make its cloacal area accessible—that is, no cloacal to cloacal contact took place). The mounting bird then stepped off of its mate's back onto the west side of the nest and then slipped onto the water, remaining at the west side of the nest while eyeing its mate. After several seconds its mate stood up, remained standing over the eggs for about 5 seconds and then walked off the east side of the nest into the water. Shortly thereafter, the other member of the pair climbed onto the nest, went over to the eggs and after briefly repositioning and/or rotating them, settled onto the eggs and began incubating. By this time its mate had made its way farther out into the river and was busy vigorously preening itself.

For a second consecutive day the parent that wasn't incubating at the time sounded the characteristic drawn-out, whooping pied-billed grebe call while well out in the river (in this context it clearly appeared that it was serving as a territorial call).

**6/16/17:** When this observation began only one of the pair was seen. This individual was well-settled on the nest incubating the eggs. A few minutes later the other member of the pair appeared at the nest and began diving in its vicinity, collecting vegetation during the dives and adding this onto the nest. Almost immediately the incubating bird left the nest and joined its mate in diving in the nest's vicinity, collecting vegetation and adding this onto the nest. The pair engaged in this task of nest maintenance/enhancement (hereafter denoted as "nest M/E") together over the next ~ 8 minutes before the bird that had originally been incubating went back on the nest and resumed incubation (its mate continued with nest M/E). (While the pair was off the nest and simultaneously engaged in nest M/E I had my best look yet at the eggs because they were more spread out than in any of my previous observations. I could now see that there were definitely 4 eggs present and probably more.) Upon resuming incubation duty it spent much time fussing with the eggs, repeatedly repositioning and rotating them—it would sit on them for only about a minute or less before standing and again fussing with them. In the period just before its mate ceased with its nest M/E the sitting bird left the nest to again briefly join its mate in this activity. It then climbed back onto the nest, fussed with the eggs a bit more, settled itself onto them and, while sitting on the eggs, engaged in an extensive self-preening bout. As it did so, its mate left the nest area, paddled well out into the river SE of the nest where it parked and also engaged in vigorous self-preening.

About 10 minutes later this member of the pair made its way back to the nest where it again engaged in nest M/E. As before, the incubating bird promptly left the nest and joined its mate in this activity. After a few minutes of this it climbed back onto the nest and fussed with the eggs a bit before settling onto them. As happened earlier, its mate continued with its nest M/E activity. Observations of the pair concluded at this point.

On a few occasions while sitting on the clutch the incubating parent left one (and possibly a second) egg uncovered for a short time (about one minute or less). Given how the parent had spread the eggs out just before sitting on the clutch on these occasions it appeared that this may have been deliberate on the part of the parent and the egg(s) left uncovered was specifically selected for this.

**6/17/17:** Only a brief (~ 20 minutes) check of the nest was conducted today. The pair continues to incubate their clutch. For the first three quarters or so of this observation period one of the pair remained well-settled on the nest while the other was mainly well out in the river SE and east of the nest, either parked or cruising about. At one point it voiced one of the species' characteristic drawn-out, whooping calls, this almost certainly functioning as a territorial call in this instance. (Such calls henceforth will simply be referred to as a "call' or it will be stated that the individual "called." In other words, if it is merely stated that the individual called, this means that it voiced one of these stereotypic drawn-out, whooping calls. Any other type of call or vocalization will be specifically described or otherwise distinguished. This call appears to have two functions, serving mainly as a territorial call but also as a contact call when the members of a pair are well separated or otherwise out of touch.)

A very probable nest exchange took place in the latter part of this observation period. I couldn't be 100% certain that an exchange occurred because I missed the moment when the apparent switch took place. At that particular moment one of the pair was on the nest and I had been visually scanning the general area trying to locate its mate. When I switched my attention back to the nest I saw that both individuals were there. One of the individuals was standing on the nest and the other was just beginning to move away from it. The grebe that was standing on the nest was wet and had obviously just been in the water. It spent ~ 1.5 minutes repositioning/rotating the eggs before finally settling onto them. It then engaged in a lengthy self-preening bout while sitting on the eggs. In the meantime, its mate continued moving SE farther out into the river, vigorously preening itself as it moved along. It proceeded south, going under the RR bridge and into the West Lagoon where it began foraging along the dense reed beds that line the west shore of the lagoon. Though I didn't actually see the birds switch, the one that swam away from the nest didn't appear to be same one that had been off the nest prior to the apparent switch (the plumage of the sexes is similar in pied-billed grebes and I am still in the process of learning to distinguish the members of this pair from each other). [*Looking back in hindsight, given what I subsequently learned I can state with certainty that a nest exchange did occur and that the female is the one who had initially been on the nest and was later relieved.*]

Once again, I could see that four eggs were definitely present in the nest and I couldn't rule out the possibility that there could be more.

**6/18/17:** The nesting pair was observed for ~ 35 minutes today. They continue to incubate their clutch. As observations began, one of the pair was well-settled on the nest, incubating the eggs. While sitting on the eggs it regularly engaged in short preening bouts. Its mate was busy foraging in the eastern half of the river (sometimes quite close to the eastern bank), east and E-NE of the SC. This situation remained so for ~ 25 minutes. At that point the incubating bird called once. I then immediately switched my view to its still foraging mate to see if and how it responded to this call. At the time, this individual was foraging close to the east bank of the river, E-NE of the SC. By the time I trained my binoculars on it, the indication was that it had heard and acknowledged the call. It was sitting still in the water near the east bank, looking out in the general direction of the nest. Approx. 5 minutes later it began making its way west across the river passing under and through the numerous interspersed algal mats that currently are present in the river and SC. It soon became apparent that this member of the pair was heading for the nest. As it worked its way across the river toward the nest I lost track of it after one of its dives. I then searched the water along its anticipated route but was unable to spot it. After ~ 10 seconds I shifted my attention back to the nest and discovered that both of the pair were in the water near the nest gathering vegetation and adding it onto the nest (the bird whose progress I had been tracking apparently had traveled a considerable distance underwater from where I last saw it dive). Soon after, one of the pair climbed onto the nest, spent ~ 30-40 seconds repositioning and rotating the eggs, settled onto them and began preening while incubating. Its mate continued diving around the nest, collecting vegetation and adding it onto the nest. It continued with its nest M/E activity for several more minutes before slowly moving out to mid-river, SE of the nest, preening itself as it moved along. Because my views of the two birds had been distant and in the case of the bird that had been incubating, too limited, I couldn't be 100% certain that a nest exchange had taken place but it appeared most probable that one had occurred. If this was indeed the case then the observations strongly suggest that this exchange had been precipitated by the call of the incubating bird.

The free swimming bird eventually made its way back into the SC where it soon went out of my view. Shortly after, it reappeared by the nest where it engaged in nest M/E while its mate remained well-settled on the eggs. This remained the situation when observations were concluded a few minutes later.

At this point in my observations of this nesting pair it is already quite clear that their floating nest requires constant attention and much effort to maintain. The pair devotes considerable time toward nest M/E with both participating in this task.

**6/19/17:** The pair was observed for one hour today as they continue to incubate their clutch. As observations got underway one of the pair was well-settled on the nest, incubating the eggs. Its mate was parked in the mid-river area SE of the nest, preening itself. It eventually moved west into the SC, making its way all the way to the dense reed beds along the mid-area of the west bank of the SC where it began foraging. After a while I lost sight of it and by the time I switched my full attention back to the nest (~ 12 minutes later) I noted that a nest exchange had just occurred. At this point I saw that the bird that had initially been on the nest was now moving away from it heading SE, farther out into the river. It vigorously preened itself and called once as it moved along. This member of the pair soon parked in the mid-river area SE of the nest and engaged in a lengthy preening bout. In the meantime, its mate had climbed onto the nest to relieve its partner, fussing with the eggs for ~ 30 seconds before settling down on them. Approx. 5 minutes after its previous call, the bird that had been relieved of incubation duty (still parked and preening in the mid-river area) called again, this call being particularly drawn out. Shortly after, it moved NW to the reeds along the west bank of the river just north of the NE point of the SC and began foraging. I soon lost sight of this bird as it foraged.

Approx. 4 minutes later the incubating bird stood up over the eggs, rearranged some of the nest's vegetation around them and then walked off the nest and dove at its east side. Shortly after, one of the pair surfaced next to the nest with some vegetation and added this onto it. It then climbed onto the nest, rearranged some of the vegetation on it, repositioned and/or rotated the eggs and then settled onto them. It wasn't until a short time later (when I was able to observe that the member of the pair that was now moving about, free of incubation duty, was the one that had been sitting on the nest prior to this recent activity there) that it became apparent that another nest exchange had taken place. The bird that was now back on the nest thus had taken only a brief respite from incubation duty. By ~ 46 minutes into this observation period the now free swimming member of the pair was busy foraging in the mid-river area east of the SC amidst the extensive floating algal mats situated there. (Note that I'm using the term 'algal mat' as a generic term. These mats of floating vegetation also generally consist of various types of plant and plant-like organisms other than just algae.)

A few minutes later the incubating bird stood up, walked off the nest, dove in its vicinity and engaged in nest M/E for several minutes before climbing back onto the nest and resuming incubation of the eggs (its mate was well out in the river SE of the nest during this time). When observations were concluded the same individual was still incubating the eggs and its mate was out in the river SE of the nest.

**6/20/17:** Time: 1313-1425.

The pair continues to incubate their clutch. Two nest exchanges were observed and the pair was seen engaging in their usual nest M/E activity.

At the time that observations began, one of the pair was well settled on the nest incubating the eggs. The other was in the western third of the river well north of the SC, preening itself. About 5 minutes into the observation period the bird on the nest stood up, rearranged the vegetation on the nest for ~ 20 seconds and then settled back onto the eggs. Its mate was now foraging in the area just north of the NE point of the SC. Approx. 4 minutes later the bird on the nest again stood up, spent ~ 15 seconds rearranging the nest material and repositioning/rotating the eggs and then settled back onto them. A few minutes later its mate went to the nest and dove near it, collected vegetation during its dive and added this onto the nest. It then dove again and I wasn't able to re-sight it at this time. The incubating bird remained settled on the nest. Ten minutes later the member of the pair that was currently off incubation duty was again seen in the water next to the nest. At this point the incubating bird stood up, walked off the nest onto the water at its east side and dove. Both birds were now in the water, leaving the nest unoccupied and the eggs uncovered. Within 30 seconds the member of the pair that had been freely swimming about climbed onto the nest, briefly fussed with the eggs and then settled onto them (a nest exchange thus having occurred). Its mate (now relieved of incubation duty) then engaged in a long bout of nest M/E while its partner remained settled on the eggs. At one point during its nest M/E activity this individual went ~ 15-20 m from the nest and collected a long reed stem which it dragged back to and worked into the nest. After ~ 15 minutes it finally ceased its nest M/E, moved out to open water SE of the nest, parked and engaged in a lengthy preening bout. It then moved north and engaged in some brief foraging before returning to the nest ~ 11 minutes after it had left the nest area. At the nest it again began diving in its vicinity and conducting nest M/E while its mate remained settled on the nest, incubating the eggs. Approx. 7 minutes later it ceased its nest M/E activity and moved away from the nest to the SE into the mid-river area. It parked there and extensively preened itself. Nine minutes later it began moving back toward the nest, calling once as it did so. It stopped when it got near the bank at the eastern edge of the fairly large algal mat that currently extends out from the east side of the nest. There, it did some final preening before diving under the algal mat and resurfacing at the north side of the nest. A nest exchange then ensued: as it climbed onto the nest from the north side its mate stood up and walked off the SE side of the nest. The nest-relieving bird then fussed with the eggs briefly before settling onto them. Its partner moved out to the mid-river area SE of the nest where it parked and vigorously preened itself for several minutes. It then moved north into the mid-river area E-SE of the nest and began foraging there. Observations were concluded at this point.

**6/21/17:** Time: 1255-1345.

Disaster has struck the nesting grebe pair. Since yesterday's observation period the water level in the river and lagoons that it feeds has risen dramatically and their floating nest wasn't able to cope with this situation. The nest is still floating within the clump of reeds where it was constructed but its above-surface section is now very low and barely above the surface (also, it looks as though the nest has begun breaking apart). In addition, the vegetation that makes up the above-surface portion of the nest is wet and saturated. Sadly, it appears obvious that the eggs were fatally inundated by the rising water—none of them were seen in the nest or anywhere else. It is thus probable that they may have been swept out of the nest or fallen through it. The compromised nest is now floating considerably higher up within the reed clump as one third to one half the length of the reed stems that were above the surface yesterday are now underwater. The floating algal mats that had surrounded the nest and the area around it have been swept away except for the mat in the small gap between the reed clump that contains the nest and the main reed bed that lines the bank.

Most telling of all is that neither member of the pair was seen on the nest. It was thus obvious that the nest (and this nesting attempt) has now been abandoned, indicating that the eggs have definitely been lost. Both individuals were seen (often simultaneously) moving about within their territory but over the 50 minutes that they were observed neither of them went any closer than ~12 m from the nest. It was abundantly clear from their ranging and behavior that the nest is no longer the focus of their attention.

On two widely separate occasions one of the pair voiced a call.

It remains to be seen if the pair will make a second nesting attempt, either rebuilding this nest or constructing a new one. The cause of this sudden rise in water level is not apparent to me at this time so I will have to investigate it. One potential cause that can immediately be ruled out is rainfall because we haven't had any rain in this region for several months.

**6/22/17:** Time: 1240-1330.

The water level has risen even higher and the lagoons have also increased a bit more in size. The nest is no longer present (or at least, no longer visible) within the clump of reeds where it was built nor is it (or any remains of it) seen anywhere else. It's most likely that it has disintegrated or sank, or was dislodged and carried away by the current.

Both members of the pair were observed parked, moving about and foraging within their territory. During the observation period one of them was mainly out in the northern half of their territory and the other was in the southern half. The latter called on three separate occasions. There were as yet no indications that they are currently preparing for another nesting attempt. I would guess that if they are going to do so they will probably wait until the water level situation has stabilized. I would also expect that if they are going to make a second nesting attempt they will want to do so as soon as possible so as to have as much of what remains of the breeding season available to them as they can salvage.

In investigating the various possible causes for the sudden dramatic rise in the water level I have been able to determine what appears to be the responsible agent. Among the possibilities for creating such a sudden rise the one that most readily comes to mind is rainfall. As mentioned in yesterday's notes, this possibility can be immediately dismissed because it has been several months since any rain has fallen in this region. Another credible possibility would be the release of a large amount of water from the reservoir that lies well upstream along the river. However, after the severe drought that has plagued the region over the past years the water level in the reservoir remains low despite the record rainfall that much of the state experienced this past rainy season. Thus, there is no chance that the authorities would have conducted a release at this time. It now appears most likely that the water that is responsible for this sudden rise came not from the atmosphere or a land-based water source but from the ocean. The past few days have experienced periods of very high tides which flowed completely over the beach and into both the East and West Lagoons. In fact, while walking along the beach today on my way to my SC observation point I noticed that the beach sand was wet from the surf line of the receding tide all the way to the south end of both lagoons.

**6/23/17:** Time: 1330-1410.

The rise in water level has continued as it has risen a bit more since yesterday. Besides the water level, something in the conditions (wind and/or water flow) has also affected the pair's territory by clearing away the great majority of the numerous floating algal mats that were present.

It appears that this resilient pair is wasting no time getting started on a second nesting attempt. At the beginning of this observation period the two of them were quite distant from each other. One of them was at the northern end of their territory while the other was in its southern end. However, they later got together in the SC and both then went into the dense reed beds that line the mid-section of its western bank. While there they were observed carrying some vegetation and adding it onto a new floating nest that they are constructing. The nest is well concealed within a clump of emergent reeds that extends slightly out from and is connected to the main reed bed. Quite a bit of construction has already been done on this nest and it appears that it could soon be ready for use. The pair spent only about 5 minutes working on the nest before heading out into the river. The area immediately outside the reeds that is adjacent to the nest within is surrounded by a fairly extensive floating algal mat. Though it's not known if such was the intent, this mat effectively enhances the concealment of the nest.

Having now observed the SR pair over a period of 11 consecutive days the boundaries of their territory have become clear. Basically, their territory consists of the southern segment of the river (the entire portion of this segment of river, including both the east and west banks), extending from the immediate north side of the RR bridge at its south end (i.e., the north end of the West Lagoon) to a point along the river approx. halfway between the RR bridge and the Highway 101 bridges at its north end. This, of course, includes the SC, which comprises the SW section of their territory (the north end of the territory is a fair distance from the north end of the SC). It also includes the short segment of the river at its very southern end. This is the ribbon of river that runs from the north end of the West Lagoon over to the north end of the East Lagoon where it feeds into it. This short south ribbon of the river runs west to east and follows adjacent to the RR bridge (actually, part of this narrow ribbon runs underneath the RR bridge with the rest of it extending along the bridge's north side). Recall that this short end section of the river was a common area used by all three neighboring pairs last season during the period that I observed the West Lagoon family. At this time, the SR pair has claimed it. Henceforth I will refer to this short end segment as the "south ribbon" of the river.

Currently, the West Lagoon is a common area. Thus far, the West Lagoon pair has not returned to nest there this season and the lagoon remains unclaimed at this time. The members of the SR pair occasionally utilize it as does the East Lagoon pair. The vital lush, dense reed beds that lined the entire east shore of the West Lagoon during last year's breeding season are now gone. They were devastated, having been completely eroded away this past rainy season during the period when the river was running very strongly (the east shore comprised the outside bank of the river during that period when the West Lagoon was transformed from a lazy lagoon back to a proper river and as such, took the brunt of the river's erosive force; the west shore/bank was essentially unscathed and still retains its dense, extensive reed beds). It seems quite possible that this is the reason why the West Lagoon pair isn't nesting here this season. While feeding/rearing their chicks last season the West Lagoon pair did their foraging almost exclusively along the edges of and within the reed beds. The West Lagoon is fairly small and it is quite probable that it took the reed beds lining both the east and west shores of this small lagoon to provide sufficient prey/foraging habitat for rearing a brood. In other words, the lagoon with only the west shore reed beds (as the situation currently stands) probably does not contain sufficient resources for rearing a brood.

**6/24/17:** Time: 1329-1450; Brief re-check of nest at 1520.

The water level appears to have stabilized in that it seems to be at the same height as yesterday.

The pair has apparently kept busy at the task of constructing their new nest. It appears to be completed or nearly so. When this observation period began, one of the pair was sitting well-settled on the nest. The other was parked inside the SC ~ 22 m east of the nest, preening itself. At that point I couldn't see whether or not any eggs were present in the nest. At 1344 the bird on the nest stood up in place, preened itself for about 35 seconds and then settled back down on the nest. No eggs were seen, however my view of the nest was rather limited. Two minutes later it stood and walked off the nest (and was quickly lost from view), leaving it unoccupied. Once again, no eggs were seen in the nest. Its mate wasn't seen at this time.

At 1350 the member of the pair that had not been on the nest was sighted near the NE edge of the SC where it called once, dived shortly after and was lost from view at that time. Three minutes later a member of the pair was sighted moving west near the north bank of the SC. It soon engaged in a dive during which it collected some vegetation. Upon resurfacing it headed in the direction of the nest, carrying the vegetation in its beak. When it got to within ~ 18 m from the nest it dove with the vegetation and covered the remaining distance to the nest underwater. It resurfaced near the nest and added the vegetation onto it. After doing so, it dove, resurfaced well away from (and east of) the nest and moved E-NE near and along the north bank of the SC, calling once as it moved along. Soon after calling, it dove and repeated the above noted sequence of collecting vegetation, taking it to and placing it on the nest. After doing so, it dove and was lost from sight at this time (1401). It was resighted at 1408 near the dense reed beds along the north bank of the SC. After calling once it dove a few seconds later and was again lost from view. At this point the nest had already been unoccupied for 22 minutes. Thus, though my view of the nest was too inadequate to be 100% certain that no eggs were present, this long period of inattention to nest sitting clearly indicated that it was unlikely that any eggs have been laid yet.

At 1415 this same individual was resighted, this time parked out in the mid-area of the SC. While parked there it engaged in a lengthy preening bout, briefly breaking from its preening to call on two separate occasions (at 1421 and 1425). Its mate was not seen at this time.

By 1431 it had slowly made its way into the NW corner of the SC. It then began moving east along the north bank. As it moved along it was keenly looking about, very likely for its mate. Just as it reached the NE end of the SC it called again, dove soon after and wasn't resighted at this time.

At 1435 the other member of the pair was sighted moving into the mid-area of the SC. It parked there and vigorously preened itself (its mate wasn't seen at this time). It ceased preening at 1442 and went to the nest. At this point I could see that its mate was also at the nest engaged in nest construction activity (it had gone there unseen by me while I was watching its partner). The individual that had just been preening then climbed onto the nest, sat and settled onto it. Its mate remained next to the side of the nest where it grabbed some vegetation and placed this onto the nest. It did this for several minutes before ceasing and just sitting in the water at the side of the nest while its mate remained well-settled on it.

At 1449 the nest-sitting bird stood up, walked off the east side of the nest, moved out ~ 15 m to the east of it and began preening. Its mate then climbed onto the nest, settled onto it and preened itself while sitting there. At this point (1450) the observation period was ended, its mate now calm and parked ~ 30 m east of the nest.

I made a brief re-check of the pair at 1520 and found that they were together at the nest, one on it and the other in the water next to it. In regard to which of the pair was now on the nest, their positions had switched from when last seen at 1450. The individual on the nest had assumed an unorthodox posture. It was lying flat, stretched out on its ventral side with its head and neck extended fully forward (oriented north). Its mate was sitting in the water at the south side of the nest next to the rump of its stretched out partner, its attention mainly focused on its mate. The pair remained like this for several minutes before the bird in the water climbed onto the nest, and walked onto (i.e., mounted) its mate's back. At this point a copulation occurred, lasting ~ 4-5 seconds. Both birds then moved off the nest and engaged in nest construction activity.

***From this point on**, in order to facilitate note-taking the members of the pair have been given names. The names are simply derived from using the letters ("R" for one and "SB" for the other) that I have assigned to each in my raw notes. These letters are based on the main physical characteristics that I use to distinguish them from each other. One individual has a notably more rounded forehead than its mate and a less robust beak. I have denoted this individual as "R" (for rounded forehead) in my raw notes. Its mate is fairly well distinguished from R by its larger, stouter beak and flatter forehead and top of head. This individual has been denoted as "SB" (for stout beak) in my raw notes. The names that were picked were based on these letters (for the sake of continuity in my notes) and the fact that, having now witnessed a copulation, I know the sex of each individual. SB is the one who did the mounting and is obviously the male of the pair. In considering a male name that contains the letters S and B the one that immediately came to mind was Sinbad. This seemed to fit this individual so it was chosen as his name. R is the one who was mounted and is obviously the female. I was very impressed by the tenacity with which she clung to her plan to breed, getting started on preparing for a second clutch right after losing her first (or at least, the first of the season that I know of). In my mind I thought that she clung to her will to breed as tenaciously as a remora clings to its host. So, she has been named Remora (she will be called "Remy" for short in the notes).

As is typical of grebes in general, the plumage of males and females of this species is similar so the sexes can't be distinguished based on this prominent physical characteristic. Thus, prior to witnessing this copulation I had no idea as to which of the SR pair was the male and which the female. Actually, the benefit from my good fortune at having witnessed this copulation is more far reaching than just this particular pair: it has now provided me with the means to visually distinguish males and females of this species in general. I have noted that the differing physical characteristics described above that exist between the SR pair members (i.e., the more rounded forehead and somewhat smaller, less robust beak of the female Remy versus and the flatter forehead and dome and larger, more robust beak of the male Sinbad) are consistent and exist in each of the PB grebe pairs that I have seen. That is, the individuals in each of these pairs can be distinguished from each other by these same physical characteristics (i.e., one member of the pair has the more rounded forehead and smaller, less robust beak, and one has the flatter forehead and dome and larger, more robust beak). In fact, I used these same physical characteristics to distinguish the members of the West Lagoon pair whom I observed last season. Of course, in their particular case, having not had the benefit of witnessing a copulation I had no idea as to whether these differences were gender related or just due to normal individual variability in general. Given the consistent nature of this set of differing physical traits within pairs this obviously is not a random phenomenon. Therefore, the most logical conclusion is that they are gender related. As expected, there is variability in the degree to which these dimorphic traits are expressed in various individuals and in particular cases this can make gender ID by this means more difficult. I have also noted that on average, males are larger than females though it appears that perhaps there is some overlap (that is, with the largest females being larger than the smallest males). *[Postscript: Further support for the apparent gender related physical differences noted above was obtained in 2018 while observing one of the nesting PB grebe pairs at the water treatment plant noted in Chapter 6. While observing them I had the good fortune of being able to witness a copulation by this pair. Consistent with what was seen with the SR pair it was the individual with the apparent male features of the head and beak described above who assumed the male role in this copulation and the individual with the apparent female features who assumed the female role. A further noteworthy point is that as with all of the observed copulations by the SR pair this copulation also took place on the nest.]*

**<u>6/25/17</u>:** Time: 1103-1133.

The water level appears to be about the same as on the previous two days, though perhaps just a bit higher.

Only a short check on the pair was conducted today. At the start of the observation period Remy was in the mid-river area east of the SC. From there she eventually made her way to the area near the west bank of the river north of the NE point of the SC. Sinbad wasn't seen at this time.

Sinbad was finally sighted at 1107 in the NW section of the SC carrying vegetation toward the nest. When he got to within ~ 15 m of the nest he dived, resurfaced next to the nest and added the vegetation onto it. Sinbad then dived, resurfaced ~ 12 m east of the nest and proceeded to the mid-area of the SC where he parked and vigorously preened himself. At this time Remy was foraging in the northern half of the pair's territory.

The nest appears to be ready for use or nearly so. No eggs were seen in the nest.

Over the next 13 minutes Sinbad remained in the SC, moving about a bit and usually continuing to preen. During this period he voiced a call on two separate occasions (at 1114 and 1118). Remy continued to forage in the northern half of their territory. At 1120 Sinbad went to the nest, sat down and settled onto it and preened while sitting. When observations were concluded at 1133 he was still well-settled on the nest and Remy was still foraging and sometimes preening in the northern half of the territory.

With the nest now looking ready for use and the fact that a copulation has been observed it appears that egg laying by the pair is imminent. Therefore, the "formal" observation periods will be initiated tomorrow. From that point, these formal observation periods will be conducted on a daily basis until at least several days after the chick rearing period has concluded.

**6/26/17:** Time: 1335-1435. [*Formal observation periods will be initiated today.*]

The water level appears to have stabilized over the past four days.

1335 - As this observation period begins, Remy is well-settled on the nest and Sinbad is sitting in the water at the south side of the nest (Remy is facing north).

1340 – Remy stands and walks off the nest (no eggs are seen in the nest). She moves to the north bank of the SC, heads east along the bank, rounds the NE point of the SC and proceeds north off the west bank of the river. While moving north she dives (probably beginning a foraging session) and isn't resighted at this time. Seconds after Remy left the nest Sinbad followed her for ~ 18 m east of the nest but remained in the western half of the SC. After a dive ~ 2 minutes later he isn't resighted at this time.

1346 – Remy is sighted foraging in the northern half of their territory off the west bank of the river. Sinbad still hasn't been resighted.

1353 – The pair is sighted swimming closely together, moving west along the edge of the reeds that line the north bank of the SC. One after the other they dive and are temporarily lost from view. Approx. 12 seconds later Sinbad is sighted next to the nest holding a large load of vegetation in his beak. He adds this onto the nest and then dives. He is soon resighted moving into the mid-area of the SC where he parks and preens.

A few minutes later Remy is sighted near the reeds just north of the nest. Soon after, she moves into the mid-area of the SC and it is obvious that she is going to Sinbad. As Remy draws near, Sinbad spots and moves toward her. They meet en route to each other and greet briefly by brushing breast to breast as Remy continues past. Remy keeps moving farther east out into the river and begins foraging in the SE section of their territory. Sinbad eventually moves to the area of the river off the NE point of the SC and remains parked there, sometimes preening briefly and often looking about.

1405 – Sinbad moves east across the river to the area near the east bank and begins foraging there.

1409 – Sinbad swims west back across the river, enters the SC along its north bank, dives soon after and isn't resighted at this time.

1412 – Remy is sighted entering into the SC from the east, parks in its mid-area and preens. Seconds later Sinbad is sighted next to the nest. He soon climbs onto the nest and settles himself on it. Several minutes later Remy ceases her preening, moves to a point ~ 18 m east of the nest and parks there.

1420 – Remy goes to the nest and sits in the water at its south side. Sinbad is still well-settled on the nest (facing north), rearranging the nest vegetation around his body as he sits there.

1421 – Remy is still sitting in the water at the south side of the nest near Sinbad's rump. She reaches her head forward and gently pokes Sinbad's rump once and then continues to wait in her place. Approx. 15 seconds later Remy again pokes Sinbad's rump once. Sinbad responds to this 2 seconds later by standing and walking off the nest at its north side. Remy then climbs onto the nest from its south side and settles onto it. With Remy sitting on the nest Sinbad then engages in nest construction activity.

1426 – While still sitting on the nest, Remy assumes what apparently is a receptive copulatory posture (the same as seen on 6/24): lying flat on her ventral side with head and neck stretched fully forward and also lying flat on the nest. Sinbad had ceased his nest construction activity and has been sitting in the water at the west side of the nest next to Remy's rump (Remy is facing east). Sinbad's attention is clearly focused on Remy. Approximately 12 seconds later Sinbad climbs onto the nest, mounts Remy's back and copulates with her. While the copulation is in progress Remy raises her head and neck (keeping them fully stretched out) at an angle of ~ 45 degrees or greater and rhythmically sways her head/neck from side to side. After ~ 5-6 seconds Sinbad ceases copulating and leaves the nest with Remy following at his heels. Remy moves only ~ 20 m east of the nest where she parks and vigorously preens (this includes some very focused preening on her rump area). Sinbad ventures out into the northern half of their territory and forages.

1435 – As this observation period is concluded, Remy is parked ~ 12-15 m east of the nest (intermittently preening) and Sinbad is still foraging in the northern half of their territory.

*Comments:* Having now observed a second copulation this solidly confirms the fact that Remy is the female of the pair and Sinbad is the male.

No eggs have been laid yet, however the indications are that laying will begin soon: the nest is now ready for use or nearly so and two copulations have been observed over the past three days. The pair has clearly invested a strong commitment to this second nesting attempt.

The pair has exhibited some behaviors that seem rather curious considering that no eggs have yet been produced. Both individuals spend time lying well-settled on the nest and nest exchanges between the two have been observed. One thus can't help but wonder why they are engaging in such behaviors when no eggs are in the nest?

**6/27/17:** Time: 1210-1320.

The water level is about the same as yesterday. It thus appears that, at least for now, the water level has stabilized.

1210 – As observations begin, Sinbad is standing on the nest preening and Remy is parked in the mid-area of the SC preening herself. No eggs are seen in the nest.

1217 – Remy (still in the mid-SC area) ceases preening and begins moving north. When she reaches the area off the NE point of the SC she dives and isn't resighted at this time (she has probably begun foraging). Sinbad (still standing on the nest and preening) sits and settles on the nest and continues preening.

1220 – While still well-settled on the nest, Sinbad calls once.

1222 – Remy (who has been foraging) is sighted swallowing a crayfish at the edge of the reeds that line the NE point of the SC. After eating the crayfish she moves to the mid-area of the SC where she parks and preens. Sinbad is still well-settled on the nest.

1232 – Sinbad (still well-settled on the nest) voices an especially long, drawn-out call. Remy (still parked in the mid-SC area) promptly responds to his call by going to the nest and a nest exchange then ensues: Sinbad stands and walks off the north side of the nest, immediately dives and is temporarily lost from view. Remy then climbs onto the nest and settles herself onto it.

1241 – Remy (while still well-settled on the nest) assumes a posture that indicates she is presenting herself for a copulation: she lies flat on her ventral side with her head and neck stretched fully forward and also lying flat on the nest (this is the same posture that she assumed immediately prior to the two copulations that I previously observed). Sinbad is sitting in the water at the west side of the nest (Remy is facing east) next to Remy's rump, his attention clearly focused on her. Approx. one minute later Sinbad climbs onto the nest, mounts Remy's back and they copulate. As seen yesterday, while the copulation is in progress Remy raises her head/neck (fully stretched out at an angle of ~ 45 degrees) and rhythmically sways her head/neck from side to side. The copulation lasts ~ 5 seconds and then both leave the nest. Remy goes to the mid-SC area where she parks and preens. Sinbad goes to the NW section of the SC and heads east along the north bank, dives and isn't resighted at this time.

1244 – Sinbad is resighted—he is now standing on the nest. He briefly preens himself before lying down and settling himself on the nest. Remy is still preening in the mid-SC area.

1300 – Remy goes to the NW corner of the SC amid the floating algal mats near the shore. Sinbad comes into the area and meets her there. Both then dive and go to the nest together. At the nest Remy climbs onto and settles herself on it. While she lies on the nest Sinbad engages in nest construction activity.

1306 – Remy stands and leaves the nest but remains in the SW section of the SC, SE of the nest. Sinbad then ceases his nest construction activity, climbs onto the nest but remains standing. About a minute later he walks off the nest and moves out near the reeds along the north bank in the mid-north section of the SC. There, he dives and collects some vegetation during his dive, takes this back to the nest and adds the vegetation onto it. Remy is now preening in the mid-SC area.

1313 – Sinbad goes out near the reeds at the NE point of the SC, collects a long reed stem from that area, takes this back to the nest and works the stem into it. He then climbs onto and settles on the nest. Remy is parked in the mid-SC area, intermittently preening herself.

1320 – Remy (still parked in the mid-SC) goes to the nest and a nest exchange takes place: Sinbad walks off the nest and Remy settles onto it. With Remy settled on the nest, Sinbad again engages in nest construction activity. At this point the observation period is concluded.

*Comments:* Three copulations have now been observed and all were conducted on the nest.

There seems little doubt that Sinbad's long, drawn-out call is what precipitated the chain of actions that ultimately led to the copulation.

Remy assumed the same characteristic posture prior to being mounted by Sinbad in all three of the copulations that have been observed. Though these are the first pied-billed grebe copulations that I have seen (and I have no prior knowledge of this species' reproductive behavior), given the above observations it is probably a safe bet that this is the standard "present" posture utilized by females of this species. Similarly, the rhythmic swaying of the head/neck from side to side during a copulation is also likely a stereotypical behavior exhibited by females.

In none of the observed copulations did Sinbad mount Remy immediately upon seeing her present to him. Instead, in each case he remained poised in the water next to the nest behind her rump (his attention focused on her) for lengthy periods of from ~ 40 seconds to more than one minute before finally mounting her.

Once again today, though no eggs have yet been laid both members of the pair spent time lying well-settled on the nest and several nest exchanges were observed.

At this point in their preparation for this second nesting attempt Sinbad has done noticeably more nest construction work and nest sitting than Remy, while Remy has spent noticeably more time foraging than her mate. Considering the far greater nutritional demands faced by females because they are the ones that have to produce the eggs, this seems quite logical from a biological perspective.

**6/28/17:** Time: 0835-0935.

0835 – As this observation period begins Sinbad is at the nest engaged in nest construction activity. Remy was initially ~ 1.5 m east of the nest before then climbing onto and settling on the nest. No eggs are present.

0843 – Sinbad is still busily engaged in nest construction activity. Remy stands, walks off the nest, stops ~ 2 m to the east of it and preens. Approx. 40 seconds later she moves farther east ~25-30 m from the nest where she parks and preens.

0847 – Sinbad ceases his nest work and swims east toward Remy (still in the general location noted at 0843). When he reaches Remy he brushes against her left flank but doesn't stop. Instead, he just continues on past her. He proceeds to the area by the reeds at the mid-north bank of the SC where the now obliterated nest from their failed nesting attempt is located. There, he dives, collects vegetation during his dive, takes this to the nest and adds this material onto it. He then makes 5 more successive trips to the same general spot where he again collects vegetation which he takes back and adds onto the nest. While Sinbad was engaged in this nest construction work Remy continued preening herself in the western half of the SC.

0858 – Remy moves east out to the mid-eastern edge of the SC, dives there and isn't resighted at this time (she has probably begun a foraging session).

0900 – After placing the vegetation from a 6$^{th}$ collecting trip onto the nest, Sinbad swims out to the mid-SC area where he parks and begins preening. He preens for ~ 45 seconds then moves farther east, calling twice (~ 10 seconds apart) as he reaches the east edge of the SC. He then moves a little farther to the north and parks just east of the NE edge of the SC, often looking about while intermittently preening.

0909 – Sinbad moves a little farther north in the area just east of the NE edge of the SC and dives. He resurfaces holding a prey item in his beak and quickly swallows it (it appears to be a fish but my view is too inadequate to be certain). He then begins moving slowly west, re-entering the SC and proceeding along the near-shore area of the north bank.

0911 – Sinbad has made his way back to the nest and upon arriving he calls once while in the water at its east side. Several seconds later he climbs onto the nest and preens for several minutes while standing before finally lying down and settling himself. Remy still hasn't been resighted.

0928 – Sinbad is still well-settled on the nest. Remy is sighted entering the SC from the east. She makes her way directly to the nest and mills around it, apparently waiting for Sinbad to move off the nest. Approx. 40 seconds later, Sinbad stands and walks off the nest. Shortly after, Remy climbs onto the nest and settles herself on it. After leaving the nest, Sinbad immediately engaged in nest construction activity.

0935 – As this observation is concluded, Remy is still settled on the nest and Sinbad is still engaged in nest construction activity;

*0940 – Just before leaving the area after concluding the observation period I took a final look at the nest and discovered that a copulation was in progress on it. Sinbad was mounted on Remy and

Remy's head/neck was stretched out and raised and swaying from side to side (just as she was seen doing in the other three copulations that were observed). This is now the third consecutive day and the fourth day in the last five that a copulation has been seen.

**6/29/17:** Time: 1223-1332; 1401-1418.

1223 – As the observation period begins Remy is well-settled on the nest and Sinbad is parked and preening near the reeds at the mid-section of the north bank of the SC.

1226 – Sinbad (in the same general location noted at 1223) breaks from his preening to call once and then resumes preening. He soon moves to the mid-area of the western half of the SC. Remy is still well-settled on the nest (because she has remained settled on the nest I haven't been able to see if any eggs are present).

1227 – Remy stands, walks off the nest and goes to the mid-area of the SC where she parks and preens. I can now see that at least one egg is present on the nest! Approx. 12 seconds after Remy leaves the nest Sinbad goes to it, climbs on, fusses briefly with the egg(s) and then settles onto the clutch and incubates it.

1230 – Remy moves out near the NE point of the SC, dives there and isn't resighted at this time (she has probably begun a foraging session). Sinbad is still well-settled on the nest, preening himself while incubating the clutch (while preening he often rubs his folded wings together over his back—this is something that he has often been observed doing).

1242 – Remy is sighted foraging in the western half of the river, just north of the NE point of the SC. Sinbad is still incubating and still often rubbing his folded wings together.

1246 – Sinbad stands, walks off the nest and parks ~ 1 m away from the side of the nest. There, he sits and intently looks about, as if waiting for his mate.

1247 – Sinbad leaves the nest area (which is then left unoccupied), goes out near the north bank of the SC and moves east along it. While doing so, he calls once, dives soon after and isn't resighted at this time.

1250 – Sinbad is resighted, now back at the nest. He climbs onto the nest, rearranges the vegetation on it for ~ 12 seconds and then settles onto the clutch and preens himself while incubating. Remy is foraging in the northern half of their territory.

1258 – Remy is sighted entering the SC from the E-NE and slowly makes her way directly to the nest (as usual, she dives when she gets within ~ 15 m of the nest and covers the remaining distance underwater). She then sits in the water at the east side of the nest and soon reaches her head out and gently pokes Sinbad's rump with her beak once. Seconds later Sinbad stands and walks off the nest. Soon after, Remy climbs onto the nest, fusses briefly with the egg(s) and the nest vegetation around it, then settles onto the clutch and incubates it. Sinbad engages in nest M/E. He makes 3 trips to gather vegetation for the nest from the area at the location of their first nest (i.e., from their failed nesting attempt—henceforth this will be referred to as the "old nest site"). After the third trip he just collects his nest material from the vicinity around the nest.

1308 – Sinbad ceases his nest M/E activity and sits in the water at the west side of the nest facing Remy (who is still incubating). Approx. one minute later Remy stands, walks off the nest and moves out to the mid-SC area where she parks and preens. Sinbad initially appears as though he is going to go onto the nest and assume incubation duty but instead moves all the way out to the mid-river area east of the SC and begins foraging. He soon works his way north into the north half of their territory.

1313 – Remy (still in the mid-SC area) ceases preening and goes back to the nest, which is still unoccupied. However, instead of going onto the nest she sits in the water at its east side and intently looks to the east (it appears clear that she is looking for Sinbad). It seems obvious by her behavior that she doesn't want to assume incubation duty at this time and would like Sinbad to do so (but at this time he is far away, foraging in the north half of their territory). Remy continues on like this, parked by the nest and intently looking east for Sinbad until 1320. At this point she abandons this vigilance for Sinbad and engages in nest M/E, repeatedly gathering vegetation from the area around the nest and adding this onto it.

1326 – Remy finally resigns herself to incubation duty. She ceases her nest M/E, climbs onto the nest, spends some time rearranging the nest material around the clutch then settles onto it and incubates. Sinbad is still foraging in the north half of the territory.

1328 – Sinbad suddenly arrives at the nest with a load of vegetation held in his beak. He adds this onto the nest and engages in nest M/E, repeatedly gathering vegetation from the area around the nest and adding this onto it—Remy sometimes rearranges this new material while sitting on the clutch.

1332 – Remy stands, walks off the nest and moves out to the mid-SC area where she parks and preens. Soon after, Sinbad climbs onto the nest, briefly fusses with the egg(s) and the nest material around it, then settles onto the clutch and incubates. Remy soon ceases her preening, moves NE toward mid-river, dives and begins foraging. End observation period.

*1401 – Brief re-check of the nest. Sinbad is well settled on the nest. Remy suddenly appears at the nest and another nest exchange ensues. Sinbad leaves the nest and both briefly engage in nest M/E together. Remy soon climbs onto the nest, settles onto the clutch and incubates. As she incubates, Sinbad continues with his nest M/E activity. This includes making a trip out to the reeds at the NE point of the SC where he collects a long reed stem which he hauls back to the nest and works into it. He continues with his nest M/E work until the pair again exchanges incubation duty at 1418. Observations are concluded at this time.

*Comments:* Egg laying has begun! This has thus occurred on only the eighth day after their first clutch was lost (this is even more remarkable when considering that they also had to build a new nest). One egg was definitely seen in the nest today but my view was too inadequate to be able to confirm whether or not this was the only one (similar to the situation with the first nest, because this nest is tucked within the reeds my view is akin to peering through a picket fence with various-sized gaps between the slats; in addition, my best view is essentially from water level—I don't have any good vantage points that would allow me to look down on the nest and gain a better view into the nest bowl). If serious incubation has begun today then the hatching window for any egg laid today opens on July 21st.

Both individuals took turns incubating the egg(s). Five nest exchanges (or shifts in incubation duty) occurred during the 1 hr. 26 mins. that the pair was observed, including the two that were seen during the nest re-check period.

When going to the nest both members of the pair habitually dive when still a considerable distance away (including when carrying nest material). They then cover the remaining distance to the nest while underwater and resurface within the reeds by the nest where they are much more difficult to see. Aside from the fact that it is easier to go under the algal mats that surround the approach to the nest area rather than through them, it is quite probable that they do this as means of trying to keep from readily giving away the location of the nest (they did this when going to their first nest also). They follow a similar pattern when leaving the nest: they dive in the nest area (within the reeds) and don't resurface until well away from the nest outside the reeds.

**6/30/17:** Time: 1310-1411; Brief re-check at 1425.

1310 – As observations begin Sinbad is well-settled on the nest incubating the clutch. Remy is foraging in the northern half of their territory.

1320 – Sinbad is still on the nest incubating the clutch. Remy has ceased foraging and is now parked in the river ~ 1/3 of the way across from the west bank, east of the NE point of the SC. While incubating, Sinbad has often been rubbing his folded wings together over his back in his typical manner. He now habitually does this. When Remy does so, she does it only briefly. [This behavior has been a useful aid for identifying the individual who is on the nest during times when that individual is facing away from my vantage point and I can only see the individual's back, and the other member of the pair isn't visible. In such cases, if the individual on the nest is rubbing its wings together in this manner for an extended period then it is a virtual certainty that it is Sinbad (in fact, this has proven to be the case in every such instance thus far).]

1324 – Remy begins moving west. She soon enters the SC and continues heading west, moving slowly along the north bank, sometimes stopping briefly and preening, then continuing on (she appears to be making her way to the nest).

1326 – Remy is still slowly approaching the nest. When she reaches within ~ 40 m Sinbad leaves the nest and heads for her. The two meet en route and Sinbad just continues on past Remy to the area near the NE point of the SC where he dives and isn't resighted at this time (he has probably begun foraging). In the meantime, Remy goes to the nest, climbs onto it, briefly repositions/rotates the eggs, then settles onto and incubates them. Two eggs were definitely seen, indicating that at least one more has been laid since yesterday's observations (though it is also possible that the second egg was present yesterday but I just wasn't able to see it; similarly, my view isn't perfect so I can't say with certainty that only two eggs are present today).

1332 – Sinbad has returned to the nest and sits in the water next to it. Remy stands, walks off the nest, moves out ~ 10-12 m to the east of it and parks there and preens. Sinbad then engages in nest M/E activity, including two trips to collect vegetation from the vicinity of the old nest site. Because he has consistently come to this location to collect vegetation for their current nest it appears rather certain that he is collecting vegetation from their previous nest that has sunk here and thus provides a ready supply of nest material that they had already collected (i.e., they are re-using material from their first nest). Remy eventually moves a little farther out to the mid-SC area where she parks and resumes preening.

1337 – Remy goes to the nest just before Sinbad returns with another load of vegetation that he had collected from the old nest site area. However, Remy doesn't go onto the nest. Instead, she lingers by it for about a minute, then goes back out to the mid-SC area and again parks there and preens. Sinbad continues with his nest M/E activity, including making more trips to collect nest material from the old nest site area.

1348 – Remy goes to the nest and joins Sinbad in conducting nest M/E. After only ~ 1 minute of this she goes back out to the mid-SC area where she again parks and preens. Sinbad soon climbs onto the nest and fusses with the eggs and rearranges the nest material around them for ~ 45 seconds but doesn't settle on the clutch. Instead, he walks off the nest and resumes his nest M/E activity, including more trips to collect vegetation from the old nest site area.

1354 – Remy goes to the nest and joins Sinbad in conducting nest M/E.

1358 – Remy ceases her nest M/E work and climbs onto the nest. She fusses with the eggs and the nest material around them for ~ 7 seconds, then settles onto the eggs and incubates them. Sinbad continues with his nest M/E activity, including additional trips to the old nest site area.

1405 – While continuing with his nest M/E work Sinbad makes another trip to the old nest site area but this time he parks near it and preens himself.

1408 – Sinbad (still parked in the old nest site area) ceases preening, dives and collects vegetation, takes this back to and places it on the nest and continues with his nest M/E.

1411 – Remy stands, walks off the nest, moves out ~ 15 m to the east of it, parks there and preens. Sinbad is still engaged in nest M/E activity. End observation period.

*1425 – Brief re-check of the nest. Remy is parked ~ 1 m east of the nest, intently looking to the east (apparently looking for Sinbad). Sinbad is foraging in the northern half of their territory.

*Comments:* There were definitely at least two eggs seen in the nest today. It thus appears that at least one more has been laid since yesterday's observations.

Though at least one member of the pair was next to or in the general area of the nest for the entire observation period, the pair spent what seemed a relatively short amount of the period incubating the eggs (only just over half of the period—35 of the 61 total minutes) and this appeared to be deliberate on their part.

Sinbad was very intent on conducting nest M/E work and did so for much of the period.

A stranger pied-billed (PB) grebe was sighted today in the river to the north of the SR pair's territory. Since I began observations on this pair (6/9/17) I haven't seen any other PB grebes enter into their territory. There is currently a PB grebe pair in the East Lagoon with 4 chicks that are well-along in the chick rearing process. I have occasionally seen members of their family foraging in the West Lagoon but haven't seen any of them in the territory of the SR pair. Unlike last year, there currently is no PB grebe pair residing in (i.e., having claimed) the West Lagoon (this matter was discussed in the notes for 6/23). As previously discussed in the 6/23 notes, the topography and general habitat of this lagoon was altered considerably during the lengthy period during the rainy season when the river flowed very strongly through and essentially obliterated the West Lagoon, this area having then been transformed from a lazy lagoon into a proper river (with the onset of the dry season the West Lagoon has since reformed). Recall that the west side of the lagoon actually remained essentially intact and it was the east side (which is the outside bank of the river here) that bore the brunt of the river's erosive force.

Calls have regularly been heard by the SR pair—thus far I have only observed Sinbad making these calls and not Remy (I have also regularly heard calls made by the East Lagoon pair). These calls definitely appear to play a role in the establishment and maintenance of a territory (they also appear to have an intra-pair communicatory function).

**<u>7/1/17</u>:** Time: 1250-1352.

1250 – As observations begin, Sinbad is well-settled on the nest, incubating (he is rubbing his folded wings together over his back as he typically does). Remy isn't seen at this time.

1312 – Remy is finally sighted. She is foraging in the northern half of their territory. Sinbad is still incubating the clutch as he has solidly been doing since observations began.

1315 – Remy has worked her way south. She enters the SC from the NE and goes to the mid-SC area where she parks and preens. Sinbad (still well-settled on the nest) apparently spots her and keenly gazes in her direction (he appears eager for a nest exchange).

1318 – Remy has moved a little closer toward the nest but is still parked and preening in the mid-SC area. Sinbad (apparently losing patience) stands and walks off the nest. He moves directly toward Remy and meets and greets her on the move as he continues past her. He proceeds to the area near the old nest site, dives and collects vegetation, takes this back to the nest and places this material onto it. While he is engaged in this, Remy goes to the nest but instead of climbing onto it she just parks in the water by its east side.

1320 – Remy finally climbs onto the nest, rearranges the nest material around and over the eggs for ~ 17 seconds, then walks off the nest and parks in the water by it (only a small patch of one of the eggs is visible and this is all that I have been able to see of the clutch thus far). Sinbad has continued with his nest M/E work, making another trip to collect nest material from the old nest site area.

1324 – Remy moves out ~ 12 m east of the nest and parks and preens there.

1325 – Remy moves back to the nest and joins Sinbad in nest M/E activity. Sinbad is still making trips to the old nest site area for nest material; Remy gathers hers locally from around the nest.

1326 – Sinbad ceases his nest M/E activity, moves out to the area just off the NE point of the SC and begins foraging.

1330 – Remy ceases her nest M/E work, climbs onto the nest, briefly rearranges the nest material around the eggs, then settles onto them and incubates. Sinbad is now foraging in the northern half of their territory.

1336 – Remy is still settled on the nest incubating the clutch. Sinbad has worked his way south and is now foraging in the river east of the SC. While there he calls once and then resumes foraging.

1340 – Sinbad enters the SC, goes to the nest and sits in the water next to it (apparently waiting for Remy to come off the nest). Shortly after, Remy stands and walks off the nest. Seconds later Sinbad climbs onto the nest, fusses with the eggs and the nest material around them for ~ 25 seconds, then settles onto the eggs and incubates them. Remy moves out to the mid-east end of the SC where she parks and preens.

1346 – Remy is still parked at the east edge of the SC. She dives and begins foraging.

1350 – A PB grebe is sighted at the far north end of the SR pair's territory. The grebe suddenly takes a long flight, flying low over the water to the south end of the territory. It lands near the east bank of the river in the area well east of the SC, dives soon after and isn't resighted at this time. [This grebe dove before I had the chance to identify whether or not it was Remy. Though I know that she had been foraging in the northern half of their territory the bird that took flight was too distant for me to be able to positively identify. Thus, especially considering that a stranger PB grebe was seen in the area north of their territory yesterday, I can't be certain about whether or not the flying grebe was Remy. Perhaps lending some support that it was Remy is the fact that a few minutes later she was seen foraging in the area where the bird had landed. However, countering this is the point that in those few minutes Remy would have had plenty enough time to have moved (in the water) from the north end of the territory to the south end. Furthermore, if the flying bird was a stranger/interloper whom Remy had charged and caused to take flight in order to flee from her, it would have been typical for her to come to the area where it had landed to search for it. Though here again, one must then consider that when a fleeing bird in such territorial defense situations takes flight, the pursuing bird typically (but not always) takes flight after it. However, in this case only the one bird was seen in flight. Finally, there is the fact that PB grebes are seldom seen in flight. So, if that was Remy one must then contemplate what possible reason she would have had to suddenly undertake what would then appear to be a seemingly unprovoked flight? All speculation aside, in the end all that can really be stated is that it simply isn't known whether or not that flying grebe was Remy.]

1352 – Sinbad is still well-settled on the nest, incubating. Remy is foraging in the SE section of their territory (i.e., east of the SC). End observation period.

*1407 – Brief re-check of the nest. A nest exchange occurs just as I check the nest: Sinbad stands and walks off the nest. Remy then climbs onto the nest, rearranges and rotates the eggs, then settles onto them and incubates (two eggs were definitely seen and almost certainly a third). Sinbad engages in nest M/E.

*Comments:* For the entire observation period the eggs were obscured by the incubating parent or the nest material on and around them except for a brief moment when a small patch of one egg was visible. It wasn't until the re-check that I was afforded a decent look at the clutch. Two eggs could be confirmed and almost certainly a third. It thus appears that at least one more egg has been added to the clutch since yesterday's observations.

If the PB grebe that was seen flying was indeed Remy then this would be the first time that I have seen either of the pair in flight. Regardless, it's one of the few times that I have ever seen an adult PB grebe in flight.

**7/2/17:** Time: 1040-1130.

1040 – As observations begin, Sinbad is engaged in nest M/E activity and Remy is entering the SC near its NE point, apparently making her way toward the nest.

1043 – Remy arrives at the nest. She soon climbs onto it, fusses with the eggs and the nest material around them for ~ 35 seconds, then settles onto the eggs and incubates them (only a small patch of one of the eggs was briefly visible). Sinbad is busy making trips to the old nest site area to collect vegetation which he then adds onto the nest.

1050 – Remy is still incubating the eggs. Sinbad is now foraging near the reeds at the NE point of the SC, working his way north.

1054 – Sinbad has parked in the mid-river area in the northern half of their territory and is preening. Remy is still on the nest incubating the eggs. After preening for ~ 3 minutes Sinbad resumes foraging and works his way south into the southern half of their territory, east of the SC.

1104 – Sinbad is foraging by the reeds at the NE point of the SC. He surfaces from a foraging dive with a prey item and quickly swallows it (my view was too inadequate to identify the prey item). Remy is still well settled on the nest incubating the eggs.

1106 – Sinbad enters the SC and moves west to the area near the old nest site where he parks and preens. Remy is still incubating the clutch.

1115 – Sinbad is still parked in the same general area, no longer preening but rubbing his folded wings together over his back in his usual manner. He soon begins moving toward the nest.

1116 – Sinbad arrives at the nest and sits in the water at its north side next to Remy's rump (Remy is sitting on the eggs facing south). Sinbad remains there and ~ 20 seconds later Remy assumes and holds the 'present' posture: she lies flat on her ventral side with her head/neck fully stretched out and also lying flat on the nest. Sinbad remains sitting, poised in the water behind Remy's rump. He remains there for more than a minute before he finally climbs onto the nest, mounts Remy's back and they then copulate (the copulation lasts ~ 9 seconds). During the copulation Remy displays the same behavior that she exhibited in each of the previous copulations that I have observed: she raises her stretched out head/neck to an ~ 45 degree angle and rhythmically sways her head/neck from side to side, turning her head slightly outward at the apex of each side swing. When they conclude the copulation Sinbad immediately dismounts Remy and walks off the nest. Remy remains sitting on the nest ~ 29 seconds longer before standing and walking off the nest (time: 1120). She then moves to the mid-SC area where she parks and preens.

1121 – Sinbad climbs onto the nest, fusses with the eggs and the nest material around them and then settles onto the clutch.

1130 – Sinbad is still well-settled on the nest incubating the eggs. Remy has drifted into the NE section of the SC and is still preening. End observation period.

*Comments:* I wasn't afforded an adequate enough look at the nest today to be able to see how many eggs were present (the most that I was able to see was a small portion of one of the eggs). I thus don't know if any more eggs were added to the clutch since yesterday's observations.

It's interesting that it is now the fourth day since the first egg was laid and the pair is still copulating.

All four of the East Lagoon pair's chicks were sighted today in the northern half of the East Lagoon.

**7/3/17:** Time: 1246-1350; Brief re-check of nest at 1402.

1246 – As observations begin Sinbad is well-settled on the nest incubating the eggs and Remy is parked and preening near the reed beds at the NE point of the SC.

1251 – Remy is still parked near the reeds at the NE point of the SC. She ceases preening, dives and begins foraging. Sinbad is still well-settled on the nest incubating the clutch. An osprey has been perched on the top of a tall snag that is located on the east bank of the river, east of the SC since before this observation period began (this is a regularly used perch by this individual and a few other ospreys).

1300 – Sinbad is still incubating the eggs. Remy is still foraging along the edge of and within the emergent reeds at the NE point of the SC.

1302 – Sinbad stands, briefly fusses with the eggs and the nest material around them and then walks off the nest (a small portion of at least two eggs are seen). Sinbad then starts swimming east in the north section of the SC and calls once as he moves along. At the same time Remy is moving in (i,e., heading west) from the NE point of the SC, also within the north section. They meet up en route near the old nest site. Both then set out for the nest, Remy first with Sinbad following shortly after at a distance. Remy arrives first just ahead of Sinbad. They mill around the nest for several seconds before they both dive. Remy moves out ~15-20 m east of the nest, mills about in that area for ~ 35 seconds, then goes to the SW corner of the SC and begins foraging among the reeds there. Sinbad engages in nest M/E activity.

1315 – Sinbad is still engaged in nest M/E activity. After placing yet another load of vegetation on the nest Sinbad heads east to the old nest site area, calling once en route. He collects a long reed stem from that area, hauls it back to and works it into the nest and then continues with his nest M/E work.

1317 – Remy is still foraging in the SW section of the SC. While she is on the surface within a floating algal mat just SE of the nest the osprey (having targeted a fish while "hovering" over the west end of the SC) suddenly swoops down toward her general area. Startled by its sudden appearance she immediately responds to this by submerging in a rapid, sinking dive. A moment after she submerges, the osprey strikes the water with its talons ~ 12-15 m to the NE from where she dove in an unsuccessful attempt to capture a fish. Remy resurfaces soon after and goes to the nest (the osprey has left the SC area). She then climbs onto it, rearranges the nest material around the eggs for ~ 12 seconds but doesn't settle onto the clutch. Instead, she walks off the nest and moves out to the mid-SC area where she parks and preens. Sinbad is still engaged in nest M/E.

1327 – Remy (still in the mid-SC area) begins slowly moving toward the nest. Sinbad (still engaged in nest M/E) goes to the old nest site area and collects another long reed stem which he hauls back to and places on the nest.

1328 – Remy arrives at the nest. She lingers at its east side for ~ 20 seconds before finally climbing onto it. She rearranges the nest material around and over the eggs for ~ 45 seconds, then settles onto the eggs and incubates them.

1330 – Sinbad moves to the area a little south of the old nest site, parks there and preens.

1343 – Sinbad is still parked and preening in the general area of the old nest site. He briefly pauses from his preening to call once. Remy is still well-settled on the nest, incubating the clutch.

1350 – Sinbad is still parked and preening in the area south of the old nest site. Remy is still well-settled on the nest incubating the eggs. End observation period.

*1402 – During a brief re-check of the pair a nest exchange occurs. Remy comes off the nest and Sinbad goes on and assumes incubation duty.

*Comments:* Once again I was not afforded a clear look at the eggs in the nest. At best, only a few small patches of two eggs were visible to me. Thus, at this time I don't know how many eggs are present. The most that have been visible to me at any one time are the three that were seen on 7/1. It seems a safe bet that more eggs have probably been added to the clutch since then.

**7/4/17:** Time: 1205-1315.

1205 – As observations begin Sinbad is well-settled on the nest incubating the eggs (while incubating he is rubbing his folded wings together in his usual manner). Remy isn't seen at this time.

1226 – Remy is finally sighted—she is foraging along the reeds that line the west bank of the river at the north end of their territory. Sinbad is still incubating the clutch as he has been doing since before this observation period began.

1228 – Remy (still foraging along the reeds of the west bank at the north end of their territory) has surfaced from a foraging dive with a crayfish which she vigorously works before eventually swallowing it. Sinbad is still incubating the clutch.

1232 – Remy parks in the mid-river area at the north end of their territory and preens.

1234 – Remy begins swimming south. She eventually rounds the NE point of the SC and enters into it.

1236 – Remy moves into the mid-SC area where she parks and preens. Sinbad (still on the nest incubating) obviously has sighted Remy. He stands and walks off the nest and moves directly toward her. Remy dives well before Sinbad can reach her and heads toward the nest.

1237 – Remy arrives at the nest and lingers in the water by it for ~ 20 seconds before climbing onto it. She briefly rearranges the nest material around the eggs, then settles onto and incubates them. In the meantime, Sinbad has moved out to the edge of the reed beds at the NE point of the SC. He dives there and begins foraging, slowly working his way north along the reeds of the west bank.

1245 – Remy is still incubating the eggs. Sinbad is still foraging along the reeds of the west bank of the river, north of the NE point of the SC.

1247 – Sinbad has reversed his course and is now working his way south. He rounds the NE point of the SC, enters the SC and goes to the nest. He then briefly engages in some nest M/E activity before moving to the mid-SC area where he parks and preens. Remy continues incubating the clutch.

1256 – Remy stands, repositions/rotates the eggs for ~ 8 seconds, then settles back onto them and resumes incubation. Sinbad is still parked and preening in the mid-SC area.

1257 – Remy is still incubating the clutch. Sinbad goes to the nest and engages in nest M/E activity.

1302 – With Sinbad in the water at the south side of the nest, Remy stands, walks off the nest, moves to the mid-area at the east end of the SC and parks there and preens. Sinbad climbs onto the nest, fusses with the eggs for ~ 12 seconds (parts of at least 2 eggs were briefly visible), then settles onto and incubates them.

1315 – Sinbad is still well-settled on the nest incubating the clutch. Remy (still parked at the east edge of the SC) ceases her preening, goes over near the old nest site, dives and collects some vegetation which she carries to and places on the nest. She then continues with her nest M/E work. End observation period.

*Comments:* It was a rather quiet, uneventful observation period today. The eggs were solidly incubated for nearly the entire 1 hr. 10 min. observation with both of the pair taking their turns on the nest.

Yet again I was not afforded a decent look at the eggs in the nest so I still don't know how many the clutch currently contains.

One of the chicks belonging to the East Lagoon family ventured a short distance into the SE edge of the SR pair's territory (it went up to the end of the east bank of the river, just north of the RR bridge). Neither of the SR pair was near it at the time: Sinbad was on the nest incubating and Remy was way out at the north end of their territory. This was the first time that I have seen any of the East Lagoon family enter into the SR pair's territory.

**<u>7/5/17</u>:** Time: 1152-1255.

1152 – The nest is unoccupied as observations begin (my view of the nest bowl is quite limited but parts of at least two eggs are visible). Sinbad is busily engaged in nest M/E activity. Remy isn't seen at this time.

1155 – Remy is finally sighted. She has come into the mid-SC area where she is parked and preening. Sinbad is still engaged in nest M/E.

1157 – Remy goes to the nest, climbs onto it, rearranges the nest material around the eggs for ~45 seconds, then settles on the eggs and incubates them. Sinbad continues with his nest M/E work.

1200 – Sinbad (still engaged in nest M/E activity) goes out to the reeds near the old nest site, collects a fairly long, stout reed stem, hauls it back to and places it on the nest. He then moves out to the mid-SC area where he parks and preens. Remy is still incubating the eggs.

1205 – While continuing to preen, Sinbad has slowly drifted to the mid-area along the east edge of the SC. Remy is still incubating the clutch.

1207 – Sinbad moves east toward mid-river, east of the SC, dives and begins foraging, slowly working his way north. Remy continues incubating the eggs.

1212 – Sinbad has continued working his way north and is now foraging along the reeds that line the west bank of the river at the north end of their territory. Remy is still incubating.

1215 – Sinbad has reversed his course and is now working his way south along the reeds of the west bank of the river. He surfaces from a foraging dive by the reeds at the NE point of the SC with a fish, swallows it, dives again and isn't resighted at this time.

1217 – Sinbad is resighted near the old nest site area carrying a load of vegetation toward the nest. He goes to the nest, adds this material onto it and continues with his nest M/E activity. Remy is still incubating the clutch.

1221 – Sinbad ceases his nest M/E work and moves out to the mid-SC area where he parks and preens. Remy continues incubating the clutch.

1229 – Sinbad goes to the nest. When he arrives, Remy stands and walks off the nest. Seconds later Sinbad climbs onto the nest, fusses with the eggs and the nest material around them for ~32 seconds, then settles onto the eggs and incubates them. Remy slowly makes her way out to the NE end of the SC, dives and begins foraging.

1234 – Remy surfaces by the edge of the reeds at the NE point of the SC with a fairly large fish. She works it for ~ 12 seconds before swallowing it and then resumes foraging. Sinbad is still incubating the eggs.

1240 – Remy has worked her way into the mid-river area east of the NE point of the SC and continues her foraging there. Sinbad is still incubating, periodically preening and rubbing his folded wings together.

1241 – Remy parks mid-river in the area just north of the NE point of the SC and preens. Sinbad is still incubating.

1246 – Remy (still in the same mid-river area) ceases preening and resumes foraging, working her way south. By 1250 she has worked her way into the mid-river area east of the SC. Sinbad is still incubating the clutch.

1252 – Remy ceases her foraging and makes her way to the nest. Upon her arrival Sinbad stands and walks off the nest. Remy briefly engages in some nest M/E work before she climbs onto the nest, briefly fusses with the eggs and the nest material around them, settles onto the eggs and incubates them (time: 1255). While she has been doing this, Sinbad has set off along the north side of the SC toward the old nest site, calling once en route. When he reaches that area he dives, collects a load of vegetation, carries it back to the nest and adds this material onto it. He then embarks on another collecting trip to the old nest site area. Remy continues incubating the eggs. Observations are concluded at this time (1255).

*Comments:* As yesterday, it was another routine, fairly uneventful observation period. Though there were some short intervals when the nest was unoccupied, the eggs were solidly incubated for nearly the entire observation period.

At this point there is little doubt that the pair (notably Sinbad) is re-using nest material from their previous nest (i.e., the one used in their failed nesting attempt). They have repeatedly dove and collected vegetation from that specific area and added this to the new nest. They have done so from the time when they began constructing the new nest to the present when engaging in nest M/E.

**7/6/17:** Time: 1200-1300; Brief re-check of nest at 1326.

1200 – As observations begin, Sinbad is well-settled on the nest incubating the eggs and Remy is parked and preening in the NW section of the SC.

1206 – Remy (still in the NW section) dives and isn't resighted at this time. Sinbad is still incubating the clutch.

1209 – Remy is resighted as she pops up next to the nest. Sinbad stands, walks off the nest, goes out to the old nest site (calling once en route), dives and collects some vegetation, then takes this back to the nest and places this material onto it. Remy lingers at the east side of the nest but doesn't go onto it.

1212 – Sinbad (still engaged in nest M/E) makes another trip to the old nest site area where he collects more vegetation that he adds to the nest. Remy climbs onto the nest, briefly rearranges the nest material around the eggs, then walks off the nest, parks ~ 2-3 m to the east of it and intently gazes to the east. It seems rather clear that, for some reason, she doesn't want to assume incubation duty at this time.

1215 – Remy moves out to the mid-SC area where she parks and preens. Sinbad is still engaged in nest M/E activity.

1217 – Sinbad ceases his nest M/E work, climbs onto the nest, briefly fusses with the nest material around the eggs, then settles onto them and incubates the clutch. Remy (still in the mid-SC) dives and isn't resighted at this time.

1219 – Remy is resighted as she parks ~18 m east of the nest. Sinbad stands, walks off the nest, goes to the old nest site area, dives and collects some vegetation which he brings back and adds to the nest.

1220 – Remy goes to the nest, climbs onto it, briefly rearranges the nest material around the eggs, then settles onto and incubates them. Sinbad continues with his nest M/E activity.

1221 – Sinbad arrives at the nest with another load of vegetation that he had collected by the old nest site and adds this onto the nest. As he does, Remy stands, walks off the nest and moves out to the mid-SC area where she parks and preens.

1230 – Sinbad ceases his nest M/E activity and heads east toward the area where Remy is (she is still parked and preening in the mid-SC). As he approaches, Remy moves in his general direction. They pass close to each other and continue on in opposite directions. Sinbad continues on to the mid-area at the east end of the SC where he parks and preens. Remy keeps moving toward the nest.

1231 – Remy arrives at the nest and climbs onto it. She initially just stands and preens for ~ 17 seconds, then walks off and engages in nest M/E activity. Sinbad is still preening in the same general location noted at 1230.

1238 – Remy ceases her nest M/E work, climbs onto the nest, briefly fusses with the eggs and the nest material around them, then settles onto the eggs and incubates them. Sinbad is still parked and preening at the east end of the SC.

1245 – Sinbad (still at the east end of the SC) ceases preening, dives and begins foraging in the mid-river area east of the SC. Remy is still incubating the clutch.

1252 – Sinbad ceases foraging and goes to the mid-area along the east end of the SC where he parks and preens. Remy is still incubating the eggs.

1255 – Sinbad begins slowly moving west toward the nest, preening as he moves along.

1256 – Sinbad arrives at the nest. Remy then stands, walks off the nest, moves out to the SE edge of the SC, preens herself there briefly, then dives and begins foraging in the SE section of their territory. Meanwhile, Sinbad has gone onto the nest. He fussed with the eggs and nest material around them for ~ 40 seconds (parts of at least two eggs were briefly visible) before settling on them (he preens himself while incubating).

1300 – Sinbad is still well-settled on the nest incubating the clutch. Remy is foraging in the SE section of their territory. End observation period.

*1326 – During a brief re-check of the pair a nest exchange takes place: Sinbad goes off the nest and immediately engages in nest M/E activity. Remy goes onto the nest and assumes incubation duty. During this exchange I was afforded the best look at the clutch that I have had up to this point. Though my view was still quite limited, parts of at least three eggs and probably a fourth were visible (my view was too inadequate to be able to determine if more than four eggs were present).

*Comments:* The eggs were incubated for less than half of the total time during today's observation period though much nest M/E work was done.

At this time the lives of the pair generally revolve around four main activities: incubating the eggs, nest M/E, foraging and territorial maintenance.

Both members of the pair take their turns incubating the eggs. At this point incubation duty has generally been equitable among the pair though Sinbad has spent a bit more time at this task. Typically, 1-3 nest exchanges may take place within an hour.

When relieved of incubation duty, that individual will typically (but not always) use the break to do some foraging. Along with spending a little less time incubating than Sinbad, Remy appears to spend more time foraging. Besides foraging, other activities during incubation breaks include nest M/E work, territorial maintenance, and much preening.

Both members of the pair engage in nest M/E work though overall, Sinbad does notably more of this than Remy.

Territorial maintenance appears to involve calls, evicting intruders who venture into the territory, and maintaining a visible physical presence throughout the territory. Regular foraging through the various sections of the territory plays a role in the latter. In addition, Sinbad and Remy regularly park themselves out in open, conspicuous places within various parts of the territory where they would be easily visible to other PB grebes. This appears to serve to intimidate and advertise to any other grebe that this territory has been claimed. With Remy spending a bit less time incubating and engaging in nest M/E and more time foraging and moving about the territory than Sinbad, she is currently more involved in presenting a physical territorial presence than he is. As for the calls, thus far I have only observed Sinbad calling.

Probably the most obvious element of territorial establishment and maintenance is to physically exclude interlopers from the territory by means of aggression (including chases) and/or displays. Up to this point the pair has not had to engage in any such behavior while I was observing them since no significant interloper situations have occurred during my observation periods (with the possible exception of the "flying grebe" incident noted on 7/1).

**7/7/17:** Time: 1150-1320.

1150 – As observations begin, Remy is well-settled on the nest incubating the eggs and Sinbad is parked and preening at the NE end of the SC.

1205 – While continuing to preen, Sinbad has drifted south to the mid-area along the east end of the SC. Remy is still incubating the clutch.

1207 – Sinbad begins moving toward the nest and calls once as he moves along.

1208 – While Sinbad is still ~ 15 m away as he approaches the nest, Remy (who has obviously spotted him) stands, walks off the nest and moves out to the mid-SC area where she parks and preens. Sinbad arrives at the nest a few seconds after Remy had left it. He climbs onto the nest, fusses with the eggs and the nest material around them for ~ 18 seconds, then settles onto the eggs and incubates them.

1212 – Remy moves east out of the SC and begins foraging, working her way north. Sinbad is still incubating the clutch.

1218 – Off the east bank of the northern half of the territory 3 mallards who have been ranging closely together there suddenly take flight as a tight group and land ~ 15 m farther north in the river. I immediately sight Remy in the area where the mallards had taken off from, peering intently in their direction. I only took note of this incident when the ducks had just taken flight and thus drew my attention. Therefore I didn't see what had caused them to suddenly fly away. However, based on the circumstantial evidence there seems little doubt that Remy had probably charged at or otherwise driven them away. Sinbad is still incubating the eggs.

1222 – Remy has parked in the mid-river area in the northern half of their territory. The mallards have moved north out of the SR pair's territory. Sinbad is still well-settled on the nest, incubating.

1225 – Remy (still in the northern half of their territory) begins foraging in the western half of the river. Sinbad continues incubating the clutch.

1231 – Sinbad stands and walks off the nest. He lingers at the side of the nest for several seconds, then climbs back on, rearranges the nest material around the eggs for ~ 8 seconds, walks off the nest again and engages in nest M/E activity. Remy (still foraging) has worked her way south and is now foraging by the reeds along the west bank of the river, north of the SC.

1237 – Sinbad is still engaged in nest M/E work. Remy is still foraging along the reeds of the west bank of the river, north of the SC.

1239 – Sinbad climbs onto the nest, briefly rearranges the nest material around the eggs and then settles onto and incubates them. Remy moves into the mid-SC area where she parks and preens.

1245 – Remy begins moving toward the nest. While she is still ~ 10 m away on her approach, Sinbad stands, walks off the nest, goes to the mid-SC area, parks there and preens. Remy arrives at the nest ~ 3 seconds after Sinbad had left it. She climbs onto the nest, fusses with the eggs and the nest material around them for ~ 30 seconds, then settles onto the eggs and incubates them. At least 3 eggs and possibly a 4th were briefly visible.

1247 – Sinbad ceases preening, goes to the old nest site area, dives and collects some vegetation, takes it to the nest and adds this material onto it. He then moves out to the east edge of the SC where he parks and preens. Remy is still incubating the clutch.

1300 – Sinbad (still at the east edge of the SC) ceases preening, moves east to the mid-river area east of the SC, dives and begins foraging. Remy continues incubating the clutch.

1307 – Sinbad resurfaces from a foraging dive ~ 17 m off the NE point of the SC with a fairly large fish. He vigorously works the fish for ~ 38 seconds before finally swallowing it. He then collects a long reed stem from near the reed bed there and begins hauling it back to the nest. Remy is still incubating the clutch.

1309 – When Sinbad arrives at the nest with the reed stem Remy stands and walks off the nest. She moves out to the mid-SC area where she parks and preens.

1310 – Sinbad climbs onto the nest, rearranges the nest material for ~ 17 seconds, then walks off the nest and engages in more nest M/E activity. Remy (still in the mid-SC) dives and isn't resighted at this time (she has probably begun foraging).

1319 – Sinbad climbs onto the nest, briefly fusses with the eggs, then settles onto and incubates them. Remy is foraging in the north half of their territory. End observation period (time: 1320).

*Comments:* Regarding the incident at 1218 in which the 3 mallards suddenly took flight there seems little doubt that Remy had chased them away. Her reason for doing so was not apparent—the mallards had merely been foraging and were doing nothing of note that would warrant this.

Prior to when the observation period began a chick from the East Lagoon family was again seen briefly foraging in the south end of the SR pair's territory, just north of the RR bridge. Neither Sinbad nor Remy was in the area at the time or in a favorable location where they would be likely to catch sight of the young interloper.

**7/8/17:** Time: 0912-1025.

0912 – As observations begin, Sinbad is well-settled on the nest incubating the eggs. Remy isn't seen at this time.

0916 – Remy is sighted as she moves into the mid-SC area where she parks and preens. Shortly after, Sinbad (having spotted Remy and obviously eager to be relieved of incubation duty) stands, walks off the nest and moves toward her. Remy catches sight of Sinbad as he approaches at a distance and moves toward him. They meet en route, greet briefly (brushing their breasts together as they pass) and then continue past each other in opposite directions. Sinbad goes to the old nest site area where he parks and preens. Remy continues on toward the nest. When she arrives at the nest (time: 0917) she lingers in the water by it for ~ 30 seconds, then climbs onto it, briefly fusses with the nest material, then settles onto the eggs and incubates them.

0924 – Sinbad moves out to the mid-area at the east edge of the SC, parks there and preens. Remy is still incubating the eggs.

0926 – Sinbad (still at the east edge of the SC) calls once, moves to the mid-river area east of the SC, dives and begins foraging, working his way north into the northern half of their territory. Remy is still incubating the clutch.

0932 – While foraging in the western half of the river at the north end of their territory Sinbad calls once, then resumes foraging. He works his way over to the reeds along the west bank. As he forages there, another PB grebe is sighted foraging along the reeds just north of Sinbad at the territorial boundary. This grebe is quickly lost from my view amid the reeds but Sinbad probably saw it.

0933 – Sinbad moves to the mid-river area at the northern boundary of their territory and parks there, his attention mainly focused to the north. The other grebe suddenly flies out from the west bank reeds and lands mid-river ~ 25 m north of Sinbad and then closes ranks a bit with him (it is now just north of the territorial boundary). What appears to be a territorial standoff then ensues as each very conspicuously sits in mid-river on their respective sides of the border, facing each other down at a distance (Sinbad also begins preening). [It is most likely that the other grebe belongs to a pair that has apparently established a territory in the section of the river immediately north of (bordering) the SR pair's territory.] Remy is still incubating the clutch.

0935 – Sinbad takes a short, low flight (just south of the territorial boundary) from mid-river and lands near the reeds of the west bank at the north end of their territory, then goes back out to the mid-river area and parks there (I can no longer see the other grebe).

0939 – Sinbad (still mid-river at the north end of their territory) dives and begins foraging. The other grebe remains unseen by me. Remy is still incubating the clutch.

0943 – Sinbad is still foraging, working his way south from the northern end of their territory. Remy is still incubating the eggs.

0945 – Sinbad moves to the mid-river area east of the NE point of the SC where he parks and preens.

0951 – Sinbad (still parked mid-river, east of the NE point) begins making his way west toward the SC. He soon rounds the NE point, proceeds along the reeds of the north bank, then angles SW toward the nest.

0953 – While Sinbad is still ~ 22 m away Remy stands, walks off the nest, goes out to the mid-area at the east end of the SC, parks there and preens. Sinbad arrives at the nest ~ 4 seconds after Remy had left it. He climbs onto the nest, fusses with the eggs and the nest material around them for ~ 41 seconds, then settles onto and incubates the eggs.

0958 – Remy (still at the east end of the SC) ceases preening and begins foraging, initially working her way north in the mid-river area, then toward the east bank. Sinbad is still incubating the clutch.

1005 – Sinbad is still incubating the eggs (he is rubbing his folded wings together in his usual manner as he incubates). Remy is still foraging in the northern half of their territory.

1016 – Remy is foraging near the east bank of the river in the northern half of their territory (she is near the head of the small side channel that branches off from the river there). She resurfaces from a foraging dive with a very large fish (i.e., for a PB grebe to swallow). Remy vigorously works the fish for ~ 70 seconds (including diving with it briefly several times) before finally abandoning it—she releases it from her beak and just lets it drop away in the water (apparently, the fish is just beyond the upper size limit of what she can swallow). Sinbad is still incubating the eggs.

1025 – Remy surfaces from a foraging dive in the same area noted at 1016 with another very large fish. She works it for ~ 40 seconds and again finally abandons it (it is quite possible that this was the same fish she had previously abandoned and was giving it a second try—considering the vigorous manner with which she had worked that fish she most likely had killed or seriously incapacitated it, in which case it would then probably have still been lying in the place where she had let it drop away). Sinbad is still well-settled on the nest, incubating. End observation period.

*Within a few minutes after this observation period was concluded a person (one of the many homeless people in this area) on a makeshift paddleboard entered the water at the south end of the SR pair's territory. He proceeded to noisily paddle his way through the heart of their territory and continued north beyond it. On his way he passed right by the area where Remy had been foraging. I no longer saw her there—the paddleboarder undoubtedly had caused her to flee from that area. This major intrusion into their territory had also caused Sinbad to come off the nest and probably to hide among the reeds (I wasn't able to sight him at that time and the nest remained unoccupied). The paddleboarder doubled back when upriver and ~ 5 minutes later he came back into their territory from the north. He lingered in mid-river, east of the NE point of the SC for several minutes before finally paddling into the side channel on the east bank (the head of this channel is where Remy had been foraging). [This side channel begins on the east bank roughly across the river from the NE point of the SC and wraps around a small island. The west side of this island forms the east bank of the very south end of the river (this comprises the east bank segment that lies across the river from the SC). The side channel flows along the north side of this island, then swings south along its east side and eventually empties into the south ribbon of the river which runs along the south end of the island (recall that the "south ribbon" location was defined in the 6/23 notes).] The paddleboarder continued on into the side channel and moved out of the area. I finally resighted Sinbad at 1045. It was clear that he was reluctant to get back on the nest and instead sat in the water at its east side, his gaze nervously fixed to the east where the paddleboarder had paused earlier. At 1050 I caught sight of Remy: she was foraging along the reeds of the west bank of the river in the northern half of their territory. By this time Sinbad still hadn't gone back onto the nest, which at this point had remained unoccupied since the intruder had spooked Sinbad off the nest at ~ 1027.

**7/9/17:** Time: 1200-1315.

1200 – As observations begin, Sinbad is well-settled on the nest incubating the eggs and Remy is foraging in the northern half of their territory. This situation remains so with nothing of note occurring until 1216.

1216 - While sitting on the eggs Sinbad calls once. Seconds later he stands, fusses with the eggs and the nest material around them for ~ 32 seconds and then settles back down on the eggs. Sinbad seems eager to be relieved of incubation duty. Remy is still foraging in the northern half of their territory.

1219 – Sinbad stands and walks off the nest and engages in nest M/E activity, collecting the new nest material locally from the general area around the nest (notably, he makes no collecting trips to the old nest site area).

1221 – Remy has made her way to the mid-area at the east edge of the SC where she parks and preens. Sinbad is still engaged in nest M/E work.

1224 – Remy (still at the east edge of the SC) dives and isn't resighted at this time. Sinbad continues with his nest M/E work.

1226 – Sinbad climbs onto the nest, fusses with the eggs and surrounding nest material for ~ 22 seconds, then settles onto and incubates the clutch.

1229 – Remy is resighted near the NE point of the SC, slowly making her way west toward the nest. While she is still ~ 15 m away as she approaches, Sinbad stands and leaves the nest. He moves east just off the north bank and rounds the NE point of the SC where he dives and begins foraging. Meanwhile, Remy arrives at the nest ~ 5 seconds after Sinbad had left it, climbs onto it (time: 1230), fusses with the eggs and nest material for ~ 11 seconds, then settles onto the eggs.

1237 – Sinbad (still foraging) works his way to the mid-river area east of the SC. Remy is still incubating the clutch.

1240 – Sinbad (still foraging mid-river, east of the SC) begins working his way north.

1242 – Remy stands and walks off the nest. She lingers in the water near the nest, mainly looking about but also doing some light nest M/E work.

1247 – Remy goes back onto the nest, fusses with the eggs and the nest material around them, settles onto the eggs and resumes incubation. Sinbad is still foraging, now mid-river east of the SC.

1249 – Sinbad moves to the mid-area along the east edge of the SC and calls once, then parks there and preens. Well in the distance to the north a stranger PB grebe is foraging just inside the north end of their territory but Sinbad doesn't appear to notice it. [It is probable that a pair has established an adjacent territory to the north of the SR pair's territory (recall the apparent territorial standoff observed on 7/8), though I haven't fully checked this out yet. If so, this stranger grebe is likely a member of that pair.] Remy continues incubating.

1253 – Sinbad is still preening himself at the east end of the SC. He calls once and then continues preening. I can no longer see the other grebe. Remy is still incubating the clutch.

1256 – Sinbad begins moving very slowly west toward the nest.

1259 – Remy stands and leaves the nest while Sinbad is still ~ 8 m away as he approaches. She goes out to the mid-area at the east end of the SC, preens herself there for several minutes, then moves east to the mid-river area and begins foraging. In the meantime Sinbad arrived at the nest seconds after Remy had left it, climbed onto it, stood on it preening himself and also fussing with the eggs and surrounding nest material for ~ 68 seconds, then finally settled onto the eggs (time:1300).

1305 – Sinbad is still incubating the eggs (he is preening himself while sitting on the eggs). Remy is still foraging in the SE section of their territory.

1307 – Sinbad stands on the nest, vigorously flutters his slightly spread wings for ~ 35 seconds, then settles back down on the eggs and resumes incubation.

1315 – Sinbad is still well-settled on the nest incubating the clutch. Remy is still foraging in the SE section of their territory. End observation period.

**7/10/17:** Time: 1159-1311; Brief re-check of nest at 1324.

1159 – As observations begin, Sinbad is engaged in nest M/E activity and Remy isn't seen at this time. One of the East Lagoon family's chicks is foraging at the very south end of the SR pair's territory, just north of the RR bridge by the SE point of the SC. It apparently is unseen by either of the pair (Sinbad is gathering his nest material from the general area around the nest). Also, there is a stranger PB grebe foraging in the river north of the SR pair's territory (probably a member of the pair that has apparently established a territory there).

1202 – Remy is finally sighted as she enters the SC from the east and slowly makes her way toward the nest. While she is still ~ 22 m away from the nest Sinbad ceases his nest M/E work, leaves the nest area and moves east toward his approaching mate. They meet and greet briefly en route as they pass, heading in opposite directions. Sinbad goes to the mid-area at the east end of the SC where he parks and preens. Remy arrives at the nest ~ 5 seconds after Sinbad had left the nest area. She then climbs onto the nest, fusses with the nest material around the eggs for ~ 17 seconds, settles onto the eggs and incubates them. The East Lagoon chick is no longer seen.

1213 – Sinbad (still at the east end of the SC) ceases preening, moves east toward mid-river, dives and begins foraging, working his way north. Remy is still incubating the clutch.

1220 – Sinbad has worked his way to the north end of their territory where he forages by and along the reeds of the west bank of the river. Remy is still incubating the clutch.

1223 – A now fairly mature chick from the East Lagoon family is foraging along the edge of and within the reeds at the NE point of the SC (this is well within the SR pair's territory). Sinbad has worked his way south along the reed beds of the west bank of the river and is now just north of the chick. From ~ 18 m away he spots the chick and immediately charges at it in a rapid, skipping, "half-flight" along the surface of the water. The chick quickly catches sight of Sinbad and promptly dives, disappearing under the surface before he reaches it. Sinbad dives after the youngster as soon as reaches the area where it dove, resurfaces seconds later and lingers in the area, keenly looking about for the chick. (I didn't resight the chick—it most likely either fled the area by moving south for a long distance underwater where it could remain out of sight, or resurfaced within the dense reeds where it could hide from Sinbad.) Sinbad remains parked in this area, continuing to look about and he also begins preening.

1228 – With Sinbad still stationed at the NE point of the SC, Remy (still incubating) stands and leaves the nest and goes directly to Sinbad (I don't know whether or not she had observed the earlier incident with the interloper chick but she obviously had sighted Sinbad, given that she moved straight to him from the nest). She meets and greets Sinbad at the NE point and they linger there together for several seconds (giving the impression that she had seen the incident with the chick). Sinbad then sets out directly toward the nest. Remy preens for about one minute and then begins foraging.

1230 – Sinbad arrives at the nest, climbs onto it, fusses with the eggs and the nest material around them for ~ 51 seconds, then settles onto the eggs and incubates them. Remy is foraging in the western half of the river, east of the NE point of the SC.

1245 – Sinbad is still incubating the clutch. Remy isn't seen at this time but is probably still foraging, most likely in the SE section of their territory or to the south (areas that are out of view from my usual observation point on the mid-south bank of the SC).

1249 – Sinbad stands and vigorously flutters his slightly spread wings over his back for ~ 12 seconds, then rotates/repositions the eggs for ~ 23 seconds before settling back down on them (at least three eggs were briefly visible and it appeared that there are probably more). Remy still hasn't been resighted.

1254 – Sinbad stands, walks off the nest and engages in nest M/E activity, gathering his new nest material locally from the general area around the nest.

1255 – Remy is finally resighted as she moves to the east edge of the SC from the east. She parks there and preens. Sinbad continues with his nest M/E work.

1259 – Remy goes to the nest, climbs onto it, spends ~ 72 seconds rotating/repositioning the eggs and rearranging the nest material around them before settling on the eggs and incubating them (time:1301). Sinbad continues with his nest M/E activity, including collecting a fresh, long reed stem for the nest.

1311 – Sinbad has ceased his nest M/E work and is now parked and preening ~ 15 m from the old nest site. Remy is still incubating the eggs. End observation period.

*1324 – During a brief re-check of the nest another nest exchange takes place: Remy comes off the nest and Sinbad goes on and assumes incubation duty.

*During continued observations about 10 minutes after this re-check Remy is sighted foraging approx. midway along the east shore of the West Lagoon. At this time one of the East Lagoon pair's chicks was foraging just inside the south end of the SR pair's territory near the SE point of the SC. A second East Lagoon chick was parked at the NW end of the West Lagoon at the edge of the dense reed beds that line its west shore. Minutes later the first chick moved south into the West Lagoon and joined its sib by the reeds at the NW end. In the meantime Remy had crossed the lagoon in its mid-section and was now foraging along the reeds of the west shore, working her way north toward where the chicks were. She spotted the chicks when she neared their location and immediately reacted to their presence. Remy initially displayed briefly at them (puffing out her cheeks and throat) and then charged each one in turn (charging twice at one of them). When charged by Remy the chicks took immediate evasive action, fleeing and diving with Remy diving at their heels in pursuit (it appeared that she may have physically contacted one of the chicks as it dove). Remy succeeded in driving the chicks out of the area or causing them to go into hiding—after the final dive by each to escape Remy they were not resighted thereafter in the West Lagoon. Shortly after this incident Remy left the lagoon, moved north under the RR bridge back into the pair's territory and eventually relieved Sinbad from incubation duty.

*Comments:* As previously noted, the West Lagoon currently has not been claimed as a territory by any PB grebe pair—it isn't part of the territory of either the SR pair or East Lagoon pair but instead is a common area used by both (including the East Lagoon chicks). [Recall that the West Lagoon pair hasn't returned to nest here this season (see below and the 6/23 notes.)] Especially over the past several weeks, members of the East Lagoon family have regularly been seen in the West Lagoon, particularly one or more of the chicks. Remy uses the West Lagoon less frequently (mainly for occasional foraging) but nevertheless claimed "squatters rights" when she encountered the chicks today. To date I haven't yet sighted Sinbad in the West lagoon but I would guess that he probably does venture into it at least occasionally.

The PB grebe pair (i.e., the West Lagoon pair) that I observed rearing their chicks in the West Lagoon (which they had claimed as their territory) last year abandoned it (and apparently the general area) during the ensuing rainy season after the river, fed by record rainfall, flowed very strongly and drastically changed the West Lagoon. During that period the West Lagoon was transformed from a lazy lagoon to a proper river as the river forcefully broke through the south end of the lagoon, cut through the beach and ran straight into the ocean. After the rainy season the river's flow into the ocean soon became reduced to a fairly insignificant little channel and this allowed the West Lagoon to reform. However, in regard to the West Lagoon pair's needs the damage had already been done.

In the years of severe drought prior to this past rainy season the rains were too insufficient to cause the river to flow strongly and only small outflow channels occasionally opened up at the south end of the lagoon and flowed into the ocean. This meant that the lagoon persisted and remained relatively stable over these years. This in turn allowed the reeds to flourish and lush, dense growths came to ring the entire lagoon except the south shore which is part of the wide, sandy ocean beach here. This made for great habitat for the grebe pair, offering lots of cover and concealment, prime nesting habitat, and the extensive reed beds supported an abundance of prey (notably crayfish). But the West Lagoon is relatively small and it appears likely that in order to provide sufficient suitable habitat for rearing a brood of chicks the lush reed beds along both the east and west shorelines are required. Unfortunately for the West Lagoon pair, when the river was flowing strongly through here it applied great erosive force to the outside bank (the east shore) and eroded away much of the east bank, totally obliterating its lush reed beds. The dense reed beds along the inside bank (the west shore) remained essentially unscathed but apparently this isn't sufficient by itself to support a chick rearing effort (most notably in regard to providing a sufficient amount of prey—recall that the provisioning parents and the developing chicks of the West Lagoon family did essentially all of their foraging along the edges of and within the reeds). Thus, if the pair wanted to engage in another reproductive effort this season they would have to seek another place to do so (and it seems probable that this may be what they have done). Their only alternative would have been to attempt to expand their territory by usurping sections of adjacent territory from the SR and/or East Lagoon pairs, a tough task.

It appears that the East Lagoon chicks have achieved independence though they still reside in their natal range. Lately when I see them they are always ranging/foraging independently or with one or more sibs. It has been a number of days since I last saw any of the chicks accompanying and taking food from a parent. At this time I have seen three of the chicks but can't confirm whether or not all four are still alive or present—note that at this stage it is also eminently likely that one of them may have left the area.

In regard to my observations of the West Lagoon family last year, there is an interesting tidbit that ties in with my current observations of the SR pair. Having become familiar with Sinbad I now realize and can identify him as being the adult grebe who attacked one of the West Lagoon chicks in the common (overlap) area (the south ribbon of the river) last season on 10/6/16.

**7/11/17:** Time: 1310-1420.

1310 – As observations begin, Sinbad is well-settled on the nest incubating the eggs and Remy is foraging in the West Lagoon.

1320 – Sinbad is still incubating the clutch. Remy isn't seen at this time.

1334 – Sinbad stands, walks off the nest and engages in nest M/E activity, collecting his new nest material from the general area around the nest (in the minutes prior to this it was obvious that Sinbad was getting restless and probably was wanting to be relieved from incubation duty).

1336 – Remy is resighted as she moves into the mid-area of the east end of the SC (having come from the east or SE) where she parks and preens. Sinbad continues with his nest M/E work.

1340 – Remy begins moving toward the nest while Sinbad moves east away from it. They meet and briefly greet in the mid-SC. Sinbad then heads toward the north bank of the SC, dives en route and isn't resighted at this time. Remy continues on toward the nest.

1341 – Remy arrives at the nest and briefly engages in some nest M/E work. She then climbs onto the nest and fusses with the eggs for ~ 19 seconds before settling on and incubating them. Sinbad appears by the nest and is engaged in nest M/E activity.

1345 – Remy is still incubating the clutch. Sinbad arrives at the nest with a fresh, long reed stem. He works it into the nest and then continues with his nest M/E work.

1347 – Sinbad ceases his nest M/E activity and goes out to the mid-area along the east end of the SC where he parks and preens. Remy continues incubating the eggs.

1353 – Sinbad (still at the east end of the SC) ceases preening, moves NE to the mid-river area east of the NE point of the SC and begins foraging, working his way north. Remy is still incubating the eggs.

1400 – Sinbad is still foraging, now in the western half of the river in the northern half of their territory. Another PB grebe is foraging in the river to the north (outside) of the SR pair's territory (this is probably a member of the pair that has established a territory there). Remy is still incubating the clutch.

1402 – Sinbad works his way across the river to the east bank area near the mouth of the side channel.

1405 – Sinbad collects a long reed stem from the area noted at 1402 and begins hauling it all the way to the nest.

1408 – Sinbad finally arrives at the nest with the reed stem and works it into the nest. Remy then stands, walks off the nest and goes to the mid-area at the east end of the SC where she parks and preens. As she moved away from the nest Sinbad climbed onto it, fussed with the eggs for ~ 13 seconds, then settled onto them.

1412 – Sinbad (still incubating) stands and preens himself for about one minute, then rotates/repositions the eggs for ~ 10 seconds before settling back down onto them and resuming incubation (at least two of the eggs were briefly seen). Remy is still preening herself at the east end of the SC.

1420 – Sinbad is still well-settled on the nest incubating the clutch. Remy is still parked and preening herself at the east end of the SC. End observation period.

*Within a minute after this observation period concluded Remy (while continuing to preen herself at the east end of the SC) gave a series of 8 single, short, one note calls over a period of ~75 seconds. The first 4 were more closely spaced and the other 4 were widely gapped (note that these were not the long, drawn-out calls that Sinbad regularly makes). Her reason for making these calls wasn't apparent nor was it clear if they were made out of distress or alarm or something other than these. Regardless, one can probably assume that the vocalizations were meant to communicate something to Sinbad, otherwise why bother vocalizing?

*Comments:* The last two days both Sinbad and Remy have become noticeably more keen in regard to rotating and repositioning the eggs. I still have not been afforded a definitive look at the nest that would allow me to see just how many eggs are in the clutch. At this point all I can say is that there are at least 4 and probably one or more others.

One is left to wonder about what was so exceptional about the reed stem that Sinbad collected from all the way over on the east bank of the river that made it worth the effort of having to haul it for such a long distance to the nest. It wasn't green and fresh. However, though it was somewhat discolored it still appeared to be pretty fresh and possessed good, firm girth. Perhaps it had very favorable buoyancy qualities (certainly a desirable feature for a floating nest) and this is what made it worth the extra effort?

All 4 of the East Lagoon chicks were sighted within a period of ~ 5 minutes. One chick was in the north end of the East Lagoon. The other 3 were in the West Lagoon: two of these were in company resting at the edge of the reeds along the west shore in the north section of the lagoon.

**7/12/17:** Time: 1200-1315.

1200 – When observations begin, the nest is unoccupied. Remy is parked at the east end of the SC preening herself. Sinbad initially isn't seen but is soon sighted moving east through the mid-SC area toward his mate. As he nears Remy she moves a short distance to the SE and continues preening. Sinbad then changes course toward the north bank and ends out engaging in nest M/E activity.

1206 – Sinbad ceases his nest M/E work, climbs onto the nest, fusses with the eggs and the nest material around them for ~ 49 seconds, then settles onto the eggs and incubates them. In the meantime, Remy had moved deep into the SC and parked ~ 12 m east of the nest. She dove soon after and wasn't resighted at that time. (It's quite likely that Remy had been incubating the eggs just before I arrived and that a somewhat drawn-out nest exchange has just taken place.)

1215 – Sinbad is still incubating the clutch. Remy hasn't been resighted (she is probably foraging).

1220 – Sinbad is still incubating the eggs. Remy still hasn't been resighted. Engaging in a quick search for her I move to a different observation area (the west end of the RR bridge) where I can see the West Lagoon and check to see if she is there. I spot an adult PB grebe foraging in the south end of the lagoon. It appears as though it probably is Remy but the grebe is too distant to positively identify (recall that the RR bridge is at the north end of this lagoon). [Note: With the nest being located in the SC, the SC is thus the hub of the pair's activities. This being the case, my usual observation station is on the mid-south bank of the SC. This particular spot happens to be the only good observation point in the SC. From it I can also see the northern section of their territory. Unfortunately, due to the topography and vegetation I'm not able to see the great majority of the SE section of their territory (including none of the south ribbon of the river) or the West Lagoon.]

1228 – The grebe sighted at 1220 in the south end of the West Lagoon has been confirmed as being Remy. She is now foraging in the mid-area of the lagoon. Sinbad is still incubating the clutch.

1236 – Remy is still foraging in the West Lagoon, now in the middle third of the lagoon. Another adult PB grebe (probably a member of the East Lagoon pair) is foraging in the vicinity of Remy. When Remy sights the other grebe she immediately charges at it in a skipping flight along the surface of the water. The other grebe quickly spots Remy and promptly flees to the east, also in a skipping flight over the surface (the two are separated by ~ 3 m). The fleeing grebe soon lands and immediately dives. Remy then also lands and dives after it. She quickly resurfaces, keenly looks about for the other grebe for ~ 7 seconds and then resumes foraging. The other grebe isn't resighted—it appears that after it dove it probably had fled underwater for a considerable distance where it could remain unseen by Remy. Sinbad is still incubating the eggs.

1245 – Sinbad stands, walks off the nest and moves east toward the mid-SC, calling once en route. When he reaches the mid-SC area he pauses there briefly and then dives. He is soon seen back at the nest engaged in nest M/E activity. Remy isn't seen at this time from my vantage point on the south bank of the SC (I have moved back to this observation point; Remy is most likely still foraging in the West Lagoon).

1250 – Sinbad ceases his nest M/E work, climbs onto the nest, briefly fusses with the eggs and the nest material around them and then settles onto the eggs. An immature double-crested cormorant is currently foraging in the SC. Remy still hasn't been resighted (she is probably still foraging in the West Lagoon).

1300 – Sinbad is still incubating the clutch. Remy is finally resighted—she is at the east end of the SC. She slowly moves to the mid-SC area. Sinbad spots Remy, leaves the nest and moves out to her. They meet and briefly greet, emitting a soft, continuous chattering vocalization (sounding somewhat like a series of continuous short "honks" strung together) as they meet. They then continue on in opposite directions. Sinbad moves to the area east of the NE point of the SC and begins foraging, working his way north. Remy goes to the nest, climbs onto it, rotates/repositions the eggs for ~ 17 seconds and then settles onto and incubates them (time: 1301).

1305 – Sinbad is still foraging—he has worked his way to the north end of their territory. Remy is still incubating the eggs.

1310 – Sinbad is still foraging in the north end of their territory. Remy continues to incubate the clutch.

1315 – Remy is still incubating the clutch. Sinbad is now parked and preening mid-river in the north half of their territory. While preening he briefly spreads and flaps his wings several times and it is noted that he appears to be molting his flight feathers (unless something else is going on with them). He is missing quite a few of his flight feathers on both wings (both primaries and secondaries) and others appear to still be growing, having not yet achieved their full length. Given the rather frayed, somewhat sparsely feathered appearance of his wings the indication is that he probably isn't as proficient a flier at this time. The observation period is concluded at this point.

*Comments:* If Sinbad is indeed molting his flight feathers then this could perhaps be the reason why he has been observed rubbing his folded wings together over his back so often. I know for certain that he has only recently shed these feathers. This is because on 7/7 I was provided an opportunity to view Sinbad's wings when he spread and briefly flapped them while preening. At that time he pretty much still possessed his full, intact compliment of flight feathers (at least, no conspicuous gaps were present). However, he has been habitually rubbing his folded wings together since the last days of June suggesting that perhaps the process leading to the actual shedding of the feathers began at around that time.

Once again a stranger PB grebe was seen foraging in the section of the river north of the SR pair's territory (it was seen during casual observations after the formal observation period had concluded). It is looking increasingly more as though a pair has indeed established a territory there.

Lately, Remy has been foraging more regularly in the West Lagoon. In addition, she has now been observed charging at and chasing away members of the East Lagoon family on two of the last three days, including chicks and probably a parent. It was particularly interesting seeing how she chased the adult (almost certainly one of the East Lagoon pair) away. This other adult put up no defense whatsoever—it simply immediately fled when it sighted Remy charging at it. Perhaps the SR pair is intent on annexing the West Lagoon with a particular eye on securing exclusive access for when they are rearing the chicks from the clutch that they are currently incubating?

**7/13/17:** Time: 1123-1225.

1123 – As observations begin, it appears that a nest exchange has probably just occurred: Sinbad has just settled on the nest and begun incubating the eggs and Remy is parked and preening in the mid-SC area.

1130 – Sinbad is still incubating the clutch and Remy is still parked and preening in the mid-SC area.

1134 – Remy moves to the east end of the SC, dives and isn't resighted at this time (she has probably begun foraging). Sinbad is still incubating the eggs.

1137 – Remy is resighted—she is foraging just east of the NE point of the SC. Sinbad continues incubating the clutch.

1142 – Remy resurfaces from a foraging dive just east of the NE point of the SC with a fairly large fish (i.e., large for a PB grebe). She vigorously works the fish (including briefly diving with it 3 times) for ~37 seconds before finally swallowing it. She then continues foraging. Sinbad is still incubating the clutch.

1150 – Remy has ceased foraging and is now parked and preening near the reeds along the west bank of the river in the northern half of their territory. Sinbad is still well-settled on the nest, incubating the eggs.

1156 – Remy (still in the same location as at 1150) ceases preening and begins foraging again. Sinbad remains on incubation duty.

1205 – Remy is now foraging just off the NE point of the SC. Sinbad is still incubating the eggs but appears to be getting a bit anxious to be relieved of incubation duty: he is keenly peering to the east while incubating, apparently looking out for Remy.

1210 – Sinbad is still incubating and still appearing anxious to be relived of incubation duty. Remy is still foraging.

1212 – Sinbad, now looking restless while settled on the eggs, stands, walks off the nest, parks at its east side and intently gazes to the east and NE for ~ 6 seconds (apparently looking for Remy). He then begins nest M/E activity, collecting his new nest material locally from the area around the nest. Remy is foraging near the NE point of the SC.

1216 –Remy moves to the mid-SC area where she parks and preens. Sinbad is still engaged in nest M/E activity.

1217 – Sinbad has spotted Remy. He ceases his nest M/E work and moves east in the direction of Remy, who is still parked and preening in the mid-SC area. He dives well before reaching Remy and resurfaces well beyond her. He steadily makes his way all the way to the north end of their territory (intermittently preening himself en route) where he soon begins foraging in the west half of the river (time re. Sinbad: 1220).

1218 – Remy continues to preen herself for ~ 30 seconds more after Sinbad moves past her. She then goes to the nest, climbs onto it and briefly fusses with the nest material around the eggs but then walks off the nest. She moves out ~ 6-8 m east of the nest, parks there and intently gazes to the east and north. Over the next four minutes she alternates between sometimes going back to the nest (walking onto it briefly several times), intermittently doing a touch of nest M/E, and then going back out to the area just east of the nest (generally from 3-8 m away) and keenly looking about. Clearly she has perceived something of note that has grabbed her attention (I'm unable to discern what it is) to the point where she is reluctant to go onto the nest and incubate and thus give up her focused vigilance for whatever it is that has made her so concerned.

1223 – With Remy parked ~ 3-4 m east of the nest a stranger adult PB grebe suddenly bursts from the dense reeds along the west shore of the SC from somewhere a little to the north of the nest (the reason for Remy's unrest is now clear!). In an obvious attempt to flee from the area the grebe rockets away in a low, speedy flight, lands by the dense reeds at the NE point of the SC, immediately dives and isn't seen again by me. Remy doesn't give chase and instead remains in the nest area. Apparently Remy had perceived the presence of this intruder grebe in the area and this is what had made her so ill at ease and reluctant to settle on the nest. Sinbad is still foraging at the north end of their territory.

1225 – Remy has continued the pattern of repeatedly going just east of the nest and intently looking about, then returning by the nest and intermittently doing some nest M/E (she briefly climbed onto the nest at least once). Sinbad is still foraging at the north end of their territory but has begun working his way south along the reeds of the west bank. End observation period.

*Remy continued as noted above, apparently still reluctant to settle on the eggs because of her concern regarding the presence of the intruder grebe. At 1242 she finally settled on the nest and assumed incubation duty. I had no further sightings of the intruder grebe. Nor on Sinbad's return to the nest did he give any indications that he saw it when he passed close to the area at the NE point of the SC where the grebe had dove and disappeared from sight.

*Comments:* When the stranger grebe dove after its flight it probably did one of the following to escape the wrath of the home pair (it probably took flight only after it perceived that Remy had detected it or that such was imminent). It most likely either resurfaced well within the dense growth of emergent reeds near where it landed and remained well hidden within or it fled the area by moving a long distance while remaining underwater (and unseen), resurfacing only when it was clear of the area.

I had been completely unaware of the stranger grebe's presence. I never saw it until it took flight so I thus have no idea of where it had come from or how long it had been there. All indications are that Sinbad had also been unaware of it. Notably, he gave no indication whatsoever while doing his nest M/E work or during his incubation period and the subsequent nest exchange that he was aware of its presence. If he had been, he undoubtedly would have charged at and chased it out. Only Remy had somehow perceived or become suspicious of its presence. The grebe had apparently been very well hidden within the dense reeds somewhere to the north of the nest.

**7/14/17:** Time: 1245-1400.

1245 – As observations begin, Remy is well-settled on the nest incubating the eggs. Sinbad is parked at the east edge of the SC. He soon begins foraging, working his way north along the reeds of the west bank of the river.

1250 – Sinbad is now foraging in the mid-river area at the north end of their territory. Remy is still incubating the clutch.

1300 – Sinbad is still foraging at the north end of their territory. Remy is still incubating the eggs.

1310 – Sinbad is still foraging in the northern half of their territory. He has been slowly working his way south. Remy is still incubating the eggs. A stranger PB grebe is foraging in the river north of the SR pair's territory.

1314 – Sinbad goes to the mid-area along the east end of the SC where he parks and preens. Remy continues incubating the clutch.

1320 – Sinbad has begun foraging near the reeds along the north bank of the SC. Remy is still incubating the clutch.

1323 – Sinbad parks near the reeds in the NW corner of the SC and just looks about.

1325 – Sinbad (still in the same location) dives and isn't resighted at this time. Remy is still incubating the eggs.

1328 – Sinbad is resighted—he is now parked at the edge of the reeds at the west end of the north bank of the SC.

1330 – Sinbad (still in the same location) dives and isn't resighted at this time. Remy continues to incubate the clutch.

1336 – Sinbad is resighted—he is parked just east of the NE point of the SC. Remy is still incubating.

1342 – Sinbad (still just east of the NE point) dives and begins foraging, working his way west into the SC near its north bank. Remy is still incubating the eggs.

1345 – Sinbad goes to the nest and sits in the water at its north side. Remy stands, walks off the nest and moves out to the mid-area at the east end of the SC where she parks and preens. About 12 seconds after she had left the nest Sinbad climbed onto it, fussed with the eggs and the nest material around them for ~ 17 seconds and then settled onto the eggs and began incubating them.

1348 – Remy (still at the east end of the SC) moves east toward mid-river, dives and begins foraging. Sinbad is still incubating the clutch.

1350 – Remy is now foraging in the West Lagoon. Sinbad continues incubating the eggs.

1400 – Remy is still foraging in the West Lagoon. Sinbad is still incubating the eggs. End observation period.

*Shortly after this observation period concluded the same paddleboarder who came through the pair's territory on 7/8 paddled into their territory at 1404. This time he was even more intrusive. He didn't just stick to the mid-river area as he went through but instead ringed (i.e., paralleled the shoreline) through the entire SC, passing within ~ 25-30 m of the nest. Upon sighting the paddleboarder entering the SE section of the SC Sinbad immediately fled off the nest and hid within the dense reeds in the nest area. Even long after the paddleboarder had left the pair's territory Sinbad remained too apprehensive to go back onto the nest (which thus was left unoccupied). Finally, at 1423 (after returning from the West Lagoon) Remy went to the nest and assumed incubation duty.

*Comments:* Remy incubated the eggs for a solid hour from the time that observations began. Sinbad didn't appear to be having much luck foraging and it seemed that he deliberately took a longer break than usual from assuming his turn at incubation duty in order to give himself more foraging time.

**7/15/17:** Time: 1205-1300.

1205 – As observations begin, Remy is well-settled on the nest incubating the eggs and Sinbad is parked at the east end of the SC.

1210 – Sinbad has moved into the SC and is now parked at the edge of the reeds on the north bank ~ 14 m east of the old nest site, just looking about. Remy is still incubating the clutch.

1217 – Sinbad (still in the same location) dives and begins foraging around the NE point of the SC. Remy continues incubating the eggs.

1220 – Sinbad is now foraging just off the NE point of the SC. Remy is still incubating the eggs.

1224 – Sinbad resurfaces from a foraging dive by the reeds along the NE point, just inside the SC, with a large fish. He works the fish (including diving several times with it) for ~ 38 seconds before diving with it a third time. At this point he isn't resighted until ~16 seconds later at which time he no longer has the fish—I thus don't know whether he swallowed the fish or abandoned it. Remy is still incubating the clutch.

1226 – Remy (appearing anxious to be relieved of incubation duty) stands, walks off the nest and moves east in the general direction of Sinbad. Sinbad spots Remy approaching and moves toward her. They meet and greet en route and just continue past each other in opposite directions. Remy keeps moving east, dives as she nears the east end of the SC and isn't resighted at this time (she has probably begun foraging). Sinbad continues on toward the nest.

1227 – Sinbad arrives at the nest. He lingers briefly and then climbs onto it, spends ~ 68 seconds rotating/repositioning the eggs and rearranging the nest material around them before settling on the eggs and incubating them (time: 1228).

1240 – Sinbad is still incubating the clutch. Remy still hasn't been resighted.

1245 – Remy is finally resighted—she has moved to the mid-area along the east edge of the SC where she is parked and preening. Sinbad is still incubating the eggs.

1247 – Remy dives and moves to the edge of the reeds at the NE point of the SC where she parks and preens.

1251 – Sinbad stands, walks off the nest and engages in nest M/E activity. Remy is still parked and preening at the NE point of the SC.

1252 – Sinbad ceases his nest M/E work, leaves the nest area and moves east toward Remy. Remy soon sights him and moves toward her mate. They meet and greet on the go in the mid-SC and continue on in opposite directions. Sinbad proceeds east to the area just off the NE point of the SC where he dives and begins foraging, working his way north along the edge of the reeds of the west bank of the river. Remy continues on toward the nest.

1254 – Remy arrives at the nest, climbs onto it, fusses with the eggs and rearranges the nest material around them for ~ 65 seconds, then settles onto the eggs and incubates them (time: 1255).

1300 – Sinbad is now foraging in the northern half of their territory along the reeds of the west bank of the river. Remy is still incubating the eggs. End observation period.

*As I was leaving the area about 5 minutes after the observation period had concluded the same paddleboarder as in the previous two incidents (this time with a passenger) entered the pair's territory from the SE end. He was headed for the SC but on this occasion I was in a location where I could ask him to please not enter into the SC because he would disturb the nesting grebes. He kindly complied but in regard to disturbing the nesting grebes it was too late: his presence just outside the SE end of the SC had already caused Remy to flee off the nest and hide in the reeds. This is now the third time that I have seen the pair get disturbed from their incubation duty by this paddleboarder. In these cases the individual who was incubating wasted no time leaving the nest and hiding in the dense reeds with the apparent intent of not wanting to give away the location of the nest (as well as out of pure alarm!).

*Comments:* Two chicks and one parent from the East Lagoon family were seen in the West Lagoon. All three were independent of each other (i.e., they were not keeping company). The indications seem certain that the chicks have achieved independence from the parents though chicks may sometimes keep company amongst themselves in various combinations and they are still utilizing their natal range (which includes the West Lagoon which, at this point, is still a common area used by both the East Lagoon family and the SR pair).

**7/16/17:** Time: 1152-1305. [*Note: Observations of the SR pair's nest had to be conducted from the more distant vantage point located on the west end of the RR bridge. This is because my usual observation point on the mid-south bank of the SC (which is much closer to the nest) was occupied by some people who were fishing there. The greater distance meant that observations could not be as detailed. Also, I had to retreat from the bridge every time a train approached. A few comments regarding these observation points that are available to me are relevant here. Unlike with my SC observation point, all of the SR pair's territory as well as the West Lagoon are visible from the RR bridge. However, the bridge is not suitable as a regular observation station. The regular train traffic makes it a hazard, keeping vigilant for approaching trains means that I can't fully focus on the observations, and should I be in the middle of an interesting observation I would have to abandon it if a train is approaching. On the other hand, the all-encompassing view makes the bridge ideal for brief checks for individuals within the pair's complete territory and surrounding range (notably the West Lagoon).*]

1152 – As observations begin, Sinbad is well-settled on the nest incubating the eggs and Remy is foraging in the northern half of their territory. An osprey is perched on the snag that it (and a few other ospreys) typically uses when it perches in this area (the snag is on the edge of the east bank of the river, across the river from the SC).

1200 – Remy (still foraging in the northern half of their territory) resurfaces from a foraging dive with a fairly large fish. She vigorously works the fish for ~ 43 seconds (including diving with it once) before finally swallowing it. She then begins slowly heading south. Sinbad is still incubating the clutch.

1203 – Remy has now reached the area just east of the NE point of the SC. Sinbad continues incubating the eggs.

1205 – Remy moves to the mid-area of the SC where she parks and preens. Sinbad is still incubating the eggs.

1213 – Remy goes to the nest. Sinbad then stands, walks off the nest and moves out to the area off the east side of the NE point of the SC (calling once while en route) where he parks and preens. Remy remains by the side of the nest and does some light nest M/E.

1215 – Remy climbs onto the nest, settles on the eggs and incubates them.

1217 – Sinbad (still parked off the NE point) ceases preening, dives and begins foraging, working his way north.

1220 – Sinbad is now foraging in the northern half of their territory. Another PB grebe is briefly sighted foraging in the river just north of the SR pair's territory. Sinbad gives no indication that he has seen it, though it didn't encroach on the par's territory (this is probably a member of the pair that appears to have established a territory immediately north of the SR pair's territory). Remy is still incubating the clutch.

1225 – Remy is still incubating the eggs. Sinbad isn't seen at this time but is probably still foraging.

1230 – Sinbad has moved just inside the SC near the reeds west of the NE point where he is parked and preening. Remy is still incubating the clutch.

1235 – Sinbad ceases preening and slowly moves south along the east end of the SC. He stops along the northern third of the east end and just sits there and looks around. Remy is still incubating the eggs.

1240 – Sinbad moves west into the northern half of the mid-SC area and again parks and just looks about. Remy continues incubating the clutch.

1245 – Sinbad begins foraging near the reeds in the NW section of the SC, eventually working his way into the NW corner. Remy is still incubating the clutch.

1250 – Sinbad goes to the nest and sits in the water at its north side. Remy soon stands, walks off the nest and goes out to the mid-SC area where she parks and preens herself. Sinbad engages in some brief nest M/E activity and then climbs onto the nest. He rearranges the nest material around the eggs for ~16 seconds but doesn't settle on them. Instead, he walks off the nest, briefly sits at its east side and then engages in nest M/E.

1255 – Sinbad is still engaged in light nest M/E interspersed with moments of just sitting in the water next to the nest. Remy is now parked and preening in the north section of the SC just west of the NE point.

1300 – Sinbad finally goes onto the nest and incubates the eggs. Remy is still parked and preening just west of the NE point of the SC.

1305 – Sinbad is still incubating the eggs. Remy is still parked and preening just west of the NE point of the SC. The osprey is still perched in the same spot as at the start of observations. End observation period.

*Comments:* The osprey remained perched in the same spot on the snag on the east bank of the river throughout the entire observation period. Neither of the SR pair paid any noticeable attention to it (I don't know for certain whether or not they were aware of its presence but it seems most likely that they were).

Three chicks and one parent of the East Lagoon family were sighted in the East Lagoon. The chicks were independent of the parent though two of the chicks were in close company with each other (it was obvious that this was intentional on their part). At one point one of these chicks captured a very large crayfish that clearly looked to be well beyond the size limit that the chick would be able to eat. It nevertheless made an attempt, working the crayfish for several minutes before finally abandoning it. Other than some missing appendages the crayfish looked relatively unscathed (of particular note, there were no noticeable breaks in its exoskeleton).

**<u>7/17/17</u>:** Time: 1205-1315.

1205 – A significant disturbance in the SC area was in progress as observations began. One of the homeless women in this general area was apparently experiencing some sort of mental episode. She was on the SE shore of the SC continuously shouting very angrily (there were no other people present at the time). She had been doing so for at least two minutes before my arrival (I could hear shouting while en route to the SR pair's territory). When I arrived at the SC I saw that the nest was unoccupied and neither of the pair was seen. There seems little doubt that this loud disturbance had probably caused the pair to flee from the nest and go into hiding. Within a few minutes the woman leaves the area and ceases her shouting.

1210 – Remy and Sinbad are finally sighted. They are close together in the mid-area along the east end of the SC. They mill about together for a few seconds, then Sinbad moves east toward mid-river, dives and isn't resighted at this time (he has probably begun foraging). Remy begins slowly moving in the direction of the nest.

1212 – Remy arrives at the nest, sits in the water at its north side for ~ 20 seconds (she appears reluctant to go onto the nest), then briefly engages in some light nest M/E work.

1213 – Remy finally climbs onto the nest. She rearranges the nest material around the eggs for ~64 seconds and then settles onto the eggs and incubates them (time: 1214).

1215 – Sinbad is resighted—he is foraging just east of the NE point of the SC. He resurfaces from a foraging dive with a fairly large fish, works it for ~ 37 seconds (including briefly diving with it once) before swallowing it. He then parks and preens there. Remy is still incubating the clutch.

1220 – Sinbad is still parked and preening just east of the NE point. Remy is still incubating the eggs.

1228 – Sinbad begins foraging in the area around the NE point of the SC. Remy continues incubating the clutch.

1231 – Sinbad is now foraging just out from the reeds along the north bank of the SC, slightly west of the NE point.

1235 – Sinbad is still foraging, now along the edge of and within the reeds that line the NE point. Remy is still incubating the eggs.

1236 – Sinbad surfaces from a foraging dive with what appears to be a small crayfish (he is just off the NE point of the SC). He quickly swallows it, then moves east toward mid-river, dives and isn't resighted at this time.

1241 – Sinbad is resighted as he moves from the east to the mid-area at the east end of the SC where he parks and preens. A stranger adult PB grebe is now foraging within the north end of the SR pair's territory (this grebe is fairly distant from Sinbad who appears to have sighted it). Remy is still incubating the eggs.

1243 – The other grebe has now moved north, out of the SR pair's territory. Sinbad is still parked and preening himself at the east end of the SC.

1244 – Sinbad ceases preening and begins making his way toward the nest. Remy is still incubating the clutch.

1245 – Sinbad arrives at the nest. Remy then stands, walks off the nest and goes out to the mid-area along the east end of the SC where she parks and preens. In the meantime, Sinbad climbs onto the nest, rotates/repositions the eggs for ~ 17 seconds (it appeared that at least 4 eggs were briefly visible) and then settles onto and incubates them (time: 1246).

1248 – Remy (still at the east end of the SC) ceases preening, dives and begins foraging, working her way north in the mid-river area.

1255 – Remy is now foraging in the north half of their territory. Sinbad is still incubating the clutch.

1300 – Remy (while foraging mid-river in the northern half of their territory) resurfaces from a foraging dive with a large fish. She vigorously works it for ~ 76 seconds (including taking a number of brief dives with it) before finally swallowing it. She then continues foraging. Sinbad is still incubating the eggs.

1306 – Remy ceases foraging, parks mid-river in the northern half of their territory and preens. Sinbad is still incubating the eggs.

1308 – Remy ceases preening and resumes foraging, slowly working her way south.

1310 – Remy is now foraging just east of the east end of the SC. Sinbad is still incubating the clutch.

1315 – Remy moves to the mid-SC area where she parks and preens. Sinbad is still incubating the clutch. (The indications are that another nest exchange is imminent.) End observation period.

*While departing the area minutes after the observation period had concluded I noted that a nest exchange had occurred: Sinbad off the nest and Remy on. After leaving the nest Sinbad moved out to the area just off the NE point of the SC (calling once while en route) and began foraging there.

*Comments:* Up to this point the pair has generally been quite diligent in regard to alternating their shifts of incubation duty.

**7/18/17:** Time: 1155-1310.

1155 – As observations begin, Sinbad is well-settled on the nest incubating the eggs and Remy is foraging in the area just east of the SC.

1158 – Remy moves to the mid-area of the SC where she parks and preens.

1159 – Sinbad has spotted Remy. He leaves the nest and moves toward her. When he reaches her they briefly greet and then proceed in opposite directions. Sinbad goes to the area off the NE point of the SC and immediately begins foraging. Remy goes to the nest and briefly engages in some nest M/E.

1200 – Remy climbs onto the nest and quickly settles onto the eggs and incubates them.

1205 – Sinbad is foraging in the mid-river area east of the NE point of the SC. Remy is still incubating the clutch.

1207 – Sinbad resurfaces from a foraging dive in the mid-river area east of the NE point of the SC with a large fish. He vigorously works it for ~ 73 seconds (including diving with it briefly 4 times) before finally swallowing it. He then resumes foraging and works his way north into the northern half of their territory. Remy is still incubating the eggs.

1212 – Sinbad parks and preens himself mid-river at the northern end of their territory.

1215 – Sinbad (still in the same location) ceases preening and begins foraging in the western half of the river at the north end of their territory. Remy is still incubating the clutch.

1216 – Sinbad (at the edge of the reeds along the west bank at the north end of their territory) resurfaces from a foraging dive with a fish. He works it for ~ 53 seconds before swallowing it, then dives and isn't resighted at this time.

1220 – Remy is still incubating the clutch. Sinbad still hasn't been resighted.

1225 – Sinbad is resighted—he is parked and preening himself at the edge of the reeds of the north bank of the SC, just west of the NE point. Remy is still incubating the eggs.

1227 – Sinbad moves to the mid-SC area where he parks and preens.

1230 – Remy (apparently having sighted Sinbad) stands, walks off the nest and moves toward Sinbad. Sinbad soon sees Remy approaching and moves toward her. They meet and greet briefly on the go and continue on in opposite directions. Remy goes to the east end of the SC, dives and isn't resighted at this time (she has probably begun foraging). Sinbad goes to the nest, climbs onto it and rearranges the nest material around the eggs for ~ 33 seconds, then (while remaining standing on the nest) continuously flutters his partially spread wings over his back for over a minute and preens himself for a while longer before finally settling onto the eggs and incubating them (time: 1233).

1237 – Sinbad stands, walks off the nest, sits in the water at its east side and just gazes to the east for one minute. He then engages in nest M/E activity. Remy hasn't been resighted.

1242 – Sinbad climbs onto the nest, flutters his partially spread wings (as before) for ~ 11 seconds but then goes off the nest—he walked off the nest just as an osprey swooped down in the mid-SC area, pulling up just before striking the water in an aborted fishing attempt. It thus wasn't clear if Sinbad had walked off the nest of his own accord or in response to the osprey's sudden, swooping appearance. Soon after, he resumed his nest M/E activity.

1246 – Sinbad ceases his nest M/E work, moves out to the mid-SC area (calling once en route), and stops there briefly. He then goes near the reeds in the area north of the nest. Soon after, he again engages in nest M/E activity.

1248 – Remy is resighted as she moves to the mid-area along the east end of the SC where she parks and preens. Sinbad climbs onto the nest, stands there and vigorously flutters his partially spread wings over his back for ~ 23 seconds before finally settling onto the eggs and incubating them.

1255 – Remy is still parked and preening in the same location. Sinbad is still incubating the clutch.

1257 – Remy is still parked and preening at the east end of the SC. Sinbad is still incubating the eggs. The osprey (which has been searching for fish prey over the SE side of the SR pair's territory) suddenly launches into a swooping strike into the water at mid-river, east of the SC in an unsuccessful attempt to catch a fish. In apparent response to this Remy emits 4 closely spaced single, short one note calls. Sinbad appears to respond to her calls as ~ 1 ½ seconds later he stands, walks off the nest and sits in the water at its east side and keenly looks to the east.

1259 – Remy suddenly takes a low flight from the east end of the SC to its mid-area. Sinbad then moves out to Remy. They briefly meet and greet and Sinbad continues on to the east end of the SC where he parks and preens. Remy sets out toward the nest.

1300 – Remy arrives at the nest, climbs onto it and rearranges the nest material around the eggs for ~ 11 seconds, then walks off the nest and engages in nest M/E activity.

1302 – Remy moves out ~ 6 m east of the nest. She parks there and gazes to the east for more than a minute before resuming her nest M/E work.

1305 – Sinbad is still parked and preening at the east end of the SC. Remy is still engaged in nest M/E activity. At one point she climbed onto the nest and spent time fussing with the eggs and the nest material around them. It appeared as though she was getting ready to settle onto the eggs but instead she walked off the nest and resumed her nest M/E work.

1310 - Sinbad is still parked and preening at the east end of the SC and has drifted to within ~ 20 m of its NE point. Remy has continued with her nest M/E activity. During this activity she has twice more gone onto the nest and spent time rotating/repositioning the eggs and rearranging the nest material around and over them. As before, it appeared that she was getting ready to settle onto the eggs but instead walked off the nest. For some reason she appears somewhat reluctant to settle onto the eggs and incubate them as had Sinbad earlier. End observation period.

*Before departing the area I noted that Remy finally settled onto the eggs at 1314. She spent considerable time rotating/repositioning the eggs before settling onto them.

*Comments:* The osprey was active in and around the pair's territory for much of the observation period and for a while thereafter. It spent periods during this time cruising over their territory in search of fish prey. During the observation period it made an aborted attempt (pulling up before hitting the water) to catch a fish in the mid-SC area and an unsuccessful strike in the area east of the SC. It wasn't clear if the former had drawn a reaction from Sinbad but Remy almost certainly had reacted to the latter. In each case it seems most likely that it was just reflexive reactions by the grebes to the sudden dramatic action of the osprey (i.e., they were startled) rather than their having viewed the osprey as a threat to themselves. Not long after the observation period was concluded the osprey swooped down and captured a fish only ~ 30-35 m east of the nest. The nest was unoccupied at the time and Sinbad was engaged in nest M/E work which he continued with.

The piled vegetation on the nest was higher than I have ever seen it to date. Thus, the top of the nest (and the nest bowl) was sitting higher above the surface of the water than I have previously seen it.

The pair's incubation behavior and their behavior around the nest were notably different today. Other than Remy's first incubation stint in the first half of the observation period, the incubation periods by both of the pair were truncated. In addition (i.e., in the second half of the period), both exhibited a curious reluctance to just settle onto the eggs. For example, there were a number of instances when either Remy or Sinbad went onto the nest and spent time rotating/repositioning the eggs and rearranging the nest material around and over them, giving every indication that they were about to settle onto the eggs, but instead walked off the nest. Also, the pair was notably more preoccupied with rotating/repositioning the eggs, rearranging the nest material around them and tending to nest M/E activity (and curiously interspersing periods of just sitting in the water next to the nest while engaged in the latter).

In general, they seemed to be behaving as though something was different or had changed regarding one or more of the eggs and they were reacting to this. The hatching window opens in just a few days. Could it be that perhaps the parents are perceiving movements by the well-developed chicks inside the eggs? Or is it possible that an early hatching was underway? In any event, the pair is entering into a crucial and exciting period regarding their clutch.

**7/19/17:** Time: 0829-0935.

0829 – As observations begin, Sinbad is well-settled on the nest incubating the eggs and Remy is foraging in the SE section of their territory.

0835 – Remy has worked her way north and is now foraging in the western half of the river within the northern half of their territory. Sinbad is still incubating the eggs.

0838 – Remy has ceased foraging and is parked and preening in the western half of the river within the northern half of their territory.

0841 – Remy (still parked in the same location) ceases preening and begins moving south. She stops along the east end of the SC ~ 30 m south of the NE point and preens herself. Sinbad is still incubating the clutch.

0845 – Remy moves to the mid-SC area. Sinbad (who has obviously sighted her there) stands, walks off the nest and heads toward Remy. Remy soon spots Sinbad approaching and moves toward him. They briefly meet and greet en route and continue on in opposite directions. Sinbad continues east to the area east of the NE point of the SC where he parks and preens. Remy proceeds toward the nest.

0846 – Remy arrives at the nest, climbs onto it, fusses with the eggs for ~ 18 seconds and then settles onto the eggs and incubates them.

0850 – Sinbad (still parked east of the NE point) dives and begins foraging near the dense reeds at the NE point. Remy is still incubating the clutch.

0855 – Sinbad is still foraging along the edge of and within the reeds along the NE point of the SC. Remy is still incubating the eggs.

0858 – Sinbad (still foraging along the reeds at the NE point) resurfaces from a foraging dive with a fairly large crayfish. He works it for ~ 15 seconds before swallowing it (he was able to swallow it surprisingly quickly considering the size of the crayfish).

0901 – Sinbad moves east ~ 25 m from the NE point, parks there and preens (breaking briefly from his preening to call once). Remy is still incubating the eggs.

0905 – Sinbad has moved back to the NE point of the SC where he is parked and just looking about. Remy continues incubating.

0910 – Sinbad moves south to the mid-area along the east end of the SC, parks there and just looks about, occasionally preening briefly as he does so. Remy is still incubating the clutch.

0912 – Sinbad (still at the east end of the SC) moves to the north bank of the SC west of the NE point and begins foraging along the reeds there.

0915 – Sinbad goes to the nest. Remy then stands, walks off the nest and moves out to the mid-area at the east end of the SC where she parks and preens. In the meantime, Sinbad climbs onto the nest, fusses with the eggs for ~ 18 seconds, then settles onto and incubates the eggs.

0918 – Remy (still at the east end of the SC) dives and begins foraging, working her way east toward mid-river, east of the SC.

0920 – Remy is still foraging mid-river, east of the SC. Sinbad is still incubating the clutch. Another adult PB grebe is foraging just beyond the north end of the SR pair's territory. A second adult grebe is seen near it soon after the first was sighted. There is essentially no doubt that these two comprise the mated pair that has established a territory there.

0925 – Remy has ceased foraging and is now parked and preening just east of the NE point of the SC. Sinbad is still incubating the clutch.

0927 – Remy (still just east of the NE point) dives and begins foraging, working her way east toward mid-river.

0930 – Remy is slowly working her way south and is now foraging mid-river, east of the SC. Sinbad is still incubating the eggs.

0935 – Remy has continued working her way south and is now foraging in the West Lagoon. Sinbad is still incubating the eggs. End observation period.

*Comments:* Unlike yesterday, today's observation period was routine and rather uneventful with normal incubation stints.

Sightings of one and sometimes two adult PB grebes (including in close proximity to each other) in the area immediately north of the SR pair's territory have occurred regularly. It now appears certain that these two (a male and a female) are a mated pair that has established a territory in the section of the river north of (bordering) the SR pair's territory. Henceforth I will refer to this pair as the North River (NR) pair. As yet I haven't found a vantage point from which I can obtain an adequate enough view to determine whether or not this pair is nesting or rearing chicks (no chicks have been sighted).

At this point, among the SR pair only Sinbad has been observed giving the long, drawn out call that is characteristic of this species. I have heard numerous such calls from the East Lagoon pair but have only had the opportunity to observe which of the pair was making the call on a few occasions. In each case it was the male. It will be interesting to see if this pattern holds within the SR pair and the other pairs in the area.

**<u>7/20/17</u>:** Time: 1152-1305; Brief re-check of nest at 1315.

1152 – As observations begin, Sinbad is engaged in nest M/E activity and Remy is foraging in mid-river, east of the SC, working her way north.

1159 – Sinbad ceases his nest M/E work, climbs onto the nest, rotates/repositions the eggs and rearranges the nest material around them for ~ 49 seconds, then settles onto the eggs and incubates them. Remy is foraging in the north end of their territory.

1204 – Remy (foraging mid-river at the north end of their territory) resurfaces from a foraging dive with a large fish. She vigorously works it for ~ 73 seconds (including diving with it several times) before finally swallowing it. Sinbad is still incubating the eggs.

1207 – Remy (foraging mid-river in the northern half of their territory) surfaces from a foraging dive with another fish. She vigorously works it for ~ 31 seconds before swallowing it. She soon begins making her way south. Sinbad is still incubating the clutch.

1210 – Remy rounds the NE point of the SC and moves to its mid-area where she parks and preens.

1211 – Sinbad (who clearly has caught sight of Remy and appears eager to be relieved of incubation duty) stands, walks off the nest and (in a very low swimming profile) heads directly toward Remy. Remy soon spots Sinbad as he approaches and moves toward him. They meet en route and greet briefly (they brush breasts as they pass) and proceed in opposite directions. Sinbad moves out to the area east of the NE point, dives and begins foraging. Remy continues on toward the nest.

1212 – Remy arrives at the nest, climbs onto it, fusses with the eggs and the nest material around them for ~ 20 seconds, then settles onto the eggs and incubates them.

1215 – Sinbad is now foraging at the edge of the reeds at the NE point of the SC. Remy is still incubating the eggs.

1220 – Sinbad has worked his way inside the SC and is now foraging near the reeds along the north bank, just west of the NE point. Remy is still incubating the clutch.

1225 – Sinbad moves out to the mid-area along the east end of the SC where he parks and preens. Remy continues incubating the eggs.

1227 – Sinbad (in the same location) ceases preening, dives and begins foraging. He soon resurfaces from one of his foraging dives with a large fish. Surprisingly (given its large size), he works it for only ~ 13 seconds before he swallows it. He then parks and preens himself again in the mid-area of the east end of the SC.

1233 – Sinbad (still parked and preening in the same location) calls once and then resumes preening. Remy is still incubating the clutch.

1235 – Sinbad begins moving west into the SC along its north bank. When he reaches the NW section he dives and isn't resighted at this time.

1237 – Sinbad arrives at the nest. Remy stands, walks off the nest and moves out to the mid-SC area where she parks and preens.

1239 – Sinbad climbs onto the nest and takes ~ 82 seconds before finally settling on the eggs and incubating them (time: 1241). He spent most of the 82 seconds standing on the nest and rapidly fluttering/shaking/rubbing together his very slightly spread wings close over his back.

1246 – Remy, still in the mid-SC area, ceases preening and begins foraging there. She soon surfaces from a foraging dive with a large fish and works it for a lengthy ~ 92 seconds (including diving with it several times) before finally swallowing it. She then parks and again preens herself.

1250 – Remy is still parked and preening but has drifted a little to the east near the east end of the SC. Sinbad is still incubating the eggs.

1255 – Remy begins moving to the nest. She arrives there with some vegetation she has collected along the way and adds this onto the nest. Sinbad then stands, walks off the nest and goes out to the area off the NE point of the SC where he begins foraging. Remy briefly engages in some nest M/E before climbing onto the nest. She then fusses briefly with the eggs, settles onto and incubates them (time: 1256).

1257 – Sinbad (foraging just east of the NE point of the SC) surfaces from a foraging dive with a rather large fish. He vigorously works it for ~ 43 seconds (including diving with it several times) before swallowing it. He then resumes foraging.

1300 – Sinbad has worked his way north and is now foraging mid-river at the north end of their territory. Remy is still incubating the clutch.

1305 – Sinbad is still foraging in the north end of their territory. Remy is still incubating the eggs. End observation period.

*1315 – During a brief re-check of the nest another nest exchange occurred: Remy came off the nest and Sinbad went on and assumed incubation duty.

*Comments:* The pair seemed particularly "workman-like" with their nest exchanges and incubation stints today. If effective incubation was begun on the day the first egg(s) was laid then the hatching window opens tomorrow. Frustratingly, I still have not been afforded a definitive look at the nest which would allow me to see how many eggs are in the clutch. I have been able to determine that the clutch contains more than four eggs but I don't how many more.

The East Lagoon pair has produced a second brood for this season: one of the pair was seen today with 5 small, very young chicks. They were at the dense reed bed situated along the mid-section of the east shore of the East Lagoon. There is notable size disparity among these chicks.

In addition to the new chicks, three of the chicks from their first brood (now fledged and independent but still residing in their natal range) were sighted foraging independently in the southern third of the East Lagoon. With the second brood now present it will be interesting to see if they will leave their natal range or perhaps be driven out by the parents. One of these chicks was twice seen vigorously exercising its wings—it vigorously flapped them for ~ 5-6 seconds while sitting in place.

**7/21/17:** Time: 1150-1315.

1150 – As observations begin, Sinbad is engaged in nest M/E activity and Remy isn't seen at this time.

1153 – Sinbad climbs onto the nest, briefly fusses with the eggs and the nest material around them and then settles onto the eggs and incubates them.

1155 – Remy is finally sighted—she is foraging in the north end of their territory. Sinbad is still incubating the eggs.

1205 – Remy has worked her way a little to the south and is now foraging mid-river in the northern half of their territory. Sinbad is still incubating the clutch.

1206 – Sinbad calls once while still well-settled on the nest and incubating the eggs. He appears to be getting anxious to be relieved of incubation duty.

1207 – A stranger adult PB grebe (most likely one of the NR pair) is foraging just inside the north end of the SR pair's territory. Remy is foraging ~ 30-40 m to the south of this grebe. She soon spots the intruder and immediately charges at it. The other grebe quickly sights Remy and promptly begins fleeing to the north. A chase then ensues with both birds in flight low over the water. The intruder is able to maintain a sizeable gap of ~ 15 m as Remy gives chase. The interloper rapidly flees out of the north end of the SR pair's territory, lands well north of the territorial boundary, immediately dives upon landing and isn't resighted by me. Simultaneously, Remy lands at the north end of their territory, immediately dives and quickly resurfaces. She then parks mid-river at the north end of their territory and intently looks to the north for the now ousted intruder.

1210 – Remy is still parked at the north end of their territory, her attention mainly focused to the north as though standing guard at the north boundary. Sinbad is still incubating the clutch.

1213 – Remy (still parked at the north end of their territory) dives and isn't resighted at this time (she has probably begun foraging).

1215 – Remy is resighted—she is foraging in the northern half of their territory near the east bank of the river. Sinbad is still incubating the eggs.

1220 – Sinbad is still incubating the clutch. Remy isn't seen at this time.

1223 – Sinbad stands, walks off the nest, goes to the mid-SC area, calls once, then dives and returns to the nest area where he lingers in the water by the nest (it's quite obvious that Sinbad is eager to be relieved of incubation duty). Remy still hasn't been resighted.

1226 – Sinbad climbs onto the nest, fusses with the eggs and the nest material around them for ~24 seconds, then settles onto the eggs and incubates them.

1228 – Remy is finally resighted—she is foraging in the northern half of their territory ~ 15 m from the reeds along the west bank of the river. She soon works her way E-SE to the mid-river area east of the SC. Sinbad is still incubating the clutch.

1232 – Remy moves to the mid-area along the east end of the SC where she parks and preens.

1236 – Remy moves farther into the SC to its mid-area and parks and preens herself there.

1237 – Sinbad (who is still eager to be relieved of incubation duty and has clearly caught sight of Remy) stands, walks off the nest and heads directly toward his mate while swimming in a very low profile. Remy spots Sinbad approaching and moves toward him. They meet en route, greet briefly as they pass and proceed in opposite directions. Sinbad continues on to the area off the NE point of the SC and begins foraging, working his way north. Remy goes to the nest and briefly engages in some light nest M/E. She then climbs onto the nest and fusses briefly with the eggs and the nest material around them before settling onto the eggs and incubating them (time: 1239).

1242 – Sinbad ceases foraging, parks mid-river at the north end of their territory and preens.

1250 – Sinbad is still parked and preening at the north end of their territory. Remy is still incubating the eggs.

1253 – Sinbad ceases preening and begins foraging, working his way south. Remy suddenly stands, walks off the nest, parks ~ 1-1.5 m east of the nest and facing it, and focuses her attention on the nest bowl (I'm not able to determine just what her attention is focused on but it appears to be something regarding the eggs).

1257 – Remy is still parked ~ 1.5-2 m east of the nest, now gazing mainly to the east. Sinbad is now foraging just east of the NE point of the SC.

1259 – Remy moves back to the nest and lingers by it, sometimes gazing at it and sometimes doing a touch of light nest M/E.

1301 – Remy climbs onto the nest and engages in very focused manipulations of the eggs.

1302 – Sinbad arrives at the nest, sits in the water at its north side and watches Remy as she continues her manipulations with the eggs.

1304 – Remy walks off the nest. Both of the pair are now behind (i.e., on the west side of) the nest where my view of them is mostly or completely obscured by the nest and the reeds around it. Occasionally I catch sight of a small fragment of one of them, still at the west side of the nest. It appears that at least one of them may be monitoring whatever it was on the nest that had drawn their attention.

1310 – Sinbad comes into view as he climbs onto the nest, gives focused attention to the eggs for ~ 45 seconds and then settles onto the eggs and incubates them. Remy isn't seen at this time.

1315 – Remy is resighted—she is foraging at the north end of their territory. Sinbad is still incubating the eggs. No chicks seen. End observation period.

*Comments:* The hatching window opened today. No chicks were seen. Incubation proceeded as usual for the first hour of the observation period but things changed dramatically during the final segment of the period. Something regarding the eggs suddenly caught and keenly held Remy's attention and also Sinbad's when he arrived at the nest later on. My view of the nest has always been rather limited and I couldn't see or determine just what it was that had grabbed their attention. However, given the time that they subsequently spent focused on and manipulating the eggs there seems little doubt that something regarding the eggs had changed. Perhaps pipping had occurred in one or more of the eggs or maybe Remy felt some notable movement that was made from inside an egg by a well-developed chick? There seems an excellent chance that I will see at least one hatched chick tomorrow!

Remy again chased away an intruder grebe today. She has shown herself to be quite feisty and a determined defender of the pair's territory.

All 5 of the East Lagoon pair's second brood of the season were seen. In addition, two and probably a third grown chick from their first brood were seen in the southern third of the East Lagoon.

**<u>7/22/17</u>:** Time: 1155-1310.

1155 – As observations begin, Remy is well-settled on the nest incubating the eggs and Sinbad is foraging at the edge of the reeds at the NE point of the SC. No chicks seen at the nest but my view of it is rather limited so I don't know at this time if any have hatched out.

1205 – Sinbad is now parked and preening ~ 8 m east of the NE point. Remy is still incubating the eggs.

1208 – Sinbad ceases preening, dives and begins foraging along the reeds at the NE point of the SC, working his way west along the north bank. Remy is still incubating the eggs.

1211 – Sinbad goes to the nest. Remy then stands, walks off the nest and moves to the area east of the NE point of the SC where she begins foraging, working her way north.

1212 – Sinbad climbs onto the nest, fusses with the eggs and the nest material around them for ~22 seconds, then settles onto the eggs and incubates them. I still haven't had an adequate enough view to be able to determine if any chicks are present.

1220 – Sinbad is still incubating the clutch. Remy isn't seen at this time.

1223 – Remy is resighted—she is foraging at and just beyond the north end of their territory. She soon ceases foraging, parks at the north end of their territory and preens herself. Sinbad continues incubating the eggs.

1226 – Sinbad suddenly stands, rotates/repositions the eggs, then walks off the nest and sits in the water by it (occasionally doing a touch of nest M/E) for ~ 2 minutes.

1228 – Sinbad climbs back onto the nest, fusses with the eggs for ~ 17 seconds, then settles onto them briefly before again standing and walking off the nest.

1229 – Sinbad climbs back onto the nest, fusses with the eggs for ~ 12 seconds, then settles onto and incubates them. Sinbad's behavior since 1226 indicates that something is likely happening with one or more of the eggs but my view of the nest is too inadequate to be able to determine what this might be. Remy is still parked at the north end of their territory.

1233 – A paddleboard (manned by a different person than in the previous episodes) enters into the SC from its SE point and moves west along the south shore. Sinbad immediately flees off the nest and hides (out of my view) in the dense reeds. Remy had still been at the north end of their territory but is no longer in sight (she probably saw the paddleboard and is most likely hiding in the reeds at the west bank of the river). When the paddleboard nears my observation point on the mid-south shore I inform the person on it about the nesting grebes and ask him to please leave the SC. I also ask if he would please stay out of the SC for the next few months. He kindly complies (and seems genuinely sympathetic about having disturbed the nesting pair), turns around and exits the SC. He then proceeds north through the pair's territory into the river beyond (he passed right by the area where Remy had been) and soon passes from view (he did not double back through the pair's territory—I suspect that he parked and dismounted his paddleboard somewhere in the section of river north of their territory).

1237 – Sinbad is finally resighted—he is parked next to the reeds north of the nest, intently looking to the east (probably looking out for the paddleboard). Remy isn't resighted at this time.

1240 – Sinbad is still as at 1237 (he is clearly still spooked regarding the paddleboard incident). Remy is finally resighted—she is parked and intently looking about next to the reeds along the west bank of the river at the north end of their territory (clearly, she is also still spooked regarding the paddleboard).

1241 – Sinbad goes to the nest and sits in the water at its NW side but appears reluctant to go onto it.

1244 – Sinbad climbs onto the nest and then immediately walks off it at its north side. **One very young chick** is briefly seen in the water by him and possibly a second one on the nest (my view of the nest at this moment is far too inadequate and it is most likely that what initially appeared as though it might have been movement by a chick on the nest was probably just a pattern of shifting sunlight falling on the nest through the reeds as they were being moved by the wind).

1246 – Sinbad climbs onto the nest and picks up a large, intact lower two thirds portion of empty eggshell with his beak (the eggshell is obviously from a recent hatching—the open edge of the eggshell shows that it has been characteristically "ringed", that is, it has been neatly cut through in a circular pattern along its circumference by an emerging chick). He then walks off the SE side of the nest with the eggshell, dives with it ~ 1.5 m SE of the nest and soon resurfaces without the eggshell approx. one meter north of where he dove (he apparently had disposed of the eggshell while underwater—disposing of and hiding the eggshell away from the nest so that it can't potentially serve as something that could give away the nest's location is almost certainly the main reason why Sinbad took the eggshell off the nest and stashed it away). Remy hasn't been resighted since 1240.

1251 – Since 1246 Sinbad has been sitting by or moving around the nest (on a few occasions he also walked onto the nest, fussed with the eggs a bit and then walked off). He again climbs onto the nest, fusses with the eggs for ~ 13 seconds, then finally settles onto and incubates them. One chick is seen at his side. This chick soon climbs onto his back. Remy still hasn't been resighted.

1300 – Sinbad is still incubating the eggs with the chick on the nest with him. Remy still hasn't been resighted.

1303 – Remy is finally resighted—she is parked and preening herself in the mid-north area of the SC ~ 7 m south of the reeds that line the north bank. Sinbad is still incubating the eggs and minding the chick that is on the nest with him.

1306 – Remy moves to the area ~ 9 m east of the nest, parks there and gazes to the east.

1307 – Remy moves a little farther east into the mid-SC area where she parks and preens. Sinbad continues with his incubation/nest sitting duty.

1310 – Sinbad and Remy are the same as at 1307. End observation period.

*As of my final look at the nest when I left the area at ~ 1500 only the one chick was seen.

*Comments:* It was an exciting day with the first confirmed chick hatching from the clutch. This occurred 24 days from the day the first egg was laid (counting that first day, 6/29, as day one). It's perhaps possible that a second chick was present but I am fairly certain that only one chick had hatched out as of my last check of the nest at 1500.

In regard to when laying actually began it is important to correctly interpret the point that 6/29 was the first day that I saw an egg in the nest. In reality, what this means is that the first egg was laid sometime between the end of my 6/28 observations (which concluded at 0940) and the beginning of my 6/29 observations (which commenced at 1223). Thus, it is certainly possible that the first egg could have been laid on 6/28 sometime after my observations concluded (at which time the majority of the day still remained since observations ended early) and therefore wouldn't have been seen by me until my 6/29 observations began. Given the latter, the best that can actually be stated is that the first egg of the clutch was laid on either 6/28 or 6/29 but was confirmed on 6/29.

Similarly, while it is possible that the lone chick seen today may have hatched yesterday during the period after I had concluded my observations, this seems very unlikely due to what was observed at 1246 today. At that time Sinbad was seen removing the larger portion of eggshell from a hatching from the nest and disposing of it. Given the typical promptness with which nesting birds remove eggshell from a hatching from the nest it seems extremely unlikely that the pair would have left that eggshell in the nest for such a long time (i.e., since yesterday). The above, of course, holds true only if this was indeed the only chick that had hatched up to that point, as appears to be the case.

The chick mainly stayed on the nest but was also seen in the water at the side of the nest on a few occasions. At one point it was seen climbing onto Sinbad's back while he was lying on the nest incubating the remaining eggs.

This is now the fourth time that a paddleboard has been observed entering the SR pair's territory and caused the incubating bird to flee from the nest and hide in the reeds (I don't know if the chick had hatched out yet at that time—the chick wasn't seen until 1244).

All 5 of the chicks from the East Lagoon pair's second brood were seen. As on 7/20 they were in and just beyond the dense reed bed at the east shore of the mid-segment of the East Lagoon. Two of the now grown and independent chicks from their first brood were seen in the southern third of the lagoon.

**<u>7/23/17</u>:** Time: 1153-1310; 1345-1530.

1153 – As observations begin, Sinbad is parked and preening in the mid-SC area. A pair of PB grebes (the NR pair) is foraging near and east of the NE point of the SC. Sinbad appears to be aware of these two but makes no move toward them (it is quite probable that he is being cautious about leaving the area because of the newly hatched chick(s) and the unhatched eggs). Remy is settled on the nest incubating the eggs. No chicks are visible.

1157 – Sinbad (still parked in the mid-SC area) calls once, dives and quickly resurfaces on his way toward the NE point where the NR pair had been (I don't see either of them at this moment). As he approaches the NE point one of the intruder grebes surfaces near there. Sinbad moves toward the intruder who (having sighted Sinbad) immediately dives. Sinbad also dives while continuing in the intruder's direction. Sinbad resurfaces soon after; the other grebe isn't resighted at this time. Sinbad briefly moves about in the area of the NE point, apparently searching for the intruder, then goes to the mid-SC area, parks there and gazes to the east.

1201 – Sinbad calls once while still parked in the mid-SC area. One of the NR pair (probably the male) is sighted moving north along the reeds of the west bank of the river in the northern half of the SR pair's territory. He calls once as he moves along.

1202 – Sinbad goes to the nest and engages in some nest M/E activity. Remy is still incubating the eggs.

1205 – One of the NR pair (probably the male) calls once from the northern half of the SR pair's territory.

1208 – Sinbad goes to the mid-SC area, parks there and preens while also looking about. Remy is still incubating the eggs. No chicks have been sighted yet.

1211 – Sinbad (still parked in the mid-SC area) calls once. Remy is still incubating and one chick is now seen lying on the nest next to her.

1213 – Sinbad goes to the nest. Remy then stands, walks off the nest and moves out to the east end of the SC where she parks and preens. Sinbad climbs onto the nest (3 chicks are clearly visible on it), fusses with the nest material and probably the unhatched eggs (none of them were visible) for ~ 39 seconds, then settles onto the eggs and incubates them. The chicks have remained on the nest with him.

1216 – Remy (now SE of the NE point of the SC) dives and begins foraging. One of the NR pair is sighted foraging in the northern half of the SR pair's territory.

1220 – Sinbad is still incubating the eggs. Remy isn't seen at this time.

1222 – Remy goes to the nest and engages in some light nest M/E. Sinbad continues incubating the eggs and nest sitting the chicks.

1223 – The member of the NR pair that was sighted at 1216 (still in the northern half of the SR pair's territory) calls once.

1229 – Remy is now foraging along the reeds in the SW section of the SC. She resurfaces from a foraging dive with a large crayfish and works it for ~ 49 seconds before swallowing it (tail first as usual with the tail partly doubled over on itself; in contrast the grebes swallow fish head first). She then resumes foraging. Sinbad is still incubating the eggs (all 3 of the chicks are probably on the nest with him though only one is visible).

1232 – Remy moves to the mid-area along the east end of the SC where she parks and preens. While preening, she voices one short, one note call.

1234 – Both members of the NR pair are in the western half of the river—one is just north of the NE point of the SC and the other is foraging near the reeds of the NE point (in other words, both are deep inside the SR pair's territory). Remy is still parked in the mid-area of the east end of the SC. She undoubtedly is aware of the intruders but stays put, making no move toward them. Sinbad is still incubating the eggs. Thus far both Remy and Sinbad have been remaining within the SC or only just beyond it during today's observation period. It is quite likely that with the serious encroachment into their territory by the NR pair they want to stay in the general nest area so as not to leave their newly hatched chicks and remaining unhatched eggs vulnerable.

1238 – Same situation as at 1234 for all 4 adult grebes.

1240 – The NR pair begins slowly working their way north in the northern half of the SR pair's territory.

1241 – Remy (still parked at the east end of the SC) dives and isn't resighted at this time.

1244 – Remy is resighted—she is moving in from the east end of the SC and goes to the area just south of the NE point and begins foraging there. Sinbad is still well-settled on the nest incubating the eggs.

1246 – One of the NR pair (probably the male, who is now parked just off the reeds at the west bank of the river in the northern half of the SR pair's territory) calls once. Within 2 seconds Sinbad answers this call with one of his own while sitting on the nest and incubating the eggs.

1249 – Remy goes to the nest. Sinbad then stands, walks off the nest and goes to the mid-SC area where he parks and preens. Three chicks are seen on the nest.

1250 – Remy climbs onto the nest, fusses with the eggs and the nest material around them for ~14 seconds and then settles onto the eggs and incubates them.

1251 – The NR male is still in the same location in the northern half of the SR pair's territory as at 1246. He calls again, this time with an especially long, drawn out call. Sinbad (still parked in the mid-SC) promptly answers this with a call of his own. Remy continues to incubate the eggs and nest-sit the chicks.

1253 – Remy stands, fusses with the eggs and the nest material around them for ~ 45 seconds, then resettles on the eggs and resumes incubation (the chicks are on the nest with her). Sinbad (still in the mid-SC area) dives and isn't resighted at this time.

1300 – Sinbad is resighted—he is foraging just south of the NE point of the SC. Remy is still incubating the remaining unhatched eggs. One of the NR pair (probably the female) is foraging in the area just north of the SR pair's territory. Its mate isn't seen at this time.

1305 – Sinbad goes to the nest, briefly lingers by it and then moves out to the mid-SC area where he parks and preens (I wasn't able to see if perhaps he had administered a feeding to the chicks). Remy is still incubating the eggs. Both members of the NR pair are in the western half of the river in the northern half of the SR pair's territory.

1310 – Sinbad is still parked in the mid-SC area. Remy is still incubating the eggs. One of the chicks is visible lying at her side; the other two are almost certainly either on the nest or on Remy's back. Both members of the NR pair are still in the north end of the SR pair's territory. Remy and Sinbad have been aware of and concerned with the presence of the NR pair throughout much of this observation period. End observation period.

*On a re-check of the nest at 1345 I found a dramatic event in progress. One of the SR pair was locked in a fierce, savage battle with the NR male at and around the nest (mainly within and at the edge of the reeds). The two were often locked together with the fight involving grappling, tumbling, spinning and biting with wings often flapping or flailing. In particular, the combatants sought to grab and hold the back of the neck of their opponent with its beak and vigorously shake it. While employing this hold the bird that was doing so also sometimes attempted to force and hold its opponent under the water. Most of the latter two tactics were being utilized by the NR male who was mostly getting the better of the SR pair member. Because the battling duo were so often locked together and my view was usually obscured by the reeds I was unable to confirm with certainty which of the home pair was involved in the fight, though it appeared that it was probably Remy (only one of the SR pair was seen until the very end of the fight). On several occasions the two fought right on top of the nest. None of the chicks were seen during or shortly after the fight. Unless injured or worse by the interloper they most likely were hiding within the dense growth of emergent reeds. The NR male was very persistent and the fight raged on for about 5-6 minutes. Finally, the two combatants emerged from the reeds well separated and no longer fighting. The other member of the SR pair was now present, the two of them in proximity in the area east of the nest. The NR male left the SC and joined his mate in the northern half of the SR pair's territory where he periodically called. Sinbad then spent much time on and around the nest. Remy appeared to be somewhat frazzled and for the first 9 minutes or so after the fight she repeatedly voiced fairly soft, short one note calls, probably calling for the chicks. She did this first from the edge of the reeds ~ 7 m north of the nest and then from the edge of the reeds in the SW corner, south of the nest. Physically she looked as though she had been the one involved in the fight. She appeared somewhat disheveled and most notably, the feathers on the back of her neck looked to be disturbed and tousled (i.e., as though this area had been grabbed, held and shaken). Note that such an inference would not exclude the possibility that both of the pair were involved in the fight (for example, this wouldn't exclude a possible scenario of Sinbad having taken her place sometime after the fight had begun). Neither of the pair appeared to be bleeding or to have any noticeable injuries.

There were a few moments during the fight when the NR male nearly had the home pair grebe subdued but the SR grebe then rallied and fought back. The SR grebe often uttered short, single note, distressful sounding calls during the fight.

To my great frustration, my view of how the fight finally came to a close was obscured by the dense reeds. Thus, I don't know if perhaps the other member of the SR pair had come to the aid of its mate and turned the tide.

Nor do I know how the fight began since it was already in progress when I arrived. Considering the location of the fight it is quite probable that the initial encounter occurred at or near the nest. Given that only one of the SR pair was seen at and engaged in the fight until its end, it seems likely that the other member of the pair probably wasn't in the area when the fight began. This seems especially so if Remy was the one involved in the fight—in such a case it is extremely improbable that Sinbad wouldn't have taken on the NR male in Remy's place. On the other hand, it could be that both were present and one of them carried the chicks away on its back to safety and protected them by hiding in the reeds with the youngsters while its mate engaged the interloper. If such were the scenario it would then be most likely that Sinbad was the one involved in the fight. Another possible scenario is that Remy was initially the one who was involved in the fight and Sinbad later came onto the scene and took on the NR male in her place (Remy is clearly smaller than Sinbad and the NR male). Unfortunately, all that I can ever do is speculate as above because my view of the fight was too inadequate to be able to confirm which one of the SR pair was involved in the fight. In fact, because there were moments when the fight moved completely out of my view within the reeds I can't even rule out the possibility that one of the SR pair had taken the other's place during such a moment.

At ~ 1415, just as things had calmed down (though I still hadn't sighted any of the chicks) and with Remy and Sinbad just east of the nest, the NR male entered deep into the SC and again engaged the SR pair in a fierce, savage, prolonged fight. The great majority of the fight took place within the southern half of the mid-SC area. With my view unobstructed for most of the fight I could confirm that it was Sinbad who engaged the intruder. On two occasions during the initial stage of the fight Remy physically joined the fray. However, in each instance the NR male then turned his attacks fully onto Remy and quickly gained the upper hand on her until Sinbad promptly re-engaged him, causing the NR male to immediately switch his full attention back to him. Thereafter, Remy continued to support Sinbad through the remainder of the fight by following the combatants closely and sometimes taking a firm peck or two at the NR male. Nevertheless, the battle was essentially fought between Sinbad and the NR male. Their fighting was particularly fierce and quite serious. They often pushed hard against each other, sometimes chest to chest with wings flapping. The pair furiously grappled with each other, locked together, pecking and tumbling on the surface. On a number of occasions they dove together, briefly carrying on their battle underwater. However, as in the first fight the main aim seemed to be to grab and hold the back of the opponent's neck with the beak (savagely shaking the beak back and forth while maintaining the hold) and controlling the opponent and the fight in this manner, including attempting to force and hold the opponent underwater. The battle between these two ebbed and flowed, the advantage constantly shifting back and forth (though the NR male appeared to be doing slightly better)., first one, then the other getting a neck hold on its opponent. Finally, the battle moved to a part of SW section of the SC that was out of my view (with Remy still following the combating males) and I was unable to see what transpired. Shortly after, the NR male emerged from that area alone and made his way out of the SC to the northern half of the SR pair's territory where he joined his mate (who had taken no part in either of the fights). Remy and Sinbad moved together to the area east of the nest where they lingered together briefly before engaging in separate tasks. The pair again had driven off the intruder but as before he had been very persistent, the fight having lasted ~ 6 minutes. In general, the NR male had proved himself to be quite formidable.

In the aftermath of this second altercation Sinbad again spent much time on the nest fussing with the nest material and apparently also the unhatched eggs. No eggs were visible from my vantage point but his actions (and later, the fact that Remy settled onto the nest in a typical incubation posture) indicated that the eggs were still on the nest (at least, one or more of them). With some of the first fight having taken place on top of the nest there is the possibility that some or all of the eggs could have been damaged or knocked into the water. In regard to possible damage, it is perhaps in their favor that the chicks within the (viable) unhatched eggs are ready to hatch out, at least three of their clutch mates having already done so. The nest itself appeared to have suffered slightly from the fighting (though nothing major)—it seemed to be sitting just a bit lower in the water.

Remy apparently had dealt with collecting the chicks. About 10 minutes after the fight, as Remy moved toward the nest from the reeds in the SW corner, I sighted one chick for certain and probably a second riding on her back, tucked under her wings. She then unloaded the chick(s) onto the nest. I thus can only confirm that one and probably at least two of the chicks had survived the fights and the serious encroachment by the NR male. The chicks are extremely vulnerable at this early stage and could easily be killed with one crushing bite.

Three chicks were definitely seen today. It thus appears that two more chicks have hatched since the end of yesterday's observations. This inference is based only on the number of chicks that I have been able to confirm. It should be kept in mind that my view of the nest is limited and it is certainly possible that there are more chicks present which I just haven't been able to see.

I observed the pair until 1530. By then things had calmed down though the pair was still vigilant. When I departed, Remy was well-settled on the nest and apparently incubating the eggs. Sinbad was parked in the mid-area of the SC. The NR pair was situated deep within the SR pair's territory, just north of the SC, the NR male periodically calling (as did Sinbad). The SR pair was still obviously reticent about leaving the SC and after witnessing the violent incursions by the NR male it was easy to understand why.

*Comments:* Sinbad, Remy and their brood are now facing a critical situation. It's clear that this matter is not over and that they are involved in a very serious territorial conflict. The NR pair appears intent on trying to take over the SR pair's territory or a significant portion of it. At this point I don't know whether Sinbad and Remy will still be here when I return tomorrow and if their chicks (hatched and unhatched) will still be alive (I don't even know for certain if more than one chick survived today). This situation could hardly have happened at a worse time for the SR pair—a time when they have newly hatched chicks to care for and the unhatched remainder of the clutch that still needs to be incubated.

**7/24/17:** Time: 1150-1315.

1150 – As observations begin I am relieved to see that the SR pair is still in the SC. Sinbad is parked and preening in the NW section of the SC and Remy is well-settled on the nest incubating the eggs. One of the NR pair (probably the female) is foraging near the west bank of the river in the northern half of the SR pair's territory.

1157 – Sinbad goes to the nest and engages in nest M/E activity, staying near the nest as he does so (one chick can be seen in the water by him). He also appears to be doing some foraging in the area around the north side of the nest. At one point it appeared as though he captured a prey item and after working it, may have handed a scrap to a chick (my view is mostly obscured by the reeds).

1200 – Remy stands, walks off the nest (at least one egg is visible) and joins Sinbad on the north side of the nest in conducting some light nest M/E as well as probably doing some foraging there.

1203 – Sinbad climbs onto the nest (followed soon after by a chick), briefly fusses with the egg(s) and nest material around them, then settles on the remaining clutch and incubates it. In the meantime, Remy emerges from the nest area heading east with the remains of a crayfish that she had apparently been working and feeding to the chicks. She quickly swallows it en route as she moves out to the mid-area along the east end of the SC where she dives and isn't resighted at this time.

1204 – Remy is resighted—she is foraging off the mid-north bank area of the SC.

1205 – A paddleboarder launches his board from the SE point of the SC and paddles it north on the river through the entire length of the SR pair's territory and beyond. Remy is spooked into a dive by the presence of the paddleboard and isn't resighted at this time. Sinbad tenses and intently watches the paddleboard but remains on the nest.

1212 – The paddleboarder returns south on the river through the SR pair's territory to the SE point of the SC. This time Sinbad is spooked off the nest by the paddleboard as it passes east of the SC and he sits in the water at the north side of the nest, keenly watching the paddleboard. Remy hasn't been resighted yet (she is probably hiding in the reeds).

1216 – Sinbad finally climbs back onto the nest and quickly settles onto the egg(s) and resumes incubation. Remy is resighted—she is parked just out from the reeds at the mid-north bank of the SC.

1218 – The paddleboarder makes another trip north through the SR pair's territory. As he passes east of the SC he spooks Remy into a rapid sinking dive and she isn't resighted at this time. Sinbad tenses and keenly watches the paddleboard but remains on the nest and continues incubating the clutch,

1223 – The paddleboarder makes a return trip south on the river through the SR pair's territory, this time causing Sinbad to flee off the nest and hide in the reeds. One chick is briefly seen on the nest. Remy still hasn't been resighted.

1231 – Sinbad is resighted—he is within the reeds next to the south side of the nest, still reluctant to go back onto it due to his apprehension regarding the paddleboard (which is now back at the SE point of the SC). Remy still hasn't been resighted. The paddleboard also spooked the NR pair into hiding. Neither of them have been sighted since the first paddleboard trip, though a call was heard at ~ 1227.

1235 – Sinbad is still sitting in the water at the side of the nest (now at its west side), intently looking to the east. Remy still hasn't been resighted.

1237 – Remy is finally resighted—she is in the water by the nest with Sinbad. Sinbad soon climbs onto the nest, fusses with the egg(s) and nest material for ~ 8 seconds, then settles onto the clutch and incubates it. Two chicks are briefly seen on the nest, one of them while climbing onto Sinbad's back. Remy moves away from the nest within the reeds. She passes out of view and isn't resighted at this time.

1242 – Remy is resighted—she is parked in the NW corner of the SC and is just looking about. Sinbad is still incubating the clutch.

1250 – Sinbad is still incubating the clutch and nest-sitting the chicks. Remy isn't seen at this time. One of the NR pair is seen foraging in the northern half of the SR pair's territory.

1253 – Remy is resighted—she is working a fairly large crayfish by the reeds at the mid-south bank area of the SC. She works it for ~ 13 seconds more, dives with it and soon resurfaces with just a large posterior section of the tail which she carries to the nest to feed to the chicks. She arrives at the nest ~ 1.5 m from its south side where my view of her is nearly completely obscured by the dense reeds. At least one chick is briefly seen going to her. Though my view is mostly blocked I can see signs of vigorous movement indicating that she is working the crayfish, probably detaching morsels to feed to the chick(s).

1256 – Sinbad stands, walks off the nest and engages in light nest M/E.

1301 – Remy climbs onto the nest, briefly fusses with the egg(s) and nest material around the clutch, then settles onto it and begins incubating. Sinbad moves to the NW corner of the SC where he parks and preens.

1305 – Sinbad is still parked in the NW corner of the SC. Remy is still incubating the clutch and nest-sitting the chicks.

1315 – Sinbad and Remy are the same as at 1305. One of the NR pair is foraging near the reeds along the west bank of the river in the northern half of the SR pair's territory. End observation period.

*Comments:* Resolution of the territorial matter is ongoing. No encounters between the two pairs occurred during today's observation period (which was significantly disrupted by the paddleboarder in this regard). However, the situation remains tense. Just as yesterday, Sinbad and Remy confined their movements and activities to within the SC (and most of this was confined within the western half of the SC). They are obviously still very concerned about ensuring that they don't leave their fragile newly hatched chicks and remaining unhatched eggs vulnerable to the intruder (NR) pair. The NR pair seems to be getting more comfortable with being in the northern half of the SR pair's territory. Unfortunately for Sinbad and Remy, they find themselves in a quandary at this point. They need to defend the northern half of their territory but it appears as though they feel that it's too risky to go out and attempt to drive the intruders away since this would entail leaving their chicks and unhatched eggs more vulnerable to the intruders.

As usual, I was not afforded a definitive look at the nest and thus don't know just how many chicks and unhatched eggs are present. All that I could confirm today was at least two chicks and at least one unhatched egg. I feel certain that there is more of each but I just couldn't see them.

It was confirmed today that the parents are already provisioning their chicks.

Hatching for this clutch has been staggered. Thus far it has extended over a period of three days.

Up to this point the chicks have mainly stayed on the nest (this includes climbing onto the parent's back while that parent is sitting on the nest incubating the eggs). Of particular note, when a parent is on the nest the chicks are essentially always on the nest with the parent. The chicks do sometimes go onto the water around the nest. Such instances mainly occur during nest exchanges between the parents or other times when the nest is unoccupied (such as when one or both of the parents are engaged in nest M/E) and when a parent comes to the nest area to administer a feeding (in the latter case, even if the chicks are on the nest with a parent they may come off the nest and go to the feeding parent—such feedings thus far have been administered close to the nest).

One of the now independent chicks from the East Lagoon pair's first brood of this season briefly foraged just inside the SR pair's territory in the area east of the SE point of the SC (just north of the RR bridge). Neither Remy nor Sinbad were in that vicinity at the time nor did they appear to be aware of the juvenile's presence.

For a fifth time now a paddleboarder has been observed to disrupt the SR pair's nesting.

**<u>7/25/17</u>:** Time: 0746-0835.

0746 – When observations begin, Sinbad is well-settled on the nest incubating the clutch and Remy is foraging along the north bank of the SC. Two and possibly 3 chicks are briefly seen and at least one egg. Neither of the NR pair is sighted at this time.

0750 – Remy has shifted her foraging to the mid-SC area. Sinbad is still incubating the clutch. One of the NR pair is sighted in the north end of the SR pair's territory.

0755 – Sinbad is still incubating the egg(s). Remy isn't seen at this time.

0757 – Remy is resighted—she is foraging in the mid-area of the east end of the SC.

0803 – Remy surfaces from a foraging dive just east of the east end of the SC with a fairly large fish. She works it briefly and then heads in the direction of the nest, carrying the fish in her beak. When she reaches the mid-area of the SC she stops there, works the fish for ~ 17 seconds and then swallows it. (It appeared that her initial intent was to take the fish to the chicks for a feeding but then acknowledged that it was much too large for any of the chicks to swallow nor would she be able to tear pieces off to feed to them.) Sinbad is still incubating the egg(s).

0807 – Sinbad stands, briefly repositions himself over the egg(s), then settles back onto the clutch and resumes incubation. Two and probably three chicks are seen on the nest. Remy isn't seen at this time.

0809 – One of the NR pair (probably the male) is in the north end of the SR pair's territory and calls once.

0811 – Sinbad is still incubating the clutch. Remy is briefly seen by the nest engaging in some nest M/E and probably also doing some foraging in that area.

0815 – Remy moves out to the mid-SC area, dives and begins foraging. Sinbad continues incubating the egg(s).

0817 – Remy resurfaces from a foraging dive in the mid-SC area and begins heading for the nest as though she has a prey item for a feeding. However, I don't see anything in her beak (perhaps she has a small prey item that I'm not able to see). As usual, she dives when still well out from the nest and covers the remaining distance to it underwater (and thus, out of sight). She resurfaces by the NW side of the nest where she remains briefly (unfortunately, my view is blocked by the nest so I'm not able to see if she administers a feeding), then dives, goes back out to the mid-SC area and resumes foraging (her actions over this sequence indicate that she probably did conduct a feeding). Sinbad is still incubating the egg(s).

0821 – Remy goes to the nest. Sinbad stands, walks off the nest, moves to the mid-SC area (calling once while en route) and begins foraging, working his way to the reeds at the south side of the NE point of the SC. Remy engages in light nest M/E activity (two chicks are briefly seen in the water by her and one on the nest; at least one egg is visible on the nest).

0825 – Remy climbs onto the nest, quickly settles onto the egg(s) and incubates the clutch. Sinbad isn't seen at this time (he is probably still foraging).

0830 – Sinbad is resighted—he is foraging in the river just east of the SC. Remy is still incubating the egg(s).

0831 – Sinbad moves to the mid-area of the east end of the SC where he parks and preens.

0833 – Sinbad is still parked in the same area at the east end of the SC. He intermittently preens himself while alertly looking to the north and east (he apparently is looking out for the NR pair who are not seen by me at this time). While doing so he gives a short, single, one note call (similar to the call that Remy has been observed emitting). He voices another one of these calls ~12 seconds later. Remy continues incubating the egg(s).

0835 – Sinbad (still parked at the east end of the SC) calls once (his usual long, drawn-out territorial/contact call). Remy is still incubating the egg(s). End observation period. [*Note: I had to make a trip to Los Angeles today so the observation period was earlier and a bit shorter than usual.*]

*Comments:* The territorial matter between the SR and NR pairs is still ongoing and remains tense. Remy and Sinbad are obviously still concerned about and preoccupied with the NR pair. Both of them foraged briefly in the river section just east of the SC. Other than these brief interludes they once again restricted their movements and activities to the SC. When sighted during today's observation period the NR pair was in the north end of the SR pair's territory and the section of river to the north of this. Perhaps it was because the NR pair was so distant that Remy and Sinbad felt secure enough to briefly extend their foraging to the area a little to the east of the SC?

As usual, I wasn't afforded a definitive view of the nest. That being the case, today I was able to confirm the presence of at least 3 chicks and at least one unhatched egg. Hatching of the clutch has now extended to a fourth day.

As has been the case since hatching began, the chicks are almost always on the nest (including being on the back of an incubating parent) but do sometimes go onto the water near the nest.

**7/26/17:** Time: 1155-1315; Brief re-check of nest at ~ 1325.

1155 – As observations begin, Remy is well-settled on the nest incubating the egg(s). Sinbad is briefly seen by the nest before he dives and is lost from view at this time.

1158 – Sinbad is resighted—he is foraging in the mid-SC area. An osprey calls while "hovering" over this area and attempting to get a fix on a fish. Sinbad suddenly becomes keenly alert to the osprey's presence and nervously mills about the mid-SC area, voicing 4 successive short, single, one note calls (these appear to be alarm calls and are probably being given out of concern for the safety of the chicks). Almost certainly in response to his calls, Remy comes off the nest and sits at its north side, alertly looking to the east. The osprey "hovers" a bit lower but aborts its hunting effort without attempting a strike and moves away from the area. Soon after, Sinbad parks in the NW section of the SC and preens himself. Remy remains in the water by the nest, still keyed-up and alert. One of the NR pair is sighted in the north end of the SR pair's territory.

1202 – Remy climbs back onto the nest, fusses with the egg(s) and the nest material around it for ~ 29 seconds, then settles onto the clutch and incubates it.

1205 – Remy stands, walks off the nest and sits in the water at its NW side and gazes to the east.

1208 – Remy is still sitting in the water by the nest. Sinbad is sighted moving toward the nest from the NE point of the SC (I'm not able to see anything obvious in his beak but it's possible that he has a small prey item). He goes directly to the nest but my view is too inadequate to be able to see if he conducted a feeding (at least 3 chicks are seen in the water near the north side of the nest). Sinbad then leaves the nest area and goes back to the NE point of the SC and begins foraging (his actions during this sequence suggest that he did conduct a feeding).

1214 – Remy climbs onto the nest, fusses with the egg(s) and nest material for ~ 19 seconds , then walks off the nest and sits at its north side. Sinbad isn't seen at this time.

1216 – Remy is still sitting in the water by the nest. Sinbad suddenly appears briefly by the nest (I'm not able to see what he does there—of particular note, I'm unable to see whether or not he conducts a feeding), then moves out to the mid-SC area where he dives and isn't resighted at this time (he is probably foraging).

1220 – Remy goes onto the nest and settles onto the clutch. Approx. 1 ½ minutes later she rises and stands over the clutch for several minutes during which time she preens herself and fusses with the egg(s) and nest material, and then settles back onto the clutch and resumes incubation (time: 1225). One chick is briefly seen in the water by the nest. Sinbad isn't seen at this time.

1230 – Sinbad is resighted—he is foraging in the SE section of the SC. Remy is still incubating the egg(s). Three chicks and probably a fourth are briefly sighted in the water ~ ½ m north of the nest, well hidden within the reeds.

1232 – Sinbad appears at the nest with what appears to be a small fish which he eventually hands off to one of the chicks near the north side of the nest. He then moves away from the nest and is lost from view. Remy is still incubating the egg(s).

1234 – Sinbad again appears at the nest. Remy then stands, walks off the nest and moves out to the mid-SC area where she begins foraging. Sinbad climbs onto the nest, settles onto and incubates the clutch.

1235 – Remy surfaces in the mid-SC area with a small fish. She carries it to the area near the north side of the nest, hands it off to one of the chicks, then goes back out to the mid-SC area and resumes foraging.

1236 – Sinbad stands, walks off the nest and is soon lost from view. Some chicks are briefly seen in the water ~ ½ - 1 m north of the nest.

1238 – Remy is foraging in the SE section of the SC when she suddenly assumes an alerted state and in a frantic manner repeatedly voices short, single, one note alarm calls. She has done this in response to an osprey which has just begun hovering over the east end of the SC, shifting its hover point around over that area (drifting in toward the mid-SC) as it attempts to get a fix on a fish. While continuing to call, Remy circles back through the mid-SC area to the nest area in a fairly frantic manner, taking three short, exaggerated, very conspicuous flights as she does so. (These flights appear to be an attempt on her part to distract the osprey and draw its attention to her rather than the chicks. The osprey, of course, has no interest in the chicks and is just trying to target a fish. However, Remy obviously perceives it as a potential threat to the chicks and is taking no chances.) Sinbad is hiding in the reeds near the nest with the chicks. Remy joins him and both then alertly look out from there. The osprey soon abandons its hunting effort, again without attempting a strike (it's even possible that Remy's frantic movement through the area inadvertently spooked away the fish that the osprey had been targeting), and leaves the area. Despite the osprey having flown away, Sinbad and Remy remain in the nest area with the chicks. Both members of the NR pair are sighted (by me—they can't be seen by the SR pair from their present location by the nest) in the northern half of the SR pair's territory.

1245 – Sinbad and Remy are still in the nest area with the chicks, alertly looking out to the east. One of the NR pair has moved to the area just north of the NE point of the SC (Sinbad and Remy would not be able to see it from their present location).

1247 – Remy finally leaves the nest area. She goes to the mid-SC area and begins foraging. Sinbad climbs onto the nest, settles onto the clutch and incubates it.

1250 – Remy resurfaces from a foraging dive in the east end of the SC with a small fish. She takes it back to the nest area, feeds it to one of the chicks, then goes out to the SE section of the SC and resumes foraging. Sinbad is still incubating the remaining unhatched egg(s).

1300 – Sinbad is still incubating the egg(s). Remy is still foraging in the SE section of the SC. One of the NR pair is now foraging just east of the NE point of the SC. Neither it nor Remy appear to be aware of each other (or they are just choosing to ignore each other).

1301 – Remy goes to the nest. Sinbad then stands, walks off the nest and makes his way out to the NE point of the SC (calling once en route) where he begins foraging (the NR grebe had already gone back into the northern half of the SR pair's territory by then). Remy engages in nest M/E.

1304 – Remy climbs onto the nest, settles onto the egg(s) and begins incubation. Sinbad is still foraging at the NE point of the SC.

1306 – Remy stands, fusses with the egg(s) and the nest material around it for over one minute, then settles back onto the clutch and incubates it. One chick is visible on the nest.

1310 – Remy is still incubating the egg(s). Sinbad appears at the nest and seems to be administering a feeding to some chicks that are around him but my view is too poor to be certain of this.

1315 – Remy is still incubating the egg(s). Sinbad is parked and preening in the mid-north section of the SC. End observation period.

*On a brief re-check of the nest at ~ 1325 Sinbad was well-settled on the nest incubating the egg(s) and Remy was foraging in the eastern half of the river east of the NE point of the SC.

*Comments:* The territorial matter is still ongoing but didn't appear to be quite as tense during today's observation period. This is perhaps largely due to the fact that Remy and Sinbad are now having to devote much of their attention and activity toward foraging for and feeding their newly hatched chicks. Nevertheless, they are still obviously concerned about and preoccupied with the NR pair. Once again they restricted their movements and activities to the SC. There was one notable exception to this. On a brief re-check of the nest ~ 10 minutes after the formal observation period concluded Remy was sighted foraging in the eastern half of the river in the area east of the NE point of the SC (the NR pair was at the north end of the SR pair's territory at the time). This is the farthest out of the SC that I have seen either of the home pair go since the territorial conflict began. For the most part, the NR pair remained in the northern half of the SR pair's territory and in the section of river north of it (i.e., their own territory). However, one of the pair did venture down to the NE point of the SC and spent a little time there.

Three chicks were definitely seen today and probably a fourth, though as usual I was not afforded a definitive look. I thus am not certain of exactly how many chicks were present today. The chicks spent much of the limited time when they were visible on the water near the nest, particularly in the area up to about 1 ½ m to the north of the nest where all of the observed feedings occurred. They remained within the reeds: I have not yet seen any of the chicks venture out from the reeds. The chicks continue to climb onto the backs of the parents.

One egg was definitely seen on the nest today and it appears that it may be the only remaining unhatched egg. As in several previous instances when it appeared that the hatching process for one of the eggs may have begun or had been underway, the incubation periods were often truncated and the pair spent considerable time fussing with the egg. Perhaps this is an indication that pipping has occurred in what appears to be the final unhatched egg? If so, then by tomorrow the pair may finally be finished with incubation duties.

In each of the feedings today in which the prey item captured and proffered by the parents could be identified, it was a small fish. This is interesting because in each of the numerous instances in which I have observed Remy and Sinbad capture a fish for their own consumption the fish was of a large or fairly large size (i.e., for a PB grebe)—in all cases, a fish much too large for these small, very young chicks to be able to swallow (and the physical characteristics of these large fish appear to be such that the parents aren't capable of tearing into the body and detaching small morsels that can be fed to the chicks). The indication here is that the parents are deliberately and selectively seeking out small fish of a size that the chicks can swallow.

The SR pair's reaction to ospreys that are hunting over the SC area is notably different now that they have young chicks (note that ospreys are commonly seen in this area and regularly perch on either of two snags on the east bank of the river across from the SC). When with no chicks they basically were indifferent to, or at most paid only casual attention to ospreys. The only exceptions to the latter that I have observed were a few instances when an osprey made a diving attempt to capture a fish and struck the water (or pulled up just before doing so) fairly close by one of the pair. Even in these cases the brief reaction by Sinbad or Remy appeared to be just a reflexive action to the sudden and dramatic appearance of the osprey (i.e., a reaction due to being startled and surprised) rather than out of fear of the osprey. In general, they don't appear to regard ospreys as a threat to themselves. However, judging by the nature of the strong reactions they exhibited today they do apparently perceive the ospreys as a possible threat to the chicks.

All 5 of the East Lagoon second brood chicks were seen accompanying one of the parents in the far north section of the East Lagoon. The other parent was nearby foraging for the chicks. One of the now independent chicks from their first brood was seen in the West Lagoon.

**7/27/17:** Time: 1136-1245.

1136 – As observations begin, Sinbad is well-settled on the nest and Remy isn't seen at this time. Both members of the NR pair are foraging along the reeds of the west bank of the river in the northern half of the SR pair's territory.

1140 – Sinbad is still well-settled on the nest (incubating or brooding?). Two chicks are visible on the nest. Remy still hasn't been sighted.

1144 – Remy is sighted moving from the NE point of the SC toward the nest (I'm unable to see if she is carrying a prey item in her beak). She arrives near the north side of the nest but my view is too inadequate to see whether or not she conducts a feeding (one chick is visible in the water by her). She then goes out to the mid-SC area where she parks and preens.

1145 – Sinbad stands, walks off the nest, moves out to the SE section of the SC and begins foraging, working his way north along the east end of the SC.

1146 – Remy goes to the nest, climbs onto it, fusses with the nest material (and egg?—no eggs have yet been seen on the nest today), then lies down and settles on the nest. One chick is seen climbing onto her back.

1150 – Sinbad is foraging by the reeds at the NE point of the SC. Remy is still settled on the nest.

1155 – Sinbad surfaces from a foraging dive at the edge of the reeds at the NE point of the SC with a very large fish. He works it for ~ 58 seconds (including diving with it 3 times) before finally swallowing it. Remy is still settled on the nest (one chick is visible on her back).

1157 – Sinbad goes to the mid-SC area where he parks and preens. One of the NR pair is sighted near the west bank of the river at the north end of the SR pair's territory. Remy is still settled on the nest.

1200 – Sinbad begins moving toward the nest. Remy spots him approaching, then stands, walks off the nest and meets and greets him en route ~ 10 m east of the nest. They continue on in opposite directions. Remy dives when she reaches the mid-SC area and isn't resighted at this time. Sinbad makes his way to the nest.

1201 – Sinbad climbs onto the nest, briefly fusses with the nest material (and egg?—I still haven't seen any eggs), then lies down and settles on the nest.

1204 – Remy is resighted—she is foraging along the reeds in the SW section of the SC. At least one chick is visible in the water ~ ½ - 1 m south of the nest within the reeds. One of the NR pair is still inside the north end of the SR pair's territory.

1210 – Remy is now foraging in the mid-SC area. Sinbad is still settled on the nest.

1211 – A turkey vulture that is soaring fairly low flies over the west end of the SC in the direction of the nest. Sinbad sights it approaching the nest area and suddenly stands, walks off the nest and hides in the reeds just north of the nest (at least 2 chicks were seen following him there). The turkey vulture passes over the nest area, just continuing on its way, and is quickly gone from the area (the turkey vulture was quite obviously oblivious to and had no interest in the grebes). Sinbad returns to the nest soon after it passes and engages in nest M/E and perhaps also some foraging in the area around the nest.

1215 – Remy appears at the nest near Sinbad. Both Remy and Sinbad appear to be administering feedings at the north side of the nest but my view is too poor to be able to see just what is going on. At least 3 chicks and possibly 4 are sometimes visible in the water by them. Remy soon leaves the nest area and heads east.

1219 – Sinbad climbs onto the nest, fusses with the nest material for ~ 25 seconds, then lies down and settles on the nest. Remy is parked and preening ~ 14 m south of the NE point of the SC.

1223 – A turkey vulture soars slowly over the east end of the SC heading in a direction (west) that will take it over the nest area. Sinbad spots it and cranes his neck as he watches the turkey vulture approach and appears poised and ready to get up and leave the nest. However, this time he remains on the nest as the turkey vulture passes by and moves out of the area (this turkey vulture was flying higher than the individual in the previous incident).

1226 – Remy (still near the NE point of the SC) ceases preening, dives and begins foraging. Sinbad is still settled on the nest.

1230 – Sinbad is still settled on the nest. Two chicks are visible on the nest with him; one of them is preening itself. Remy isn't seen at this time.

1235 – Sinbad stands, walks off the nest, sits in the water by it and calls once. He then engages in nest M/E. Remy still hasn't been resighted (she is probably foraging).

1237 – Sinbad moves out to ~ 6 m east of the nest and calls once while looking east (probably for Remy). He lingers there briefly, then goes back to the nest and resumes his nest M/E work and perhaps also does some foraging in the nest area.

1238 – Sinbad climbs onto the nest, briefly fusses with the nest material, then lies down and settles on the nest. Remy still hasn't been resighted.

1242 – Remy is briefly sighted by the nest (I didn't see whether she had conducted a feeding—no chicks were visible around her) before she dives and moves out to the mid-SC area where she parks and preens.

1245 – Remy is still parked and preening in the mid-SC area. Sinbad is still settled on the nest. Neither of the NR pair is seen at this time. End observation period.

*Comments:* The territorial situation was similar to yesterday. It is still ongoing but appeared less tense (here again, much of the SR pair's attention and activities have been diverted to foraging for and feeding the chicks). Still, as has been the case since this territorial conflict began, Remy and Sinbad restricted their movement and activities to within the SC. It's clear that they remain concerned about and preoccupied with the NR pair. The NR pair ranged only within the northern half of the SR pair's territory and to the north of it in their own territory during today's observation period. Sinbad and the NR pair (probably only the male) continue to call periodically but regularly. No physical confrontations have been observed since the first day (7/23) of the conflict. At this point it appears that both pairs would prefer to keep it so but I wonder what has happened between them during all of the hours when I haven't been here to observe them. It remains to be seen what turn this situation will take when the chicks become mobile (something that will be happening fairly soon).

I had no confirmed sightings of any eggs today or even of a probable one. I thus don't know for certain one way or the other whether the pair was still incubating an unhatched egg(s) or just brooding and protecting the newly hatched chicks. In general, the pair behaved as though incubating: they remained tied to the nest; settled on the same spot on the nest; often fussed with (manipulated) the area that they sat upon; laid on the nest in an incubation posture; and continued to engage in nest M/E activity. In addition, one or the other of the pair was settled on the nest for a combined 56 of the 69 minutes of the observation period. On the other hand, it could be that they were just brooding and very protectively nest sitting the chicks. Unfortunately, without knowing for certain whether or not an egg(s) was present (once again, due to my limited view of the nest) I can do no better than speculate at this point. My best guess is that I didn't see any eggs because none were present and all of the chicks have hatched out. Perhaps there will be something tomorrow that will help clarify this matter?

It was interesting seeing Sinbad react to those turkey vultures today. Prior to the presence of the chicks I had never observed either of the pair pay any noticeable attention to turkey vultures that were passing overhead (turkey vultures are common in this area). Obviously, as with the ospreys, the concern here is for the newly hatched chicks. Both Sinbad and Remy are quite sensitive regarding anything that they perceive as a possible threat to the chicks.

All 5 of the East Lagoon second brood chicks were again seen in the far north section of the East Lagoon. When they were sighted a feeding was in progress (a crayfish) with 4 of the chicks swarming around the feeding parent (the other parent was also with the group) and competing with each other. The 5$^{th}$ chick for some reason did not join in during the feeding but instead just sat by itself ~ 12 m away.

**7/28/17:** Time: 1045-1155.

1045 – As observations begin, Sinbad is parked and preening ~ 20 m south of the NE point of the SC. Remy is parked and looking about (mainly to the east) ~ 5 m east of the nest. Some chicks are visible in the water by the nest. No eggs are visible on the nest.

1047 – Remy moves to the nest. She lingers by it momentarily before eventually climbing onto it, lying down and settling on it (some chicks are briefly sighted joining her on the nest). Remy appears to be brooding or looking after the chicks rather than incubating an egg. It appears that hatching has indeed concluded and very likely had done so yesterday or perhaps the day before.

1049 – Sinbad (still by the NE point of the SC) begins foraging near the reeds at the NE point.

1055 – Sinbad is still foraging by the reeds along the NE point. Remy is still settled on the nest (all of the chicks are probably on the nest with her).

1058 – One of the NR pair (probably the female) is sighted near the west bank of the river in the north end of the SR pair's territory.

1100 – Sinbad surfaces from a foraging dive near the reeds at the NE point with a medium-sized crayfish and begins making his way to the nest with it. When he reaches the area ~ 1 m north of the nest (well within the reeds) some chicks are seen streaming off the nest to him. Sinbad then works the crayfish for ~ 59 seconds as he feeds the chicks. It appears that he is probably repeatedly tearing off morsels from the crayfish and handing these to the chicks but my view is much too inadequate to make out the details (unfortunately, the spot just north of the nest where the feedings have predominantly been taking place lies within the reeds behind an extremely dense cluster of closely packed reeds that essentially blocks my view). At least 4 chicks and almost certainly a 5$^{th}$ are intermittently seen swarming around Sinbad as he works the crayfish. When Sinbad completes the feeding he moves back out to the NE point area (calling once en route) and resumes foraging.

1102 – Remy walks off the nest and moves out ~ 3 m to the east of it where she briefly lingers and looks to the east. She then returns to the nest area and begins foraging there. As before, no eggs are seen on the nest.

1108 – Sinbad surfaces from a foraging dive just SW of the NE point of the SC with a fairly large fish. He works it for ~ 20 seconds before swallowing it. He then starts moving toward the nest. From the nest area Remy spots him approaching and heads out toward him. They meet and greet en route and continue on in opposite directions. Remy moves out to the area around the NE point and begins foraging. Sinbad moves to the nest .As he reaches a point ~ 1 ½ m north of the nest (this is within the reeds) some chicks rush over to him looking for a feeding but Sinbad has nothing for them. Sinbad then engages in light nest M/E. Some of the chicks stay in the water just north of the nest and others go onto it.

1113 – Sinbad climbs onto the nest, lies down and settles on it. Some of the chicks are briefly seen going onto the nest with him.

1120 – Sinbad is still settled on the nest (all of the chicks are probably on the nest with him). Remy isn't seen at this time (she is probably foraging).

1121 – Remy is resighted—she is engaged in a fierce fight with the NR male in the mid-river area east of the SC! The two were already well-engaged when first sighted and the fight had probably already been going on for a while. At this initial sighting Remy is badly losing the battle and is in a difficult situation. The NR male has a firm grip on the back of her neck with his beak with his body positioned directly behind her and partly on her back. He has her firmly under control and she is unable to fight back under such a hold. It appears that her only recourse at this point is to try to break free from his grasp. She attempts to wriggle her neck and pull away while moving forward but he maintains his controlling grip, then rides onto her back and forces her under the surface. Both then disappear under the surface for several seconds. When they resurface nothing has changed: the NR male still has his controlling grasp on her neck. Within seconds he forces her under a second time and they again disappear below the surface with the male locked onto her. This time Remy was obviously able to break free (or the male released her) as she resurfaces ~ 6 m to the west of him. The NR male remains where he resurfaced and doesn't pursue her. Remy (looking frazzled and somewhat disheveled, notably along the feathers on the back of her neck where the male had grasped her) moves farther west to the mid-area along the east end of the SC where she parks, preens and composes herself. The NR male moves north into the northern half of the SR pair's territory. No bleeding or other injuries could be seen on Remy (in this regard, she fully spread her wings several times while preening and they looked to be in good shape). Throughout this fight Sinbad was on the nest with the chicks. He gave no obvious signs that he had been aware of the fight (there were no obstructions other than the reeds around the nest that would have blocked his view but the fight occurred a considerable distance away from the nest).

1124 – Remy (still parked at the east end of the SC) voices one short, single, one note call and resumes preening. Sinbad is still settled on the nest with the chicks. The NR male is in the northern half of the SR pair's territory.

1128 – A call is heard coming from the northern half of the SR pair's territory. Though neither of the NR pair is sighted at this time the call undoubtedly is coming from one of them (probably the male).

1131 – Remy ceases preening and, apparently undeterred by the savage fight, moves to the mid-river area east of the SC and begins foraging (it is most likely that the NR male had attacked her while she was foraging in this general area). The NR male is now foraging along the reeds of the west bank of the river in the northern half of the SR pair's territory and is a fair distance away from Remy. He is soon joined by his mate.

1134 – Remy surfaces from a foraging dive in the mid-river area east of the SC with a small to medium-sized fish and carries it to the nest (Sinbad is still settled on the nest with the chicks). When she reaches the area ~ 1 m north of the nest chicks are seen coming off of the nest and swarming around her. Though, as usual, my view of the feeding was largely blocked it appeared that Remy probably had been able to break through the skin into the body of the fish and was detaching morsels and handing these off to the chicks. This feeding took ~ 2.5 minutes (time: 1137). When Remy concludes the feeding she lingers there briefly, then dives and isn't resighted at this time. While Remy was busy feeding the chicks Sinbad walked off the nest, moved out to the mid-SC area (calling once en route) and began foraging.

1142 – While foraging at the east end of the SC, Sinbad surfaces from a dive with a small prey item (I'm unable to positively identify it but it appears to be a small fish). He carries it to the nest and conducts a quick feeding (probably just handing it to one of the chicks). He then moves out to the mid-SC area (calling once en route) and resumes foraging. Remy isn't seen at this time. One of the NR pair is sighted near the west bank of the river in the north end of the SR pair's territory.

1145 – Some of the chicks are visible on the nest by themselves. Neither of the parents is sighted at this time (both are probably foraging).

1151 – Sinbad is resighted—he is foraging in the SW corner of the SC, not very far from the nest. Remy still hasn't been resighted.

1155 – Remy is resighted—she is foraging near the east bank of the river east of the SC. Neither of the NR pair is sighted at this time—they are most likely at the north end of the SR pair's territory or to the north inside their own territory. Sinbad is foraging in the SW section of the SC. Some of the chicks are visible on the nest and probably all of them are on it. End observation period.

*Comments:* The ongoing territorial conflict heated up today and another fierce fight was observed, this time between Remy and the NR male. Though I only saw the tail end of the fight it seems most likely that it was started by the NR male attacking Remy while she was foraging in the mid to east areas of the river, east of the SC. The NR male was easily the winner of the fight and Remy looked to be no match for him. Nevertheless, she is quite spunky and appeared undeterred: despite the thrashing she received she returned to that area just ten minutes later and resumed foraging there (the NR male was some distance away in the northern half of the SR pair's territory at the time). She appears quite determined in her task to feed her new brood.

Sinbad was settled on the nest with the chicks during the time that the fight occurred. He gave no noticeable indications that he was aware of the fight (it took place a considerable distance away from the nest). However, even if he had seen it taking place it seems questionable as to whether he would have joined the fray and left the chicks so unprotected and vulnerable with both he and Remy so far away from the nest with an intruder present.

Though the SR pair (mainly Remy) did spend a bit more time foraging out in the river east of the SC, they still mostly restricted their movements and activities to within the SC (as they have done since the territorial conflict began on 7/23—I haven't seen them venture into the northern half of their territory since then).

The indications are that this conflict is only going to intensify. With all of the eggs having now hatched the SR pair will continually have to step up their foraging for the chicks (in fact, they already noticeably did so today). In addition, the chicks will soon be mobile. There is no getting away from the reality that the SR pair is going to have to expand out of the SC to other parts of their territory in order to meet the needs of their growing chicks. Thus, at some point soon this territorial conflict is going to have to come to a decisive resolution.

There are a number of possible ways that this situation can end. The most obvious are as follows: the SR pair could drive the intruders out and retain their full territory; the NR pair could oust the SR pair and usurp their territory (and possibly kill some or all of the chicks?); the NR pair could take over just part of or all of the northern half of the SR pair's territory, thus adding a range expansion from their current territory which is to the immediate north of the SR pair's (added to their current territory this would probably comprise a sufficiently resourced area in which to rear a brood—their current territory doesn't seem sufficient by itself). This latter scenario seems to be the most likely of the possible outcomes.

If the NR pair does usurp most or all of the northern half of the SR pair's territory Remy and Sinbad will still have a possible option for obtaining sufficient habitat in which to successfully rear their chicks (the southern half of their territory by itself doesn't contain a sufficient amount of habitat). As previously noted, the West Lagoon is currently unclaimed by any PB grebe pair. If

Sinbad and Remy then expanded their territory by annexing the West Lagoon to the southern half of their current territory (which, of course, includes the SC) this would likely be sufficient for rearing their brood. It remains to be seen how this situation will play out. It currently is heading in the direction of this latter scenario.

There is still enough time remaining in this breeding season to produce and rear a clutch and it seems that the NR pair is bound and determined to claim sufficient territory in which to do so. On several occasions I have observed both Remy and Sinbad easily chase an intruder out of their territory. However, the situation with those intruders was quite different from that of the NR pair. They were just looking to do a bit of foraging, something that isn't worth the risk of engaging in a fight, especially a sustained one. On the other hand, attempting to obtain a breeding territory (as the NR pair is attempting to do) is worthy of significant risk and makes for a much more determined foe.

It was obvious today that hatching has concluded: once again, no eggs were seen. In fact, after observing the behavior of the pair at the nest today I feel certain that hatching had already concluded before I began my observations yesterday. It thus seems that the last egg probably hatched sometime during the period after my 7/26 observation and before my 7/27 observation. Hatching therefore was staggered over a period of from 5 to 7 days (recall that the first egg hatched on either 7/21 or 7/22, though it appears virtually certain that it occurred on 7/22).

Overall, incubation of the eggs had been equitably shared between Sinbad and Remy, with Sinbad doing a bit more than his mate. Considering 7/26 as the last day of incubation, the pair logged a combined total of 1,560 minutes (26 hours) incubating the eggs during the formal observation periods that were conducted over the course of the incubation period (observation periods generally were from 1 hr. to 1hr. 15 mins. in length). Individually, Sinbad incubated for 826 minutes (53% of the total minutes) and Remy did so for 734 minutes (47% of the total minutes). Sinbad thus spent about 6 % more time incubating the eggs than did Remy. Because this is the only PB grebe incubation period that I have ever observed I don't know how representative this is for the species as a whole.

Both of the pair also shared in the nest M/E work. However, though I didn't specifically keep track of the amount of time that each of the pair engaged in nest M/E during the incubation period it was nevertheless extremely obvious that Sinbad did notably more of this than did Remy.

The pair remains very protective of the newly hatched chicks. Except for a few brief periods when both parents were away foraging, they kept to a strategy of keeping one parent at the nest guarding and looking after the chicks while the other foraged (for itself as well as for the chicks). The chick-sitting parent generally spent its shift lying on the nest with the chicks on its back and/or on the nest around it. As with incubation, the pair regularly exchanged chick-sitting duty. I still haven't been afforded a definitive view of the brood but at least 5 chicks were definitely seen. With all of the chicks now hatched and hungry, feedings were noticeably stepped up today.

**7/29/17:** Time: 1038-1145; Re-check of the pair: 1159-1215.

1038 – As observations begin, there is a major disturbance in the area. A helicopter has been very loudly flying over and around the SC as it attends to a small grass fire far to the west of the SC. Sinbad is settled on the nest with the chicks. Remy isn't seen at this time.

1045 – Remy shows up ~ 1 m north of the nest and several of the chicks are seen going to her. It appears that she probably administers a feeding to the chicks but my view is mostly blocked by the reeds so I can't be certain of this. Sinbad is still settled on the nest. After the probable feeding, Remy dives and isn't resighted at this time. The helicopter has moved out of the area.

1047 – Remy is resighted as she moves from the nest area into the mid-SC where she begins foraging. Remy still appears to be a bit spooked as a carryover from when the helicopter was in the area: she isn't fully resurfacing from her dives and is employing sinking dives.

1049 – The helicopter makes another pass through the area. Remy is sighted resurfacing briefly from a dive near the SE point of the SC. She voices a short, single, one note call (probably an alarm call) and then dives again.

1050 – Remy surfaces from a dive in the mid-river area east of the SE section of the SC and emits another short, single, one note call. She parks there and preens herself for ~ 45 seconds, then moves farther to the east where she passes out of my view. Sinbad is still settled on the nest with the chicks.

1055 – Remy is now parked and preening in the mid-river area east of the SC. Sinbad is still settled on the nest with the chicks.

1057 – Remy ceases preening and moves to the SE section of their territory where she soon passes out of my view.

1106 – Remy is resighted—she is at the nest (she may have conducted a feeding there before I sighted her). Sinbad stands, walks off the nest, moves out to the east end of the SC, dives and isn't resighted at this time. Remy remains by the nest with the chicks.

1108 – Remy climbs onto the nest, lies down and settles onto it with the chicks. Neither of the NR pair has been sighted yet today. When the helicopter was operating in this area it was passing directly over the area where the pair typically ranges. It is probable that this has affected their ranging and behavior up to this point in the observation period.

1112 – Sinbad is resighted—he is entering the east end of the SC from the east with a medium-sized fish. He makes his way to the nest, stopping several times to work the fish. Sinbad arrives at the usual chick feeding location ~ 1 m north of the nest (henceforth this will be referred to as the "north feeding station") and chicks are seen leaving the nest (and Remy, who was still sitting with them), going to Sinbad and swarming around him (time: 1115). My view is mostly blocked but it appears that Sinbad is working the fish, apparently trying to break through its outer surface into its body so that he can detach bite-sized morsels to feed to the chicks (the fish is much too large for the chicks to swallow). However, after ~ 20 seconds he decides to just swallow the fish himself. He then goes out to the mid-north area of the SC (calling once en route) and resumes foraging.

1117 – Sinbad surfaces from a foraging dive just west of the NE point of the SC with a large fish and heads toward the nest with it. When he arrives at the north feeding station some of the chicks stream off the nest toward him. However, just as they reach him Sinbad begins working the fish down his throat and soon swallows it (the fish was far too large for the chicks—he apparently had taken it to the nest with the intention of feeding it to the chicks but quickly realized this; it's somewhat surprising that he even bothered taking it to the nest in the first place).

1119 – Sinbad goes out to the mid-SC area where he parks and preens. Remy is still settled on the nest with the chicks (all of them went back onto the nest when Sinbad left the area).

1125 – Sinbad is still parked and preening in the mid-SC area. Remy is still settled on the nest with the chicks.

1128 – Sinbad ceases preening and goes to the SW section of the SC and begins foraging.

1130 – Remy stands, walks off the nest and is quickly lost from my view among the reeds.

1133 – Remy appears at the north feeding station. Chicks swarm around her and it appears that she is conducting a feeding (as usual, my view is mostly obscured).

1135 – Just as Remy appears to be finished with her probable feeding Sinbad arrives at the nest with a medium-sized crayfish. He goes to the north feeding station and begins working the crayfish. My view is again mostly blocked but it appears that he is detaching morsels from the crayfish and handing these off to the chicks. While he is busy with the feeding, Remy moves out to the mid-SC area and begins foraging.

1140 – Sinbad climbs onto the nest, lies down and settles onto it with the chicks. Remy is foraging in the SE section of the SC.

1142 – Remy is briefly sighted at the nest—it appeared that she had handed off a small prey item to one of the chicks on the nest. She then goes to the NW section of the SC and begins foraging.

1145 – Sinbad is still settled on the nest with the chicks. Remy is still foraging, now in the mid-SC area. End observation period.

*Just after the formal observation period was concluded both members of the NR pair were seen together in the western half of the river in the north end of the SR pair's territory. They were working their way south. As I was leaving the area I noted that the NR male had moved south to the area east of the NE point of the SC. I also noted that Sinbad and Remy were exchanging chick-sitting duties: Sinbad left the nest and Remy settled onto it.

*With the NR male having been just east of the NE point of the SC and Sinbad having just been relieved of chick-sitting duty I realized that a situation had arisen that could potentially result in another physical confrontation between the two males. I thus returned soon after at 1159 to check on the situation and discovered that this had indeed occurred. When I arrived I found Sinbad and the NR male locked in a fierce fight just east of the NE point of the SC. The NR female was nearby but didn't join in. Remy was settled on the nest with the chicks. The two battling males broke apart soon after I had first sighted them and from this point on the confrontation became mostly a territorial standoff between the two, interspersed with brief fighting. For the most part, the standoff consisted of what are probably stereotypical behaviors, displays and postures for this species when engaged in territorial conflict. (Note that I'm not familiar with this species so much of what I observe is new to me. My only previous notable experience with PB grebes is the specific case study that I conducted on the West Lagoon family last breeding season.) The two males remained close to each other (sometimes touching), confronting each other either face to face/breast to breast or broadside (side by side with their heads either at the same or opposite ends and intermittently rotating from one orientation to the other). They did this with their heads held in a stereotypic manner: head held up stiffly but at a normal angle, with cheeks/upper throat puffed up and intermittently executing a series of repeated, short, side to side swings of the head. While engaged in these standoffs the pair either remained stationary or moved slowly along a "boundary line." In either case each of the opponents remained on his respective territorial side: thus Sinbad stayed south of the NR male and the NR male stayed north of Sinbad (i.e., unless the two were entangled while fighting). Periodically, one of them (usually the NR male) would strike at and/or charge at the other and a brief fight would ensue. These brief fights involved any or all of the following: pecking; biting; bumping and/or shoving with the body, often with wings flapping (the latter appeared to be mainly for maintaining balance or generating force while pushing). These brief scuffles generally lasted only from 2 to 4 seconds. The last of these scuffles was particularly noteworthy. The two faced each other breast to breast, each furiously working their feet and flapping their wings to raise themselves as tall as they could to the point where they were essentially "standing" on the surface of the water or nearly so. At the same time they were pressed together breast against breast, each pushing at the other in this fashion, attempting to stand taller than their opponent in order to gain leverage (this clash lasted only a few seconds). After each of these brief fights they would resume their standoff.

About 45 seconds after the last brief scuffle the NR male just began somewhat casually moving away from Sinbad who was still sitting steadfast at his "boundary." The NR male moved to his mate who was nearby to the north. The pair lingered briefly and then slowly began making their way north along the west bank of the river. As his opponent began moving to his mate, Sinbad watched him and then (apparently when he felt assured that his opponent wasn't going to come back at him) he left his "station" on the east side of the NE point of the SC and made his way back into the SC (calling once when he reached the NW section) and to the nest.

This confrontation was quite lengthy. It lasted approx. ten minutes from the time when I first sighted the two locked in battle (I have no idea how long it had been going on prior to this). In general, it was the NR male who was taking to the offensive. He was obviously attempting to get the better of Sinbad; to intimidate him and force him to yield and back away. However, Sinbad steadfastly held his ground and refused to yield after a prolonged battle.

*Comments:* From a standpoint of the location where this confrontation occurred it was basically a stalemate. However, within the larger picture of this territorial conflict was this an indication of a significant loss for Sinbad and Remy? Did this confrontation (and standoff location) signify the establishment of a new territorial boundary between the two pairs, extending across the river from just north of the NE point of the SC? Or is it that Sinbad took his stand at that point simply because this was where the two just happened to cross paths? If it does represent the establishment of a new territorial boundary this would mean that Sinbad and Remy have lost the entire northern half of their territory to the NR pair (it now seems particularly foreboding that neither of the SR pair has been seen venturing into the northern half of their territory since this territorial conflict began on 7/23). It remains for observations over the upcoming days to reveal just what the significance of this incident was.

The NR female remained nearby throughout the conflict (sometimes fairly close) but never joined her mate in the fight. In fact, she hasn't been involved in any of the observed fights between members of these two pairs.

Remy remained on the nest with the chicks during the confrontation. She gave no clear indications as to whether or not she saw or was otherwise aware that it was taking place. It was fairly distant from the nest though at least for some periods, her sightline to it wouldn't have been blocked. I wasn't able to hear any vocalizations over a substantial portion of the conflict because a super-long train was very noisily passing by. However, I did see Sinbad's beak move several times indicating that he was calling, though it would have been near impossible for Remy to hear any calls over the noise from the train. If Remy had been aware of the fight then her movements not long after indicated that she apparently wasn't too put off by it. Sinbad relieved her at the nest shortly after the confrontation and she went out and foraged in the mid-river area east of the NE point of the SC, near where the fight had occurred [the NR pair was at the north end of the SR pair's (former?) territory at the time]. At times during the conflict Remy (while sitting on the nest) was seen alertly looking in that direction making it seem likely that she was aware of it.

As for the chicks, they again remained either on the nest (and/or on the nest-sitting parent's back) or on the water close to it. I still have yet to see any of the chicks come out (not even very briefly) from within the reeds. Only 4 of the chicks were visible at any one time today. I haven't yet been afforded a view that will allow me to see just how many chicks are in the brood. All I can say at this point is that there are at least 5.

**7/30/17:** Time: 1151-1305.

1151 – As observations begin, Remy is settled on the nest with the chicks. Sinbad isn't seen at this time.

1155 – Remy stands and walks off the nest to its north side. Chicks surround her but I'm not able to see what she's doing, other than some light nest M/E. Sinbad still hasn't been sighted.

1159 – Remy climbs back onto the nest and settles on it with the chicks

1201 – Sinbad is sighted—he is foraging by the reeds at the NE point of the SC.

1205 – Sinbad is still foraging at the NE point of the SC. Remy is still settled on the nest with the chicks.

1206 – The NR female is sighted foraging in the west side of the river in the northern half of the SR pair's (former?) territory.

1210 – Remy is still settled on the nest with the chicks. Sinbad isn't seen at this time. The NR female is still foraging in the northern half of the SR pair's territory near the west bank of the river.

1212 – Sinbad is resighted—he is foraging in the NW section of the SC along the north bank.

1214 – Sinbad moves to the mid-area along the east end of the SC where he parks and preens. Remy is still on the nest with the chicks. The NR female is still in the northern half of the SR pair's territory. Sinbad flaps his wings briefly and this reveals that his wings are still raggedy with a number of missing and growing flight feathers on each wing.

1217 – Sinbad (still at the east end of the SC) begins moving toward the nest. Remy sees him approaching from a distance, stands and walks off the nest and heads directly for him. They meet and greet on the move in the mid-SC area and continue on in opposite directions. Remy goes to the SE section of the SC and begins foraging. Sinbad continues on toward the nest.

1218 – Sinbad arrives at the north side of the nest. He briefly preens and then engages in some light nest M/E and is soon lost from view within the reeds (he is perhaps foraging in the vicinity of the nest).

1220 – Neither of the parents is seen at this time. Both are probably foraging (Sinbad is probably doing so in the general vicinity of the nest).The chicks are thus by themselves at the nest for the moment.

1229 – Sinbad appears by the nest with a medium-sized fish. He goes to the north feeding station and is surrounded by chicks (at least 3-4 are seen). My view is mostly blocked but it appears that he works the fish for ~ 30 seconds, perhaps trying to break through its skin so that he can tear morsels from inside it for the chicks (the fish is much too large for the chicks to swallow). Regardless, he ends up swallowing the fish himself. He then resumes foraging in the general area around the nest. The NR female is no longer seen. [Note on the north feeding station: The pair most likely predominantly uses this particular location because it's the only spot close to the nest within the reeds that has enough space to maneuver and conduct feedings.]

1231 – While foraging in the SW corner of the SC Sinbad calls once and then resumes foraging. Remy still hasn't been resighted.

1235 – Sinbad appears briefly at the north side of the nest but I'm unable to see whether or not he delivers a feeding. He then quickly goes out of view amid the reeds.

1236 – The NR male is sighted foraging near the reeds along the west bank of the river ~ 35 m north of the NE point of the SC. He is working his way north.

1238 – Sinbad moves east past the SE point of the SC to the mid-river area and passes from my view. Remy is now in the general area of the nest (but not at the nest).

1242 – Sinbad captures a small fish in the SE section of their territory, takes it to the nest area and hands it off to one of the chicks. He then goes out to the northern section along the east end of the SC where he parks and preens. Remy is sighted at the nest and appears to be doing some light nest M/E. The NR male is at the north end of the SR pair's territory.

1245 – Remy climbs onto the nest and settles on it with the chicks. Sinbad is parked and preening in the NE section of the SC.

1247 – The NR male is no longer seen (he was last seen at the north end of the SR pair's territory).

1251 – Sinbad (now parked in the mid-north section of the SC) calls once, then goes near the NE point of the SC and begins foraging, working his way to the SE section of their territory. Remy is still settled on the nest with the chicks.

1255 – Sinbad is still foraging in the SE section of their territory. Remy is still settled on the nest with the chicks.

1300 – Sinbad is still foraging in the SE section of their territory (east of the SE point of the SC). Remy is still settled on the nest with the chicks (at least one of them is visible on her back). Neither of the NR pair is seen at this time.

1303 – Sinbad again catches a small fish in the SE section of their territory. He takes it to the nest, hands it off to a chick and then resumes foraging. Remy left the nest when she saw Sinbad approaching and began foraging.in the mid-SC area, working her way SE.

1305 – Remy heads for the nest from the SE section of their territory with a small fish that she captured there. She takes it to the nest and feeds it to a chick. Sinbad is at the nest when she arrives. Both then leave the nest area and begin foraging in the mid-SC area, working their separate ways east. The chicks are thus by themselves at the nest while both parents are a fair distance away. This is the first such instance that has occurred during my observations of the brood. End observation period.

*Comments:* Today the pair appear to have begun the process of "loosening the apron strings" regarding the chicks. During most of the observation period they continued with their pattern of greater protection for the chicks—keeping one parent at the nest to guard and look after the chicks while the other is foraging. However, in the closing segment of the observation period the chicks were left by themselves at the nest while both parents simultaneously foraged a fair distance away. This is the first time that I have seen them do this since the time when the first chick hatched. This is probably largely a reflection of the fact that the chicks are growing and the demands on the parents to feed them are increasing.

The general situation regarding the chicks was basically the same as it has been since hatching began. They were a bit more active and appeared to spend a little more time off the nest (which they need to do during feedings) but they still remained on or close to the nest and stayed within the reeds. In addition, they continued to stay on the nest with the parent (or on the parent's back) when the parent was settled on the nest.

There were no further confrontations regarding the territorial conflict during today's observations. However, there were some interesting and suggestive observations that perhaps may relate to the standoff that took place yesterday between the two males. While Remy and Sinbad conducted the majority of their movements and activities within the SC, they also spent time in the river section of their territory that lies east of the SC and they appeared to do so more readily and seemed less tense (of course, this could just be solely related to the increasing demands for feeding their chicks). I didn't see them go north of the "line" extending east across the river from just north of the NE point of the SC where the standoff occurred. In addition, each of the NR pair briefly came down to that "line" (separately) today. Are these observations an indication that a new territorial boundary was indeed established there yesterday? Or were these observations just coincidental? Obviously, further observations are required to clarify this matter.

**7/31/17:** Time: 1109-1220.

1109 – As observations begin, Sinbad is en route to the NE point of the SC where he dives and begins foraging. Remy is briefly seen at the north side of the nest before she dives and isn't resighted at this time. The NR male is near the reeds along the west bank of the river in the north end of the SR pair's (former?) territory.

1115 – Neither of the parents is seen at this time. The chicks are alone at the nest (all are either on the nest or in the water at its side).

1117 – Sinbad is resighted—he is foraging near the reeds at the NE point of the SC.

1120 – Remy is sighted briefly at the nest engaged in some light nest M/E work before she moves out to the mid-SC area where she parks and preens.

1122 – Sinbad is sighted heading for the nest from the SE section of the SC with a small fish. He goes to the north feeding station, hands the fish off to a chick and then goes to the SW corner and resumes foraging.

1123 – Remy (still parked in the mid-SC area) ceases preening, moves to the east end of the SC, dives and begins foraging.

1127 – Sinbad is seen by the north side of the nest—he appears to be doing some light nest M/E and/or foraging in the nest area.

1130 – Sinbad leaves the nest area, goes out to the mid-SC area (calling once en route) and begins foraging. Remy isn't seen at this time.

1132 – Sinbad parks ~ 30 m east of the nest and preens. The NR male is foraging along the reeds of the west bank of the river in the north end of the SR pair's territory, working his way south.

1135 – Sinbad ceases preening and goes to the nest. He then climbs onto it, lies down and settles on it with the chicks.

1136 – The NR male is now foraging along the reeds only ~ 12 m north of the NE point of the SC.

1137 – Remy goes to the nest from the SE section of their territory with a small prey item (I'm not able to identify it). She hands it off to a chick and then goes out to the north section along the east end of the SC where she parks and preens. The NR male isn't sighted at this time but Remy's location isn't far from where he was last seen (just north of the NE point of the SC).

1145 – Remy is still parked and preening in the same location. Sinbad is still on the nest with the chicks.

1150 – Remy is now parked in the mid-area along the east end of the SC. The NR male is foraging in the mid-river area just north of the SC. Remy appears to be monitoring his movements. He makes his way over near the west bank of the river just north of the NE point of the SC and soon heads north. When he reaches a point about halfway between the NE point and the north end of the SR pair's (former?) territory Remy moves to the NE point and begins foraging by the reeds there. Sinbad is still settled on the nest with the chicks.

1155 – The NR male, now just south of the north end of the SR pair's territory, calls once. Sinbad is still settled on the nest with the chicks. Remy isn't sighted at this time (she is probably foraging).

1157 – Remy is sighted at the north feeding station with some chicks around her (she probably had just given a brief feeding). Sinbad stands, walks off the nest, moves out to the mid-SC area and begins foraging. Remy goes to the NW corner of the SC and forages there.

1159 – Sinbad surfaces from a foraging dive in the mid-SC area with a small fish. He goes to the nest area, hands it off to a chick (who quickly swallows it), then moves back out to the mid-SC area (calling once en route) and resumes foraging.

1202 – Remy is sighted at the north feeding station with chicks swarming around her as she works a large crayfish. Approx. 20 seconds later she is seen swallowing the crayfish but I wasn't able to see if she had torn off any morsels to feed to the chicks before she swallowed it.

1205 – Remy climbs onto the nest and preens herself, then lies down and settles on the nest with the chicks.

1207 – Sinbad goes to the nest from the SE section of the SC with a small fish, feeds it to a chick and then goes to the mid-SC area and resumes foraging.

1210 – Sinbad resurfaces from a foraging dive at the NE point of the SC with a small fish. He goes to the north feeding station, feeds it to a chick, then goes to the NE section of the SC and resumes foraging. Remy is still settled on the nest with the chicks.

1212 – Sinbad surfaces in the NW section of the SC with a medium-sized fish and goes to the north feeding station with it. Chicks excitedly swarm around him as he works the fish. At this point my view is mostly blocked but I'm able to see some of what transpires. On two separate occasions a chick is seen trying to swallow the fish (I don't know whether or not it was the same chick in both instances). In each of these cases the chick is unsuccessful and spits the fish out into the water where it is retrieved by Sinbad (the fish was obviously a little too large for the chick(s) to swallow). The action then moves behind an especially dense cluster of reeds and my view is completely obstructed. After ~ 40 seconds of working this feeding attempt Sinbad emerges from the nest area without the fish (time: 1214), goes to the north section of the SC and resumes foraging (because my view was blocked I don't know whether one of the chicks was finally able to swallow the fish or if Sinbad ended up doing so—I suspect the latter because the fish appeared to be a little too large for the chicks to swallow). Remy is still settled on the nest.

1220 – Remy is still settled on the nest with the chicks. Sinbad is foraging in the north section of the SC. The NR male is parked and preening just off the west bank of the river at the north end of the SR pair's (former?) territory. End observation period.

*Comments:* This was a fairly routine observation period with the pair quite busy feeding and watching over their chicks. The pair continued to restrict their movements and activities mainly within the SC, though (as yesterday) they did spend some time foraging in the river area east of the SC. There were again some observations suggesting the possibility that a new territorial boundary between the SR and NR pairs has been established on a line across the river at or just north of the NE point of the SC. The NR male ventured down to this area and foraged in a line from the west bank across to mid-river along this probable boundary line and didn't go south of it. Interestingly, while he was doing so Remy was parked in the mid-area of the east end of the SC. Though it seems a virtual certainty that the NR male saw her, he didn't cross the "boundary" to go after her. For her part, Remy appeared to be monitoring his movements and not until he had eventually moved well north of the NE point of the SC did she go over and forage by the reeds there. This matter requires further observations for unequivocal confirmation but it appears highly likely that this is the new territorial boundary that has been established and that Sinbad and Remy have lost the entire north half of their territory to the NR pair.

Five chicks were definitely seen today and it appears that this may be the full complement of this brood. The chicks again remained on or near the nest but were noticeably more active and spent more time off the nest. A few were even seen diving briefly. Also, for the first time some of the chicks were observed going to the inside edge of the reeds by the nest (they often looked east out into open water while they were there, probably watching for incoming feedings). Though the chicks are still quite small they have noticeably grown some since hatching.

The parents continue to take turns sitting on the nest guarding and babysitting the chicks. However, they appeared to spend less time doing so today. In addition, the chicks spent more time at the nest by themselves while the parents were simultaneously out foraging for them.

**8/1/17:** Time: 1045-1210; 1215-1245.

1045 – As observations begin, a light drizzle is falling. Sinbad is settled on the nest with the chicks. Remy isn't seen at this time.

1050 – The light drizzle continues. Sinbad is still settled on the nest with the chicks. Remy still hasn't been sighted. The NR male is sighted by the reeds along the west bank of the river at the north end of the SR pair's (former?) territory.

1052 – Sinbad stands, walks off the nest and moves to the mid-SC area where he begins foraging, working his way out toward the SE point. Remy still hasn't been sighted. The drizzle has stopped. The NR male isn't seen at this time.

1056 – Remy is finally sighted—she is foraging in the mid-SC area. Sinbad is foraging near the SE point of the SC.

1058 – Remy parks and preens in the mid-SC area. Sinbad goes to the nest area from the SE point of the SC with a small fish. He hands it off to a chick and then goes back out near the SE point and resumes foraging.

1101 – Remy (still in the mid-SC area) ceases preening and goes to the nest. She then climbs onto the nest, lies down and settles on it with the chicks (one chick remains on the water at the north side of the nest).

1105 – Remy is still settled on the nest with the chicks (all of the chicks are now on the nest with her). Sinbad isn't seen at this time (he is probably foraging in the SE section of their territory).

1108 – Sinbad goes to the nest from the SE section of their territory with a large crayfish. He goes to the north feeding station and the chicks swarm around him. He then works the crayfish for ~ 107 seconds, tearing morsels from the crayfish and handing these off to the chicks. When he concludes the feeding he goes out to the NE point of the SC and resumes foraging. After the feeding the chicks go back onto the nest with Remy (who had remained settled on it).

1115 – Remy is still settled on the nest with the chicks. Sinbad is still foraging at the NE point of the SC.

1116 – Sinbad surfaces from a foraging dive just west of the NE point with a very large fish. He works the fish for ~ 62 seconds (the fish flaps strongly three separate times and appears to almost escape but Sinbad dives briefly and secures a better grip and/or recaptures the fish) before finally (and with much effort) managing to work the fish down his throat (this large fish appeared to be at or near the upper size limit of what he can swallow). Sinbad then parks and preens in the mid-north section of the SC.

1120 – Sinbad is still parked and preening in the same location. Remy is still settled on the nest with the chicks. The NR male is sighted foraging in the mid-river area at the north end of the SR pair's (former?) territory.

1130 – Sinbad has moved to and parked in the mid-SC area and is vigilantly gazing from NE to SE (as though on "sentry duty"). Remy is still settled on the nest with the chicks.

1132 – Sinbad (still parked in the mid-SC) calls once and remains parked there, looking about. Remy is still settled on the nest with the chicks. The NR male is near the west bank of the river at the north end of the SR pair's territory and moves a little farther south.

1135 – Sinbad is still parked in the mid-SC area looking vigilantly from NE to SE (it's not clear as to what has been causing him to look about so intently). Remy is still settled on the nest with the chicks. The NR male has moved a bit farther south in the northern half of the SR pair's territory.

1137 – Sinbad moves out to the east end of the SC, north of the SE point and alertly looks about for ~ 40 seconds, then dives and begins foraging. Remy is still settled on the nest with the chicks.

1140 – Remy stands, walks off the nest and goes to the mid-north area of the SC where she parks and preens.

1142 – Remy (still in the same location) dives and begins foraging. Sinbad isn't seen at this time nor is the NR male (he was foraging in the mid-river area in the northern half of the SR pair's territory when last sighted at 1139).

1144 – Sinbad is sighted moving west in the SE section of the SC. While moving along he voices ~ 6 short, sharp, single one note calls and then dives (~ 5 of the short calls were emitted successively and one shortly after these). He resurfaces soon after with a small fish which he takes to the nest area and hands off to one of the chicks. He then moves out to the mid-SC area (calling once en route) and resumes foraging. The NR male is foraging mid-river in the northern half of the SR pair's territory.

1145 – Remy is sighted at the north feeding station with chicks swarming around her. My view is mostly obstructed but it appears that she conducts a feeding. She then dives and isn't resighted at this time.

1147 – Sinbad surfaces from a dive in the mid-SC area with a small fish. He goes to the nest, hands it off to a chick and then goes back out to the mid-SC area and resumes foraging.

1150 – Neither parent is sighted at this time. The chicks are at the nest by themselves. The NR male has moved to the north end of the SR pair's territory.

1155 – Remy is sighted at the north feeding station amid swarming chicks as she conducts a feeding. My view is mostly obscured and I'm not able to identify the prey item or see the details of the feeding. When the feeding is concluded Remy goes to the mid-SC area and begins foraging.

1200 – Sinbad surfaces from a foraging dive just west of the NE point of the SC with a very large crayfish and works it briefly as he makes his way to the nest. He goes to the north feeding station where chicks swarm around him. He works the crayfish for ~ 49 seconds before swallowing it himself (once again, my view was mostly obstructed so I couldn't see whether or not he had torn off any morsels for the chicks before he swallowed it). Sinbad then engages in some light nest M/E. After doing so he goes out ~ 8 m east of the nest, calls once, lingers there briefly and returns to the nest (time: 1205). He then climbs onto and settles on the nest. Remy forages in the SW and mid areas of the SC, then moves out to the mid-area along the east end of the SC. While there, she moves slowly along the east end of the SC, looking about vigilantly and voicing a series of separate short, sharp, single, one note calls (time: 1210). Her manner is as though she is checking to see if all is safe and clear in the general area. Perhaps in response to her calls Sinbad has come off the nest and is sitting at its north side with the chicks, alertly looking to the east toward Remy's direction. End observation period.

*1215-1245: Something curious seemed to be imminent at the conclusion of the observation period so I checked the family again about 5 minutes later. I discovered that for the first time during my observations of the hatched chicks they had moved out of and beyond the dense reed bed in which the nest is located! Five chicks were seen sitting around Sinbad in the floating algal mat in front of (i.e., clear of) the edge of the reeds, east of the nest. This was my first definitive view of the brood and (as suspected yesterday) I can now confirm that the brood consists of 5 chicks. The chicks remained outside the reeds for about a half hour. They generally stayed within the algal mats but did sometimes venture just beyond their outer edges (mainly during feedings). [These algal mats extend a short distance out from the edge of the reeds forming an essentially continuous carpet on the surface along the entire west end of the SC.] The chicks ranged as far as ~ 20-25 m north of the nest in the algal mats. They were sometimes there by themselves while their parents foraged. At least 5 feedings were conducted out here, the first ones that I have observed taking place away from the immediate area around the nest (and out of the reeds). Some food solicitations by the chicks to the parents were also observed (these solicitations were the same as those described for the young West Lagoon chicks when I observed them last year). The chicks had some trouble moving through the algal mats, sometimes appearing to be half walking as they moved along. Several of the chicks were seen diving briefly (including a sinking dive). At 1245 Sinbad had the chicks gathered around him and he slowly made his way back to the nest with the chicks following behind him. Upon reaching the nest Sinbad climbed onto it, settled himself on it and was joined there by the chicks.

*Comments:* Some notable size disparity was seen among the chicks but not nearly to the extent of that seen among the West Lagoon brood last year and at this point none of the chicks look runty.

In light of the fact that the chicks came out from the reeds right after Remy's vigilant looking about and short calls while at the east end of the SC, it thus seems quite possible that she had indeed been checking to make sure that "the coast was clear" for the chicks to come out (and probably in particular, ensuring that the NR pair wasn't in the area). It is also then plausible that Sinbad had responded to her calls and taken these as an "all clear" signal to lead the chicks out. Of course, it's possible that this could all have just been coincidental but it didn't seem so.

This marks an important milestone for the chicks. It signifies that they will no longer be strictly tied to the nest and its immediate area and will very soon be accompanying the parents around as they forage for them.

Observations were again consistent with the probability that a new territorial boundary has been established between the SR and NR pairs just north of the NE point of the SC. The situation is still unclear. However, with the chicks now about to become mobile things are bound to come to a head and reveal just where the situation stands (including whether it is still in flux or has already been resolved as appears to be the case).

The period ahead will be nerve-wracking (for me as well as for Sinbad and Remy) because it is particularly dangerous for these young, fragile chicks as they begin to move around in their world. The inexperienced chicks are naïve about their environment and have much to learn (and their increasing mobility means they will be spending less time under the protective close presence of the parents). Also, at this early stage of their lives their physical skills are as yet poorly developed. For example, they are not as proficient at swimming (i.e., paddling with their feet while on the surface) and their swimming speed is much slower, their diving skills are at an early stage of development as is their endurance, and of course, they aren't yet able to fly. All of these things make it that much more difficult to avoid or escape predators and other hazards. In addition, their small size and undeveloped bodies make them physically more vulnerable. They are far less capable of physically defending themselves, they can more easily be injured or killed (including by adult grebes), and they are susceptible to a greater variety of predators (this of course includes predators that aren't capable of preying on adult PB grebes but can handle small chicks). On top of all of this is the turbulent territorial matter and the possible danger from the NR pair lurking in the area. In regard to the latter, I don't know if adult PB grebes have been known to kill the chicks of neighboring pairs. However, what is obvious is that an adult would easily be able to kill a young chick with one firm bite to the appropriate anatomy (as evidenced by the way adults are able to crack through the exoskeletons of crayfish).

**8/2/17:** Time: 1045-1200; Brief re-check at ~ 1210.

1045 – As observations begin, neither of the parents is seen at this time. The chicks are at the nest—some of them are on the nest and some are in the water by it.

1050 – Sinbad is sighted foraging in the NW corner of the SC. Remy still hasn't been sighted. The chicks are at the nest.

1055 – Sinbad is now foraging along the north bank of the SC, working his way east. Remy still hasn't been seen. The chicks are at the nest—some are on it and some at its side.

1056 – The NR male is parked and preening in the western half of the river at the north end of the SR pair's (former?) territory.

1101 – Sinbad is foraging just south of the NE point of the SC. Remy still hasn't been sighted. The NR male is foraging along the reeds of the west bank of the river in the northern half of the SR pair's territory.

1103 – Sinbad and Remy are both sighted at the north feeding station with the chicks gathered around them. My view is mostly obscured but it appears that one or both of them is conducting a feeding. Sinbad spends ~ 40 seconds there with the chicks, then moves out to the mid-SC area and begins foraging, soon working his way to the NE point of the SC (time: 1105). Remy moves out ~ 20 m east of the nest where she parks and preens. The NR male isn't seen at this time.

1110 – Remy is still parked and preening in the same general location. Sinbad isn't seen at this time. The chicks are at the nest.

1115 – Remy ceases preening and goes to the nest. She then climbs onto the nest, lies down and settles on it (some of the chicks are on the nest with her and some are in the water at its side). Sinbad still hasn't been resighted.

1117 – All of the chicks are now on the nest with Remy.

1120 – Sinbad is resighted as he moves to the mid-area at the east end of the SC from the east. He parks there and preens himself. Remy is still on the nest with the chicks.

1124 – Sinbad moves to the mid-SC area where he parks and intermittently preens. He also alertly looks about and soon voices 4 consecutive single, short, sharp, one note calls and a fifth one a few seconds later. Apparently in response to these Remy walks off the nest and the chicks move with her to the edge of the reeds east of the nest. Sinbad then begins moving toward the nest and Remy begins moving toward him (the chicks remain at the edge of the reeds). They meet and greet on the go and proceed in opposite directions. Remy goes out to the east end of the SC, parks there and vigilantly looks about, intermittently voicing 4-5 single, short, sharp, one note calls (2-3 of these were given successively). In the meantime, Sinbad has moved to ~ 10 m east of the nest and has parked there. The chicks move into the algal mat just beyond the reeds, east of the nest. Sinbad moves to a point just beyond the edge of the algal mat, east of the nest (and adjacent to the chicks). The chicks soon move near (and within) the edge of the algal mat near him. Sinbad eventually goes into the algal mat with the chicks. All 5 of the chicks are seen and at some point all of them are observed engaging in brief dives (time: 1130). Sinbad soon parks within the algal mat at a point just beyond the reeds ~ 7 m north of the nest and sits there with the chicks. Remy (still at the east end of the SC) begins foraging and is lost from my view.

1135 – Remy is resighted—she is by Sinbad and the chicks (I didn't see if she had conducted a feeding). She then goes out to the mid-SC area and resumes foraging. Sinbad moves out ~ 6-8 m east of the algal mat into open water with the chicks following along and he parks there with the chicks. The chicks initially remain very close around him (this is easily the farthest from the nest into open water that I have seen them venture).

1137 – Remy moves to the SE section of the SC and forages there. Sinbad is still parked in the same open location with the chicks.

1140 – Sinbad (now preening) is still parked in the same open location with the chicks. Remy isn't sighted at this time.

1142 – Sinbad moves near the edge of the algal mat with the chicks, then dives and begins foraging. The chicks then move well into the algal mat ~ 18 m north of the nest near the edge of the reeds.

1146 – The chicks move just beyond the edge of the algal mat in the NW corner of the SC. Neither parent is seen at this time (both are probably foraging).

1148 – Sinbad surfaces from a foraging dive (unsuccessful) just south of the chicks. Some of the chicks rush to him looking for a feeding but he has nothing to offer them. He then dives and resumes foraging.

1149 – Sinbad surfaces from an unsuccessful foraging dive, calls once and then resumes foraging.

1150 – Sinbad surfaces from a foraging dive in the NW corner with a small prey item (I'm unable to confirm what it is though it's probably a small fish). He hands it off to a chick and resumes foraging. The chicks are still just off the edge of the algal mat in the NW corner.

1152 – Remy is sighted moving toward the nest area from the east end of the SC with a small fish. As she nears the general area she visually searches for the chicks (not having known that they had moved north into the NW corner). She soon spots the chicks and veers toward them. Four of the chicks see her approaching and rush toward her. The largest of the chicks wins the race and takes the fish. Remy lingers with the chicks for ~ 10 seconds, then dives and resumes foraging.

1155-1200: Remy goes to the chicks in the NW corner with a large crayfish that she had captured in the mid-north section of the SC. The chicks excitedly swarm around her as she begins working the crayfish. She often bites on and shakes the crayfish, creating numerous breaks in its exoskeleton from which she tears out morsels that she hands to the chicks. She frequently lays the crayfish on the water while tearing out these morsels and a number of times has to dive completely under the surface to retrieve the crayfish after it sinks (it appears that some chicks may also have torn some morsels from the carcass as she worked it). As she does this she is progressively reducing the bulk of the carcass so that one of the chicks will eventually be able to assume control of and swallow it. On several occasions a chick takes the carcass and attempts to swallow it but is unsuccessful because it's still too large. After Remy had worked the crayfish for ~ 2 minutes 44 seconds its bulk had finally been reduced to the point where one chick was able to take control of and swallow the remaining bulk of the carcass. Just before this feeding had concluded, Sinbad came to this intensely competing group of chicks with a small fish which was quickly taken by one of the chicks. Remy and Sinbad then resumed foraging. (These feedings took place in open water ~ 3-4 m beyond the algal mat in the NW corner.) The NR male is sighted at this time foraging in the western half of the river in the northern half of the SR pair's (former?) territory. End observation period.

*~1210 – Brief re-check. Remy is sighted ~ 4 m beyond the edge of the algal mat NE of the nest with the chicks. One of the chicks is on her back tucked under her left wing (with its head sticking out from the anterior edge of her wing) and the other 4 are on the water around her. She soon leads the chicks to the nest but doesn't stay with them. Instead, she heads out to do more foraging while the chicks remain at the nest.

*Comments:* Remy and Sinbad were quite busy with the task of feeding their chicks during today's observation period. They also continued with the process of "loosening the apron strings" regarding the chicks, while the chicks took another small step toward becoming mobile.

Remy and Sinbad combined spent only about 9 minutes sitting on the nest with the chicks during the observation period. For the majority of the time the chicks were by themselves while the parents were out foraging. The chicks remained at the nest (on it and in the water near it) for the first half of the observation period. After the parents led them away from the nest the chicks spent the second half of the period outside of the reeds, mainly in the algal mat in the area from the nest north to the NW corner (they also spent time in open water beyond the edge of the algal mat). When Sinbad led the chicks out ~ 8 m from the edge of the algal mat this was easily the farthest out in open water that I have seen the chicks venture. All 5 of the chicks were seen and at some point all of them were seen engaging in brief dives.

Remy and Sinbad did the same general thing today as they did yesterday before leading the chicks out from the nest beyond the reeds. They again had one parent with the chicks at the nest while the other was out vigilantly looking about and then voiced a series of single short, sharp, one note calls—the parent at the nest then apparently responding to the calls by leading the chicks away from the nest to the edge of and beyond the reeds. It thus appears quite probable that, in these introductory forays out from the reeds, they are indeed checking to see that all is clear and then giving the "all clear" signal before leading the chicks out from the relative security and concealment of the nest area.

Today's observations were again consistent with the probability that a new territorial boundary has been established between the SR and NR pairs on a line across the river just north of the NE point of the SC.

**8/3/17:** Time: 1031-1150.

1031 – As observations begin, Sinbad is parked and preening in the mid-north section of the SC. Remy is settled on the nest babysitting the chicks (some of the chicks are on the nest with her and some are in the water at the north side of the nest).

1035 – Sinbad is now parked in the mid-SC area. Remy is still on the nest watching over the chicks (some are still in the water near the nest).

1037 – The NR male is sighted foraging in the western half of the river in the northern half of the SR pair's (former?) territory.

1039 – Sinbad has moved to a position ~ 18 m east of the nest and is vigilantly looking about. Remy is still on the nest watching over the chicks. The NR male isn't seen at this time.

1045 – Sinbad is still parked in the area ~ 18 m east of the nest. Remy is still on the nest watching over the chicks.

1046 – Sinbad moves a few meters closer to the nest, parks there and continues to vigilantly look about. Soon after, Remy stands, walks off the nest and heads out toward Sinbad who then starts moving toward her. They meet and greet ~ 9 m from the nest and proceed in opposite directions. Remy makes her way to the NE point of the SC where she first spends some time vigilantly looking about and then begins foraging (time: 1048). In the meantime, Sinbad moves to a point ~2 m out from the edge of the algal mat, east of the nest, calling once as he does so. Soon after, 3 of the chicks emerge from the reeds and join him there. Shortly thereafter, a fourth chick joins them and then finally, the fifth (the smallest of the brood). Sinbad sits there only a short time with the chicks around him before he dives and begins foraging in the area east of the nest and then in the SW corner of the SC (as a reminder here, recall that the nest is located in the reeds at the mid-west end of the SC and thus lies between the NW and SW corners).

1051 – Sinbad surfaces from a dive in the SW corner with a small fish. He moves to the chicks, hands it off to one of them and resumes foraging. Remy isn't seen at this time.

1052 – Sinbad surfaces from a foraging dive in the SW corner with another fish. This time it's a medium-sized fish that looks as though it is right at or perhaps a bit beyond the size that any of the chicks will be able to swallow. He goes to the chicks (who are at and just beyond the edge of the algal mat east of the nest) and they immediately swarm around him. Sinbad vigorously works the fish and it appears that he quickly incapacitates it. He seems bound and determined to get this fish fed to one of the chicks. As he works it he often holds it out for the chicks (apparently offering it up for a chick to take from him). The chicks sometimes peck at it while he holds it up for them or while he lays it on the water in front of them. On several occasions a chick takes the fish from him and makes a brief, pathetic attempt to swallow it before dropping it onto the water. Sinbad isn't deterred and continues to thrash and work the fish. He is incredibly persistent and keeps at this for nearly 5 minutes until the largest chick finally swallows the fish. This outcome was somewhat surprising as usually it would be expected that he would have just decided to swallow the fish himself long before the time that this feeding finally ended.

1102 – While foraging in the NW corner Sinbad calls once, then moves east to the mid-north section of the SC, dives and resumes foraging. The chicks move back to the edge of the algal mat, just north of the nest. Remy isn't seen at this time.

1105 – The chicks are still in the same location. Neither parent is sighted at this time (both are probably foraging).

1111 – Remy appears at the north feeding station with a prey item (probably a fish but my view is too inadequate to be certain). The chicks go to Remy and swarm around her as she works the prey item. My view is mostly obscured so I'm not able to see the details of the feeding. She works the prey item for ~ 70 seconds and then moves out ~ 9 m east of the nest (no longer with the prey item) where she parks and preens. The chicks remain at the north feeding station.

1115 – Remy is still parked and preening in the same location (she is ~ 3 m beyond the algal mat in open water). One of the chicks (the largest) has come out to Remy and sits by her side. The other chicks are at the edge of the reeds, east of the nest. The chick that is with Remy occasionally solicits her for food.

1120 – Sinbad is sighted moving to Remy and the chick that is with her from the SE section of the SC with a small fish. As he nears the pair Remy moves away to the east. Sinbad then hands the fish off to the chick (who had no competition because it had been the only one of the brood who had ventured out to Remy). Both parents move out to forage: Sinbad goes to the SW corner and Remy goes out to the east end of the SC. The chick who had been with Remy moves back to the edge of the algal mat near the other chicks.

1123 – Sinbad is on the surface ~ 22 m east of the nest as a Cooper's hawk who is being chased and harassed by several common crows rapidly approaches the nest area from the east. He quickly catches sight of the hawk and voices one single, short, sharp, one note call (apparently an alarm call for the chicks who are at the inside edge of the reeds, east of the nest and thus are already under cover). The hawk and pursuing crows rapidly pass over and out of the area to the west. Sinbad then parks and preens ~ 11 m east of the nest (he is in open water ~ 5 m beyond the edge of the algal mat), Soon after, one of the chicks (once again, the largest) comes out and joins him there. The other chicks remain at the edge of the reeds, east of the nest. Remy isn't seen at this time.

1128 – Remy is sighted as she moves toward Sinbad and the chick from the east end of the SC with a small fish. Sinbad calls once as she nears the pair and begins moving away from the chick. Remy hands the fish off to the chick (who again had no competition for the fish because it had been the only one of the brood who came out and stayed with Sinbad), then moves out near the NE point of the SC and resumes foraging. Sinbad initially heads toward the SE section but doubles back. The chick who had been with Sinbad withdraws back to the edge of the reeds, east of the nest and joins the other chicks there. Sinbad continues moving toward the nest, dives when ~ 8 m to the east of it and resurfaces at the north feeding station (he has no prey with him). Some of the chicks move to him. He lingers there for ~ 20 seconds and then begins foraging near the nest and in the SW corner.

1135 – Remy is sighted moving toward the nest from the NE point with a medium-sized crayfish. She goes to the north feeding station where the chicks swarm around her as she works the crayfish. As usual, my view is mostly obscured so I'm not able to see the details of the feeding. Finally, after ~ 109 seconds one of the chicks is seen swallowing the remaining reduced bulk of the carcass. Sinbad had come to the nest and joined the group just before the crayfish feeding concluded (I wasn't able to note whether or not he had brought a prey item with him for the chicks). After Remy's feeding, both parents move out from the nest. Sinbad calls once as he moves away from the nest area and goes to the area off the NE point of the SC. Approx. 2 -3 seconds after Sinbad had called, the NR male (apparently in response) called from his location in the west half of the river in the northern half of the SR pair's territory. Remy moves out to the NW section where she parks and preens. The chicks remain at the north feeding station.

1142 – Remy moves to the nest, climbs onto and settles on it. All of the chicks eventually join her on the nest (with one or more going onto her back).

1145 – Remy is still settled on the nest with the chicks. Sinbad (parked south of the NE point) begins foraging. The NR male is still in the northern half of the SR pair's (former?) territory.

1050 – Sinbad is still foraging at the NE point of the SC. Remy is still settled on the nest with the chicks. The NR male isn't seen at this time. End observation period.

*Comments:* Remy and Sinbad continued guiding the chicks on their progression toward becoming fully mobile. The pair and the chicks remained in a cautious phase today. As in the previous several days, the parents led the chicks out into the algal mat and only a short distance beyond, staying within the west end of the SC in the general area of the nest. The chicks spent most of the time that they were out from the nest within the algal mat or at the edge of the reeds. They again spent much time by themselves at or near the nest while both parents were out foraging. Nevertheless, the parents still spent some time settled on the nest with the chicks (and chicks are still seen climbing onto the backs of the parents). The chicks didn't venture as far out from the algal mat as they did yesterday.

The largest chick is already noticeably ahead of its sibs. Its larger size is certainly a factor as it allows this chick to outcompete its sibs and it is also able to swallow larger food items than them (as exemplified with the feeding of that larger fish at 1052). In addition, this chick is also farther ahead behaviorally. For instance, it was the only chick today who felt secure enough to go out and sit with Remy and Sinbad when they were parked in open water a fair distance (i.e., for the chicks) beyond the algal mat. It definitely benefitted from this as it was thus able receive two uncontested feedings by virtue of being so far out and the first chick that the incoming prey-bearing parent met. In essence, this chick essentially intercepted the parent well before the parent neared any of the other chicks (i.e., its competitors). It's unfortunate that I wasn't able to observe the hatching order of the chicks over the 5-7 day period during which hatching occurred. It's probably a good bet that this largest chick was the first to hatch and thus got a head start over its brood mates (probably receiving feedings while its sibs were still in their egg).

Observations today were once again consistent with the probability that a new territorial boundary has been established between the two pairs just north of the NE point of the SC. Regardless, it has been clear for some time now that the NR pair is an established fixture here and not mere transient intruders. At this point the observations strongly suggest that the NR pair has indeed usurped the northern half of the SR pair's territory. Given this likelihood, I've been trying (unsuccessfully) to find a vantage point from which I can adequately check to see if the NR pair is currently nesting or working toward this end.

All 5 chicks of the East Lagoon second brood were seen today in the north end of the East Lagoon. Three of the now independent chicks from the first brood were also sighted: two in the south end of the East Lagoon (these two were again seen associating) and one in the West Lagoon.

**8/4/17:** Time: 1036-1145.

1036 – As observations begin, Sinbad is foraging near the SE point of the SC and Remy is parked ~ 10 m NE of the nest (in open water). The chicks are at the nest—some are on the nest and some are in the water just north of it.

1038 – Remy (still in the same location) dives and isn't resighted at this time.

1040 – Sinbad goes to the nest from the area off the SE point of the SC with a small fish. He hands it off to a chick, then moves out to the mid-SC area and resumes foraging, soon working his way into the SE section.

1045 – The chicks are at (but not on) the nest. Neither parent is seen at this time (both are probably foraging).

1047 – The NR male is seen parked and preening ~ 30 m north of the NE point of the SC off the west bank of the river.

1048 – Sinbad is sighted moving away from the south bank in the SE section of the SC. As he moves along he gives a series of ~ 5-6 single, short, sharp one note calls. The calls are apparently signaling alarm or distress but I'm not able to determine the source of this (it is possible that he caught sight of me at my observation point on the mid-south bank but I never saw him look my way). He dives soon after and resumes foraging.

1050 – Remy is sighted foraging in the SE section of the SC near the area where Sinbad is foraging. The chicks are at the nest. The NR male moves to the north end of the SR pair's (former?) territory.

1052 – Remy and Sinbad are both foraging in the SE section of the SC. While there, Remy spots Sinbad moving north and then heads toward him. When she gets to within about one meter from him she ceases her pursuit (perhaps she had just wanted to confirm that it was Sinbad?) and stops there. Sinbad continues on his way to the NE point of the SC and begins foraging there.

1054 – Remy moves to a point ~ 15 m east of the nest where she parks and preens. Eventually all 5 chicks stream out to Remy and sit close around her (time: 1055). The NR male is no longer seen.

1057 – Sinbad heads toward Remy and the chicks from the NE point of the SC with a very small fish. However, when he gets to within ~ 11 m of the chicks he decides to swallow the fish, then does an about face and heads back to the NE point where he resumes foraging. Just before he had reversed course several of the chicks had begun moving toward him. When they saw him turn around (they never got close to him) they returned to Remy.

1100 – Remy is still in the same location with the chicks. Sinbad is foraging near the NE point of the SC.

1102 – At least 3 of the chicks have been intermittently (but frequently) soliciting Remy for food (the chicks do this by moving in front of or at the side of Remy's head and extending their head and beak to Remy's beak, either touching it or nearly doing so—this is the same as described for the West lagoon chicks that I observed last year).

1105 – Remy is still in the same location with the chicks (one of the chicks has climbed onto her back and is tucked under her left wing with its head protruding from the anterior edge of her folded wing). Sinbad isn't seen at this time.

1110 – Remy has moved farther out into the SC. She is now ~ 20 m east of the nest with the chicks who continue to regularly solicit her for food. She soon dives and begins foraging. The chicks remain where she left them.

1112 – While foraging in the mid-SC area Remy sights a great-blue heron that is on the north bank of the SC just west of the NE point. She begins repeatedly voicing short sharp, single one note calls, these apparently being alarm calls given for the chicks. Remy moves a short distance toward the heron (but doesn't approach anywhere near it) and then circles back to the mid-SC. By this time the chicks have already begun moving toward the nest and quickly make their way into the reeds, east of the nest. Remy goes to them and lingers there for a while. She then goes to the SW corner of the SC with the chicks following along and begins foraging (time: 1114). Sinbad is soon sighted in the SE section of the SC. He repeatedly voices short, sharp, single one note calls (apparently alarm calls) for several seconds, then goes to the NE point, dives and isn't resighted at this time. It wasn't clear as to whether the heron or something else undetected by me was the cause for the alarm calls. (At the same time that Remy began reacting to the heron a train was noisily passing by. Thus, it is possible that Remy's actions could have been in response to the train instead. However, trains regularly pass by here periodically throughout the day and I have never seen the SR pair respond in such a strong way to them—they generally appear to be habituated to them. On the other hand, this was the first time that I have seen the chicks out in open water so far away from the nest when a train was passing by. It may be that, with the chicks now venturing out from the reeds, the parents are more sensitive to the trains, similar to how they have become more sensitive to the presence of raptors and large birds in general since hatching began.)

1115 – Remy is foraging in the SW corner of the SC. The chicks are in the algal mat in the SW corner, awaiting the next feeding. Sinbad isn't seen at this time.

1117 – Remy is sighted in the SW corner with a small to medium-sized fish that she gives to one of the chicks (my view of this is mostly obstructed). As she conducts this feeding, Sinbad is spotted moving toward Remy and the chicks. Sinbad and Remy meet and greet by the chicks. Remy then goes to the mid-SC area and begins foraging, soon working her way to the SE section of the SC. Sinbad remains with the chicks in the SW corner and soon begins foraging there.

1122 – Remy is foraging in the SE section of the SC. Sinbad is foraging in the SW corner. As he works his way a little to the east along the south bank the chicks follow, keeping within the algal mat or just beyond. Sinbad soon reverses course and works his way back into the SW corner; the chicks follow him there.

1225 – Sinbad is still foraging in the SW corner. The chicks are still in the SW corner at and just beyond the edge of the algal mat. The heron is still at the same spot on the north bank at the edge of the water (it has mostly been "ambush" hunting for fish there). Remy isn't seen at this time.

1132 – Sinbad is sighted with the chicks in the SW corner where he evidently has just delivered a medium-sized fish—one of the larger chicks is seen swallowing it. Sinbad then moves out to the mid-SC area (calling once en route) where he parks and preens. The chicks remain in the SW corner.

1134 – Remy goes to the chicks (still in the SW corner) from the SE section of the SC with a fairly large crayfish. When she nears within ~ 5 m of the chicks they rush out and swarm around her as they meet her en route. She vigorously works the crayfish for ~ 98 seconds. In working the crayfish she often holds it in her beak while biting down and shaking her head from side to side. Typically, after from just a few to 4-5 such shakes the crayfish is flung onto the water as it either slips from her grasp or is bitten through. In this manner Remy works the carcass into a state in which she can detach morsels of a size that the chicks can swallow. She tears morsels from the carcass both while it is on the surface of the water and submerged. (The carcass sank below the surface numerous times as she handed morsels to the chicks and she often had to dive completely below the surface to retrieve it. On a number of these dives it was obvious that she had detached a morsel or a larger piece of the carcass while underwater because she resurfaced with the morsel rather than the full carcass. She then dove back under and resurfaced with either the full remaining carcass or another morsel or larger piece.) During this feeding Remy tore off and fed many morsels from the crayfish and it appeared that all 5 chicks got at least one morsel from it (Remy has proven herself to be quite proficient at tearing morsels and detaching larger segments from crayfish). When the feeding is concluded Remy parks with the chicks ~ 8 m out from the algal mat (in open water) and preens herself.

1140 – Remy is still parked in the same location with the chicks. Sinbad isn't seen at this time.

1145 – Remy is still parked in the same location but only 2 of the chicks are with her now (one of them tries unsuccessfully to climb onto her back). The other 3 chicks are sitting at the edge of the algal mat in the SW corner. Sinbad isn't seen at this time. The heron is still in the same location (it has remained there since it was first sighted). End observation period.

*Comments:* The chicks continued on their progression toward becoming fully mobile. To this point it has been a gradual but steady process once it got underway. Today the chicks took their biggest step yet and the parents loosened the apron strings to the greatest extent thus far. This was the first observation period during which the parents spent no time sitting on the nest with the chicks. In addition, it was the first one in which the chicks spent the majority of the time out away from the nest. In a restricted sense, the chicks were actually fairly mobile today (restricted in that they stayed within the SW quadrant of the SC and especially in the SW corner). They ventured about as far out into open water as they've ever been observed doing and they took most of their feedings away from the nest area and beyond the reeds. It appears that it won't be long now before the chicks are accompanying the parents on their foraging forays.

At this point it appears most probable that the SR pair has indeed lost the northern half of their territory to the NR pair. As previously noted, if not unequivocally clarified beforehand it will certainly become evident just where this territorial situation stands when the chicks become mobile.

An adult PB grebe pair was sighted together in the West Lagoon today. I thus can't help but wonder if this is just a pair passing through the area or perhaps the West Lagoon pair that I observed rearing their chicks here last season. It will be interesting to see whether this pair hangs around.

**8/5/17:** Time: 1030-1150; Brief re-check at ~ 1200; Second re-check: 1250-1300.

1030 – As observations begin, all 5 chicks are in the algal mat in the SW corner of the SC. Neither parent is seen at this time.

1032 – Remy is sighted—she is ~ 12 m east of the nest (in open water) and is soon joined by the chicks who go to her from the SW corner.

1033 – Remy goes to the nest with the chicks. She then climbs onto and settles down on it. Some of the chicks join her on the nest and some remain in the water at its north side.

1034 – Sinbad is sighted moving toward the nest from the SE section of the SC with a very large fish (it is far beyond the size that the chicks can swallow). He stops briefly ~ 8 m east of the nest and works the fish, then goes to the north feeding station. Some of the chicks go to him but after working the fish for ~ 11 seconds Sinbad swallows it (time: 1035). He then goes out to the mid-SC area (calling once en route) where he parks and preens. Remy is still settled on the nest.

1040 – Sinbad is still parked and preening in the mid-SC area. Remy is still settled on the nest babysitting the chicks—some of them are on the nest with her and some are in the water next to it.

1045 – Sinbad is still parked in the mid-SC area. He is no longer preening and is vigilantly looking about (as though on "sentry duty"). Remy is still settled on the nest (at least one chick is in the water at its south side and it appears that all of the others are on the nest with her).

1046 – Remy stands, walks off the nest and heads toward Sinbad who then moves toward her. They meet and greet en route and continue on in opposite directions. Remy goes out to the NE point of the SC and begins foraging. Sinbad moves to the nest. As he approaches to within ~ 6 m away, one and then a second chick go to him. He stops at the edge of the algal mat east of the nest where the other 3 chicks also join him. He lingers there briefly, goes to the nest with the chicks and lingers by it for ~ 20 seconds. He then moves out just beyond the algal mat east of the nest, calls once and soon begins foraging in that area. The chicks eventually move to the algal mat in the SW corner.

1050 – The chicks are in the algal mat in the SW corner. Sinbad is foraging in the SW corner. Remy is foraging SE of the NE point of the SC.

1052 – Sinbad moves out ~ 22 m east of the nest (well out in open water) and parks there. Three chicks soon stream out to him from the SW corner. Shortly after, a fourth chick goes out and joins them. A short time later the fifth chick finally joins them and all 5 chicks sit close around Sinbad as he preens himself.

1055 – Sinbad is still parked and preening himself with the chicks in the same location. Remy isn't seen at this time. The NR male is sighted in the west side of the river in the northern half of the SR pair's (former?) territory.

1056 – While preening himself Sinbad plucks a small feather from his breast and hands it to a chick who immediately swallows it.

1059 – Sinbad (still in the same location) dives and begins foraging. The chicks remain by themselves in this open water location.

1100 – Remy heads toward the chicks from the SE section of the SC with a small fish. As she nears the chicks several of them rush toward her and she hands the fish off to the first one that reaches her. She then resumes foraging.

1102 – Remy and Sinbad are both foraging in the SE section of the SC. The chicks (staying together) have drifted out to ~ 35 m east of the nest (about as far out as I've ever seen them go).

1103 – Remy goes to the chicks with a small to medium-sized fish She works the fish for ~ 39 seconds (during which time two chicks took the fish and made unsuccessful attempts to swallow it) before the largest chick finally swallows it. Remy then resumes foraging in the NW section of the SC. The chicks remain in the mid-SC area.

1105 – The chicks are now a bit spread out in the west end of the mid-SC area. Neither of the parents is seen at this time (both are probably foraging).

1109 – Remy goes to the chicks from the NE section of the SC with a large crayfish. The chicks swarm around her as she works the crayfish for ~ 148 seconds, tearing morsels and larger chunks from it and handing these to the chicks. Numerous times she dives completely under the surface to retrieve or tear pieces from carcass after it sinks while she's working it. While this feeding is still in progress Sinbad goes to the feeding group from the SE section of the SC with a small fish (time: 1111). As he nears the group two of the chicks rush toward him and he hands the fish to the first one that reached him. The other chick (who was trailing right behind the first) unsuccessfully attempts to take the fish away from the first, who quickly swallows it. Sinbad then joins Remy and assists her feeding by picking up and retrieving small scraps that detach from the crayfish as she works it and handing these to the chicks. Before the feeding concludes, Sinbad moves out to the NE section of the SC and resumes foraging. After completing the feeding, Remy remains with the chicks in the west portion of the mid-SC area.

1113 – A train very loudly passes by the area (including repeatedly tooting its whistle) while Remy and the chicks are well out in open water in the mid-SC area. This spooks the group: Remy moves to the area near the reeds just north of the nest but the chicks don't follow her. Instead, 4 of them retreat to the algal mat in the SW corner and one chick remains out in open water in the mid-SC (this chick soon also goes to the SW corner).

1115 – Remy and all of the chicks have gathered together in the SW corner.

1120 – Sinbad joins Remy and the chicks in the SW corner at the edge of the algal mat. Remy and Sinbad appear to be concerned about something and they are alertly looking toward the reeds near and north of the nest (I'm unable to see or determine what has caught their attention).

1125 – All are still together at the edge of the algal mat in the SW corner. Sinbad and especially Remy are still vigilantly looking to the reeds.

1126 – Remy continues looking alertly to the reeds, now specifically to those in the SW corner. Sinbad also continues vigilantly looking about. The chicks are still with them.

1129 – Sinbad (still in the SW corner with Remy and the chicks) dives and isn't resighted at this time. Remy is ~ 5 m beyond the algal mat with 4 chicks; the other chick is within the algal mat.

1132 – The fifth chick moves out and joins Remy and its sibs.

1135 – Remy is still parked in the same location with the chicks, still alertly looking about.

1136 – While briefly preening herself, Remy plucks out a small feather from her breast or flank and hands it to a chick who immediately swallows it. She then plucks another small feather and hands this to a different chick who also quickly swallows it. Approx. 4 seconds later Remy dives and begins foraging, working her way to the SE section of the SC. The chicks remain where she left them and drift a little farther to the east. (I wasn't able to see or determine what it was that had caught and held Remy's and Sinbad's attention since 1120.)

1142 – Remy heads toward the chicks from the SE section of the SC with a small fish. Several chicks rush out to her as she nears them and she hands the fish to the first chick that reaches her. Remy then stays with the chicks in the SW section.

1144 – Remy moves away from the chicks to just beyond the edge of the algal mat in the SW corner and alertly looks toward the reeds there. One chick begins to follow her but then sees Sinbad approaching the chicks with a medium-sized fish. The chicks swarm around Sinbad and compete intensely for the fish. Sinbad intermittently works and retrieves the fish as various chicks attempt to control and swallow it, the fish thus changing hands among the chicks a number of times. At least twice a pair of chicks engaged in a tug-of-war over the fish and at least twice one chick was seen taking it from another. Finally, after ~ 30 seconds of this one of the larger chicks swallows the fish (time: 1145). Sinbad and the chicks then join Remy (still alertly looking to the reeds) just beyond the algal mat in the SW corner.

1148 – Remy dives and isn't resighted at this time.

1150 – Sinbad moves to the mid-SC area with the chicks and parks there with them. Remy isn't seen at this time. End observation period.

*1200 – Re-check of the family: Sinbad has taken the 5 chicks far out to the western portion of the east end of the SC (or in other words, the far east end of the mid-SC area). He left them within a modest sized algal mat where the chicks generally remained, waiting for and receiving feedings from the parents. This is the farthest out (by a considerable margin) in open water and from the nest that I have seen the chicks venture.

*1250-1300: A second re-check of the family: On this re-check I found Sinbad and the NR male engaged in another territorial standoff at the same location (just north of the NE point of the SC) as on 7/29. Unlike on that previous occasion, no physical fighting occurred between them despite the fact that they were very close to each other most of the time and sometimes touching. Instead, this lengthy confrontation consisted solely of posturing/displaying by the two opponents. Over this entire encounter the two males engaged in the same stereotypical posturing/displaying as exhibited in the 7/29 incident. They positioned themselves close to each other (sometimes touching) side by side (broadside), shifting between having their heads adjacent or in the opposite orientation (head adjacent to the rump of the other). While doing so they held their heads up stiffly (but at a normal angle) with the cheeks and upper throat puffed up, and intermittently turning the head from side to side. As they went about this posturing/displaying each male kept to his side of an imaginary line that stretched between them: Sinbad kept to the south (in his territory) and the NR male kept to the north side (in his territory). The pair conducted this standoff both while remaining stationary and while slowly moving along the imaginary east-west line. After a lengthy 6 minutes the confrontation abruptly ended, each male just suddenly going his own separate way. Sinbad went to the area just south of the NE point of the SC where he began foraging and the NR male headed north in his territory. During the confrontation the chicks were mainly in the mid-SC area and Remy mainly foraged (within the SC).

*Comments:* What has seemed increasingly obvious can now be confirmed. The fact that no physical fighting occurred and the confrontation was conducted in such a ritualized manner at the same boundary as on 7/29 indicates that the territorial conflict has been resolved; that a new boundary between these two pairs has indeed been established at this location (on a line across the river from just north of the NE point of the SC). In other words, that Sinbad and Remy have lost the northern half of their territory to the NR pair. Whether this situation will persist as it is remains to be seen. For example, will the SR pair be able to secure enough food for their chicks within the remaining portion of their territory (which seems unlikely), and if not, what will happen then?

It was a significant day for the chicks as they became mobile today (though they restricted their ranging to within the SC). The brood thus became mobile on the 10th day after the last egg hatched (i.e., considering 7/26 as the day of the last hatching and counting that as day one). Chick mobility, of course, means that the parents will be able to take the chicks to the general areas where they want to forage or the chicks can just accompany the parent as it forages. Obviously, this will greatly reduce the distances that the parents have to carry their captured prey to the chicks. This in turn, means that the parents will be able to expand their foraging range for chick feedings.

An American coot with 2 fairly young chicks was seen in the north end of the East Lagoon. Not too long from now this general area will be teeming with coot parents and their newly fledged chicks.

**<u>8/6/17</u>:** Time: 1045-1150; Re-check of the family: 1218-1225.

1045 – As observations begin, Sinbad is with the chicks in the mid-SC area. He soon leads them to the old nest site area and they drift to just west of this. Remy isn't sighted at this time (however, several minutes before this observation period began she was seen in the West Lagoon chasing a PB grebe).

1050 – Sinbad leads the chicks to a spot ~ 20 m NE of the nest and parks there with them. Remy still hasn't been sighted.

1051 – Sinbad and the chicks move to the edge of the algal mat east of the nest.

1052 – Remy is sighted—she is heading toward the chicks from the SE end of the SC with a medium-sized fish. As she approaches them one of the chicks rushes out and meets her en route ~ 15 m east of the nest. Soon after, the other chicks go out to Remy and all swarm around her. She works the fish amid the competing chicks for ~ 21 seconds before one of them (Phantom) finally swallows it (*note: the chicks have been given names—see post-observation period comments below*). Remy then resumes foraging in the mid-SC area. The chicks remain where they are in the mid-SC area.

1055 – The chicks are still in the mid-SC area. Sinbad is foraging in the SE section of the SC and Remy is foraging in the mid-SC.

1059 – Sinbad heads toward the chicks from the SE section of the SC with a small fish. As he approaches them one chick (Quark) and then several others string out to meet him. With its head start over the others, Quark is the first to reach him and takes and swallows the fish. Sinbad then resumes foraging. The chicks re-gather and move a little farther east in the mid-SC area (in open water) into a modest-sized algal mat and park there.

1101 – Remy heads in the direction of the nest from the SE section of the SC with a medium-sized fish. She apparently isn't aware that the chicks have moved well east from where she last fed them. She is well south of the chicks and passes west of them before she finally becomes aware of their present location. She then veers N-NE toward them and is soon met en route by the eager chicks. They swarm around her as she works the fish for ~ 25 seconds before one of them (Spot) swallows it. Remy then resumes foraging.

1105 – Sinbad is sighted (he is already with the chicks) with a medium-sized fish. The intensely competing chicks are swarming around him as he works the fish. The feeding carries on for ~ 75 seconds. The fish changes hands a number of times as various chicks attempt unsuccessfully to swallow it before the largest chick (Sparky) finally succeeds. Sinbad then resumes foraging.

1110 – The chicks have moved to the algal mat at the south side of the NE point of the SC. Sinbad (foraging near them, off the east side of the NE point) calls once and resumes foraging. Remy isn't seen at this time. The NR female is foraging mid-river, north of the NE point (a fair distance away from Sinbad).

1114 – Remy is sighted moving toward the chicks from the east with a large crayfish. Three of the chicks rush out and meet her en route and then swarm around her as she works the crayfish. In the usual manner, she repeatedly tears morsels from the crayfish and hands these off to the chicks. Approx. 50 seconds later a fourth chick joins the feeding. It isn't until ~ 95 seconds into the feeding that the fifth chick (Sparky, who had just eaten the fish from the 1105 feeding and is perhaps temporarily satiated) goes to the group but doesn't join in on the feeding. Instead, Sparky just lingers there and soon moves west to Sinbad (who is now parked in the mid-north area of the SC) and sits by him. The feeding is quite long, lasting ~ 3 minutes 22 seconds. When it concludes, Remy resumes foraging and the 4 chicks string their way over to Sinbad and Sparky.

1123 – Sinbad is now parked in the mid-SC area with the chicks. Remy isn't seen at this time. The NR female is foraging off the west bank of the river in the southern half of their territory (this is what used to be the northern half of the SR pair's territory before the NR pair took it from them).

1126 – Sinbad (with the chicks following along) has slowly made his way to the edge of the algal mat east of the nest and is parked there with the chicks. Remy is foraging at the east end of the SC.

1130 – Sinbad (with the chicks following) moves a little farther to the south to the edge of the algal mat in the SW corner. He often looks alertly to the east (perhaps for Remy?).

1135 – Sinbad is still parked in the same general location with the chicks, still intently looking to the east. Remy isn't seen at this time.

1137 – Sinbad moves out to the mid-SC area with the chicks and parks there. The chicks intermittently solicit him for food.

1140 – Sinbad (accompanied by the chicks) moves a little farther out to the east and calls once. Soon after, he dives and begins foraging. The chicks then move to a nearby modest-sized algal mat situated in open water at the far east end of the mid-SC area.

1145 – The chicks are still parked in the same algal mat. Sinbad is foraging in the mid-SC area. After resurfacing from an unsuccessful foraging dive he calls once and resumes foraging. Remy isn't seen at this time.

1147 – The chicks move to the algal mat just south of the NE point of the SC, looking very eager for a feeding.

1150 – The chicks are in the same location as at 1147. Sinbad is foraging by the NE point in the general area of the chicks. Remy isn't seen at this time. End observation period.

*1218-1225: On a re-check of the family I once again found Sinbad and the NR male already engaged in a lengthy territorial standoff. This standoff was basically a carbon copy of the one observed yesterday. There was no physical fighting; only the same stereotypical posturing/displaying as described in yesterday's notes and it occurred at the same location just north of the NE point of the SC. Also as happened yesterday, it ended abruptly with each of the males suddenly just moving away into their own respective territory. There now can be no doubt that a new territorial boundary between the two pairs has been established here. In looking back, the situation certainly has changed significantly from the time during the SR pair's early nesting period when they were observed charging at and chasing members of the NR pair out of the now former northern half of their territory.

*Comments:* The chicks were quite mobile today, though again remaining within the SC. However, they did venture all the way out to the NE point of the SC which was easily the farthest out that I have yet seen them go. This was the first observation period during which the chicks were not seen on or at the nest.

At this point the chicks still typically keep together as a group as they move about. They sometimes are a bit spread out or one or more may move a short distance away from the others with a parent but they basically stick together.

A curious observation occurred shortly after the formal observation period had concluded. Sinbad was sighted by the NE point of the SC moving in the direction of the nest while towing a long, fresh, green reed stem. This is something that he was seen doing a number of times during the periods when the pair was building the nest and incubating the eggs. Obviously, they have progressed well beyond that point now. Regardless, when Sinbad reached to within ~ 40 m of the nest he suddenly just released and abandoned the reed stem and began foraging. Though the pair no longer needs the nest for the major purpose of nesting and the chicks have just become mobile, the nest nevertheless still gets some use (though it didn't during today's observation period). The parents still periodically babysit the chicks on and next to the nest (however, they weren't observed doing so today—it remains to be seen if they will continue to do so now that the chicks have become mobile). It is still a safety/security location that the chicks may retreat to if they are alarmed or feel threatened (this includes being led there by a concerned parent). In a more general sense, the vicinity of the nest still currently serves the brood as a kind of home base area. Both parents have occasionally still been seen doing a brief touch of nest maintenance. Besides the above, though I haven't observed how the family spends its nights it seems a viable possibility that the chicks are still staying at the nest during the night.

Now that the chicks have come out from the reeds and have become mobile (and thus have become fully visible) the individual chicks can be distinguished and identified. One of the key methods for the natural visual identification of individuals within a species is to find a general physical trait that is common to all members but varies among individuals in the details of how it is expressed; in other words, its expression is distinct in each individual (for example, all Grevy's zebras possess stripes but the exact pattern of the stripes is distinct for each individual). Very young PB grebe chicks of the stage that the SR brood currently is in have white markings on the top of the head that vary in their amount, location and pattern. I have found the distinctiveness of these white head markings to work quite well for distinguishing individuals and this is the main means by which I identify the individual chicks while they are in this stage of their development (these white head markings disappear as the chick ages and it is then necessary to switch to another distinct physical feature(s) to key on—by that point I will be keying on the pattern of the black makings on a white background on the sides of the head and neck). I also find the pattern of dark markings on the back of the head and neck to be quite useful for individual identification at this stage.

In order to distinguish the individual chicks in the notes I have given them names (as a human, I find names to be much easier to remember than numbers or letters). The names given to the chicks are derived from a specific physical characteristic that distinguishes each individual at this point in time. The names are as follows:

Phantom—This chick has a very large, triangular white patch that covers most of the top of its head and its entire body is noticeably paler than the other chicks, giving it a somewhat ghostly appearance (hence the name "Phantom"). These very conspicuous features make it the most readily identifiable among the brood.

Quark—This chick is noticeably the smallest of the brood so it was just named after something small. As previously noted, though it is the smallest of the chicks it is not runty.

Sparky—The white markings on the head give the appearance of being devilish horns. I thus named this chick after the mascot (a sun devil) of my graduate school, Arizona State University. Sparky is the largest chick in the brood.

Spot—This name is a natural because this chick has just a single white spot on its head just above the base of its beak. Spot is perhaps the second largest chick in the brood.

Cobra—The pattern of the dark markings on the back of this chick's neck is reminiscent of the pattern on the hood of an Indian cobra. Cobra and Spot are very close in size and it's difficult to tell which of these two is larger. Regardless, they are the second and third largest of the brood.

**8/7/17:** Time: 1035-1135. Brief re-checks at 1142 and 1155.

1035 – As observations begin, Remy is parked with the chicks in the NW section of the SC. Sinbad isn't seen at this time. The NR female is foraging in the southern half of their range (i.e., the former northern half of the SR pair's range before it was taken from them by the NR pair).

1038 – Remy moves to the mid-SC area and parks there; the chicks follow and remain with her.

1040 – Sinbad is sighted heading toward the chicks from the east end of the SC with a small fish. One of the chicks (Cobra) spots him early, rushes out and meets Sinbad en route. Sinbad then hands the fish to Cobra who quickly swallows it. After handing it off, he goes back out to the east end of the SC (calling once en route) and resumes foraging. The chicks remain with Remy in the mid-SC area.

1043 – Remy goes to the SW corner of the SC (with the chicks following along) and begins foraging. The chicks park there at and just beyond the edge of the algal mat.

1045 – One of the chicks (Sparky) moves out ~ 8 m from the other 4 chicks who are parked just beyond the edge of the algal mat in the SW corner.

1047 – Remy (still foraging) works her way to the mid-SC area and continues foraging there.

1048 – Remy is sighted heading toward the chicks from the SE section of the SC with a small fish. As she approaches, 3 of the chicks rush out toward her. Quark (the smallest of the chicks, who managed to get a head start over the other two) reaches Remy first and takes the fish. The other two chicks arrive at Quark's heels and try to steal the fish but Quark manages to hold onto and swallow it. Remy returns to the SE section of the SC and resumes foraging.

1051 – Remy heads toward the chicks from the mid-SC area with another small fish. Spot rushes out and meets her en route and takes the fish from her. Remy then goes to the SE section of the SC and resumes foraging, The chicks re-gather and park in the mid-SC area.

1053 – While foraging in the now NE section of their territory (formerly the SE section before the NR pair usurped the northern half of the SR pair's territory) Sinbad calls once.

1056 – Remy goes to the chicks from the SW section of the SC with a small fish. Spot effectively maneuvers its way among its sibs and takes the fish from her. At this point Sinbad is sighted moving north from the area by the chicks and they follow him. Remy begins foraging in the mid-SC area. Sinbad goes to the mid-north area of the SC where he parks and preens. The chicks gather around Sinbad and remain with him.

1058 – The NR male is sighted foraging in the mid-section of their territory (formerly the north end of the SR pair's territory).

1100 – Sinbad is now parked ~19 m east of the nest with the chicks. The chicks intermittently solicit Sinbad, attempting to encourage him to go and forage for them. Remy isn't seen at this time.

1105 – Sinbad is still in the same general location with the chicks.

1108 – Remy heads toward the chicks from the east end of the SC with a small fish. Phantom (still with Sinbad and the other chicks) spots her approaching, rushes out to her and takes the fish. Remy then parks in the mid-SC area ~ 10 m from Sinbad and the other 4 chicks and preens. Phantom stays with Remy. Sparky soon leaves the others and joins these two. The other 3 chicks remain with Sinbad.

1112 – Remy (with Phantom and Sparky) moves to within ~ 6 m from Sinbad. Soon after, Sinbad moves out to the area by the NE point of the SC (calling once en route), parks to the south of it and vigilantly looks to the north and east (it appears likely that he is checking for the possible presence of the NR pair—neither of them are seen at this time). While Sinbad was moving there, Remy moved and parked ~ 21 m east of the nest and preened herself with the chicks gathered around her.

1115 – Remy is still parked with the chicks in the same location. Sinbad is still parked SE of the NE point of the SC vigilantly looking to the north and east.

1118 – Sinbad moves to the mid-river area east of the SC, calls once and begins moving SE into the NE section of their revised territory. He soon passes from my view.

1125 – Remy is still parked with the chicks in the same general location. Some of the chicks are intermittently soliciting her. Sinbad isn't seen at this time.

1126 – Remy moves out to the mid-SC area (followed by Phantom) where she dives and begins foraging.

1127 – The chicks gather together in the mid-SC area and park there. Remy works her way into the SE section of the SC and continues foraging there.

1129 – Remy moves toward the chicks from the SE section of the SC with a small fish. Phantom rushes out and meets her en route and takes and swallows the fish. Remy then resumes foraging in the SE section. The chicks re-gather in the mid-SC area.

1132 – Sinbad goes to the chicks from the NE section of their territory with a large crayfish. He vigorously works the crayfish in the usual manner for ~ 118 seconds, tearing off numerous morsels and handing these to the intensely competing chicks as they swarm around him. It appeared that all 5 chicks received at least one morsel from the crayfish. While Sinbad's feeding was in progress Remy approached the feeding group from the SE section of the SC with a small fish (time: 1133). Quark spotted Remy and rushed to her. They met en route and Quark took the fish from her and quickly swallowed it, then rejoined the feeding group around Sinbad. Remy resumed foraging.

1135 – Sinbad's feeding has concluded. Remy is foraging in the SE section of the SC. Sinbad moves to the SW section near the reeds along the south bank and the chicks follow him there. He then calls once and begins foraging, quickly working his way across to the mid-north section of the SC. The chicks remain by the reeds at the south bank in the SW section. End observation period.

*1142 – During continued observations of the family after the formal observation period had concluded Remy was sighted moving into the mid-SC area from the east with a large crayfish (this was the area where she had last seen the chicks). With the obvious intention of feeding this crayfish to the chicks, she pulled up in the mid-SC area when she wasn't able to sight them. She milled about there and intently searched visually for the chicks (with a beak-full of large crayfish she apparently wasn't able to call to them). The chicks were still by the reeds along the south bank in the SW section. Remy had been away foraging when the chicks first went there with Sinbad so she was unaware that they had moved to that location—and the chicks apparently had not spotted her, as none of them rushed out for a feeding. Having looked in vain for the chicks Remy decided that her only recourse was to eat the crayfish herself. She then reluctantly began working the crayfish, looking about for the chicks as she did. When she had consumed about two thirds of the crayfish she held the final third in her beak and milled about, giving a final visual search for the chicks. When this again failed she finally swallowed this last chunk of the crayfish and resumed foraging. The chicks apparently had remained oblivious to her presence in the mid-SC.

*1155 – On a final re-check of the family I found that mother and chicks had reunited. Remy was lying well-settled on the nest with the chicks around her (some of them were on the nest with her and some were in the water next to it).

*Comments:* It was a rather routine observation today with the parents busy feeding the chicks. The chicks were even more mobile than yesterday but again remained within the SC. They again basically stayed together but there were signs that this will soon change. For example, there were a few instances when the two parents were parked a modest distance away from each other and the brood was split amongst them. It has been several days since I last saw any of the chicks climb onto the back of a parent.

**8/8/17:** Time: 1055-1208.

1055 – As observations begin, Remy is parked with the chicks in the NW corner of the SC. Sinbad is parked in the mid-area along the east end of the SC.

1100 – Remy is still parked with the chicks in the same location. Sinbad is still parked in his same location but soon dives and isn't resighted at this time.

1101 – Sinbad is resighted—he is engaged in another territorial standoff with the NR male at the territorial boundary (just north of the NE point of the SC). As in the previous two standoffs that have been observed there is no fighting, just posturing/displaying. Remy is still parked with the chicks in the same location as at 1055 and 1100. The NR female is foraging in the mid-area of their territory, well north of the confrontation between the males.

1107 – The standoff abruptly ends as the males just suddenly move apart (the NR male first) and go their own ways. Sinbad heads west into the SC and the NR male makes his way north in his pair's territory. The behavior of the two males and the standoff in general were the same as in the previous two such encounters. By this time Remy has begun foraging and the chicks are by themselves in the NW corner of the SC.

1110 – Sinbad heads toward the chicks from the NE point of the SC with a small fish. One of the chicks (my view is too inadequate to ID it) rushes out and meets him as he approaches and takes the fish. Sinbad then resumes foraging. The chicks move east along the north bank and park west of the NE point.

1113 – The chicks move into a modest-sized, elongated algal mat that is situated just SW of the NE point. Sinbad is foraging in this general area. Remy isn't seen at this time.

1115 – While foraging, Sinbad works his way into the mid-SC area. The chicks follow him and park there (they appear to be especially eager for a feeding).

1117 – Sinbad (still foraging) works his way back to the area by the NE point and the chicks go back to the algal mat that they had gone to at 1113.

1119 – Remy is resighted—she is also foraging by the NE point near the chicks and works her way to the mid-SC.

1121 – Remy moves toward the chicks from the mid-SC area with a small fish (the chicks are still just south of the NE point). Quark spots her early as she approaches, rushes to her and meets her en route and takes the fish. Remy then resumes foraging.

1122 – Sinbad surfaces from a foraging dive near the chicks at the south side of the NE point with a large fish (it is much too large for any of the chicks to swallow). The chicks swarm around Sinbad but after working the fish for ~ 39 seconds he swallows it himself. Sinbad then moves to the mid-SC area with the chicks following. He parks there with the chicks and preens himself (some of the chicks also preen).

1125 – Sinbad is still parked with the chicks in the mid-SC area.

1130 – Remy moves toward the chicks from the SE section of the SC with a medium-sized crayfish. As she approaches, all of the chicks leave Sinbad and rush to her. With the chicks swarming around her, Remy works the crayfish (breaking off some morsels and handing these to the chicks) for only ~ 18 seconds before Spot assumes control of the carcass and swallows it. Both Sinbad and Remy then move east to forage. The chicks move east in the mid-SC area and park there.

1134 – Remy heads toward the chicks from the SE section of the SC with a small fish. As she approaches, first Sparky, and then several other chicks rush out toward her. With the head start, Sparky is the first to reach Remy. Sparky promptly takes the fish and quickly swallows it. Remy then parks in the mid-SC area and preens with the chicks gathered by her.

1135 – Sinbad is sighted with the chicks near Remy (he may have given a quick feeding just before I sighted him). He then moves out just east of the NE point. Three of the chicks follow but don't go so far out (they stop south of the NE point). The other 2 chicks stay with Remy. Sinbad soon dives and begins foraging. Shortly after, the 3 chicks who had followed Sinbad go to Remy and the other 2 chicks. Remy also soon begins foraging.

1140 – Remy heads toward the chicks (who have now moved back into the algal mat just south of the NE point) from the south section of the SC with a small to medium-sized crayfish (the crayfish is of a size that the chicks can swallow). Phantom gets a head start on the other chicks as they all rush toward Remy as she approaches. With this advantage, Phantom reaches her first, takes the crayfish and is able to swallow it. Remy then moves to the north bank near the old nest site and resumes foraging. The chicks follow and park near the old nest site.

1145 – The chicks are still near the old nest site. Remy is still foraging in that general area. Sinbad is foraging near the NE point.

1147 – Remy goes to the chicks with what appears to be a medium-sized fish. My view of the ensuing feeding is mainly blocked because the chicks and Remy are at the edge of and just within the reeds as well as behind the cluster of reeds within which the old nest had been built. Eventually, I briefly catch sight of the tail of the fish disappearing down the throat of one of the chicks. Remy then resumes foraging. The chicks stay inside the edge of the reeds just west of the old nest site.

1150 – The chicks are still in the same location. Remy is foraging along the reeds of the north bank of the SC. Sinbad isn't seen at this time.

1152 – Remy moves to the NW corner of the SC and continues foraging there. The chicks follow and park in the NW corner.

1153 – Sinbad heads toward the chicks from the SE section of the SC with a small fish. As he approaches, one of the chicks (my view is too inadequate to ID it) rushes toward him, meets him en route and takes the fish.

1154 – Remy heads toward the chicks (who are a bit spread out in the NW corner) with a small fish. Sparky quickly spots Remy (and is closest to her), rushes to her and takes the fish. Remy then resumes foraging. Sinbad parks in the mid-SC area and preens. All of the chicks eventually make their way to Sinbad and gather around him (time: 1155).

1200 – Sinbad is parked with the chicks ~ 39 m east of the nest. Sinbad is preening as are some of the chicks.

1205 – Three of the chicks (Phantom, Quark and Cobra) are with Sinbad, parked in the same area as at 1200. The other 2 chicks (Spot and Sparky) moved to the SW corner and are parked at the edge of the algal mat there. Remy isn't seen at this time.

1208 – Spot and Sparky come out of the SW corner and join Sinbad and their other sibs. Remy isn't seen at this time. End observation period.

*Comments:* Sinbad and the NR male engaged in another lengthy (6 minutes) territorial standoff at the newly established territorial border between the two pairs (the border is situated on a line across the river just north of the NE point of the SC). Once again, it was basically a carbon copy of the previous two that have been observed: no physical fighting, just posturing/displaying with each male keeping to his side of the imaginary line. The standoff also ended abruptly in typical fashion with the two just suddenly separating and going their own ways within their respective territories.

As will be the case for weeks to come, Remy and Sinbad were quite busy feeding their chicks today. Since becoming mobile on 8/5 the chicks have expanded their ranging on each subsequent day including today but again remained within the SC. The chicks still basically tended to stay together as a group but their trend toward breaking this pattern continued.

In watching the parents busily feed their chicks and contemplating all of the food that their rapidly growing chicks will require over the coming weeks, I can't help but wonder if what now remains of their territory will be sufficient for this task (it seems a virtual certainty that it is not). And if not, what course of action will the parents take? On 8/6 (prior to the observation period) Remy was sighted aggressively chasing a PB grebe out of the north half of the West Lagoon all the way to the south end in a long, low flight just above the surface of the water. Was this an indication that the pair will be looking to annex the West Lagoon as a part of their territory (or perhaps that they have already done so—the West Lagoon hasn't been claimed by any PB grebe pair up to this point)?

Several of the chicks were again observed swallowing a feather that the parent had plucked from itself and handed to the chick.

One of the chicks was seen briefly flapping its wings today. Some developmental progress was evident: the wings have elongated a bit, they don't look quite as downy and it appeared as though the buds of the flight feathers have sprouted along the posterior margin of the wings.

**8/9/17:** Time: 1025-1145.

1025 – As observations begin, neither of the parents or any of the chicks is seen (however, shortly before observations began Remy was seen foraging in the NW section of the West Lagoon).

1030 – None of the family has yet been seen.

1035 – Still no sightings of any of the family.

1037 – Sinbad is sighted moving by the SE point of the SC and heading toward the NE point. As he moves along he voices numerous short, single one note calls. These almost certainly are contact calls and he is obviously searching (and calling) for the chicks. He makes his way to the NE point, intently looking to the reeds. He then searches along the reeds of the north bank of the SC, slowly moving west. He soon dives and isn't resighted at this time. The NR male is foraging in the mid-southern half of their territory near the west bank of the river.

1043 – Sinbad is resighted—he is still searching for the chicks along the reeds of the NE point and to the west of it.

1047 – Sinbad surfaces from a dive just south of the NE point with a medium-sized fish. Holding the fish in his beak (with the obvious intent of wanting to feed it to the chicks) he circles in a wide arc through the margin of the mid-SC area, looking to the reeds that line the shore of the SC as he moves along. When he has circled back to the east end of the SC he doubles back to the mid-SC, then moves out by the SE point and finally back to the mid-SC area (time: 1051). On a quick check of the West Lagoon from the RR bridge I sight Remy near the reeds in the NW section of the West Lagoon but no chicks are seen.

1053 – Sinbad (still holding the fish in his beak) is still searching for the chicks from the mid-SC area. He soon moves to a point ~ 32 m east of the nest. There, he finally gives up on his attempt to feed the fish to the chicks and swallows it himself. He then parks there and preens (perhaps expecting that the chicks will eventually come out and join him).

1055 – Sinbad is still parked in the same location, still looking intently to the reeds that line the NW through SW shores of the SC.

1100 – Sinbad is still as at 1055, still visually searching for the chicks. The NR male is sighted foraging along the west bank of the river in the mid-southern half of their territory. [At this time I relocate my observation point from the south bank of the SC to the west end of the RR bridge from where I can see both the SC and the West Lagoon, since it now seems quite likely that the chicks are in the West Lagoon.]

1103 – Remy is sighted holding a medium-sized fish near the dense reeds of the west shore at the north end of the West Lagoon ~ 32 m south of the RR bridge. She faces the reeds, dives into them and is lost from view. It thus appears likely that the chicks are within the reeds here and she has gone in to feed them. Sinbad is still in the same general area of the SC as at 1100, looking about for the chicks—if the chicks are indeed in the West Lagoon he appears to be completely unaware of this.

1105 – Sinbad gives up his search in the SC and makes his way toward the West Lagoon. Remy isn't seen at this time and I still have had no sightings of the chicks.

1112 – Sinbad and Remy are both sighted in the NW section of the West Lagoon near the west shore ~ 21 m apart (Sinbad is south of Remy). They soon move toward each other. Both dive while en route and apparently meet up while underwater as they resurface near each other. Sinbad then moves east toward mid-lagoon (calling once) and Remy moves to within ~ 9 m of the west shore reeds. Each calls (short, single one note calls) separately, apparently for the chicks.

1117 – Remy is parked ~ 8 m from the west shore reeds ~ 36 m south of the RR bridge. She calls (again, short, single one note calls) several times for the chicks. Soon after, one by one all 5 of the chicks stream out of the reeds and go to her. Remy parks there with the chicks and preens herself. Sinbad is foraging along the reeds to the south of them.

1120 – Remy is still parked with the chicks in the same location. Sinbad isn't seen at this time (he is probably still foraging).

1123 – Sinbad goes to the chicks from the east side of the lagoon with a large crayfish (the chicks are still with Remy in the same location as at 1120). As Sinbad approaches, Quark rushes out to meet him, takes the crayfish from Sinbad and attempts to swallow it but the crayfish is too large (Quark then drops the crayfish onto the water and it is retrieved by Sinbad). Phantom soon joins them and Sinbad works the crayfish for ~ 76 seconds (tearing off morsels and handing these to the chicks) before Phantom finally assumes control of the reduced carcass and is able to swallow it. Sparky had joined this feeding briefly but for some reason didn't really get involved and left (he seemed rather uninterested—perhaps he was temporarily satiated from a recent feeding). When the feeding is concluded (time: 1125) Sinbad resumes foraging and Remy also begins foraging. The chicks gather ~ 14 m out from the reeds and soon move into them. (At one point while the chicks were still in open water and a bit spread out into groups of 2 and 3, a western gull passed fairly low over the group of 3. As it approached, the 3 chicks suddenly engaged in sinking alarm dives, instantly disappearing below the surface.)

1130 – The chicks are still in the west shore reeds (out of my view) in the NW section of the lagoon. Neither of the parents is seen at this time (both are probably foraging).

1135 – Sparky moves out ~ ½ m from the edge of the reeds, briefly looks to the east (apparently impatient for a feeding), then moves back into the reeds.

1138 – Remy moves toward the chicks with a small fish. At least 3 chicks rush out to meet her as she approaches just beyond the edge of the reeds. Quark reaches her first and takes the fish. Remy then resumes foraging and the chicks go back into the reeds.

1143 – Sinbad is sighted moving toward the chicks' location in the reeds with a large crayfish. As he approaches them, first 3, then a fourth and then all 5 chicks move to Sinbad and swarm around him ~ 2-3 m from the reeds. Remy is near the group and joins them. She assists Sinbad with the feeding as both take turns working the crayfish—tearing off morsels and handing these to the chicks, retrieving the carcass when it sinks below the surface, and picking up stray scraps that detach from the carcass as it is being worked and giving these to the chicks. After ~ 68 seconds Sparky takes control of the reduced carcass and (with considerable effort) is able to swallow it (time: 1145). Remy and Sinbad then resume foraging and the chicks go back into the reeds. End observation period.

*Comments:* This was a very significant day for the chicks as they notably expanded their mobility. For the first time they were seen ranging outside of the comparatively safe and familiar confines of the SC. It appears quite likely that they had only recently moved to the West Lagoon (probably led by Remy) at the time that the formal observation period began. I suspect this was the case because Sinbad spent the first half of the period intently searching for the chicks within the SC and was obviously completely unaware that they were in the West Lagoon. He even captured a fish in the SC and searched within it for the chicks with the obvious intent of wanting to feed it to them. Thus, the most likely scenario is that shortly before the period began Sinbad had been foraging away from the chicks while they were still in the SC and, unseen by him, Remy led the chicks (I can't imagine these young, inexperienced chicks making this move on their own) out of the SC, under the RR bridge and into the NW section of the West Lagoon. As is evident from the notes, Sinbad did eventually locate them.

The chicks were notably more apprehensive and cautious within their new environs in the West Lagoon. Lately, while in the SC they usually remained outside the reeds that ring the shores of the SC and oftentimes even in open water by themselves while the parents were foraging. In contrast, while in the West Lagoon today they basically stayed within the reeds other than during feedings or when a parent had parked nearby. After the feedings and when the parent whom they had been parked with moved away, the chicks returned to the reeds. In addition, the chicks regressed from their recent trend of progressively becoming looser in regard to their tendency to stay grouped together. Today in the West Lagoon they mainly kept together as a close-knit group, apparently seeking the greater comfort and security this offers while experiencing the uncertainties of this unfamiliar new area.

It appears that the question that I had posed in yesterday's notes (regarding what possible action the SR pair might take in the wake of having lost the northern half of their territory to the NR pair) has been answered: expansion into/annexation of the West Lagoon. In light of what happened today, the observation of Remy that occurred shortly before the 8/6 observation period now holds more significance. At that time she was seen chasing another PB grebe out of the north half of the West Lagoon all the way into the south end of the lagoon.

**<u>8/10/17</u>:** Time: 1034-1135; Brief re-check of family at 1145.

1034 – As observations begin, Remy is parked next to the chicks who are in the west shore reeds in the NW section of the West Lagoon. Sinbad isn't seen at this time.

1040 – Remy is parked ~ 1 m from the reeds (same location as at 1034). She has been joined by Phantom, Sparky and Quark. Spot and Cobra are at the edge of the reeds.

1042 – Sinbad is sighted in the mid-SC area with a large crayfish. He is making his way toward the West Lagoon.

1043 – Sinbad arrives at a point in the West Lagoon ~ 8 m north of where the chicks had been (they are no longer there—Remy had led them south along the reeds and they moved out of my view). He stops there and intently looks into the reeds for the chicks. Unable to spot them he then dives and goes into the reeds while under the surface.

1044 – Sinbad emerges to the edge of the reeds, still with the crayfish. He then slowly cruises south along the edge of the reeds, peering into them for the chicks as he moves along. At this point Remy is sighted with the chicks. They are much farther south than Sinbad—they are at the west shore reeds in the mid-segment of the lagoon.

1047 – Sinbad finally gives up on his search for the chicks and begins eating the crayfish. He eats only a small portion of it and then just releases and abandons it, allowing it sink away below the surface.

1048 – Remy leads all 5 chicks north along the west shore reeds (they are heading toward Sinbad's location). As they near, Sinbad sees them approaching and moves toward Remy. The pair meet and greet en route. Sinbad then doubles back to where he had released and abandoned the crayfish, dives and apparently retrieves it (he resurfaces with a crayfish that looks to be the same size as the one he had partially consumed and abandoned). He takes this to the chicks and conducts a feeding (had he deliberately cached the crayfish for this purpose?). Remy had begun foraging in the mid-water area during this time. She surfaces from a dive with a large crayfish claw and heads toward the feeding chicks. One of the chicks spots her approaching, leaves the feeding group around Sinbad, meets her en route and takes and swallows the claw (pincer end facing up).

1050 – Sinbad has concluded the feeding and he resumes foraging. Remy also continues foraging. The chicks retreat into the west shore reeds (in the NW section of the lagoon).

1054 – Sinbad ceases foraging, parks ~ 1.5 m out from the chicks location in the reeds and preens himself. First Phantom and then Spot emerge from the reeds, go to Sinbad and sit with him

1056 – Sinbad (with the two chicks) has drifted to the edge of the reeds by the chicks who had remained there. Remy isn't seen at this time.

1059 – Sinbad moves ~ 1 m out from the reeds and calls once. The chicks remain at and within the edge of the reeds.

1100 – Sinbad is parked ~ 1 m out from the chicks who are still at and within the edge of the reeds. A now independent chick from the East Lagoon first brood is foraging at the east side of the West Lagoon near the mouth of the small channel that connects the West and East lagoons [This channel connects in the north end of each lagoon and is located ~ 40-50 m south of the RR bridge (*see diagram of study area*). Henceforth it will be referred to as the "inter-lagoon channel." The mouth of this channel at the north end of the east shore of the West Lagoon is the only remaining section of the east shore that contains reed beds. In fact, this entire short channel is lined with lush growths of reeds and is a good foraging area for the grebes—the East Lagoon pair has often foraged for both of their broods there.] Remy isn't seen at this time.

1102 – Sinbad (still in the same location) dives and begins foraging. The chicks remain in the reeds.

1105 – The chicks are still at the edge of and within the reeds at the same location. Sinbad is foraging in the mid-water area (i.e., midway between the east and west shores of the lagoon) in the northern half of the lagoon. [Note: In the notes the term "mid-water area" will be used to denote the area midway between the east and west shores of the lagoon.]

1107 – Remy is sighted in the eastern side of the mid-section of the lagoon. An East Lagoon first brood chick (now independent) is ~ 20 m SW of her. Remy spots the grown chick and charges toward it. The chick immediately sights the charging Remy and flees. A chase then ensues with both grebes taking flight and flying just above the surface. The chick lands in the southern section of the lagoon and immediately dives. Remy lands ~ 11 m behind and dives after the chick. Both resurface soon after. The fleeing chick then dives again and isn't resighted. Remy lingers in the area briefly and then slowly begins making her way to the chicks.

1110 – Remy makes it back to the chicks and parks ~ 1.5 m from the reeds (and the chicks within them). As she preens she plucks a small feather from her breast. Sparky notes this, goes out to her and takes and swallows the feather. Phantom soon moves out and joins Remy and Sparky.

1112 – Sinbad is sighted moving toward the chicks with a large crayfish. First 2 and then 2 more chicks move out and meet him en route and Sinbad begins working the crayfish. The feeding lasts ~ 67 seconds. I was unable to see much of the feeding but did note that only 4 of the chicks participated (my view was too inadequate to be able to identify the chicks). When the feeding concludes Sinbad resumes foraging (Remy is also foraging). The chicks park at the edge of and within the reeds.

1120 – Remy has moved to the chicks' area along the west shore reeds in the NW section of the lagoon and parks ~ 2 m from the reeds near the chicks. Sinbad is foraging out from the west shore reeds in the mid-section of the lagoon, working his way south. Two of the independent East Lagoon first brood chicks are sighted on the east side of the West Lagoon near the mouth of the inter-lagoon channel.

1125 – Remy is parked ~ 1.5 m from the reeds where the chicks are located. Phantom, Quark and Sparky come out from the reeds and join her there.

1128 – While foraging in the southern half of the lagoon Sinbad calls once and resumes foraging.

1130 – Quark leaves Remy and the other 2 chicks (Sparky and Phantom) and goes back to the reeds. Sinbad is still foraging in the southern half of the lagoon.

1131 – Sparky leaves Remy and goes to the reeds. Only Phantom now remains with Remy (they are parked ~ 1.5-2 m out from the reeds).

1133 – Remy (still parked with Phantom by the reeds) spots Sinbad who is out in mid-water moving north from the southern half of the lagoon. Remy then moves out into open water toward Sinbad. Phantom follows her but trails well behind and the little chick looks very exposed in the open water. About two dozen western gulls are grouped together in the water to the east of the grebes. Remy had apparently intended to meet and greet with Sinbad. However, when she gets to within ~ 5 m of him she notes that Phantom had followed her and is trailing well behind and that an adult western gull is moving in the chick's direction from the east. Remy immediately doubles back toward Phantom. Sinbad also heads that way with the obvious intent to protect the chick. The pair quickly head off the gull and position themselves between it and Phantom; the gull then rather casually ceases advancing. It lingers briefly and then drifts back to the east, melding with the group of gulls. (The gull had not yet gotten very near Phantom and it was obviously the intent of Remy and Sinbad to keep it from doing so. The gull had definitely been moving in a direction toward Phantom but it wasn't clear if it was doing so with any intent on the chick. Regardless, the parents clearly viewed it as a threatening situation for the chick and acted accordingly.)

1135 – Remy stays with Phantom and leads the chick west toward the section of the west shore reeds in the NW section of the lagoon where the other chicks are situated. As they near that area Sparky and Quark move out and join them as they advance a little farther north, ~ 2 m out from the reeds. Sinbad is now foraging in the NW section of the lagoon. End observation period.

*1145 – In a brief re-check of the family the chicks were at the edge of and within the west shore reeds in the NW section of the West Lagoon. Sinbad was near them (it appeared that he had probably just delivered a feeding). Remy was foraging in the mid-area of the SC.

*Comments:* The chicks once again spent the entire formal observation period in the West Lagoon. As was the case yesterday, they were apprehensive and cautious and mainly kept at the edge of and within the reeds (though they were a bit looser in regard to staying close together as a group). The most striking exception was when Phantom attempted to follow Remy as she headed well out into open water. The chick then found itself in a potentially dangerous situation, at least, its parents certainly perceived it as such.

For the second consecutive day Remy was seen aggressively chasing a PB grebe while in the West Lagoon. In addition, for a second straight day Sinbad was heard voicing territorial calls while in the West Lagoon. Thus, the indications are that the pair is in the process of annexing the West Lagoon to their territory or has already done so. And though the chicks spent the entire formal observation period in the West Lagoon the parents aren't forgetting about utilizing and guarding the older part of their territory. Both of them were seen foraging in the SC (including during continued observations after the formal observation period) and thus are continuing to have a presence there. As I departed the area after concluding the day's observations I noted that Sinbad was parked in the NW section of the SC and vigilantly looking to the east. He clearly appeared to be "guarding" the area and watching out for possible intruders and probably especially for the NR pair (neither of them was seen at the time).

All 5 of the East Lagoon second brood chicks were seen today in the southern half of the East Lagoon. The chicks were very widely spread out with only 2 of them being in proximity to each other. On two occasions a parent was seen rebuffing a chick. The chicks appear to be well on their way toward achieving independence. Besides these 5 chicks, at least 3 of the East Lagoon first brood chicks (who are now independent) were seen. Three were seen in the south end of the East Lagoon and at least 2 were also seen in the West Lagoon (the latter sightings occurred much earlier than the former—i.e., there had been plenty enough time for the chicks to have gone from the West Lagoon to the East Lagoon). One of the ones seen in the West Lagoon was aggressively chased away by Remy. At one point one of the parents (the female) was foraging in the south end of the East Lagoon near a few of their first brood chicks (2 of the second brood chicks were also in the vicinity). Interestingly, the parent was quite tolerant of their presence—notably, she didn't attack, chase or otherwise attempt to drive them away.

**8/11/17:** Time: 1025-1140; Brief re-check of the family at 1150.

1025 – As observations begin, Remy is parked with the chicks near the edge of the algal mat at the west end of the SC, north of the nest. Sinbad isn't seen at this time.

1030 – Remy is still with the chicks at the edge of the algal mat. They have drifted a little to the south and are now east of the nest. Sinbad still hasn't been sighted. Remy soon dives and begins foraging.

1033 – Remy surfaces from a foraging dive (unsuccessful) ~ 18 m east of the nest where she parks and preens. Quark, Sparky and Spot move out and join her there. While preening, Remy is twice seen plucking a small feather from her breast or flank and "feeding" these to a chick.

1034 – The other 2 chicks (Cobra and Phantom) come out and join Remy and their sibs. Sinbad still hasn't been sighted.

1038 – Remy and the chicks are now parked together ~ 22 m east of the nest. The chicks are restless and obviously eager for a feeding (as a consequence, Remy has been getting solicited by some of the chicks).

1040 – Sinbad is finally sighted—he is foraging along the reeds of the north bank of the SC working his way west (he was just west of the NE point when first seen). As he works his way into the NW corner Cobra leaves Remy and goes there in anticipation of Sinbad catching a prey item. Sparky is now by itself at the edge of the algal mat east of the nest. The other 3 chicks remain with Remy.

1043 – Sinbad moves east along the reeds, goes back out near the NE point and shifts his foraging to that area. Four of the chicks follow him part of the way, stopping and parking at the old nest site. Remy begins foraging in the mid-SC area. Sparky comes out and joins his sibs at the old nest site (all 5 chicks are now together again).

1045 – The NR male is sighted foraging off the west bank of the river in the mid-segment of the southern half of their territory.

1046 – Remy surfaces from a foraging dive in the SW section of the SC with a large crayfish and heads for the chicks (they are still at the old nest site, across the SC from Remy). The chicks (very eager for a feeding) spot her early as she approaches and 4 of them stream out and meet her en route in the mid-SC. The feeding is conducted in the usual manner. Remy works the crayfish for a lengthy ~ 161 seconds, detaching morsels and larger chunks and handing these to the chicks (chicks also pick up scraps from the water that detach as Remy works the carcass). Spot didn't initially come out with its sibs but finally joins them about halfway through the feeding. While Remy's feeding is still in progress, Sinbad moves toward the swarming, feeding group from the NE point with another large crayfish (time: 1048). As he approaches, Sparky and Quark leave Remy's feeding group and meet Sinbad en route. Sinbad then begins working the crayfish and conducting a second, simultaneous feeding. When Remy's feeding concludes the 3 chicks who had remained at her feeding join in on Sinbad's feeding. Sinbad works the crayfish for ~95 seconds before Phantom finally assumes control of and swallows the remaining large segment of the crayfish (time: 1049). At the conclusion of their respective feedings both parents resume foraging. Sinbad goes to the NW corner to forage and the chicks (strung out behind him) follow him and park there (though they are a bit spread out).

1057 – I catch only the tail end of a crayfish feeding by Sinbad in the NW section—all of the chicks are swarming around him and Sparky seizes control of the reduced carcass and swallows it. Sinbad then resumes foraging

1100 – Remy has parked in the NW section of the SC with the chicks.

1101 – Sinbad (now in the SW corner) moves to ~ 5 m east of the nest with a large crayfish. Three chicks leave Remy's vicinity, go to and swarm around Sinbad (a fourth chick later joins them). Sinbad works the crayfish for ~ 55 seconds but doesn't feed it to the chicks. Instead, he retains control of the crayfish and swallows it himself (it appeared that he had intended to keep it for himself right from the start). Remy goes out near the NE point and begins foraging. Sinbad moves to the mid-SC area, briefly parks and preens there, then goes out to the east end of the SC.

1105 – The chicks have regathered (they are by themselves) ~ 31 m east of the nest.

1107 – Sinbad moves west into the SC from its east end. Cobra moves out toward Sinbad and meets him en route all the way out in the mid-SC area and solicits him. Sinbad continues moving west with Cobra tagging along and is soon met en route by Quark who had also come out a considerable distance to meet him (Quark also solicits Sinbad). Sinbad proceeds west all the way to the edge of the algal mat ~ 19 m north of the nest where he parks with the chicks (all 5 are with him).

1115 – Sinbad is still parked with the chicks in the same location.

1116 – A turkey vulture suddenly appears, flying over the group from the west. As it does, 3 of the chicks instantly disappear below the surface as they engage in alarm dives. The vulture quickly passes through the area.

1120 – Sinbad is still with the chicks in the same location. Remy isn't seen at this time.

1122 – Remy goes to the chicks with a large crayfish. The feeding proceeds in the usual manner, lasting ~ 59 seconds. Only 4 of the chicks participated (Spot didn't join in).

1124 – Remy parks in the NW corner with the chicks gathered around her. All preen themselves. Sinbad has begun foraging.

1126 – Sinbad heads for the chicks from the NE point with a medium-sized (though fairly large) fish. Four of the chicks stream out from Remy, meet him en route and swarm around him. The fish changes hands a few times as one or two of the chicks unsuccessfully attempt to swallow it. Sparky finally takes control of the fish and with considerable effort is able to swallow it. Sinbad then resumes foraging (time: 1127). The chicks go back and sit around Remy in the NW corner.

1130 – Remy is still with the chicks in the NW corner. Sinbad is foraging near the NE point and at the east end of the SC.

1131 – Remy dives and begins foraging. The chicks then go to the edge of the algal mat in the NW corner and (a bit spread out) park there.

1133 – Remy is sighted at the east end of the SC. She moves back to the NW corner where several chicks move to her, looking for a feeding (however, she has nothing to offer). She then dives and resumes foraging (time: 1135).

1140 – The chicks are a bit spread out at the edge of the reeds in the NW corner. Neither parent is seen at this time (both are probably foraging). End observation period.

*1150 – Brief re-check of the family: Four of the chicks (all but Spot) are a bit scattered in the algal mat and reeds by the nest. All four engage in what appear to be fairly brief foraging dives. All of the dives are unsuccessful, though a few of the chicks are seen dabbing at the surface of the water and briefly picking up some small strands of vegetation. Are they perhaps catching and eating insects (such as water striders) or other small invertebrates? Unfortunately, they are too distant and my view is much too inadequate to be able to make out such details. Spot is resting by itself at the edge of the reeds in the NW corner, perhaps ~ 15-16 m away from the others. Neither parent is seen at this time.

*Comments:* The chicks were back in the SC today where they spent the entire observation period. It was obvious how much more comfortable and secure they felt in the familiar confines of the SC. This was reflected in how they picked up again in their progression toward breaking from their tendency to stay together as a group. They were clearly much looser in this respect today. Overall, they oftentimes were more spread out when in the same vicinity, were more often broken into subgroups, and individual chicks were seen moving greater distances on their own toward a parent than on any previous day.

Recently (and especially today) all of the chicks have been observed engaging in at least some of what appear to be short, rudimentary foraging dives (all such dives that have been observed have been unsuccessful). Chicks have been seen diving since the time when I was able to first see them in the water near the nest. However, by their nature none of the observed dives until recently looked to be foraging dives. Besides the foraging dives the chicks have been seen dabbing at the surface and I wonder if they are catching and eating insects (such as water striders) and other small invertebrates? Unfortunately, I have always been too distant and my view much too inadequate to be able to see if this is the case.

Neither of the parents has been observed engaging in a babysitting session on or at the nest with the chicks since 8/7. It seems clear that the parents have now progressed beyond the phase in which they make use of the nest in the chick production/rearing process (unless they still do so during the nights—the study area is closed to the public after sunset so I have never observed how the family spends their nights).

**<u>8/12/17</u>:** Time: 1040-1050; Re-check of the family: 1306-1318.

1040 – As observations begin, 2 kayaks are entering the area east of the SC and move into the east end of the SC. One parent (probably Sinbad) was briefly sighted at the edge of the reeds near the old nest site just as one of the kayaks approached the east end of the SC. After that none of the grebe family is sighted at this time, all undoubtedly having been spooked into hiding by the kayaks, regardless of whether they are in the SC or the West Lagoon.

1046 – (The kayaks are gone from the area.) Sinbad and 3 chicks (Phantom, Sparky and Quark) are sighted in the NW corner of the SC. Sinbad soon begins foraging.

1048 – Sinbad is sighted moving to the 3 chicks in the NW corner with a medium-sized fish. As they swarm around Sinbad, Sparky emerges from the group with the fish and swallows it. Sinbad then resumes foraging.

1050 – Sparky, Phantom and Quark are now in the reeds and algal mat near the nest. These 3 are the only chicks that have been seen at this point. Remy also hasn't been seen.

1052 – Remy is finally sighted—she is foraging in the NE section of the SC.

1054 – A fourth chick is sighted with the other 3 in the NW corner (my view is too inadequate to ID it).

1055 – Remy is sighted in the NW corner with a large crayfish. Three chicks are gathered around her but Remy swallows the crayfish herself (as usual, tail first with the tail partly doubled over on itself). (I saw only this end portion of this episode so I don't know if she fed any morsels to the chicks before she swallowed it—probably not.)

1058 – Sinbad is parked near the reeds of the north bank in the NW section of the SC with 3 chicks (Sparky, Phantom and Quark). Remy is parked and preening ~ 21 m NE of the nest. Cobra moves out from the reeds to the west of Remy, goes to Sinbad and joins him and the other 3 chicks.

1102 – The fifth chick (Spot) is sighted in the reeds to the west of Remy.

1104 – Spot comes out of the reeds, goes to Remy and solicits her. Sinbad goes near Remy, lingers there briefly and then goes to the mid-SC area. Spot leaves Remy, follows Sinbad and solicits him.

1106 – Sinbad moves out to the east end of the SC and parks there. Spot and Phantom (who has just moved into their area) follow him there and solicit him. Remy is still parked in the same location as at 1058; Quark has joined her there. The other 2 chicks (Sparky and Cobra) are at the reeds west of the old nest site.

1107 – Spot leaves Sinbad and makes its way all the way over to Remy and solicits her (Quark is still with Remy).

1109 – Sinbad (with Phantom following) goes to the mid-SC area and begins foraging there. Phantom then moves to the NW corner.

1110 – Remy dives and begins foraging. All of the chicks gather (a bit spread out) in the NW corner.

1113 – The 2 kayaks return and pass by the area east of the SE point of the SC and go into the West Lagoon. This spooks the entire grebe family and causes them to hide in the reeds: the chicks are in the reeds in the NW corner; Remy is in the reeds at the north bank in the NW section; Sinbad is in the reeds just east of the old nest site.

1120 – All of the family is still at the edge of or tucked in the reeds. Phantom is now with Remy.

1122 – At least 4 of the chicks are now with Remy (who is still in the reeds of the north bank in the NW section of the SC). Sinbad is still at the edge of the reeds just east of the old nest site.

1124 – Remy and Sinbad (separately) go out near the NE point and begin foraging. Four of the chicks (all but Cobra) move to the area just out from the reeds, west of the NE point.

1130 – Sinbad is foraging south of the NE point. The four chicks (Sparky, Spot, Phantom and Quark) are still in the area west of the NE point.

1132 – Sinbad is sighted with a medium-sized (though fairly large) fish in the NE section of the SC. All of the chicks go to Sinbad and swarm around him as he works the fish. Spot takes control of the fish fairly early in the feeding and frantically darts about while being chased by some of its sibs who are attempting to take the fish from Spot. Sinbad follows along, overseeing the feeding. At one point Spot dropped the fish and it sank. Sinbad then dove, retrieved the fish and handed it back to Spot. After ~ 97 seconds of intense competition Spot finally swallows the fish.

1134 – Just as this feeding is ending, Remy nears the group with a large crayfish. Some of the chicks swarm around her but she quickly swallows it. Sinbad resumes foraging. Remy moves to the area just south of the old nest site, trailed by 3 of the chicks. She parks there and preens. One of the chicks (Quark) remains with her and the other 2 move to the reeds by the old nest site.

1138 – Remy moves to the old nest site and parks there. At least 3 of the chicks are in the reeds by her.

1142 – Remy moves out ~ 2 m from the old nest site. Quark goes with Remy and solicits her.

1144 – Sinbad is sighted foraging near the reeds north of the nest. Sparky and Cobra are in the area near him. Remy moves toward the mid-SC area, dives and begins foraging. Two of the 3 chicks who had been near Remy (Quark and Phantom) go to the NW section of the SC and park there.

1148 – All of the chicks but Spot are parked in open water (a bit spread out) in the mid-north area of the SC. Remy is foraging by the reeds of the south bank in the SE section. A great-blue heron (flying low at about the height of the reeds) approaches the chicks' area from the east. As it nears them they all catch sight of it and immediately engage in alarm dives, all 4 instantly disappearing below the surface. Remy happened to be on the surface as the heron flew in and she immediately rushed toward the chicks in a half-flight, skipping over the surface. The heron quickly passes over the area of the now submerged chicks, circles back and heads east out of the area (it had most likely just been looking for a spot along the shore in the SC where it could land and ambush-hunt for fish). Remy dove as soon as the heron cleared the immediate area where the chicks had gone under. She resurfaces near the chicks, who are now back on the surface. Remy goes to the 4 chicks and Spot soon comes out of the reeds by the old nest site and joins Remy and its sibs. Remy leads them to the NW corner of the SC where all of the chicks gather near her.

1150 – Remy is parked with all 5 chicks in the NW corner of the SC. Sinbad isn't seen at this time. End observation period.

*1306-1318: When passing by the area again at 1306 I arrived just in time to see the beginning of another territorial standoff between Sinbad and the NR male at their newly established territorial boundary located just north of the NE point of the SC. As with the previous standoffs it was rather lengthy, lasting for 9 minutes. Until the last ~ 50 seconds it was a carbon copy of the previous three standoffs that were observed: no physical fighting, just stereotypical posturing/displaying. However, during the final ~ 50 seconds physical fighting suddenly erupted between them. This consisted solely of the breast to breast pushing/bumping matches that were observed in their 7/29 confrontation (i.e., with the two adversaries "standing" as high as they could, wings flapping and legs churning to keep their bodies raised on the surface, shoving and bumping breast against breast, and attempting to "stand" higher in order to gain leverage over the opponent). A series of 4 brief such bouts occurred. In each case the NR male was the initiator/aggressor, attempting to beat back/shove Sinbad back and away but each time Sinbad solidly held his ground. Shortly after the final such bout, in typical fashion, the confrontation abruptly ended, each male just moving well back into their respective territories and getting on with their usual business.

*Comments:* The chicks again spent the entire observation period in the SC. They were quite mobile; more so than in any previous observation period. In addition, they were often spread out and broken into subgroups. Of particular note, on several occasions Spot was off by itself.

Boats (on this occasion, two kayaks) again disrupted and put a temporary halt to the family's activities by spooking them and causing them to hide in the reeds. This happened twice today: at the beginning of the period and at 1113. Noticeably fewer feedings were observed during today's observation period and this undoubtedly was a factor.

**8/13/17:** Time: 1040-1147; Brief re-check of the family at 1155.

1040 – As observations begin, Sinbad is parked with the chicks in the mid-water area in the northern half of the West Lagoon. Remy is foraging mid-river, east of the SC.

1045 – Sinbad has drifted west with the chicks and they are now ~ 2 m from the west shore reeds in the mid-north section of the lagoon. Remy is foraging at the east end of the SC.

1047 – Sinbad, with the chicks following along, goes to the edge of the west shore reeds in the mid-north section of the lagoon. Remy has captured a medium-sized fish at the east end of the SC and heads south toward the West Lagoon and the chicks. Sparky is parked ~ 5-6 m from the reeds (the other 4 chicks are at the reeds) and is the first to spot Remy approaching with the fish (time: 1050). Sparky rushes toward her while she is still fairly distant, meets up with her in open water and takes the fish from her. Quark, who had made a slightly later start toward Remy (these 2 were the only chicks who went to her), soon reaches Sparky and pursues its sib with the intent of trying to steal the fish. Quark does this intently for ~ 35 seconds, then appears to concede the fish to Sparky and just trails its sib in a more casual manner, making no realistic attempt to take the fish away (Quark is the smallest of the brood and Sparky is the largest). The fish is rather large for the chicks and Sparky has much trouble trying to swallow it. Sparky works the fish for a total of ~ 116 seconds before finally being able to swallow it (Sparky had engaged in a number of failed attempts before this). Remy shadowed Sparky throughout this time as she oversaw the feeding. She then resumes foraging.

1055 – Quark, Phantom and Sparky are parked ~ 6 m from the west shore reeds in the mid-north section of the lagoon (the other 2 chicks are at the reeds). Sinbad calls once while foraging near the west shore ~ 18 m south of the RR bridge.

1057 – Remy passes under the RR bridge as she heads toward the chicks from the north with a medium-sized fish (the fish is rather large for the chicks). All of the chicks are in the mid-water area of the mid-northern section of the lagoon, having followed Sinbad there as he was foraging. Quark is the first to spot Remy approaching and rushes toward her. Quark meets her en route and she hands the fish to the chick. Phantom arrives soon after and shadows Quark but makes no serious attempt to steal the fish from its sib. Remy shadows Quark and oversees the feeding. Quark works the fish for ~108 seconds before, after much difficulty, finally swallowing it (time: 1059).

1100 – Remy parks in mid-water in the mid-north section of the lagoon with the chicks. Sinbad isn't seen at this time.

1105 – Remy is still parked with the chicks in the mid-north section of the lagoon, now ~ 18 m from the west shore reeds. Sinbad still hasn't been resighted.

1108 – Remy and the chicks retreat into the reeds as a train noisily passes over the RR bridge. [Note that besides the considerable normal operating noise that the trains make they must also repeatedly toot their loud whistles while approaching and crossing the bridge. This is because the many homeless people who reside in this area regularly cross on the bridge as they come and go here so the trains sound their whistle to warn people of their presence.]

1112 – None of the family is seen at this time (Remy and the chicks are probably still in the west shore reeds where they sought cover when the train passed at 1108). Approx. 3 dozen western gulls (adults and immatures) are grouped in the water ~ 25-30 m from the location where Remy and the chicks had entered the reeds.

1115 – Sinbad is sighted foraging in the western half of the mid-north section of the lagoon. He soon parks and preens ~ 20 m from the west shore reeds.

1119 – Sinbad (still in the same location) calls once. Remy is sighted passing under the RR bridge with a large crayfish heading south toward the area where I last saw the chicks. She moves to the general vicinity of Sinbad (who then goes to the southern half of the lagoon and begins foraging). Remy lingers there, looking in vain to the reeds for the chicks, nor do any come out to her for a feeding. Giving up on her intended feeding, she reluctantly begins working the crayfish, doing so for ~ 39 seconds before swallowing it. Afterward, she parks ~ 15 m out from the reeds and preens (time: 1120).

1125 – Sinbad is sighted with a medium-sized fish near the reeds to the west of where Remy is parked. He dives, entering into the reeds while underwater and isn't resighted at this time (he likely went into the reeds to deliver a feeding to the chicks).

1127 – Sinbad is resighted—he is moving just west of Remy, heading south where he resumes foraging along the west shore reeds. The chicks haven't been visible to me since they entered the reeds at 1108.

1128 – Remy moves to the reeds and is lost from view (she is probably with the chicks there).

1129 – An immature western gull comes into the area, moves to ~ 4-5 m from the section of reeds where the chicks probably are and peers toward the reeds. Remy comes out of the reeds with her attention focused on the gull. She dives several times toward it (she was easily close enough to have contacted it while underwater but I couldn't determine whether or not she did) and this appears to drive it away as it takes flight and goes over to the large group of gulls in the mid-lagoon area. Remy then goes back into the reeds (probably to rejoin the chicks). The gull's intent wasn't clear. All that can be stated is that its behavior was consistent with the possibility that it had sighted the chicks within the reeds and came over to investigate for some reason. However, I can't be certain that this was actually the case.

1133 – Remy comes out of the reeds, dives and begins foraging, working her way out to the mid-water area. I still haven't seen any of the chicks since 1108 though Remy's behavior has indicated that they are tucked within the reeds in the area that she just left.

1135 – Remy is now foraging near the west shore reeds ~ 15 m south of the RR bridge.

1140 – None of the family is sighted at this time.

1147 – Sinbad is sighted foraging in the east side of the lagoon near the mouth of the inter-lagoon channel. He calls once while foraging there. Remy is foraging in the east side of the lagoon as well, closer to the mid-water area in the NE section. Still no sightings of the chicks (they are probably still in the same location within the reeds). End observation period.

*1155 – Brief re-check of the family: Only Sinbad is sighted. He is parked in the NE section of the SC, looking about. He apparently is standing "guard duty" and is presenting a presence within this vital portion of their territory.

*Comments:* For a third time the chicks spent an entire observation period in the West Lagoon. Though they did spend some time in open water (usually only at times when a parent was present with them) they clearly are still more apprehensive and feel less secure when there. Of particular note, they much more readily flee into the relative safety of the reeds and are more apt to remain within them (for example, they spent over half of the observation period tucked within the reeds today). It's certainly possible that the nearly constant presence of several dozen or more western gulls in the lagoon is a contributing factor to the chicks' unease.

As observed among the chicks that I studied last year in the West Lagoon (the chicks of the West Lagoon pair) and in my casual observations of the East Lagoon chicks this season, the competition between chicks during feedings is intense. Up to this point the competition has mainly consisted of muscling aside and trying to outmaneuver rival chicks, attempting to steal a food item that is in a rival's possession (this includes pressuring and hotly pursuing that rival; a number of tug-of-wars have also been observed), and being vigilant for parents approaching with a food item and outracing rival sibs to that parent (or in general, being the first to reach the food-bearing parent). This latter strategy is especially effective when the parent has a small fish or crayfish: in other words, a prey item that the chick can readily swallow without it having to be worked by the parent, which would then allow more time for other chicks to reach the provisioning parent and compete for the food item.

I still haven't been able to find an adequate vantage point from which to see if the NR pair currently has a clutch of eggs. However, there are behavioral indications which suggest that the pair is currently nesting. It has been awhile since I last saw both of the pair simultaneously. This is suggestive of what I observed during the period when Remy and Sinbad were incubating their clutch.

Within that time frame, generally only one of them at a time was out and about while the other was on the nest incubating the eggs, with the roles being reversed at each nest exchange. Is this the explanation for why lately I have typically only been seeing one or the other of the NR pair at any one time? Regardless, it is now near or at the end of the window for when a reasonable nesting attempt can be started for this breeding season.

However, I suppose that, though risky, a later nesting attempt in this area is still feasible providing that the chicks can be reared to independence before the rainy season begins. The closer to the rainy season that a nesting attempt is started in this area, the greater the risk. This is because should there be a sufficient amount of rainfall in the river's watershed to cause the river to flow strongly this will drastically alter the lagoon habitat at the terminal end of the river, making the area unsuitable for nesting and chick rearing. As previously pointed out in the notes for both the SR and West lagoon families, when the river flows strongly it breaks through the retaining (or restricting) sand bar at the end of the West Lagoon and in a wide mouth, forcefully flows through the beach and directly into the ocean. At this point the habitat here is transformed from a lazy lagoon to a proper river. The grebes in this area require and prefer the lazy, quiet waters of the lagoon habitat (with its essentially stable water level) for nesting and chick rearing. Of particular importance here are the favorable conditions that the lagoon environment creates in regard to the reed habitat. As noted previously, the emergent reed habitat is vital for the grebes in this area, particularly for nesting and chick rearing. This vital reed habitat rings the banks and shorelines along a substantial portion of this estuarine area. It exists in lush, dense, mainly continuous and extensive growths along the west bank of the river from just south of the Highway 101 bridge all the way to just north of the RR bridge (including all of the SC shoreline except for the eastern half or so of its south bank), along the north bank of the south ribbon of the river, along the northern two thirds of the west shore of the West Lagoon, throughout the inter-lagoon channel, along the entire west shore of the East Lagoon and along the northern two thirds of its east shore. For their purposes, the grebes need these reedy areas to be emergent and sufficiently submerged in order for them to be able to build their floating nests, dive and forage, and move deep within them for cover, concealment and safety. It is only when the lagoons and adjacent areas are fully formed and intact that the shoreline reed habitat is sufficiently submerged for these endeavors—and this occurs only during the dry season (or during periods of severe drought during the rainy season) when the flow in the river is too inadequate to break through the beach and reach the ocean or can only do so in a small, weak-flowing channel. It is during this period that the water then backs up and the lagoons and adjacent features are able to form. When the river flows strongly the water moves swiftly through, generally staying confined to the main river channel (and some overflow areas). The reach of the water then is not nearly as extensive as when the flow is backed up and as a consequence a considerable portion of the reed habitat is no longer sufficiently submerged, with much of it even being completely out of the water. (For example, during this past rainy season when a temporary break in the horrible drought in this area was brought on by a severe El Niño event the river flowed very strongly. During this time the two locations in the reeds in the SC where the SR pair would later build their floating nests were completely out of the water.) In addition, the flow of water to the East Lagoon through the south ribbon of the river and the inter-lagoon channel that occurs when the system is backed up is mostly or completely shunted away. As a consequence, without this flow the East Lagoon gets considerably smaller (with much of the reed habitat being left high and dry or insufficiently submerged, though unlike the West Lagoon it still remains a lagoon), the south ribbon may be left completely dry or with its flow severely reduced (with the entire north bank reed habitat completely out of the water), and the water in the inter-lagoon channel becomes quite reduced and the inter-channel connection is typically lost or greatly diminished.

Even during a period of severe drought the risk of a late nesting attempt being impacted by the occurrence of a strongly flowing river remains. This is because it takes only a single rainstorm of sufficient intensity within the watershed during an otherwise dry rainy season to cause the river to flow strongly.

Aside from the risk regarding the potential for a strong river flow event there is also risk associated with the condition of the reeds themselves. Seasonally, extensive swaths of reeds may brown and lay down, creating a near impenetrable mass. Such habitat is clearly unsuitable for chick rearing. For their purposes, the grebes need the reed stems to be vibrant and standing sturdily upright so that they can readily get in amongst them. Obviously, should a late nesting attempt carry a brood into a period in which such extensive browning develops the consequences could be severe.

All 5 of the East Lagoon second brood chicks were seen in the southern half of the East Lagoon. The chicks are now in the period during which all are doing some independent foraging yet still taking food from the parents (they are considerably farther along in their development than are the SR chicks). Some extremely intense feeding competition was seen among some of the chicks. In addition to these chicks, 3 of the now independent East Lagoon first brood chicks were seen, also in the southern half of the East Lagoon.

**<u>8/14/17</u>:** Time: 1154-1305.

1154 – As observations begin, Sinbad and 3 of the chicks (Sparky, Quark and Phantom) are parked (and a bit spread out) in the west end of the SC. Remy, Cobra and Spot aren't seen at this time.

1157 – Sinbad moves out to the east end of the SC (calling once en route) and parks there. First Quark, then Phantom and Sparky move to the mid-SC area. The NR male is sighted near the reeds along the west bank of the river in the mid-southern half of their territory.

1200 – Remy is finally sighted—she is heading toward the area east of the nest from the SE section of the SC with a small crayfish. Sparky rushes to her (arriving well before any of the other chicks), takes and quickly swallows the crayfish. Spot also came out of the reeds for the feeding but arrived too late.

1201 – Remy lingers in the west section of the SC with 4 of the chicks but soon resumes foraging.

1202 – Sinbad moves in from the east end of the SC and parks ~ 10 m east of the nest. He is joined there by Quark and Phantom. Remy, Sparky, Cobra and Spot aren't seen at this time.

1203 – Sinbad dives and begins foraging. Quark and Phantom then go into the reeds by the nest.

1205 – Sinbad goes to the NW corner and begins foraging there (on the way there he was briefly met and solicited by Sparky).

1206 – Quark, Sparky and Phantom have moved into open water in the NW corner (they are a bit spread out from each other). Cobra still hasn't been sighted but is probably in the reeds somewhere in the west end of the SC.

1207 – Sinbad is sighted moving to some of the chicks with a medium-sized crayfish. Sparky takes the crayfish from Sinbad and moves rapidly east with it, trying to escape the hotly pursuing Quark and Phantom. Sparky quickly swallows it and Quark and Phantom then break off their pursuit.

1210 – Remy is sighted foraging in the NW corner of the SC. Several chicks are spotted engaging in some fairly brief foraging dives within the reeds at the north bank in the NW corner.

1214 – Quark, Phantom and Sparky are now parked (a bit spread out from each other) in the NW section of the SC. Remy moves toward them from the SW corner with a small fish. Quark (who is nearest to Remy) sees her approaching, rushes to her, takes and quickly swallows the fish. Spot had come out of the reeds and also rushed toward Remy but was much too late. Remy then resumes foraging. All 5 of the chicks (Cobra is finally sighted) park in the algal mat east of the nest.

1219 – Remy goes to the chicks from the SW corner with a large crayfish. All of the chicks but Cobra swarm around her as she works the crayfish. Remy works it for ~ 63 seconds but retains full control of it and swallows the crayfish herself (tail first as usual, with the tail partially folded over on itself). Cobra comes out of the reeds and joins the group but only just as Remy is swallowing the crayfish (time: 1221). Remy then parks in the mid-SC area and preens. Quark and Phantom park with her and also preen. Sinbad is foraging mid-river, east of the SC. He calls once, then resumes foraging.

1225 – Remy is still parked with Quark and Phantom in the same location. Sparky and Spot come out of the reeds in the SW section, go to and join Remy and their 2 sibs (Spot solicits Remy upon arrival). Spot soon leaves the group and goes back to the reeds in the SW section.

1235 – Remy has moved to the NW corner and is parked there with all 5 chicks.

1237 – There is a loud human disturbance at the SE corner of the SC which causes Remy and all of the chicks to dive and flee into the reeds in the NW corner.

1241 – On a brief check from the west end of the RR bridge Sinbad is sighted foraging in the north half of the West Lagoon.

1242 – A person (one of the many homeless people in this area) jumps into the water near my observation point on the mid-south bank of the SC, making a large, loud splash and thus ensuring that Remy and the chicks remain hidden in the reeds.

1244 – The person comes out of the water and leaves the area.

1245 – Sinbad arrives in the NW corner of the SC and all of the chicks but Cobra come out of the reeds. Remy also comes out and parks in the NW section where she preens herself. Sinbad then moves into the mid-SC area and parks and preens. A few chicks are gathered around each of them as they preen: Spot takes and swallows a small feather from Sinbad, and Quark does likewise from Remy.

1250 – Sinbad is still parked and preening with Spot in the same location. Sparky then joins Quark and Phantom as this trio (by themselves) goes all the way out to the east end of the SC, north of the SE point. Remy is now foraging in this area and this apparently is what drew them out there.

1255 – Sinbad moves out to the SE point area where Quark, Sparky and Phantom are; Spot follows Sinbad there. Soon after, Sinbad and Remy lead all 4 of these chicks to the NW section of the SC. Cobra comes out from the reeds and joins them.

1259 – A double-crested cormorant enters into the NE section of the SC (it is just foraging). This spooks the chicks and all of them flee into the reeds north of the nest. Remy and Sinbad go to the edge of the reeds there.

1301 – Sinbad moves out ~ 37 m east of the nest where he parks and preens. Remy moves out ~1.5-2 m from the reeds (west end, north of the nest) and parks there with Phantom, Spot and Quark. Cobra and Sparky are in the algal mat just west of them.

1305 – Same as at 1301 except Sinbad has moved a little farther east in the mid-SC area. End observation period.

*Comments:* The chicks were very mobile during today's observation period. They remained within the SC though 3 of them ventured all the way out to its east end by themselves in order to be in the vicinity of Remy who was foraging in that area. They were looser in their grouping behavior today than I have yet seen them. They were often spread out or in various subgroups and sometimes even ranging alone. Cobra in particular spent considerable time ranging alone (mainly within the reeds). All 5 of the chicks spent at least some time foraging independently within the reeds and algal mats, usually in company with one or more brood mates (this was observed both during the formal observation period and in my continued observations after it was concluded). No successful foraging dives were noted but I wasn't able to see much detail because my view usually was at least partly obscured by the reeds. Besides diving, it appeared that the chicks also foraged along the surface, dabbing their beaks at and just below the surface (perhaps snapping up insects such as water striders or other invertebrates?).

As I was preparing to leave the area I saw Sinbad feed what appeared to be a dragonfly to one of the chicks. Unfortunately, I didn't get a sufficient look at the prey item to be certain that it was indeed a dragonfly.

**8/15/17:** Time: 1055-1200.

1055 – As observations begin, Sinbad is sighted with a medium-sized fish in the NW corner of the SC. At least 4 chicks can be seen swarming around him. Sparky takes the fish and quickly swallows it. Sinbad then resumes foraging. All of the chicks but Cobra move to ~ 1.5 m east of the nest and park there. Remy isn't seen at this time.

1100 – Quark, Phantom and Sparky are parked from .5-2 m east of the nest. None of the other family members are seen at this time.

1103 – First Sparky, followed soon after by Quark and Phantom dive and duck into the reeds as a double-crested cormorant approaches their area from the SW section of the SC (the cormorant is just foraging for fish in the usual manner, engaging in successive dives). It passes ~ 4-5 m east of the nest as it rounds the west end and proceeds east off the north bank of the SC and soon leaves the SC.

1107 – Sinbad is sighted just east of the nest with a large crayfish. All of the chicks but Spot are swarming around him as he works the crayfish. Sparky soon takes control of the crayfish and keeps on the move, circling about ahead of its pursuing sibs and making repeated attempts to swallow the crayfish. After over one minute of this Sparky is finally able to swallow the crayfish (time: 1109). The feeding lasted ~ 93 seconds. Sinbad then resumes foraging and all of the chicks eventually make their way into the reeds north of the nest.

1111 – Remy is sighted moving into the SC from the east. She goes to the mid-SC area where she parks and preens. Phantom, then Quark, Sparky and Cobra go out to Remy and remain with her.

1118 – Sinbad comes out of the SE section of the SC with a large fish and heads toward the 4 chicks who are with Remy. Quark, Sparky and Phantom spot him approaching from a distance. They rush toward Sinbad and meet him en route. The 3 chicks swarm around Sinbad as he begins working the fish. Though the fish is clearly too large for the chicks he nevertheless gives them a chance to swallow it. Several attempts are made by chicks but all fail miserably. Finally, after working the fish for ~ 87 seconds Sinbad swallows it (time: 1120). During this feeding Phantom physically attacked and tried to drive Quark away from the feeding by pecking at and briefly chasing Quark. However, Quark was persistent and quickly rejoined the feeding fray. When the feeding concluded, Phantom suddenly attacked Quark again. Quark didn't back down or retreat and instead engaged in a very intense fight with Phantom. The two locked together in battle, pecking and grappling, with their wings often flapping. This intense fight continued for ~ 5-6 seconds (and appeared to be intensifying) before Sinbad put an abrupt end to it. He had been ~ 1-1.5 m away when the fight began and as it continued on he quickly decided that such intense fighting among these chicks wasn't going to be tolerated and he charged toward the fighting pair. The battling chicks promptly spotted Sinbad and immediately dove, with Sinbad diving after them. Shortly after, Phantom and Quark resurfaced separately, away from each other and the fight was thus concluded (time: 1120). This was the first time that I have observed the competition between the chicks of this brood involve directed physical attacks.

1125 – Sinbad is parked and preening by himself in the mid-SC area. Remy is in the NW section of the SC. Phantom is with her and the other chicks are in the reeds in the NW section. The NR male is sighted off the west bank of the river just north of the territorial boundary between the pairs.

1129 – Remy (still parked in the NW section with Phantom; Sparky has also moved near Remy) moves toward Sinbad (still parked in the mid-SC area); Phantom and Sparky follow her. Remy nears Sinbad but then veers east and goes to the east end of the SC. Phantom and Sparky then move to Sinbad (soliciting him upon arrival) and remain with him.

1133 – Sparky is still with Sinbad in the mid-SC area (Phantom has moved away), intermittently soliciting him. The other 4 chicks are spread out in the NW section.

1135 – Remy moves toward the chicks from the NE point of the SC with a small prey item (my view is too inadequate to ID it). Sparky spots Remy early and rushes to her, meets her en route and takes and quickly swallows the prey item. Remy then goes back near the NE point and resumes foraging. Sparky goes to the reeds at the north bank, west of the NE point (one other chick is already there).

1136 – Sinbad moves out to the NE section of the SC and parks there.

1138 – Quark goes to Sinbad, solicits him upon arrival and remains with him. Soon after, Phantom and Sparky also come out and stay near him.

1141 – Remy moves toward these 3 chicks from the NE point with a small prey item (again my view is too inadequate to ID it). The 3 chicks rush toward her. Quark (who had been the nearest to Remy) arrives first, takes and quickly swallows the prey item.

1142 – Sinbad, Remy and all 5 chicks have gathered near the north bank, east of the old nest site.

1143 – Sinbad moves out to the east end of the SC, south of the NE point. The chicks remain with Remy in the same location as at 1142. The NR male is just north of the territorial boundary.

1144 – Sinbad and the NR male engage in a territorial standoff at the territorial boundary between the pairs (just north of the NE point of the SC). Remy is still with all of the chicks in the same location as at 1142 (east of the old nest site). The standoff proceeds in the usual manner.

1149 – The standoff between Sinbad and the NR male continues on in the usual manner as when no fighting occurs (i.e., just posturing/displaying). Remy moves to the NW corner of the SC; all of the chicks follow along with her except Spot, who remains at the reeds east of the old nest site.

1150 – The standoff abruptly ends as Sinbad and the NR male just suddenly move apart, each going his own way into his respective territory: Sinbad heads into the SC and the NR male moves north.

1151 – Remy goes to the mid-SC area and begins foraging. All of the chicks are now in the NE section of the SC.

1152 – Sinbad moves toward the chicks (who are still in the NE section) from the NE point with a medium-sized fish. All 5 of the chicks go to Sinbad and swarm around him as he works the fish. Spot soon takes it from Sinbad and eventually swallows it (the feeding lasted ~ 28 seconds). Sinbad then resumes foraging.

1154 – While foraging in the mid-SC area Sinbad calls once.

1156 – Sinbad parks and preens in the mid-SC area. Sparky goes to Sinbad, remains with him and intermittently solicits him.

1200 – Sinbad is still parked with Sparky in the mid-SC area. Cobra and Quark are in the NW section. Remy, Spot and Phantom aren't seen at this time. End observation period.

*Comments:* The chicks spent the entire rather eventful observation period in the SC. As is usual now, they were very mobile. They are at the point now where they are beginning to move out to areas where they see a parent foraging or will accompany a parent to the area where the parent forages for them (though they are still rather restrictive in regard to just how far they will go and where). Basically, they have progressed well beyond the stage of just staying together in a location and waiting for the parents to bring them food. In a similar vein, they have now also moved beyond the tendency and inherent desire to mainly stay together as a group. In fact, lately several of the chicks have been spending a notable amount of time ranging independently (though mostly within or along the margins of the reeds and algal mats). Specifically, Cobra and Spot are the chicks who mainly have done so (they have also been seen off together by themselves). All of the chicks have been observed doing some independent foraging.

Another territorial standoff between Sinbad and the NR male (at the territorial boundary) was observed today which involved only stereotypical posturing/displaying and no fighting. Given the number of times lately that I've seen these during my observational sampling periods it seems a likely possibility that territorial standoffs are a daily occurrence at this time, perhaps even occurring multiple times during the day.

The already intense feeding competition among the chicks ramped-up enormously today between a few of the chicks. For the first time in my observations of this brood a chick was seen taking direct aggressive physical actions against one of its sibs during (and immediately after) a feeding. During the "feeding" that occurred at 1118 Phantom very aggressively pecked at, charged and briefly chased Quark (the smallest of the chicks). Then immediately after the feeding (in which Sinbad ended up eating the large fish himself) Phantom attacked Quark and a fight ensued. Though Phantom was getting the better of the battle, Quark staunchly stood its ground, conceding nothing to Phantom. Regardless, Sinbad was having none of this sustained, extremely intense fighting among the chicks and broke it up. Clearly, Phantom was looking to establish and/or consolidate its dominance over the smaller Quark. For its part, Quark demonstrated that (at least at this point), even though subordinate it will not readily concede and will be a staunch competitor (i.e., it will not be an easy mark). Thus far Quark has proven that it is no pushover. It readily mixes it up with its sibs during feedings and despite typically being bested by its brood mates it has managed to regularly get its share of food. Somewhat surprisingly, on several occasions while competing with 2-4 sibs during crayfish feedings Quark has managed to come away with and swallow the reduced but still mostly whole carcass, which is the ultimate prize of such feedings. Quark is also proficient at vigilantly watching for parents approaching at a distance with a prey item and being the first to spot the parent, then (having gained a head start) racing to the parent, reaching it first and thus being able to take the prey item before having to compete with any sibs for it (i.e., providing that the prey item is of a size that Quark can readily swallow without it requiring extensive working). As previously noted, though the smallest of the brood, Quark is not runty.

**8/16/17:** Time: 1050-1200

1050 – As observations begin, a check of the SC reveals that none of the family is there. I thus head over to the RR bridge to see if they are in the West Lagoon.

1053 – (From the RR bridge) Remy is sighted foraging in the northern half of the West Lagoon. Sinbad isn't sighted at this time nor are any of the chicks. This indicates that the chicks most likely are at the edge of or within the west shore reeds. There are no safe locations on the RR bridge from which I can conduct lengthy observations of the West Lagoon (it has proven to be a somewhat unsatisfactory observation point during my previous observations of the family when in the West Lagoon) so I make a trip around the south shore of the lagoon to find a satisfactory vantage point on the east shore (notably, one which provides a good view of the west shore—thus far, the chicks have exclusively utilized the west shore reeds and the waters just east of them when in the West Lagoon).

1105 – (From an observation point on the mid-east shore of the lagoon—henceforth when noting that I have gone to a West Lagoon observation point it will refer to this location unless otherwise stated.) Sinbad is sighted by the edge of the west shore reeds in the mid-north section of the lagoon. None of the chicks are visible but it is likely that they are in the reeds near him.

1107 – Sinbad moves out to the mid-water area in the north end of the lagoon. Remy has parked near the reeds in the area where Sinbad had been at 1105. The chicks are finally sighted—all 5 are with Remy.

1112 – Sinbad moves to the edge of the west shore reeds and joins Remy and the chicks there. Approx. 20 western gulls (adults and immatures/fledglings) are grouped in the mid-water area to the east of the grebes. The chicks appear to be even more uneasy than usual when in the West Lagoon (with some retreating into the reeds) and the presence of the gulls is likely the cause.

1115 – Sinbad begins foraging by the reeds, working his way south. At least 2 chicks can be seen following him for part of the way, keeping to the edge of the reeds (my observation point on the east shore is fairly distant, making it difficult to readily ID the chicks). Remy is still parked in the same location; at least 2 chicks are with her.

1118 – A chick is seen foraging in the reeds by Remy.

1121 – Sinbad is sighted moving north along the reeds with a medium-sized crayfish. Two chicks spot Sinbad approaching and rush to him. Sinbad begins working the crayfish and soon one of the chicks (probably Sparky) takes control of the crayfish and swallows it (the feeding lasted ~ 25 seconds). Sinbad then resumes foraging.

1122 – Remy also begins foraging. The chicks who had been with Remy remain at the edge of and within the reeds in the general area where she left them.

1126 – Remy moves toward the chicks from along the west shore reeds with a medium sized fish (it is fairly large for the chicks). Four chicks rush toward her as she approaches. Sparky takes the fish and while being pressured and pursued by some of its sibs, works the fish for ~ 86 seconds before finally swallowing it (Remy shadowed Sparky and oversaw the feeding).

1130 – All of the chicks are within the west shore reeds in the mid-north section of the lagoon but are currently out of view. Sinbad is foraging in the mid-water area of the northern half of the lagoon near the gulls.

1132 – Sinbad parks near the gulls and preens himself.

1136 – Quark and Cobra emerge from the reeds and move out to Sinbad, soliciting him upon arrival (the gulls are still nearby). Sinbad then leads them back to the edge of the west shore reeds and begins foraging. Quark and Cobra remain at the edge of the reeds.

1139 – Remy is sighted moving along the reeds from the north toward Cobra and Quark with a medium-sized crayfish. Cobra spots Remy early, rushes to her (easily beating Quark), takes and quickly swallows the crayfish. Remy then parks by the reeds in this area in the northern half of the lagoon and preens. Four of the chicks (all but Cobra) sit around her.

1145 – Remy is still parked in the same location (by the west shore reeds in the mid-north section of the lagoon). Quark is sitting next to her. The other chicks are nearby at the edge of or within the reeds. Sinbad isn't seen at this time.

1147 – Sinbad is sighted foraging near the west shore reeds in the midsection of the lagoon, working his way south. At least 3 chicks have followed him to this general area, mainly staying within the reeds (some of them appear to be doing some independent foraging there).

1151 – Sinbad is still foraging by and within the west shore reeds in the mid-section of the lagoon. At least 3 of the chicks are in that area. The gulls have left the northern half of the lagoon and have gone to its south end and also out on the beach.

1155 – Sinbad is moving along the west shore from the south in the mid-section of the lagoon with a fairly large fish. He goes to the area where the chicks are and 2 of them come out from the reeds and go to him. He gives them a chance to swallow the fish but after ~ 55 seconds and several failed attempts by the chicks (the fish is just too large for the chicks) he swallows it himself. Sinbad then resumes foraging and the 2 chicks go back into the reeds.

1200 – Remy has parked ~ 8 m from the west shore reeds in the mid-north section of the lagoon and is preening herself. None of the chicks are seen (they are probably tucked in the reeds near her). Sinbad isn't seen at this time. End observation period.

*Comments:* The chicks spent another entire observation period in the West Lagoon. As is usual for them when in the West Lagoon, they spent little time in open water and stayed tucked within the reeds for most of the period. The contrast between their restrictive ranging today and their very free ranging while in the SC yesterday is quite striking. Clearly, they still feel notably more apprehensive and less secure when in the West Lagoon than when in the much more familiar SC. From my human perspective the chicks just seem considerably more exposed in the greater openness of the West Lagoon than in the relatively enclosed SC. I suspect that (aside from their being less familiar with the lagoon) in some manner the chicks perceive something similar.

In the last two days a few of the chicks were seen stretching and briefly flapping their wings. In regard to the feathers, the wings are still essentially undeveloped. They are still covered in down but this downy covering has thinned considerably and is clearly on its way out. Also, the buds of flight feathers have appeared along the posterior margin of the wings. However, the base structure of the wing (i.e., the limb bones and associated tissue and skin) has notably elongated and in form and length (though not yet full length) now resemble a proper wing. In general, this aspect of the wings has progressed well beyond the stunted, neonatal form.

All 5 of the East Lagoon second brood chicks were seen. They were spread out in both the northern and southern halves of the East Lagoon. Two of the now independent East Lagoon first brood chicks were seen in the south end of the lagoon.

**8/17/17:** Time: 1002-1105.

1002 – As observations begin, Quark is sighted with Remy ~ 17 m from the reeds at the west shore of the mid-north section of the West Lagoon. Quark is in possession of a fairly large crayfish that it has just taken from Remy (the crayfish has been worked a bit by Remy but it is still quite large and bulky). Sparky is approaching the pair from the south. Quark attempts to swallow the crayfish but it is too large and Quark ejects it onto the water. Sparky arrives and takes control of the crayfish and soon swallows it. Remy and the 2 chicks then remain parked there. Neither Sinbad nor any of the other chicks are seen at this time. It appears obvious that the chicks are probably in and/or along the west shore reeds (I checked the SC and found that they were not there) so I head to my observation point on the mid-east shore of the lagoon. As I am leaving my location on the RR bridge Spot and Cobra come out from the reeds and join Remy and the other 2 chicks (time: 1005).

1015 – I arrive at my West Lagoon observation point. Four chicks are seen at the edge of the west shore reeds in the mid-north section of the lagoon. At this time there are no gulls anywhere near this area—all of the gulls in the general vicinity are at the south end of the lagoon and the adjacent section of beach.

1017 – Sinbad approaches the chicks along the reeds from the south with a large crayfish. Phantom is the first to spot Sinbad and rushes to him. Sinbad then begins working the crayfish. The other 4 chicks soon join in on the feeding. Quark eventually grabs and takes control of the carcass and darts about with the crayfish ahead of its hotly pursuing sibs. Quark retains control for most of the feeding and tries several times to swallow the crayfish. However, the carcass had not been reduced enough for Quark and is still too bulky for this chick to swallow. Finally, when Quark drops the carcass after another failed attempt to swallow it, Sparky takes control of it and soon swallows the carcass (time: 1020). This feeding was quite lengthy, lasting ~ 3 minutes 5 seconds. Sinbad shadowed whichever chick had control of the carcass (diving completely below the surface to retrieve it several times when it sank after being dropped) and oversaw the feeding. At its conclusion, Sinbad moves out toward mid-water (calling once en route) and resumes foraging. Several of the chicks move to Remy who is parked in the mid-water area in the northern half of the lagoon.

1027 – One chick is in the mid-water area by itself (my view is too inadequate to ID it). Sinbad moves toward it from the east side of the lagoon with a large crayfish. The chick spots Sinbad, moves toward him and they meet en route in the mid-water area. Sinbad briefly works the crayfish and detaches what appears to be a claw, which the chick quickly takes. Sinbad then continues toward the west shore with the chick (who swallows the claw) in tow. When Sinbad reaches the edge of the west shore reeds with the chick he begins working the crayfish again. Three other chicks come out to Sinbad and he then has 4 chicks swarming around him and intensely competing for the crayfish and the scraps that he detaches from it. The feeding continues for a lengthy ~ 3 minutes 28 seconds with several failed attempts to swallow the crayfish before Sparky finally assumes control of the carcass and eventually swallows it (time: 1031). Sinbad then resumes foraging. Just as the feeding concluded several of the chicks went to Remy who was approaching in the mid-water area with another crayfish and Remy conducts a feeding there.

1036 – Remy is now parked and preening near the east shore at the north end of the lagoon. One chick is visible parked near the edge of the west shore reeds in the mid-north section of the lagoon. None of its sibs are seen at this time (they are tucked in the reeds). Sinbad also isn't seen at this time.

1038 – Sinbad is sighted moving along the west shore reeds from the south with a medium-sized fish. First one, and shortly after, a second chick spots Sinbad approaching and they rush toward him (my view is too inadequate to ID the chicks). The first chick reaches Sinbad well before the second and takes and quickly swallows the fish. Sinbad then resumes foraging. A group of about two dozen western gulls has come into and occupied the mid-water area to the east of the chicks' location (the chicks are at the west shore reeds in the mid-north section of the lagoon).

1041 – Two of the chicks (Sparky and Cobra) move south along the reeds to the mid-section of the west shore (the general area where Sinbad is foraging).

1043 – Sinbad moves toward Sparky and Cobra (a third chick has also moved near them) from the south with a medium-sized crayfish. Sparky and Cobra rush toward Sinbad and they meet him en route. Sparky takes and soon swallows the crayfish. Sinbad then goes out to the mid-water area (calling once en route) and parks by the gulls and preens himself.

1044 – Sinbad moves to the edge of the west shore reeds in the mid-section of the lagoon, parks there and looks about. Remy parks and preens by the west shore reeds in the mid-north section of the lagoon. At least 3 chicks are in her vicinity.

1049 – Sinbad (still in the same location) begins foraging by the reeds. Remy is still parked in the same general location as at 1044. At least 4 of the chicks are in her vicinity.

1056 – Remy leaves the chicks, heads east into the mid-water area and begins foraging. The chicks remain at the edge of and within the reeds where she left them.

1100 – Sinbad is sighted moving by the chicks (just east of the reeds) from the south, veers out to the mid-water area and resumes foraging. Phantom follows after him part of the way out but then retreats back to the reeds.

1103 – Remy is sighted moving toward the chicks from the NE with a medium-sized crayfish. Cobra sees her approaching, rushes out to her, takes the crayfish and quickly swallows it. Remy then goes to the mid-water area and resumes foraging. Cobra returns to the reeds.

1105 – The chicks are still at the edge of and within the west shore reeds in the mid-north section of the lagoon. Remy and Sinbad aren't seen at this time (they are probably foraging). The gulls are still grouped in the mid-water area east of the chicks. The chicks have mainly kept to the reeds since the gulls moved there. End observation period.

*Comments:* The chicks again spent the entire observation period in the West Lagoon. During the first part of the period they looked more at ease there, spending more time than usual in open water. However, when the western gulls later moved into the northern half of the lagoon the chicks once again kept mainly to the west shore reeds.

The chicks have grown noticeably over the past few weeks and all have progressed. Though there is noticeable size disparity among the brood, none of the chicks look runty. In fact, the size disparity among this brood is notably less marked than that of the West Lagoon brood that I observed here last year. Quark is noticeably the smallest of the brood but still appears to be progressing well. At this point Sparky is the largest of the chicks and is the best competitor during group feedings (relating to the latter, Sparky also appears to be able to swallow larger prey items than its sibs). All of the chicks engage in some independent foraging, mainly within the reeds and algal mats. Some of the chicks are already spending notably more time ranging and foraging independently than their sibs, most notably Cobra (Spot ranks second in this respect). On the other hand, Quark and Phantom are seen next to the parents more often than the others. Of course, all of the chicks are still completely dependent on their parents for their survival (i.e., for food, protection and guidance). This includes dependence on the ability of the parents to retain a sufficient amount of territory in which to rear the chicks. In this case, the annexation of the West Lagoon after losing the northern half of their territory to the NR pair may prove to be a decisive factor (providing that the West Lagoon can fully compensate for that loss—the current indications are that it easily can).

All 5 of the East Lagoon second brood chicks were again sighted today. They were widely spread out (with a few in proximity) in the southern half of the East Lagoon. Several of the chicks were seen following and taking food from one of the parents. In this latter situation some intense competition was observed between the chicks as one of them pecked at, charged and briefly chased its sib away from close proximity to the foraging parent (and thus was attempting to exclusively hold this favorable position for itself).

**8/18/17:** Time: 1102-1205.

1102 – As observations begin, Sinbad (foraging) is being accompanied by Sparky in the east side of the West Lagoon at the mouth of the inter-lagoon channel. At the same time, Remy is parked in the west end of the SC, NE of the nest (well out from the reeds) and 2 of the chicks are sighted in the NW section of the SC (my view is too inadequate to ID them). The other 2 chicks aren't seen at this time (they are probably in the reeds somewhere in the SC).

1103 – Sinbad leads Sparky west across the lagoon (calling once en route) to the west shore reeds. He then begins making his way north toward the SC with Sparky following along.

1107 – Sinbad (trailed at a distance by Sparky) enters the SC from the SE end, goes to the area along the north bank just west of the NE point and begins foraging. Sparky follows at a distance to that area.

1111 – Sinbad is foraging along the reeds west of the NE point. He calls once and then resumes foraging. Several of the chicks are in open water in the NW section and they duck into the reeds as a train noisily passes by the area (all of the chicks are now in the reeds).

1118 – Remy moves to the mid-SC area from the SW corner with a medium-sized crayfish. Only Quark is in the area that she is heading for. Quark spots Remy approaching and rushes to her. They meet en route ~ 16 m NE of the nest and Remy begins working the crayfish. Quark takes it soon after and unsuccessfully attempts to swallow it, dropping it onto the water after the failed attempt. Remy then grabs it and works it some more. Quark soon takes it again and after much effort (with Remy shadowing the chick and overseeing the feeding) is finally able to swallow the crayfish (the feeding lasted ~ 53 seconds). Spot came out from the reeds in the NW corner late in the feeding and rushed toward the pair looking to join the feeding. However, Spot arrived too late, reaching them just as the crayfish was disappearing down Quark's throat. After the feeding, Remy resumed foraging.

1123 – Phantom and Spot have parked in the NW section of the SC. Remy is foraging along the reeds at the NE point. None of the other family members are seen at this time.

1127 – Sinbad moves to the area ~ 10 m NE of the nest from the SW corner with a fairly large crayfish. Sparky, Quark and Phantom go to him and he begins working the crayfish. In its impatience, Sparky takes and assumes control of the crayfish too early (Sinbad allowed this)—it is obvious that the carcass hasn't yet been reduced sufficiently for the chicks to be able to swallow it. Sparky then darts about with the crayfish, keeping ahead of its hotly pursuing sibs, as it tries to swallow the crayfish. After ~ 30 seconds Quark and Phantom concede the carcass to Sparky and cease contesting it. Sparky works the carcass for over 1.5 minutes, detaching and eating a few morsels but in the end isn't able to reduce the carcass enough to swallow it and abandons it, just dropping the carcass and letting it sink away. Sinbad had shadowed and oversaw the feeding but after ~ 70 seconds of Sparky controlling the carcass (and Quark and Phantom ceasing to contest it anymore) he just left Sparky to it and resumed foraging (calling once just before doing so).

1133 – Sinbad parks at the SE end of the SC and preens. Remy is now parked with Phantom and Sparky in the mid-SC area, preening herself.

1136 – Sinbad moves a little farther into the SC from the SE end and voices a short, single, one note call (I'm not able to determine what prompted this call). One of the NR pair (probably the female) is foraging far to the north in the mid-section of the NR pair's territory.

1140 – Sinbad (still in the same location) calls once, then goes to the SW section of the SC and begins foraging. Remy is still parked in the same location in the mid-SC with Phantom. Sparky is now parked ~ 16 m to the east of Remy.

1141 – Phantom leaves Remy and moves to the SW corner where Sinbad is foraging (Phantom is obviously looking for a feeding, thus leaving an inactive parent and going to one who is foraging).

1142 – Phantom moves back to Remy. Remy and Sparky are in their same respective locations as at 1140. The other 3 chicks are sighted in the reeds of the west bank in the NW corner (2 of the chicks are foraging there).

1145 – Sparky moves to within ~ 3 m east of Remy and Phantom. Phantom is intermittently soliciting Remy. Sinbad isn't seen at this time.

1150 – Remy ceases preening, moves to the reeds at the south side of the NE point of the SC and begins foraging. Phantom has followed her there and parks by the reeds where she is foraging, monitoring her progress.

1152 – Sparky and 2 other chicks move to the algal mat and reeds in the SW corner where Sinbad is foraging.

1153 – Sinbad exits the reeds in the SW corner, calls once, goes out to the area south of the NE point and resumes foraging there. The 3 chicks remain in the reeds in the SW corner. Quark is parked and resting in the reeds in the NW corner.

1157 – Sparky and Cobra emerge from the reeds in the SW corner and move into open water ~7 m from the reeds. They stop by what appears to be a floating piece of vegetation or a branch and both repeatedly dive alongside and under it (it's not clear whether they are foraging or just diving out of curiosity).

1159 – Remy is sighted with a medium-sized crayfish by the reeds of the north bank, just west of the NE point. Three chicks (Sparky, Spot and Quark) are swarming around her. The feeding lasts ~ 53 seconds but my view is poor and I'm unable to see if the crayfish was swallowed or who swallowed it.

1202 – Remy is again sighted with a crayfish by the reeds of the north bank. This time she has a large crayfish and is east of the old nest site. Quark, Spot and Sparky are swarming around her but after working the crayfish for ~ 23 seconds she swallows it herself. Remy then moves to the NW section of the SC where she parks and preens (in open water). Some of the chicks follow her west but stay by and within the reeds.

1205 – Remy is parked in the same location as at 1202. All of the chicks are at the edge of and within the reeds that line the bank in the NW section. An osprey approaches the SC from the east and circles over the east end of the SC. As it does, Remy reacts by suddenly voicing 3-4 successive short, sharp, single one note alarm calls and poising her body for potential physical action. Since all of the chicks were already in the reeds there was no noticeable reaction to her calls. The osprey quickly executed its U-turn and moved east out of the area. Shortly after the osprey left the area Phantom came out of the reeds at the north bank, went to Remy, solicited and remained with her. End observation period.

*Comments:* When (at the beginning of the observation period) Sinbad was seen with a chick in the West Lagoon and Remy was with several chicks in the west end of the SC, this marked the first time that chicks have been observed to be split among the parents when the parents were far away from each other (notably, when one was in the SC and the other was in the West Lagoon). Another notable aspect regarding the chicks' increasing mobility is that over the past few days they have increasingly been seen accompanying foraging parents for short distances and also moving a fair distance to a parent whom they see foraging in the distance.

In the early period after the chicks hatched the parents were seen feeding fish to the chicks more often than crayfish. Over the past week or so this trend has begun to reverse, especially since the chicks began going into the West Lagoon.

**8/19/17:** Time: 1135-1250.

1135 – As observations begin, Sinbad and Remy are parked with 3 chicks ~ 3 m out from the west shore reeds in the mid-section of the West Lagoon (this sighting is from the RR bridge). Neither of the other 2 chicks is seen in the SC—they are probably in the reeds near the others in the West Lagoon. Approx. 3 dozen western gulls are in the north half of the West Lagoon, mainly in its eastern side. At this time I begin heading for my West Lagoon observation point on the mid-east shore.

1145 – I arrive at the east shore observation point. Remy and Cobra are sighted as they move north along the west shore reeds from the mid-section of the lagoon.

1150 – Sinbad is sighted foraging along the east shore, working his way north from the south end of the lagoon. As he passes close to my observation point he repeatedly voices short, single, one note calls, probably in response to my presence. Two chicks (Sparky and Cobra) are visible at the edge of the west shore reeds in the mid-section of the lagoon.

1151 – Sinbad continues on into the north half of the lagoon and then crosses the lagoon to the west shore area.

1153 – Sparky moves north along the west shore reeds to the north end of the lagoon (just south of the RR bridge). At least 3 other chicks are already there at the edge of the reeds. About 2 dozen of the gulls have left the north half of the lagoon but a dozen or so still remain in the east half.

1158 – Sparky and Quark are sighted by the edge of the west shore reeds at the north end of the lagoon. One of the parents (probably Remy) is briefly sighted foraging there.

1201 – Remy is sighted just as she is swallowing a large crayfish by the west shore reeds at the north end of the lagoon (just south of the RR bridge) with 2 chicks swarming around her. She then resumes foraging.

1203 – Remy parks and preens near the west shore reeds in the mid-north section of the lagoon. Quark, Sparky and Cobra sit in the water near her.

1205 – Remy is still in the same location preening herself. Quark is still sitting by her. The other 4 chicks are strung out along the edge of the reeds to the south of Remy.

1209 – Sinbad is sighted foraging by the west shore reeds in the mid-section of the lagoon, slowly working his way south. Phantom moves to the area where he is foraging. Remy and Quark are still parked together in the same location as at 1205.

1210 – While foraging (in the same area as at 1209) Sinbad calls once and then continues foraging.

1212 – Remy is still parked with Quark in the same location. Sparky and Spot are visible at the edge of the reeds near them.

1214 – Quark leaves Remy and heads south along the reeds toward Sinbad and Phantom. I sight the latter two just as Phantom finishes swallowing a crayfish that it undoubtedly had taken from Sinbad (they are by the west shore reeds in the mid-section of the lagoon). Sinbad then resumes foraging.

1215 – Remy (still in the same location) begins foraging.

1216 – Cobra, Sparky and at least one other chick are by the area in the mid-section of the lagoon at the edge of the west shore reeds where Sinbad is foraging.

1217 – Sinbad moves out to mid-water in the mid-section of the lagoon (calling once en route) where he parks and preens. Sparky follows him at a distance, solicits Sinbad upon reaching him and remains with the parent.

1220 – Cobra emerges from the reeds and moves out toward Sinbad (Sparky is still with him). Upon reaching Sinbad, Cobra immediately solicits him.

1221 – Sparky and Cobra leave Sinbad and move back to the reeds. Sinbad has drifted west to ~13 m from the reeds and is parked there and looking about. Remy hasn't been sighted since 1215.

1224 – Several of the chicks are briefly seen foraging within the west shore reeds in the mid-section of the lagoon by the snags that are located there.

1226 – Remy is sighted with a large crayfish just off the west shore reeds in the mid-section of the lagoon (south of the snags). Four chicks (all but Phantom) go to Remy. She works the crayfish for ~ 52 seconds as the chicks swarm around her but in the end, she swallows it herself. Remy then moves a little farther north to the mid-section of the lagoon (near the reeds) where she parks and preens (time: 1228). Quark sits by Remy and Sparky sits ~ 1.5 m to the south of her.

1235 – Remy, Quark and Sparky are still parked in their same locations. Sinbad and the other 3 chicks aren't seen at this time.

1236 – Sparky drifts over to the edge of the reeds ~ 8 m south of Remy. Quark is still with Remy and intermittently solicits her.

1239 – Cobra emerges from the reeds, goes to Remy and solicits her. Cobra remains with Remy (Quark is also still with her). The gulls are still present in the north half of the lagoon on the east side.

1243 – All 5 of the chicks are out of the reeds and are by Remy.

1245 – Four of the chicks are still by Remy; Spot is just west of them foraging at the edge of the reeds.

1250 – Same as at 1245. Sinbad isn't seen at this time. End observation period.

*Comments:* The chicks spent another entire observation period in the West Lagoon. They appeared to be a little more at ease than usual when here but still mainly stayed at the edge of and within the west shore reeds.

One of the chicks was seen briefly flapping its wings. Though short and rudimentary, the flight feathers have begun growing.

**8/20/17:** Time: 1110-1215; Brief re-check of the family at 1305; 1315-1355.

1110 – As observations begin, Remy is sighted moving out of the SW section of the SC with a fairly large fish. She goes to the reeds along the north bank just west of the NE point of the SC and looks to the reeds for the chicks (she obviously is intending to feed the fish to the chicks). She continues looking (no chicks come out to her) and finally (from near the edge of the reeds) she dives with the fish and enters the reeds while underwater. She isn't seen emerging from the reeds at this time, indicating that she probably had spotted chicks well inside the reeds and went in to deliver a feeding (she went out of my view when she entered the reeds so I can't be certain of this). Sinbad isn't seen at this time nor are any of the chicks.

1115 – None of the family is seen at this time (Remy has not been seen emerging from the reeds where she entered with the fish). Both members of the NR pair are sighted in the mid-area of their territory.

1118 – Remy, Sinbad and 3 chicks have emerged from the reeds just west of the NE point of the SC (the location where Remy had gone with the fish). Remy moves to the area just south of the NE point where she parks and preens. Phantom goes to Remy and sits by her. Sinbad goes to the mid-SC area and begins foraging. The other 2 chicks (my view is too inadequate to ID them) move a short distance to the west and go back into the reeds.

1123 – Remy moves near the reeds of the north bank of the SC with Phantom. As she nears the reeds Cobra emerges from them, goes to Remy and solicits her. Cobra then lingers with her briefly before going back into the reeds. At least 2 other chicks are briefly seen inside the reeds just west of the NE point.

1126 – Remy is parked with Phantom in front of the reeds to the west of the NE point. Three chicks are seen inside the reeds just north of her. Sinbad isn't seen at this time.

1129 – Sinbad is sighted moving into the mid-SC area where he parks and preens. Remy is in the same location as at 1126, now with Cobra and Quark by her (they intermittently solicit her). Phantom has left Remy and gone into the reeds.

1134 – Remy moves to the NW corner of the SC; Quark and Cobra accompany her there.

1136 – Sinbad moves out to the east end of the SC (calling once en route), parks there and looks about (especially to the north and east).

1140 – Sparky and 2 other chicks are in the reeds just west of the NE point. Sinbad is still parked in the same location as at 1136, continuing to look about. Quark is in front of the reeds in the NW corner. Remy isn't seen at this time.

1144 – Sparky, Phantom and Spot and (separately) Sinbad move to the west end of the SC in the general area of the nest and go into the reeds there. Sinbad begins foraging (one of the chicks solicits him just before he begins).

1148 – Remy moves toward the west end of the SC from the NE section with a medium-sized crayfish. Spot comes out from the reeds by the nest and rushes toward her. Spot meets Remy en route, takes the crayfish and swallows it ~ 15 seconds later. Spot then goes back to the reeds by the nest. Remy parks in the NW section and preens.

1152 – Sinbad, Phantom and another chick (my view is too inadequate to ID it) emerge from the reeds north of the nest. Sinbad goes to the north bank of the SC in the NE section (followed by the unidentified chick) and continues foraging there by the reeds. Phantom veers over to Remy (still parked in the NW section), solicits and remains with her.

1156 – All 5 of the chicks gather around Remy (still in the same general location as at 1152 but she has drifted to ~ 10 m from the reeds). Soon after, all of the chicks go into the reeds along the west end of the SC (the chicks had repeatedly solicited Remy but she didn't respond to this).

1200 – Sinbad is sighted foraging in the NW corner. Remy is still parked in her same location. Phantom is resting inside the edge of the reeds in the NW corner. The other chicks are tucked in the reeds.

1202 – Remy moves out to the area just south of the NE point and parks there. Two of the chicks follow her at a distance. Both eventually go to Remy and solicit her. They then go to the reeds at the south side of the NE point.

1205 – Sinbad moves out of the NW corner to the mid-SC area and continues foraging there. Phantom follows him out for only ~ 8 m and then returns to the reeds in the NW corner. Remy begins foraging at the NE point. The other 4 chicks are in the reeds at the NE point.

1207 – Remy is sighted near 3 of the chicks at the reeds just west of the NE point. One of the chicks (my view is too poor to ID it) has a small crayfish which it soon swallows (it almost certainly had obtained the crayfish from Remy). Remy then resumes foraging.

1210 – Sinbad moves east from the SW corner of the SC to the area ~ 20 m north of the SE point (calling once en route) where he parks and preens. Spot (coming from the reeds at the NE point) moves all the way across the lagoon by itself to Sinbad, solicits and remains with him. Remy is now parked ~ 12 m south of the old nest site preening herself. Quark is with her. Sparky (by itself) goes to the SW corner by the reeds.

1215 – Spot leaves Sinbad (who is still parked north of the SE point) and heads toward Remy (who is still parked with Quark in the same location as at 1210). Sparky is still parked in the SW corner. The other chicks are in the reeds, out of my view. End observation period.

*1305 – Re-check of the family: Sinbad is in the West Lagoon with Spot and Quark near the west shore reeds in the mid-north section of the lagoon. Remy is parked in the NW section of the SC with Phantom and Sparky (both solicit her often). Cobra is in the reeds in the SW corner. About 5 minutes later Sparky (who appeared to be extremely eager for a feeding) apparently became frustrated at Remy's lack of response to its repeated food solicitations. Sparky then left Remy and (by itself) made its way all the way into the West Lagoon, obviously searching for Sinbad whom it apparently hoped would (unlike Remy) be foraging for the chicks. Sparky eventually found Sinbad there and (along with Spot and Quark) accompanied him as he foraged. This was easily the farthest that I have observed any of the chicks travel by itself during my observations of the family.

*1315-1355 – As I was departing the area at ~ 1315 I sighted Sinbad in the north half of the West Lagoon heading north ~ 9 m east of the west shore reeds. At that moment I wasn't aware that this would mark the beginning of what would become a rather fraught drama. Sinbad was clearly on his way toward leaving the West Lagoon and heading back to the SC. He made his way directly toward the RR bridge leading the chicks at a steady pace. The chicks followed close at his heels however only 2 of them (Spot and Sparky) were with him. Little Quark was nowhere to be seen. Sinbad led the 2 chicks under the RR bridge all the way to the west end of the SC, calling once en route. Still there was no sign of Quark in the West Lagoon. Several minutes later Quark was finally sighted by the west shore reeds in the mid-north section of the lagoon (at some point Quark obviously had lost contact with Sinbad and its sibs). The small chick was slowly working its way north, searching along the edge of the reeds for Sinbad and its 2 sibs. When Quark reached the area ~ 25-30 m south of the RR bridge it stopped and lingered there, appearing reluctant and apprehensive about going under the bridge by itself. It then went a short way out (east) into open water, then angled back to the reeds and doubled back to the south, continuing its search. This general pattern frustratingly repeated itself several more times. At one point the chick approached to within ~ 12 m of the bridge and it appeared that Quark was finally going to cross under it. However, the chick again lost its nerve and doubled back to the south. As the minutes dragged by the situation was looking increasingly more serious. Factoring into the troubling nature of this situation was the presence of several dozen western gulls in the mid-area of the northern half of the lagoon, just a short distance south of Quark. Though the chicks have grown considerably since hatching they are still small and very vulnerable (this is especially so for Quark, who is the smallest of the brood). It was especially worrisome whenever the chick moved farther out into open water where it looked particularly exposed. Several times during this ordeal Sinbad was heard calling from the SC. I had hoped that this would give some direction to Quark but the chick made no noticeable response to these calls (these were Sinbad's stereotypical long, drawn-out territorial/contact calls). With the passing time Quark was becoming more uncertain and frantic in its searching. By about 30 minutes into this episode Quark had become noticeably more frantic in its searching and this in turn was causing the chick to become less mindful of its usual caution. At this point, after yet again turning back from the bridge the situation appeared to be taking a turn for the worse. After turning away from the bridge Quark began moving at a quickened pace more deeply into the lagoon (and thus farther away from the relative safety of the SC and the protection and security of its parents and sibs). Quark first angled SE, far out into open water (approx. one third of the way across the lagoon) before finally angling back toward the reeds. Shortly after, Quark got itself into a particularly harrowing moment. One of the adult western gulls had left the main group and was sitting by the west shore reeds. Being less cognizant of remaining cautious in its current frantic state, Quark passed within a mere 3-4 m of the gull. Fortunately, the gull paid little attention to the passing chick. Quark then continued south along the reeds, angled out to open water (about one fifth of the way across the lagoon) and continued its frantic searching southward. Just as this was happening, the situation finally began looking hopeful. While in the mid-SC area Sinbad quite clearly had come to realize that one of the chicks was missing and that it was one of the chicks who had been in the West Lagoon with him. With clear purpose he set out nonstop at a fairly quick pace, rounded the SE point of the SC, passed under the RR bridge and steamed south into the West Lagoon. As he moved south away from the bridge he called. At this point Quark was in open water in the mid-north section of the lagoon and this time the chick responded immediately to the call. Quark quickly caught sight of Sinbad and the two steamed toward each other. They met animatedly en route and Sinbad promptly led Quark over to the west shore reeds, the chick sticking close to him (and hopefully having learned its lesson!). Shortly after, Quark began soliciting Sinbad and when the pair then moved deeper into the reeds they went out of my view.

The observational evidence left no doubt that Sinbad had realized that a chick was missing and that he went to the West Lagoon specifically with the intent of finding and retrieving Quark.

This episode dramatically brings home the potential for losing a chick in such situations, either through the chick becoming permanently separated or through its becoming more vulnerable and exposed to predation. Fortunately, the chicks are on the verge of entering into a stage of somewhat lesser vulnerability.

*Comments:* The chicks are very mobile now and are entering the stage where the brood will routinely be split among the two parents. In addition, they are entering the stage in which chicks will commonly seek and accompany parents who are actively foraging (including leaving a parent who is parked and inactive, and unresponsive to food solicitations).

**<u>8/21/17</u>:** Time: 1102-1205.

1102 – As observations begin (from the RR bridge), Sparky is foraging alone by the reeds at the SE end of the river just north of the RR bridge (opposite the SE end of the SC). Three of the chicks (Cobra, Quark and one other whom I can't see well enough to ID) are by the west shore reeds in the mid-north section of the West Lagoon. Sinbad, Remy and the fifth chick aren't seen at this time. None of the family is seen in the SC so I head to my West Lagoon observation point on the mid-east shore of the lagoon.

1105 – While en route to my east shore observation point Sinbad is heard calling once from the West Lagoon.

1116 – I arrive at my east shore observation point. Sinbad is parked with Phantom and Quark in the mid-water area in the mid-north section of the lagoon. Remy is parked ~ 18 m from the west shore reeds, also in the mid-north section. The other 3 chicks aren't seen at this time (they are probably in the reeds). No gulls are present in the north half of the lagoon and only a few are in the southern half.

1123 – Same as at 1116. Remy soon dives and begins foraging. One of the now independent East Lagoon first brood chicks is in the far south end of the lagoon.

1125 – Phantom leaves Sinbad and moves to the west shore reeds. Sinbad is still parked and preening with Quark in the same location as at 1116.

1127 – Phantom comes back out to Sinbad (and Quark).

1129 – Phantom again leaves Sinbad and returns to the west shore reeds in the mid-north section of the lagoon. Sparky is at the reeds nearby, preening.

1130 – Phantom, followed shortly after by Sparky, move into the reeds and go out of view. Sinbad is still parked with Quark in the same location.

1131 – Sinbad moves to the west shore reeds in the mid-north section of the lagoon with Quark closely following along. Sinbad dives by the reeds and begins foraging.

1134 – Sparky and Cobra are sighted near the reeds in the mid-north section. Both dive into the reeds and go out of view.

1136 – Sinbad is heard calling from the west shore area but isn't seen (he's probably in the reeds).

1137 – Remy is briefly sighted foraging by the west shore reeds just north of the snags that are located by the reeds in the mid-section of the lagoon (these two snags are in the water just out from the edge of the reeds with the shoreline curving around them in a small semicircle).

1140 – Spot and 2 other chicks (my view is too inadequate to ID them) are sighted with Remy near the reeds by the snags. At the moment when I first caught sight of them one of the unidentified chicks was just finishing with swallowing a prey item which it almost certainly had taken from Remy. Remy then resumes foraging and the chicks go back into the reeds.

1143 – Quark is sighted foraging at the edge of the reeds by the snags. Remy approaches from the south with a small prey item (my view is too poor to be able to ID the prey item). Quark spots Remy approaching, rushes to her, takes and quickly swallows the prey item. Remy then begins moving slowly south along the reeds and Quark follows.

1144 – Remy parks ~ 8 m from the west shore reeds in the mid-section of the lagoon; Quark remains with her.

1146 – Remy moves east toward mid-water, dives and begins foraging. Quark then goes to the reeds.

1150 – Three chicks (too distant to ID) move south along the west shore reeds to the south end of the reed beds in the southern half of the lagoon. [The west shore reeds extend in a long lush, continuous growth along the west shore of the lagoon from just south of the RR bridge all the way to about the mid-south section of the lagoon. These end reeds were actually at the SW point (they were in the SW corner) of where the West Lagoon ended during last year's breeding season. Due to the abundant rains that followed in the subsequent rainy season the lagoon is now longer than it was during last year's breeding season, its south end having extended farther out on the beach.] They go into the reeds there and are lost from view. Remy and Sinbad aren't seen at this time.

1152 – Sinbad is sighted moving north just off the west shore from the south half of the lagoon. Quark is following him. Sinbad calls once as he moves along.

1153 – Sinbad parks and preens ~18-20 m from the west shore reeds in the lower north half of the lagoon. Quark sits by him.

1155 – Sinbad begins moving north, calling once as he goes. Quark then heads for the west shore reeds. About two dozen western gulls have moved into the west half of the mid-section of the lagoon. As Quark moves toward the reeds a gull passes low in the air over the chick as it prepares to land. Quark immediately engages in an alarm dive and instantly disappears under the surface.

1159 – Remy and 3 chicks are sighted near the west shore reeds in the mid-north section of the lagoon. Spot has just taken a fairly large crayfish from Remy. The other 2 chicks (not seen well enough to ID), who had been trailing behind Spot, reach the feeding and pursue Spot as they attempt to steal the crayfish. However, Spot manages to retain possession of it and after ~ 25 seconds is able to swallow the crayfish. Remy then parks near the reeds and preens. Several of the chicks sit around Remy and often solicit her (time: 1200).

1204 – Remy is still parked and preening in the same location. Four of the chicks (all but Phantom) are sitting around her. Phantom is resting just inside the edge of the reeds just south of them.

1205 – Remy is still parked in the same location. Cobra, Spot and Quark are still sitting with her. Sparky is now resting next to Phantom. Sinbad isn't seen at this time. The East Lagoon first brood chick is still at the far south end of the lagoon where it has been (on both the east and west sides) during the entire observation period, well away from Sinbad, Remy and the chicks. End observation period.

*Comments:* All of the chicks but Sparky spent the entire observation period in the West Lagoon and Sparky spent the majority of the period there. It was a fairly routine "West Lagoon observation period" for them. They continued with their trend of appearing progressively more at ease when there but still mainly stayed at the edge of and within the reeds. Quark looked none the worse in the wake of its ordeal yesterday.

Sparky is notably more advanced than its sibs in its progression toward achieving independence. It spends noticeably more time ranging and foraging independently and appears bolder and more confident. Sparky is the largest and most dominant of the brood and is clearly the best competitor. Probably due to its larger size, Sparky is able to swallow larger prey items than its sibs.

A prime example of the latter points was observed during my continued observations of the family after the formal observation period had concluded. Sinbad was conducting a feeding of a fairly large fish with 3 of the chicks (Sparky initially wasn't present). As he worked the fish he repeatedly offered it to the chicks. However, because of its large size the chicks appeared a bit reluctant to take it, at least at that point. Eventually, several of the chicks took a turn at attempting to swallow the fish but none of these attempts came close to be being successful. At this point (~ 30 seconds into the feeding) Sparky (who had been a little to the south when the feeding began but soon caught sight of it and rushed toward the group) reached the fray. Sparky promptly grabbed the fish and took control of it and, though with some effort, managed to swallow this large fish on its first attempt (taking ~ 10 seconds to do so).

At this time, Sparky is on track to reach independence before any of its sibs. As for the other 4 chicks, at present there is no glaring gap between them in this respect. Quark (the smallest of the chicks) is lagging slightly but not by much and seems to be doing a decent job at obtaining its share from the feedings (and overall, Quark appears to be coming along well). While two of these four (Spot and Cobra, who spend noticeably more time ranging and foraging independently than Phantom and Quark) are a bit more advanced than the other two in regard to where they currently are at in their progression toward becoming independent, these four are basically on a par with each other.

**<u>8/22/17</u>:** Time: 1103-1210.

1103 – As observations begin, Sinbad and all of the chicks but Spot are in the NW section of the SC. Remy and Spot aren't seen at this time.

1105 – Remy comes out of the SW corner of the SC with a medium-sized crayfish and moves north along the west end. Sparky spots Remy and rushes to her. They meet en route and Sparky takes and quickly swallows the crayfish.

1108 – Sinbad moves out to the NE point of the SC (calling once en route), dives and isn't resighted at this time. Remy heads out to the SE end of the SC, continues east to mid-river and passes from my view. All 5 chicks are parked from 2-4 m south of the old nest site.

1110 – Quark moves to the edge of the reeds in the NW corner and rests there. The other 4 chicks are still parked south of the old nest site. Remy and Sinbad aren't seen at this time.

1115 – All of the chicks have now moved to the NW corner of the SC and are fairly spread out there.

1116 – Sparky and Spot move east along the reeds of the north bank to the area just west of the NE point. Spot soon doubles back to the NW corner.

1117 – Sparky now also goes back to the NW corner. Remy and Sinbad still haven't been resighted.

1120 – Sparky and Spot are parked by the reeds in the mid-west end of the SC. The other chicks are in the reeds in the NW corner and west end.

1123 – Remy moves into the NW section of the SC from the NE point with a fairly large crayfish. Sparky (in the NW section) spots Remy approaching, rushes to her and takes and quickly swallows the crayfish. Remy then parks in the NW section and preens herself. Quark and Sparky sit with Remy (Quark occasionally solicits her).

1124 – Cobra is sighted with Sinbad at the far SE end of the SC. Sinbad suddenly charges at and pecks Cobra (Cobra had begun to solicit him), driving the chick farther back into the SC. Cobra then moves to the NE section.

1130 – Sinbad moves to the area south of the NE point. Cobra goes to Sinbad and solicits him. Soon after, Sinbad dives and begins foraging. Remy is still parked in the NW section with Quark (sitting right at her side) and Sparky.

1134 – Remy slowly drifts to the NW corner. Quark follows Remy closely (often soliciting her) and Sparky does so at a distance. Cobra goes to the area ~ 6 m south of the old nest site (Sinbad is still foraging in the area to the east of Cobra).

1136 – Remy moves to the edge of the reeds north of the nest and begins foraging. Quark follows her and parks there, waiting in anticipation of Remy catching a prey item. Sparky goes to the reeds in the NW corner. Cobra is still parked south of the old nest site.

1139 – Sparky moves east along the reeds of the north bank and parks near Cobra by the old nest site. Quark is still parked by the reeds in the mid-west end of the SC.

1140 – Remy (now in the SW section) moves toward Quark along the reeds east of the nest with a medium-sized crayfish. Quark spots Remy approaching and rushes to her. Remy works the crayfish briefly before Quark takes control of it (a bit prematurely). Quark drops it a few times while trying to swallow it during which time Remy takes the opportunity to further work the crayfish. Phantom sights the pair from the SW corner and rushes toward them. After ~ 30 seconds from when the feeding commenced, Quark finally swallows the crayfish, just before Phantom arrives. Remy then resumes foraging.

1141 – Sparky and Cobra move to the NW section of the SC.

1144 – Sinbad is sighted with a large crayfish by the reeds just west of the old nest site. Sparky and Cobra are swarming around him. As Sinbad works the crayfish the competition among the chicks gets quite heated. On three separate occasions Sparky aggressively pecks at Cobra while charging and briefly chasing Cobra away from close proximity to Sinbad. After each of these acts of dominance, Cobra lingers briefly at the periphery and then moves back in. Sparky clearly has the upper hand and after ~ 64 seconds seizes control of the crayfish and swallows it. Sinbad then goes to the NW corner (with Cobra following along) and resumes foraging.

1148 – Sinbad has worked his way across the lagoon to SW corner. He then moves all the way out to the east end of the SC (calling once en route) and continues on toward mid-river, passing beyond my view.

1149 – Remy is sighted with Sparky and Cobra in the NW section of the SC. Sparky has just taken a fairly large crayfish from Remy. Sparky darts about, keeping ahead of the hotly pursuing Cobra. After ~ 24 seconds of this, Sparky swallows the crayfish. Remy then resumes foraging and Cobra and Sparky park in the NW corner.

1153 – Remy is sighted with a medium-sized crayfish ~ 16 m east of the nest as she moves toward Cobra. Cobra spots her approaching and rushes to her. Remy works the crayfish briefly before Cobra takes and swallows it (the feeding lasted ~ 28 seconds). Remy then parks ~ 31 m NE of the nest and preens. Quark goes to her, solicits Remy and remains close by her. Cobra and Spot also park near her.

1200 – Remy and all 5 of the chicks are now parked in the NW section of the SC (most of them are preening). Sinbad is parked and preening at the east end of the SC. He soon moves to the mid-SC area and parks there.

1201 – Remy moves to the reeds in the SW corner and goes out of my view. Cobra, Spot, Phantom and Quark follow her there. Sparky remains in the NW section.

1205 – Quark and Spot have moved out and joined Sparky in the NW section where the three are parked a bit spread out from each other. Sinbad is now parked south of the NE point.

1210 – Remy is foraging in the SW section of the SC by and within the reeds. Sparky and Cobra are sitting just south of the old nest site. Sinbad is still parked south of the NE point. The other three chicks aren't seen at this time. End observation period.

*Comments:* Shortly after the formal observation period concluded a paddleboard (heading north) passed by in the river to the east of the SC. It again passed by to the east on its return trip ~ 7 minutes later. As usual, this disrupted the grebes' activities by spooking them and causing them all to flee and hide in the reeds until several minutes after the paddleboarder had left the area.

The chicks spent the entire observation period in the SC. As has become usual now, the chicks were quite mobile and at times were split among the two parents. In regard to their associations with the parents during the course of the period chicks were observed sitting with the parents (either with just one or split among both), accompanying the parent(s) on foraging forays, or moving to a foraging parent after spotting the parent from a distance.

More instances of specifically directed physical aggression among the chicks during feeding competition were observed today. Of particular note were some very aggressive attacks by Sparky against Cobra (involving pecking, charging and brief chasing by Sparky) in a feeding that just these two chicks participated in.

The first observed parental rebuffs of chicks were seen today. Sinbad was seen rebuffing Cobra (he pecked and briefly chased the chick) as Cobra solicited him. Also, during my continued observations after the formal observation period had concluded Sinbad was seen rebuffing Sparky as the chick was soliciting him while approaching. Is this an indication that the parents are just beginning the weaning process, with the rebuffs serving as an inducement to the chicks (or just certain ones) to engage in more independent foraging? Or were these rebuffs merely done out of annoyance? Given the general patience that the parents have typically shown for the chicks, the former seems most likely (especially considering that this is similar to what was seen during my observations of the West Lagoon family last year).

**8/23/17:** Time: 1050-1205.

1050 – As observations begin, Sinbad is parked with Quark in mid-water in the north half of the West Lagoon. Sparky and Spot are ~ 7 m from the west shore reeds in the mid-north section of the lagoon. Remy is parked with Phantom at the east end of the SC. Cobra isn't seen at this time.

1056 – Sinbad leaves the West lagoon and goes to the SC. He is trailed at a distance by Quark and trailing still farther back is Sparky. As Sinbad enters the east end of the SC he heads straight toward Remy (still at the east end, in its northern half). Remy soon spots Sinbad approaching and moves toward him. They meet and greet each other en route (voicing a fairly intense continuous, strung out, chattering vocalization as they approach and meet), linger together briefly and then Sinbad goes to the NW corner of the SC with Quark following along. Remy goes to the area NE of the SE point and intently looks to the south (she apparently is watching the late returning Spot as the chick makes its way to the SC from the West Lagoon).

1059 – Remy parks north of the SE point of the SC; Phantom, Spot and Sparky are with her.

1100 – Remy moves to the mid-SC area; Phantom goes with her.

1103 – Sinbad moves out from the SW corner of the SC with a fairly large crayfish and heads toward Quark (who is ~ 7 m east of the nest). Quark spots Sinbad approaching, rushes to him and they meet en route. Quark excitedly swirls around Sinbad as he begins working the crayfish. Spot soon joins the feeding and aggressively pecks at Quark, briefly grabs the skin on the back of Quark's neck with its beak and forces Quark to the periphery. However, Quark quickly rejoins the fray and eventually is able to sneak in, grab and take control of the crayfish. Quark then darts about, being hotly pursued by Spot while working the crayfish and attempting to swallow it. Sinbad follows Quark and oversees the feeding as it drags on but after ~ 90 seconds he leaves the two competing chicks to it and resumes foraging. Quark continues to work the crayfish and makes several failed attempts to swallow it (the crayfish is still too large and bulky for Quark to swallow). After the second failed attempt Quark ejects the crayfish onto the water and it sinks. Spot and Quark both dive after it. Spot resurfaces with the crayfish and takes control of it with Quark now doing the pursuing. Finally, ~ 2 minutes 23 seconds after the feeding began, Spot is able to swallow the crayfish (time: 1105).

1110 – Sparky is sighted at the northern end of the east end of the SC (south of the NE point), slowly making its way south. When at the mid-area along the east end an osprey who is carrying a fish in its talons suddenly flies over Sparky's location heading SW. Sparky sights it as it approaches overhead and immediately engages in an alarm dive, instantly disappearing below the surface. When Sparky resurfaces a few seconds later the osprey is long gone and the chick continues south toward the West Lagoon where it appears that some of the others have already gone.

1114 – Cobra is sighted foraging along the reeds of the north bank of the SC, west of the NE point. Cobra is the only one currently seen in the SC as it appears that all of the other family members have gone to the West Lagoon. The NR male calls once from the mid-area of their territory. At this time I change my observation point from the mid-south bank of the SC to the west end of the RR bridge from where I can see both the SC and the West Lagoon. [Only a part of the reedy west shore of the lagoon is visible from the west end of the bridge. In order to view the entire length of the west shore it is necessary to go farther out (farther east) onto the bridge. As previously noted, it is unsafe to conduct lengthy observations from the RR bridge (especially from farther out on it) because of the regular train traffic that passes over the bridge.]

1117 – Remy is sighted parked in the mid-water area of the north half of the West Lagoon; Quark and Sparky are with her.

1121 – Remy begins foraging. Quark and Sparky then move closer to the west shore reeds but remain in open water. There currently are no gulls in the north half of the lagoon; there are about two dozen in the far distance at the south end of the lagoon.

1124 – Cobra is foraging inside the edge of the reeds just west of the NE point of the SC. This chick is still the only family member in the SC; the others are in the West Lagoon.

1125 – Cobra moves south to the mid-area along the east end of the SC, parks there and preens.

1129 – Cobra is still parked in the same general area as at 1125, preening and also looking about (probably for family members).

1131 – Cobra begins slowly moving south along the east end of the SC toward the West Lagoon.

1134 – Cobra moves into the mid-river area east of the SC and NE of the SE point of the SC, mostly looking to the south toward the West Lagoon, apparently trying to spot family members. Cobra edges farther south, looking somewhat anxious, but then turns back toward the east end of the SC. It appears that Cobra wants to go to the West Lagoon but is somewhat apprehensive about crossing under the RR bridge by itself.

1136 – Cobra finally gathers its resolve and goes under the RR bridge and into the West Lagoon. The chick slowly moves south along the west shore reeds, peering into them for family members (none of them are visible from my vantage point at this time).

1139 – Cobra goes into the reeds (and out of my view) in the mid-north section of the lagoon (it is most likely that the chick found family members there).

1142 – Remy is sighted foraging in the southern half of the lagoon. She soon parks in the mid-water area there and preens herself.

1143 – Several chicks are sighted by the snags.

1144 – I begin heading over to my West Lagoon observation point on the mid-east shore.

1150 – I arrive at my east shore observation point. Remy is sighted moving north past the snags ~ 4-6 m from the west shore reeds. At least 3 chicks are seen sitting just inside the reeds at the south end of the northern half of the lagoon.

1153 – Remy parks and preens ~ 6 m from the reeds just north of the snags. Quark goes to and sits by Remy, often soliciting her.

1155 – Remy is still with Quark in the same location. Sinbad is sighted moving north along the west shore reeds in the mid-section of the lagoon with a fairly large crayfish. Spot sees Sinbad approaching and rushes to him. The chick excitedly moves around Sinbad as he begins to work the crayfish. Cobra soon spots the feeding in progress and races toward the pair. However, by the time Cobra reaches them, Spot has already taken control of the crayfish and eventually swallows it (the feeding lasted ~ 35 seconds).

1202 – Four chicks are seen by the snags. Sinbad is there also with a very large crayfish. A few of the chicks go to him but neither seem to be very keen on the prospect of eating it—they appear to be put off by its large size and the fact that Sinbad has been keeping it to himself. Sinbad eventually swallows the crayfish (which appeared to be his intention all along).

1205 – Phantom, Spot and Cobra are parked and preening themselves ~ 4-5 m from the west shore reeds in the mid-north section of the lagoon. Sinbad is parked mid-water in the north half of the lagoon. Remy, Sparky and Quark aren't seen at this time (but they are still in the West Lagoon). End observation period.

*As I was about to leave the area after the observation period had concluded I spotted Sinbad conducting a crayfish feeding with Cobra and Spot. While Sinbad was working the crayfish, Cobra briefly but aggressively chased Spot away from Sinbad several times to a peripheral location. Cobra thus effectively excluded Spot from the feeding by maintaining the favorable close-in position for itself. When the crayfish had been worked sufficiently, Cobra then was easily able to take control of the crayfish from Sinbad and swallow it.

*Comments:* In an earlier feeding today Spot had bested Quark in an aggressive episode during the feeding. Yesterday Cobra had been soundly bested by Sparky. On 8/15 Phantom bested Quark in an aggressive dominance interaction. At this point, the probable dominance hierarchy among the chicks looks to be linear and appears to be as follows (from highest to lowest ranking): Sparky, Cobra, Spot, Phantom, Quark. The situation between Cobra and Spot is not clear at this time. Though Cobra clearly bested Spot in the above noted feeding, Spot appeared to get the better of Cobra in some previous but less clear-cut instances. In any event, my assessment of the rankings must be considered somewhat tentative at this time as not nearly enough aggressive/dominance confrontations have been observed among the various combinations of chicks to be 100% certain of this order or even to be able to determine with certainty whether or not the hierarchy is linear. Nevertheless, scientific conservatism aside, with the exception of the relative positions of Cobra and Spot to each other the rankings as noted above appear to be pretty solid.

Today the chicks were observed doing their most extensive ranging to date within the course of an observation period. Three of them (Sparky, Spot and Quark) went from the West Lagoon to the SC and then back to the West Lagoon.

Once again one of the chicks was seen isolated by itself in one of the main sections of the SR pair's territory while the rest of the family was in the other. This time it was Cobra, this chick finding itself alone in the SC while the rest of the family was in the West Lagoon. Cobra eventually made its way to the West Lagoon by itself where it joined the rest of the family. Cobra looked notably more confident than Quark did in a similar situation on 8/20. However, it is perhaps of significance that Cobra's situation was the opposite of Quark's in that Cobra was isolated in the SC where the chicks obviously feel more secure. While this may be a factor I think Cobra's greater confidence is mostly down to the fact that (along with Sparky and Spot) Cobra spends notably more time foraging and ranging independently than does Quark (and Phantom).

**8/24/17:** Time: 1057-1205.

1057 – As observations begin, Remy is sighted with a medium-sized crayfish south of the NE point of the SC moving south (toward Cobra) along the east end of the SC. Cobra is foraging by itself in the SE end of the river just north of the RR bridge. Cobra spots Remy approaching and rushes toward her. They meet en route and Remy begins working the crayfish. Cobra soon takes control of the crayfish and swallows it (the feeding lasted ~ 20 seconds). Cobra immediately solicits Remy after eating the crayfish. Remy then resumes foraging, working her way toward the West Lagoon. Cobra lingers briefly, then moves across to the reeds along the east bank at the SE end of the river.

1105 – No family members are seen in the SC so I move to the west end of the RR bridge to check for them in the West Lagoon where Remy was headed.

1109 – Sinbad is sighted with Sparky and Spot. He and the chicks are parked mid-water in the north half of the West Lagoon.

1110 – At this time I begin heading over to my West Lagoon observation point on the mid-east shore of the lagoon. As I depart, I note that Spot leaves Sinbad (and Sparky) and goes to the west shore reeds. Sinbad calls once.

1117 – I arrive at my east shore observation point. About 7 western gulls are in the west side of the northern half of the lagoon. Sinbad is sighted moving south along the west shore reeds in the north half of the lagoon. He is followed at a distance by Quark and, at a still farther distance, by Sparky. He makes his way to the area by the snags where he begins foraging.

1120 – Sinbad is foraging just east of the snags. He calls once and resumes foraging.

1123 – Phantom and Sparky are sighted foraging at the edge of and within the reeds south of the snags.

1128 – Sinbad is sighted parked ~ 9 m from the south end of the west shore reeds (recall that the west shore reeds end about midway into the south half of the lagoon). Three chicks go to Sinbad and excitedly solicit him. Sinbad soon resumes foraging and the chicks go back to the reeds.

1132 – Remy is sighted parked and preening herself ~ 12 m from the west shore reeds in the north half of the lagoon. Cobra and Quark are with her and they are intermittently soliciting her (trying to coax her to begin foraging for them). Sinbad has moved north and is now parked and preening mid-water at the north end of the lagoon. One chick is with him (my view is too inadequate to ID it).

1139 – All of the family members who are in open water suddenly dive and go into hiding (probably in the west shore reeds) as a paddleboard moves from the east into the area beyond the NE end of the lagoon, just north of the RR bridge, and lingers there.

1141 – Sinbad is sighted just in front of the west shore reeds in the mid-north section of the lagoon. He is intently peering toward the paddleboarder (still in the same location as at 1139).

1143 – The paddleboarder finally begins moving north in the river and soon passes out of view.

1145 – Three of the chicks and Remy are sighted at the reeds in the area north of the snags. Remy soon dives and begins foraging.

1147 – Phantom and Spot (preening) are sighted by the reeds ~ 22 m north of the snags.

1152 – Sinbad is sighted by the snags. Two chicks rush toward him and two others come out of the reeds farther to the north but these two just remain by the edge of the reeds (my view is too inadequate to ID any of these chicks). The first chick to reach Sinbad solicits him as it approaches and is immediately rebuffed by Sinbad: he pecks at the chick and briefly chases it ~1 m away. Sinbad then begins moving south and the chick follows, staying ~ .5-1 m behind. There are no longer any gulls in the north half of the lagoon.

1156 – Sinbad is sighted moving to the snag area from the south with a medium-sized crayfish. Quark is closest to Sinbad, spots him first and rushes to him. Sparky, who is a little farther away, spots Sinbad approaching a few seconds later and also rushes toward him. With its head start and shorter distance to cover, Quark reaches Sinbad much earlier than Sparky and takes and quickly swallows the crayfish. Sinbad then resumes foraging.

1201 – Four chicks are sighted by and just south of the snags; some are foraging and some are preening.

1204 – Sinbad is sighted moving north near the snags with a medium-sized crayfish. Spot (who is nearer to Sinbad) and Phantom spot Sinbad approaching and rush toward him. Spot, taking advantage of its closer proximity, reaches Sinbad first and takes and swallows the crayfish. As Spot is moving away while swallowing the crayfish, Sinbad picks up (from the surface of the water) what appears to be a claw that had been detached from the crayfish. Phantom sees this, quickly goes to Sinbad and takes and swallows the claw (time: 1205). Sinbad then resumes foraging. None of the other family members are seen at this time. End observation period.

*Shortly after the formal observation period was concluded some of the chicks were for the first time observed in the south end of the West Lagoon. They were near the west shore south of the southern end of the west shore reeds. There were 3-4 of them there—they apparently had followed Sinbad, who was foraging in that area. Also seen during my continued observations of the family was another parental rebuff of a chick. This was observed not long after the chicks returned to the north half of the lagoon. Remy was seen suddenly jumping all over a chick that was soliciting her while approaching (my view was too inadequate to ID the chick). Just as the chick reached her she aggressively pecked at and briefly chased the chick who immediately dove briefly to escape her wrath. When the chick quickly resurfaced ~ 2 m away, Remy promptly jumped the chick a second time, again pecking at and briefly chasing the chick (who again dove to escape her). At that point Remy ended her attack on the chick and began foraging soon after this incident. This was the second parental rebuff of a chick that was observed today (Sinbad rebuffed a chick at 1152 during the formal observation period).

*Comments:* The chicks generally looked more comfortable and at ease while in the West Lagoon today. They spent more time in open water and ranged farther south than observed at any time previously.

A number of the chicks were seen spreading and/or briefly flapping their wings. In general, their wings are coming along quite well. In regard to their basic structural aspect, the wings have elongated nicely and now look like proper wings in this respect. They appear much less downy and the flight feathers are lengthening (though they are still short).

Today the chicks often were too distant or too poorly seen to be readily identified or even to allow for a rough guess at who the individual was. With the chicks now being so mobile and beginning to range so widely, such situations can now be expected to be common.

**8/25/17:** 1043-1152.

1043 – As observations begin (from the west end of the RR bridge), Phantom and Cobra are parked in the mid-water area in the north half of the West Lagoon. No other members of the family are seen in the lagoon.

1045 – (I have moved to my usual observation point on the mid-south bank of the SC.) Remy and one chick are parked together at the NE end of the SC. No others are seen at this time.

1049 – Sparky and one other chick are sighted moving south along the southern half of the east end of the SC. Remy moves to the SE section of the SC, north of the SE point.

1050 – A chick is sighted in the NW section of the SC with a crayfish that it had probably taken from Sinbad (the probable handoff from Sinbad to the chick occurred before I first caught sight of the chick). The chick is now by itself. It works the crayfish for ~ 20 seconds before swallowing it. At least one other chick is seen in the reeds at the mid-north bank of the SC.

1056 – Remy is parked with Quark north of the SE point of the SC. Spot and Sparky are parked at the east end of the SC, south of the NE point.

1058 – Remy (with Quark following along) moves to the NE point, picking up Spot and Sparky along the way, and begins foraging. The 3 chicks go to the reeds at the south side of the NE point.

1105 – Remy is sighted by the reeds at the south side of the NE point with a medium-sized crayfish. She is very slowly moving along the edge of the reeds, obviously looking for one or more chicks to feed it to (none of the chicks are visible to me but at least 3 are inside the reeds there).

1106 – Sparky, followed soon after by 2 other chicks emerge from the reeds and rush toward Remy. Sparky reaches her first and takes and quickly swallows the crayfish. Remy then resumes foraging and the chicks return to the reeds.

1110 – (Check of the West Lagoon from the west end of the RR bridge.) Sinbad is parked and preening ~ 23 m from the west shore reeds, north of the snags in the West Lagoon. Cobra and Phantom are with Sinbad and intermittently solicit him.

1115 – Sinbad has moved to ~ 15 m from the reeds, north of the snags. Cobra and Phantom are still with Sinbad and continue to intermittently solicit him.

1120 – Sinbad is still parked with Cobra and Phantom in the same location. Remy is parked at the east end of the SC, south of the NE point with Spot and Quark (both chicks are intermittently soliciting her). (At this time I move back to my observation point in the SC.)

1123 – Spot and Quark are still with Remy, south of the NE point of the SC. As Spot solicits Remy she rebuffs the chick: she aggressively pecks at Spot and drives the chick to a peripheral location around her. However, Quark remains close to Remy and she allows Quark to do so.

1126 – At least one of the two chicks from the West Lagoon (Phantom) has moved to the SC. Four chicks are briefly gathered around Remy south of the NE point but soon all but Phantom and Quark go to the reeds.

1128 – Remy moves to the mid-area along the east end of the SC. Quark and Phantom accompany her there.

1131 – All 5 of the chicks have moved to the mid-river area east of the SC where Remy has begun foraging. The chicks are within a long, patchy algal mat where they also engage in some foraging dives. Sinbad isn't seen at this time.

1135 – The 5 chicks are well strung out from the mid-river area as they begin making their way to the NE point of the SC.

1138 – All 5 of the chicks have made their way to the reeds at the NE point and the large algal mat at its east side. Remy and Sinbad aren't seen at this time.

1140 – On a quick check of the West Lagoon from the RR bridge Sinbad is sighted ~ 25 m out from the west shore reeds, very slowly moving south through the north half of the West Lagoon. He soon parks in open water in the mid-north section of the lagoon. (At this time I move back to my SC observation point on the mid-south bank.)

1141 – Quark, Phantom and Sparky are parked at the east end of the SC, south of the NE point.

1146 – Sparky and Quark move south along the east end of the SC.

1147 – Remy is sighted moving from the mid-river area toward Sparky and Quark with a medium-sized crayfish. Sparky spots Remy approaching first and rushes to her, takes the crayfish and quickly swallows it.

1149 – Remy moves toward the NE point of the SC and Quark, Sparky and Phantom follow her.

1152 – Remy is parked and preening just south of the NE point. All 5 of the chicks are gathered around in her vicinity. Sinbad isn't seen (he is probably still in the West Lagoon). End observation period.

*Comments:* The chicks were extremely mobile today and collectively did their widest ranging to date. They were literally all over the family's territory. Besides being in both the SC and West Lagoon, chicks were seen in several areas where they haven't previously been observed. These included the following. During the observation period all 5 of the chicks ventured out into the mid-river area east of the SC (including east of the NE point, just south of the territorial boundary with the NR pair). Some of the chicks have been seen moving through that area but today they lingered there and even did some foraging. During continued observations after the formal observation period had concluded Sparky and Spot were sighted together by themselves at the NE side of the West Lagoon at the mouth of the inter-lagoon channel. Later, these two were seen foraging in the mid-section of the south ribbon of the river (recall that this is the short, narrow end ribbon of the river that runs west to east under and adjacent to the RR bridge from the north end of the West Lagoon to the north end of the East lagoon where the river feeds into it).

More parental rebuffs of chicks were observed today. In addition, more specifically directed aggression by a dominant chick toward a subordinate during feeding competition and for closest access to a foraging parent was observed. Both such interactions are now being observed on a daily basis (either during formal observation periods and/or during continued observations afterward).

**8/26/17:** Time: 1117-1224.

1117 – As observations begin, Phantom is parked and preening ~ 11 m south of the old nest site.in the SC. None of the other family members are seen at this time (nor were any seen in the West Lagoon).

1121 – Cobra emerges from the reeds by the south side of the NE point of the SC and moves west (passing close to Phantom on the way) to the NW corner. Spot and Quark are sighted preening themselves near the reeds in the NW corner.

1125 – Phantom also moves to the NW corner, parks there and preens. Remy, Sinbad and Sparky haven't been sighted yet.

1127 – Sinbad is sighted foraging in the SW corner of the SC. In between foraging dives he calls once. Cobra goes to Sinbad and solicits him before he engages in another foraging dive.

1129 – Sinbad moves out of the SW corner and heads toward the mid-SC area. On his way there 4 chicks gather around him. As they move in Sinbad pecks at and pounces on one of the chicks (my view is too inadequate to ID which one), causing it to dive and move away. The other 3 chicks also part with Sinbad as he continues on to the mid-SC where he parks and preens.

1133 – Remy is sighted with Cobra in the NE section of the SC. Remy is working a fairly large crayfish as Cobra intently swirls around her. Cobra eventually takes the crayfish and continues working it. Sparky spots the pair from a distance and rushes toward them. However, Sparky reaches them too late: as Sparky arrives Cobra already has full control of the crayfish and has begun swallowing it. Remy has a scrap from the crayfish that she had detached just as Cobra had taken the carcass from her (it appears to be a claw). Sparky quickly notes this, goes to her and takes and swallows the claw. Cobra swallows the crayfish ~ 37 seconds after I had first sighted this feeding (which was already in progress at that time). Remy then heads toward Sinbad (still in the mid-SC area) and one or both of them voice extended, continuous chattering vocalizations as she approaches Sinbad. They linger together briefly and then split: Sinbad circles through the SW section of the SC and moves out to the SE section; Remy parks in the mid-SC area and preens. Quark moves out to Remy, solicits and remains close to her.

1137 – Sinbad is cruising around slowly at the SE end of the SC intently looking about (mainly to the east and SE). He appears to be looking for someone (perhaps one of the chicks?) and voices several short, single one note calls. Sparky, Cobra and Phantom are near the reeds in the NW section. Quark is still with Remy in the mid-SC area and Spot is foraging a little to the east of them. Thus, all of the chicks and Remy are accounted for so it's not clear what Sinbad is so intently visually searching for (from my vantage point my view beyond the SC to the south and east of the SE point of the SC is completely obstructed by the vegetation and topography).

1142 – All is the same as at 1137 except that Sinbad has now moved out of the SC farther out into the river to the SE and has passed beyond my view.

1145 – Quark is still with Remy, who is preening in the mid-SC area. Remy plucks a small feather from her breast and hands it to Quark who immediately swallows it. Spot, Cobra and Phantom are parked ~ 7 m south of the old nest site. Sinbad is heard calling once from the area east of the SE point (he is out of my view).

1151 – Remy is still parked with Quark in the mid-SC area. Cobra moves out toward Remy and begins soliciting her while approaching near. Remy suddenly charges and aggressively pecks at Cobra, who immediately turns away and dives to escape her wrath. Cobra resurfaces ~ 1.5 seconds later ~ 2 m from Remy and she again charges and pecks at the chick. As before, Cobra dives to escape Remy and this time stays clear of her after resurfacing. Quark remains close to Remy and is allowed to do so. Shortly after, Remy (now in the NW corner, the rebuffs of Cobra having led her there) dives and begins foraging.

1154 – Sinbad (now back in the SC) is sighted in the mid-area at the east end of the SC with a large crayfish. Spot is with Sinbad and eagerly watches as Sinbad works the crayfish. From a distance, 2 other chicks catch sight of the feeding in progress and rush toward the pair. However, Spot takes and swallows the crayfish before they arrive. Sinbad then resumes foraging.

1159 – Sinbad is now parked and preening in the NW section of the SC. Remy comes out of the NW corner (followed by Quark) and heads directly toward Sinbad. They meet and move together to the mid-section of the west end of the SC and then part company. Remy goes to the mid-SC area where she parks and preens. Sinbad goes to the SW corner and also parks and preens. Quark has accompanied Sinbad and sits with him.

1201 – Sinbad, accompanied by Quark, has moved deeper into the SW corner and the two soon go out of my view.

1203 – Remy is still parked and preening in the mid-SC. Cobra goes to Remy, solicits and remains with her. Shortly after, Phantom also goes to Remy, lingers by her briefly and then goes to the west end of the SC.

1209 – Remy (preening) is still parked with Cobra in the mid-SC area. At least 3 chicks are sighted in the reeds north of the nest. Sinbad isn't seen at this time.

1213 – Sinbad emerges from the SW corner with a large crayfish. He moves to the area east of the nest (in open water) and 4 chicks rush out to him (all but Cobra, who is still with Remy). As Sinbad works the crayfish one of the chicks leaves the feeding group. The other 3 chicks swarm around him and appear reluctant to take the crayfish because it is still too large and bulky to swallow. While working the crayfish Sinbad briefly rebuffs Sparky by giving Sparky a quick peck and backing the chick away. Sparky is not perturbed by this and remains in the feeding fray. Sinbad works the large crayfish for well over a minute before Sparky finally takes control of and swallows it (the feeding lasted ~ 90 seconds). Sinbad then resumes foraging.

1218 – Remy is still parked with Cobra in the mid-SC area. Cobra intermittently solicits her.

1222 – Quark and one other chick move out toward Remy (Cobra is still with her). As the chicks near Remy she charges and pecks at one of them (I'm unable to ID which one), causing both chicks to scatter and one of them to dive. The situation calms but only momentarily. Quark moves to Remy and she suddenly charges, pecks and pounces on Quark, forcing the chick underwater. Quark quickly resurfaces ~1.5 m away and Remy again charges, pecks and pounces on the chick who dives to escape her. Remy briefly lingers there, then moves to the NW corner and begins foraging. Cobra follows her part of the way but stops and parks in the NW section (time: 1224). End observation period.

*During my continued observations after the formal observation period had concluded it was noted that at ~1340 Sinbad and Remy led Quark, Phantom and Spot to the West Lagoon. Sparky and Cobra remained in the SC by themselves. They were in the reeds and apparently didn't see the others leaving the SC.

*Comments:* A number of parental rebuffs of chicks were observed during and in continued observations after the formal observation period. Both parents were seen rebuffing chicks. At this point it seems clear that the parents have begun the weaning process, the rebuffs by the parents apparently conveying further inducement for the chicks to do more independent foraging.

Several of the chicks were seen stretching and/or briefly flapping their wings. Though still short, growth of the flight feathers is well under way.

The chicks have grown considerably since hatching. Sparky is still notably ahead of its four sibs but all appear to be progressing well. All are spending time foraging independently but they are all (including Sparky) still clearly dependent on their parents. I still haven't had a confirmed sighting of a chick catching a prey item on its own. However, the chicks have done most of their independent foraging in the reeds where my view of them is typically quite poor or completely obscured. In a few instances I have seen what possibly may have been cases of chicks catching very small prey items but my views of these have all been too inadequate to be certain of this.

**<u>8/27/17</u>:** Time: 1147-1305.

1147 – As observations begin, Sinbad, Sparky, Phantom and one other chick are moving south along the southern half of the east end of the SC in the direction of the West Lagoon. Remy and two of the other chicks aren't seen at this time.

1152 – Sinbad and 4 chicks have gone to the West Lagoon.

1200 – I have been looking for Remy and the other chick in the SC but still haven't sighted them. I thus begin heading for my observation point on the east shore of the West Lagoon.

1206 – I arrive at my West Lagoon observation point. None of the family is seen at this time (the chicks are probably in the west shore reeds). Approx. 20 western gulls are in the mid-water area of the north half of the lagoon.

1210 – Sinbad is sighted with a crayfish—he is moving south by the reeds north of the snags. Three chicks (my view is too inadequate to ID them) rush to him. As he works the crayfish one of the chicks takes it and moves south just ahead of its 2 hotly pursuing sibs. After ~ 22 seconds the chick with the crayfish ducks into the reeds north of the snags and the other 2 chicks follow after it. Sinbad (who has been shadowing the chicks and overseeing the feeding) also follows them into the reeds. All are then lost from my view within the reeds.

1214 – Sinbad is sighted with a large crayfish by the reeds just north of the snags. Four chicks (again my view is too inadequate to ID them at this moment) are swarming around him as he works the crayfish. At this point (probably because the carcass is still too large and bulky to swallow) the chicks are reluctant to take the crayfish and one of them leaves the feeding group. Eventually, Sparky takes the large carcass and with much effort is finally able to swallow it (the feeding lasted ~ 66 seconds from the time when it was first sighted, at which point it was already in progress).

1218 – Two chicks (one of them appears to be Phantom) are sighted ~ 8 m from the west shore reeds in the mid-north section of the lagoon where they are parked and preening. Sinbad is parked with Sparky ~ 17 m north of them. Remy still hasn't been seen.

1222 – Sinbad moves to the west shore reeds in the north half of the lagoon, dives and begins foraging.

1223 – A paddleboard moves under the RR bridge (heading south) and enters into the NE end of the lagoon. The chicks and Sinbad are spooked by this and hide in the west shore reeds in the north half of the lagoon.

1228 – The paddleboard is by the NE shore of the lagoon. Neither Sinbad nor the chicks are seen (they are still hiding in the reeds).

1230 – The paddleboard exits the lagoon through its NE end, goes under the RR bridge and stops and parks just north of the bridge where it is still a very visible presence to Sinbad and the chicks.

1232 – Sinbad is sighted inside the edge of the reeds just north of the snags, intently staring toward the paddleboard.

1334 – Sinbad emerges from the reeds and moves out ~ 4 m from them, parks there and continues intently staring at the paddleboard. Two of the chicks are near him. He soon moves close to the reeds, dives and apparently begins foraging.

1237 – The paddleboarder finally leaves the area. Phantom and one other chick are parked ~ 5 m from the west shore reeds in the lower northern half of the lagoon. Sinbad is parked ~ 8 m north of them, looking toward the area where the paddleboard had been sitting.

1243 – Sinbad moves to the area by the snags and begins foraging. Sparky followed him and was lightly rebuffed by Sinbad (a quick peck) just before he dove.

1245 – Sinbad is sighted moving north by the reeds south of the snags with a large crayfish. Sparky (to the north of him) spots Sinbad approaching from a distance, races to him and takes and quickly swallows the crayfish. A second chick (probably Quark) who was a short distance farther north than Sparky had initially also started off toward Sinbad but gave up on this early as Sparky raced well ahead of it and was easily winning the race to Sinbad (this chick probably also realized there was little point in wasting any more energy and effort on this feeding since besides being well behind in the race, Sparky was also much more dominant).

1250 – Sparky and one other chick are parked and preening just east of the snags. Sinbad and the other 2 chicks who are known to be in the lagoon aren't seen at this time. Remy and the fifth chick still haven't been sighted in the West Lagoon (or at all today) so I begin heading to the SC to see if they are there.

1300 – I arrive at my SC observation point and find that Remy is parked with Cobra in the mid-SC area. Cobra is intermittently soliciting Remy.

1305 – Remy is still parked (and preening herself) with Cobra in the mid-SC area. While preening, Remy plucks 4 long flight feathers from her wing(s). On three occasions she also plucks a small feather and hands it to Cobra who immediately swallows it. Sinbad, Sparky, Phantom, Quark and Spot are still in the West Lagoon. One of the NR pair is sighted foraging in the far distance in the mid-area of their territory. End observation period.

*Comments:* Remy plucked out 4 long flight feathers in a single preening bout indicating that she is probably molting her flight feathers at this time. In this same preening bout, on three occasions she plucked a small feather from her breast or flank and handed it to Cobra who promptly swallowed it. In fact, while Remy was preening Cobra appeared to be avidly seeking these "feather meals" from her.

Over the past 10 days all of the observed feedings (i.e., in which the chicks actually ate the proffered prey item) have consisted of crayfish (the last time that a fish was observed being fed to the chicks was on 8/17). This contrasts notably with the period prior to this during which fish made up the majority of the observed feedings.

**8/28/17:** Time: 1033-1140; 1305-1325.

1033 – As observations begin, Phantom and Sparky are sighted (from the RR bridge) in the north half of the West Lagoon moving west across the lagoon from the east side (from near the mouth of the inter-lagoon channel) toward the west shore reeds.

1034 – (From my SC observation point on the mid-south bank.) Sinbad is in the SW section of the SC intently staring toward a distant paddleboarder who is moving south in the river toward the mouth of the channel on the east bank that flows around the island that comprises the east bank in the SE end of the river. Sinbad soon dives and isn't resighted at this time (he is probably hiding in the reeds).None of the other family members are seen in the SC at this time.

1041 – The paddleboarder has entered the channel and moved out of the area. Spot emerges from the reeds in the SW section of the SC, moves to the mid-SC area and parks and preens there.

1042 – Remy is sighted foraging by the reeds along the mid-south bank of the SC. Cobra goes out to the mid-SC area from the reeds at the mid-south bank. The chick lingers there briefly and then returns to the reeds.

1043 – Sinbad and Remy emerge from the reeds in the SW section of the SC. Sinbad goes to the mid-SC area where he parks and preens. Remy goes out near the north bank, west of the NE point; Spot follows her.

1045 – Remy moves to the mid-SC area; Spot accompanies her. She lingers there briefly and then moves to the area ~ 7 m east of the nest; Spot remains parked in the mid-SC area. Cobra moves toward Remy and follows her as she moves to the reeds east of the nest.

1047 – Cobra, Remy and Spot are now parked a bit spread out from each other ~ 3-7 m east of the nest. The other family members aren't seen at this time.

1050 – Phantom is sighted rounding the SE point of the SC as the chick returns to the SC from the West Lagoon. Phantom moves to the west end, passes right by Remy and goes into the reeds by the nest.

1052 – Remy has begun foraging and is sighted working a medium-sized crayfish in the mid-SC area with Quark eagerly hovering around her. Quark soon takes and swallows the crayfish.

1053 – Sinbad is sighted moving east along the reeds in the NW section of the SC with a very large fish. Spot and Phantom move to Sinbad and swarm around him as he works the fish. Quark soon joins them. Sinbad allows the chicks the chance to eat the fish. Several of them attempt to swallow it but fail because the fish is just too large for them. Finally, after ~ 108 seconds Sinbad (with considerable effort) swallows the fish himself (time: 1055).

1100 – Sinbad is parked with Phantom in the mid-SC area. Remy is parked with Cobra (who occasionally solicits her) in the SW section. None of the others are seen at this time. Sparky still hasn't returned from the West Lagoon.

1101 – Remy goes to the reeds in the SW section accompanied by Cobra and both go out of my view.

1103 – Phantom leaves Sinbad and goes to the reeds in the SW corner (where Remy had gone and is probably foraging). Sinbad remains parked in the mid-SC.

1105 – Quark emerges from the reeds at the west end of the SC and goes out to Sinbad. Quark solicits Sinbad repeatedly for ~ 11 seconds and then goes to the NW section, parks there and preens.

1109 – Sinbad is still parked and preening in the mid-SC area. Phantom comes out of the reeds in the SW corner and goes to Sinbad. The chick solicits Sinbad upon arrival and remains with him. Quark drifts to within ~ 3 m of them.

1111 – Phantom leaves Sinbad and heads toward Remy who is now in the NW end of the SC. As Phantom nears Remy she rebuffs the chick by charging at Phantom for ~ 1-2 m and driving the chick away. Phantom then moves back to Sinbad (who calls once before Phantom arrives), solicits him briefly and then moves away.

1115 – Phantom is now parked with Sinbad in the mid-SC area. Spot and Cobra are also parked in the mid-SC area. Quark is parked in the SW section. Remy isn't seen at this time. Sparky still hasn't returned from the West Lagoon.

1119 – Remy emerges from the reeds in the SW section with a medium-sized crayfish and heads toward the mid-SC area. Spot and Cobra are parked together there and Phantom is with Sinbad farther to the NE. First Spot, then Cobra and then Phantom all rush toward Remy as she approaches the general area. Spot (who was the closest to Remy) easily wins the race to her, takes the crayfish and swallows it in ~ 11 seconds, before any of the others reach Remy.

1120 – Remy then parks at the east end of the mid-SC area and preens. Phantom goes back to Sinbad (now ~ 30 m south of the NE point), solicits and remains with him. Spot and Cobra eventually go to the reeds at the west end.

1125 – Sinbad moves back to the mid-SC area. Phantom accompanies Sinbad and intermittently solicits him. Remy is still parked in the same location as at 1120.

1126 – Phantom leaves Sinbad and heads toward Remy but when the chick reaches ~ 3 m from her she charges at Phantom for ~ 1 m and drives the chick back from her (there was still ~ 2 m between them when Phantom retreated). Phantom then circles back and solicits Remy from a "safe" distance.

1127 – Remy goes to the SW corner and passes out of my view. Phantom follows her there at a distance (Remy has probably begun foraging).

1130 – Sinbad is still parked in the mid-SC area. None of the others are seen at this time. Sparky still hasn't returned from the West Lagoon.

1132 – Sinbad goes to the NW corner (he is met briefly and separately by two chicks on the way but neither stay with him) and begins foraging there.

1136 – A feeding by Remy is underway by the reeds in the SW corner but it is mostly out of my view. It appears that 3 or 4 chicks are around her. Eventually Quark emerges from the reeds with a crayfish and swallows it. All then go back into the reeds in the SW corner and pass out of my view.

1140 – Sinbad is foraging near the reeds east of the nest. While Sinbad is momentarily on the surface between foraging dives Cobra emerges from the reeds in the SW corner and heads toward him. Sinbad calls once, then dives and resumes foraging before Cobra reaches him. None of the other family members are seen at this time. Sparky still hasn't returned from the West Lagoon. End observation period.

*1209 – Sparky finally returns to the SC from the West Lagoon and joins the rest of the family there. Sparky had been in the West Lagoon by itself since Phantom left there at 1050.

*1305-1325 – An extra-long 15 minute territorial standoff took place between Sinbad and the NR male. As in all but a few of the standoffs that I have observed, no physical fighting occurred. This standoff was basically the same as the others that had no fighting: it involved only stereotypical posturing/displaying. However, there was a noteworthy difference this time. All of the previous standoffs had taken place along the newly established territorial boundary located at a line across the river that is situated a little to the north of the NE point of the SC. This one took place slightly farther south, adjacent to the NE point. Is this an indication that the NR pair has extended their territory a little farther south to the NE point? The NR female remained nearby to the north (in her territory) during the confrontation but stayed well clear of the interaction (i.e., she didn't come close to or join in on the confrontation). After 15 minutes of this standoff it abruptly ended as Sinbad just moved away and went back into the SC. Seconds later the NR male went to his mate. They greeted briefly and began foraging, working their way north farther into their territory. Remy and the chicks had remained well inside the SC during the standoff but it wasn't clear as to whether they were aware of it.

*Comments:* Remy was observed plucking another long flight feather while preening today. She was also seen briefly stretching her wings. It could then be seen that she is missing a number of flight feathers, confirming that she is indeed molting them at this time.

As has become usual, a number of parental rebuffs of chicks were observed today as well some aggressive, specifically directed competitive interactions among the chicks.

At this point all of the chicks appear to be progressing well toward achieving independence. They are generally at the stage where all of them engage in some independent ranging and foraging (either alone or more commonly with one or more of their sibs) and also accompany and take food from the parents. None of the chicks are as yet capable of adequately feeding themselves, not even Sparky. In addition, they still look to the parents for guidance, security and protection. Sparky is still notably ahead of the other 4 chicks and appears to be well on track toward achieving independence before them (of particular note today, Sparky spent well over an hour ranging/foraging by itself in the West Lagoon). There are some noticeable gaps among the other four (for example, Spot and Cobra spend considerably more time ranging and foraging independently than do Phantom and Quark, and

Spot can swallow larger prey items than the other three) but they are more or less on a par with each other. From the time of hatching, none of the chicks have been runty.

In obtaining food from the parents the chicks employ the same general strategies as those utilized by the West Lagoon brood that I observed last year. One of these strategies is to just stay in a specific area and wait for foraging parents to bring food to them. This, of course, was the initial manner in which they received their food. In the period right after hatching and before they became mobile the chicks were bound to the nest and its immediate area and the parents brought prey items to them there. With the chicks now being mobile the feasibility of this strategy varies greatly depending on the situation. If one or more chicks are accompanying a parent that is on a foraging foray the parent will give any prey item that it catches to this nearby chick(s) rather than carry it some distance to a chick(s) who is waiting somewhere for a feeding to be delivered. The same applies in situations in which one or more chicks have spotted a parent who is foraging some distance away and they go to the area where the parent is foraging and vigilantly watch for the parent to surface with a prey item. Any chick that remains behind and doesn't go to that area will lose out to the ones that did (i.e., regarding that particular parent).

When in this mode of waiting for the parents to deliver food items while in a competitive situation (i.e., two or more chicks are waiting in the same area) the chicks typically try to gain an advantage by being the first one to spot a parent who is approaching with a prey item. The chick then can hopefully get a head start in the race to the parent in an attempt to be the first to reach the parent (and thus have the first shot at taking the prey item from the parent). Note however, that getting a head start isn't the only factor involved in winning these races. Obviously, proximity to the approaching parent is a significant factor, especially if a chick is much closer than any of its sibs. Another obvious factor is swimming speed (or more appropriately for this species, "paddling" speed since they mainly move about on the surface by paddling their feet). For example, probably because it is larger and a bit more advanced developmentally, Sparky is the fastest swimmer among the brood and is very notably faster than the smallest chicks. In a number of instances Sparky has blown past sibs who had a head start and easily won the race to the parent.

Reaching a prey bearing parent first achieves its greatest advantage when the prey item is of a size that can be quickly and easily swallowed. Larger prey items that require time for working by the parent or the first arriving chick lessens or negates the advantage of primacy because the time required for working the prey allows more time for competing sibs to arrive at the scene (this, of course, is much less of a concern for a dominant chick like Sparky).

Occasionally a situation has arisen in which a group of chicks who were in the "waiting in a specific area for a delivery" mode moved a fair distance away at a time when a foraging parent was some distance away and didn't see the chicks move. In these instances, after having captured a prey item the parent returned with it to the last place that it had seen or fed the waiting chicks (i.e., the place that the chicks had moved from). The parent then slowly moved along the reeds looking for the chicks. In a few cases the parent went into the reeds to search for the chicks and later emerged from them with the prey item still held in its beak. Eventually, with the parent having no clue where the chicks were and with none having come out to it for the feeding, the parent reluctantly ate the prey item itself (in one case Sinbad just dropped a crayfish and later came back for it when he had found the chicks—it was as though Sinbad had deliberately cached the crayfish).

Often when in this particular feeding mode chicks will do some independent foraging in the area while waiting for the parents to deliver food to them.

The other two major strategies employed by the chicks for obtaining food from the parents were alluded to above. One of these is for the chick(s) to go to a parent that it has seen was engaged in foraging and remain in that area and monitor the parent, watching to see if it catches a prey item.

The other main strategy is to accompany a parent and stay with it when it forages. Also, a chick(s) will sit with or accompany a parent who is parked or otherwise not foraging and wait for it to begin foraging. In such situations the chick(s) typically will repeatedly solicit the parent in an attempt to coax it to engage in a foraging foray. The parents also benefit when the chicks employ this strategy. They can be much more wide-ranging with their foraging and it's more efficient since the chick(s) is right there and the parent thus doesn't have to waste time and energy traveling some distance in order to conduct the feeding.

As with the West Lagoon brood that I observed last year, these chicks also look for opportunities to isolate themselves with a parent or otherwise reduce competition by hooking up with a parent that is being accompanied by just one other chick (preferably a subordinate sib). The ideal such situation for a chick is exemplified by the situation observed yesterday in which Cobra had isolated itself with Remy in the SC while the other 4 chicks were with Sinbad in the West Lagoon. In other words, the 4 chicks (= competitors) were a great distance away in a location where they couldn't see Cobra isolated with Remy (or see Remy, notably when she captured prey). Cobra thus had sole access to Remy and any prey items that she captured, without even the threat of any competition from its sibs.

**<u>8/29/17</u>:** Time: 1020-1126.

1020 – As observations begin, Quark, Spot and Sparky are parked in the NW section of the SC. None of the other family members are seen at this time (including in the West Lagoon).

1023 – Quark moves to the reeds in the SW corner. Spot and Sparky move to the area SW of the NE point of the SC.

1025 – A chick is sighted with Sinbad by the reeds along the mid-north bank of the SC swallowing a crayfish. Sinbad calls once and resumes foraging. The chick goes into the reeds (my view was too inadequate to ID the chick).

1027 – Remy is sighted with a very large crayfish in the NW section of the SC. Quark stays by Remy as she works the crayfish for ~ 46 seconds before swallowing it herself (the crayfish was much too large for Quark and probably would have been too difficult and time consuming to break down for the chick).

1030 – Remy is now parked with Cobra in the mid-SC area. Spot is parked ~ 17 m south of the old nest site. Quark is now with Sinbad in the mid-area at the east end of the SC (Quark often solicits Sinbad).

1033 – Remy moves to the reeds at the west end of the SC and begins foraging; Cobra follows her there.

1035 – Sinbad is still parked (and preening himself) with Quark in the same location as at 1030. Quark continues to intermittently solicit him.

1037 – Remy is now foraging near the old nest site; Cobra is following her.

1038 – Quark leaves Sinbad and moves to the NW section of the SC. Sinbad remains in his same location. Phantom hasn't been seen yet today.

1041 – Remy is sighted with a very large crayfish by the old nest site. Cobra and Quark are with her but don't appear to be very keen on trying to get the crayfish, probably having realized that it is much too large for them and that Remy is retaining control of it. Remy works the crayfish for ~ 43 seconds before swallowing it. She then goes to the mid-SC area where she parks and preens. She is soon joined by Quark who solicits her often.

1045 – Remy is still parked with Quark in the mid-SC area. Sinbad is still parked at the mid-area of the east end of the SC. Sparky and one other chick (probably Cobra or Phantom) are foraging east of the NE point of the SC.

1047 – Sinbad (still in the same location) calls once. Both members of the NR pair are sighted foraging in the north half of their territory.

1050 – Remy is still parked with Quark in the mid-SC area. Sparky and Phantom are by the east bank of the river east of the SC. Several American coots are fairly close around them. They pay no heed to the chicks but their proximity appears to be enough to make Sinbad concerned for the chicks. He then crosses the river to the east bank area and retrieves the chicks, leading them back across the river all the way to the west end of the SC. As Sinbad re-entered the SC heading west, Quark left Remy, met

Sinbad en route and solicited him on the move as Sinbad continued to the west end. Cobra eventually goes out to Remy (in the mid-SC area) solicits and sits with her.

1055 – Remy is still parked with Cobra in the same location as at 1050. Sparky is parked ~ 9 m to the west of them.

1057 – Sinbad is sighted in the SW section of the SC. He calls once, then moves east off the south bank to the area north of the SE point. Quark follows Sinbad at a distance and eventually catches up with and stays with him as Sinbad then veers north along the east end of the SC (time: 1100). Sinbad soon stops and parks at the east end. Quark then leaves Sinbad, goes near the reeds at the north bank, west of the NE point and begins foraging.

1105 – Sinbad has moved from the east end and isn't seen at this time. Of the family, only Spot is visible—the chick is parked and preening in the NE section of the SC.

1110 – Remy is sighted working a crayfish by the reeds just west of the NE point with Phantom and Quark swarming around her. Phantom soon takes control of the crayfish and moves west, away from the hotly pursuing Quark. Phantom then ducks into the reeds and Quark follows. Both have now gone out of my view. Quark eventually emerges from the reeds with the tail section of the crayfish and soon swallows it (Phantom emerged from the reeds after Quark did but gave only token pursuit).

1114 – Sinbad is sighted with 3 chicks by the reeds at the south side of the NE point of the SC. I catch sight of them just as one of the chicks finishes swallowing a crayfish. Sinbad then resumes foraging.

1116 – Sinbad is sighted following Sparky along the reeds at the south side of the NE point. Sparky has a large crayfish that it soon swallows (Sparky apparently had taken the crayfish from Sinbad who then trailed Sparky while overseeing the feeding). Sinbad then resumes foraging. Sparky goes to the mid-SC area and parks and preens.

1120 – Sparky is still parked and preening in the mid-SC. None of the others are sighted at this time.

1124 – Remy surfaces from a foraging dive (unsuccessful) in the mid-SC area near Sparky and suddenly charges and pounces at the chick. Sparky immediately flees, diving to escape Remy. Remy then dives after Sparky. Sparky isn't resighted at this time but Remy resurfaces and cruises east along the surface at a quick pace in an aggressive looking, very low profile: she is hunkered very low on the water in a flattened, very streamlined posture, the top of her body forming a straight line from her tail to her beak, the back of her head drawn in against her shoulders, her beak stretched straight out on the surface and cutting a path through the water. She almost certainly was still looking for Sparky.

1126 – Remy finally settles down and parks and preens in the mid-SC area. A passing chick briefly solicits her and moves on. Sparky again parks in the mid-SC area well west (and thus well away) from Remy. None of the others are seen at this time. End observation period.

*1200 – During my continued observations of the family after the formal observation period has concluded Phantom is sighted by itself in the West Lagoon at a time when both parents and at least 3 other chicks (Quark, Spot and Cobra) are in the SC. Sparky isn't seen so it's not known whether or not Sparky is in the West Lagoon with Phantom. Approx. 8 minutes earlier Remy was sighted entering the SC from its SE section so it's possible that she may have been coming from the West Lagoon. If so then it is quite possible that Phantom may have initially followed or accompanied her there.

At one point when Phantom was NE of the snags a great-blue heron flew in from the NE and landed on one of the snags (the tallest of the two). As it was approaching overhead Phantom caught sight of the heron and immediately engaged in an alarm dive, instantly disappearing under the surface.

Phantom returned to the SC at 1209 and went to Sinbad (who was parked in the SE section) and repeatedly solicited him. Just prior to Phantom's return Sparky was sighted by the NE point of the SC, thus indicating that Phantom had been alone in the West Lagoon, at least from the time when first sighted there.

*Comments:* Spot and Phantom were both seen stretching and briefly flapping their wings. Spot's wing development is notably ahead of Phantom's. The flight feathers on Spot's wings are very far along and perhaps 3/4's or so grown. In contrast, Phantom's flight feathers are still fairly short (though coming along well). Spot is currently the second largest of the brood and is now only just slightly smaller than Sparky (the largest chick). Both Sparky and Spot aren't that far off from attaining Remy's size (Remy is noticeably smaller than Sinbad). Quark is still the smallest of the brood but has nearly caught up to Cobra and Phantom.

American coots were seen in the SR pair's territory for the first time this season. Last year they were common and numerous here during the period when coot parents were ranging with their fledged chicks, a period that is about to begin. In my observations of the West Lagoon family last year and the East Lagoon broods this year they have proven to be no threat whatsoever to the chicks (nor are they food competitors with the grebes since they eat plant matter) and they and the grebes basically just ignore each other as they go about their daily business. *[Note: In an intriguing deviation from their normal herbivorous diet, during the course of this study I had a one-off anomalous sighting of a coot feeding on the carcass of a large fish (probably a species of carp) that was floating on the surface of the water.]* However, Sinbad obviously was taking no chances in the episode noted at 1050. Perhaps this was just down to the novelty of the coots appearing there for the first time this season (of course, it's certainly possible and even likely that coots have already come into the SR pair's territory during the considerable time when I wasn't there to observe it but if so, this would have been a recent occurrence anyway).

**8/30/17:** Time: 1100-1212.

1100 – As observations begin, (from the RR bridge) Sinbad is sighted with Phantom and Quark ~ 10 m from the west shore reeds in the mid-north section of the West Lagoon. None of the other family members are seen at this time, including in the SC.

1105 – Spot is sighted with the other three who have been seen in the West Lagoon (Sinbad, Quark and Phantom) in the same general location. At this time I begin heading for my West Lagoon observation point on the mid-east shore.

1112 – I arrive at my east shore observation point. Sinbad is still parked with Phantom and Quark, now ~ 2 m from the west shore reeds in the mid-north section of the lagoon. None of the others are seen at this time.

1116 – Phantom leaves Sinbad and goes into the reeds. Quark remains with Sinbad and intermittently solicits him.

1118 – Sinbad begins foraging. Quark remains where Sinbad left the chick.

1120 – Cobra and Phantom come out of the reeds and park near Quark.

1122 – Sinbad is sighted moving toward the 3 chicks from the east half of the lagoon with a very large fish. Phantom and Quark rush to Sinbad as he nears their area and swarm around him. As Sinbad works the fish he retains firm control of it and it appears obvious that he intends to keep this fish for himself (it is much too large for the chicks to swallow). After ~ 33 seconds Sinbad dives into the reeds with Phantom and Quark still swarming around him and all three of them go out of my view.

1128 – Sinbad is sighted with Quark and two other chicks by the reeds just north of the snags. One of the chicks (it appears to be Cobra) has a large crayfish and is working it on the move with Sinbad following and overseeing the feeding. Cobra soon swallows the crayfish. Sinbad then resumes foraging. (Though the feeding was already in progress when I first sighted it there seems little doubt that Cobra had obtained the crayfish from Sinbad.)

1130 – None of the family members are seen at this time. Remy and Sparky haven't been seen yet today.

1133 – Quark and one other chick (it appears to be Spot) are sighted ~ 8 m from the reeds north of the snags. Both are preening.

1134 – Sinbad is sighted with 3 chicks near the reeds north of the snags. He is spotted just as he finishes swallowing a large crayfish. Sinbad then parks there and preens.

1137 – Sinbad is parked ~ 6 m from the reeds, north of the snags. Four chicks are nearby (they appear to be Sparky, Quark, Spot and the fourth isn't seen well enough to ID).

1139 – Sinbad is still parked and preening in the same location. Sparky approaches Sinbad, soliciting him while drawing near. Sinbad suddenly rebuffs Sparky by charging (for ~ 1.5 m) and lunging at the chick and thus backing Sparky away.

1142 – Sinbad is still parked near the reeds, north of the snags. One chick (my view is too poor to ID it) goes to Sinbad, solicits and sits by him.

1143 – Sinbad begins foraging. An East Lagoon first brood chick swims by right in front of me as it moves north along the east shore, well distant from Sinbad and the chicks. None of them give any indication that they are aware of this chick's presence. The chick quickly moves out of my view as it passes my location.

1146 – Sparky, Quark and Cobra are now parked together and preening themselves ~ 5-6 m from the reeds, just north of the snags.

1151 – Sinbad is sighted moving by Spot (who has come out of the reeds, north of the snags), heading south with Sparky following him. Sinbad soon dives and continues foraging. Cobra goes to the reeds near the snags.

1153 – Sinbad and the 5 chicks are still in the West Lagoon. At this time I begin heading for the SC to see if Remy is there (she still hasn't been seen today).

1200 – I arrive at my usual observation point in the SC. Remy isn't seen at this time.

1205 – Remy is finally sighted—she is foraging in the mid-river area east of the SE section of the SC, working her way west toward the SC.

1207 – Remy moves into the SC and goes to its mid-area where she parks and preens.

1209 – Remy is still parked in the mid-SC area. She heads in the direction of the SW corner, dives and isn't resighted at this time.

1212 – Remy emerges from the SW section of the SC and moves back to the mid-area where she parks and preens. Sinbad and the 5 chicks are still in the West Lagoon. End observation period.

*Comments:* This was a fairly calm and uneventful observation period. Sinbad and the 5 chicks spent the entire period in the West Lagoon, while Remy spent it elsewhere. The chicks were rather subdued in their ranging and movements. They generally remained near or within the reeds in the area by the snags and north of them. An osprey was perched on the tallest snag for about the first 15 minutes of the period but it then departed from the area. Some western gulls were in the general area of the chicks (in the north half of the lagoon) but none were close and they didn't spend much time there.

The chicks are at the stage where they are now able to swallow fairly large prey items (notably Sparky and Spot) though still not as large as what their parents are capable of handling. As a consequence they have been taking control of crayfish sooner during feedings with less working of the crayfish by the parent. However, in feedings involving very large crayfish the chicks are typically reluctant to even try taking control of the crayfish, probably realizing that it is too large for them and perhaps being a bit intimidated by the large crustacean. The chicks' enthusiasm in these feedings generally (but by no means always) wanes quickly. In these cases, eventually the parent always ends out swallowing the crayfish. It seems that carrying prey items that they have captured to the chicks is such a habitual thing for the parents that they sometimes even do so with extra-large prey that they had intended all along to keep for themselves (probably realizing from the initial capture that the prey was too large for the chicks).

As was observed with the West Lagoon brood last year, a notable change regarding crayfish feedings has set in at this stage of the chicks' development. When the chicks were smaller and the parents had to tear off scraps from the crayfish that were of a small enough size for the chicks to swallow, several chicks (i.e., in a group feeding) and sometimes all 5 would typically receive morsels from a crayfish feeding. However, with the chicks now able to swallow fairly large prey items and with their increasing proficiency at working a crayfish, the situation has evolved to the point where typically only one chick obtains food from a crayfish feeding (though another chick may obtain the odd scrap such as a detached claw or other appendage, or pluck a morsel or steal the crayfish while the chick in possession is working it).

**8/31/17:** Time: 1042-1158.

1042 – As observations begin, Remy, Sinbad, Spot, Phantom and at least 2 two other chicks are sighted (from the RR bridge) in the north half of the West Lagoon. And in a potentially troubling sighting the NR male is foraging in the mid-SC area.

1044 – Quark and Phantom are with Sinbad ~ 8 m from the west shore reeds in the north half of the lagoon. At this time I begin heading for my West Lagoon observation point on the mid-east shore of the lagoon.

1052 – I arrive at my east shore observation point. Three chicks (my view is too inadequate to ID them) are sighted in the reeds north of the snags. Approx. two dozen western gulls are in the northern half of the lagoon, in its eastern half.

1055 – Quark is sighted with a fairly large crayfish moving south past the snags. Quark keeps on the move, staying ahead of the hotly pursuing Phantom (Sinbad is also following as he oversees the feeding). After ~ 33 seconds Quark ducks into the reeds with the crayfish. Phantom and Sinbad promptly follow and all go out of view.

1057 – Sinbad emerges from the reeds south of the snags, moves to the mid-water area and begins foraging. Remy is sighted accompanied by Cobra. They are moving north in the mid-water area in the south half of the lagoon. They proceed over near the east shore in the mid-section of the lagoon where they pass out of my view.

1100 – Sparky is sighted moving into the mid-water area in the mid-section of the lagoon and parks there. The gulls have moved to the mid-water area east of the snags. (All 5 chicks have now been sighted in the West Lagoon.)

1102 – Sparky moves east toward the east shore and goes out of my view. Sinbad has parked in the mid-water area in the north end of the lagoon; Cobra is with Sinbad and intermittently solicits him.

1105 – Sinbad is still parked in the same location with Cobra. None of the others are seen at this time.

1107 – Cobra leaves Sinbad and slowly cruises toward the east shore. Cobra proceeds about halfway there, then reverses course and returns to Sinbad.

1110 – Remy is sighted moving from the mid-east shore area in the direction of Sinbad (she is fairly distant from him). Sinbad (still in the same location with Cobra) calls once.

1111 – Remy stops and parks in the mid-water area in the north half of the lagoon (still fairly distant from Sinbad).

1112 – Sparky is sighted moving toward Remy from the east. When Sparky nears to ~ 3-4 m of Remy she suddenly charges at the chick. She gets only within ~ 1.5 m before Sparky flees, diving to escape her rebuff. Sparky resurfaces away from Remy and remains clear of her a little farther to the north.

1114 – A chick goes to Remy (still parked in the mid-water area in the north half of the lagoon), solicits and remains with her (my view is too poor to ID the chick). Sinbad and Cobra have drifted to the east and have gone out of my view.

1116 – A train noisily passes over the RR bridge. This spooks Remy and the chick who is with her and both rapidly flee south and soon dive. All of the gulls in the north half of the lagoon are also spooked and flee south; some of them return soon after.

1117 – Remy again parks with a chick in the mid-water area in the north half of the lagoon.

1121 – Remy moves north with the chick accompanying her. They pass under the bridge and it initially looks as though Remy is going to go the SC. She starts off in that direction, pauses and gazes toward the SC but then moves east along the reeds of the north bank of the south ribbon of the river (time: 1123). The chick follows closely at her heels and the pair soon passes from my view (which is blocked by the base of the bridge pillars). (Remy's actions and movements lead me to wonder if the NR male is still in the SC. Was it perhaps due to his presence there that she chose not to go into the SC? Unfortunately, I can't see into the SC from my West Lagoon observation point so I can't see whether or not he is still there.). None of the others are seen at this time.

1130 – None of the family members are seen in the West Lagoon at this time. I begin heading over to the SC to check on what is going on there.

1138 – I arrive at the SC. The NR male is sighted foraging near the reeds of the north bank, west of the NE point. He soon works his way to the mid-SC area (he has probably been in the SC since he was seen there at the beginning of the observation period). None of the SR family is seen in the SC. At this point the situation seems somewhat ominous for the SR family. Have they been ousted from the SC by the NR pair who may possibly have annexed it to their territory? Or is the NR male merely taking advantage of their temporary absence to sneak in some foraging there?

1143 – The NR male is still foraging in the mid-SC.

1146 – (I have moved my observation point to the west end of the RR bridge.) Sinbad is sighted foraging inside the west end of the inter-lagoon channel. He soon moves out to the mid-water area in the north half of the West Lagoon where he parks. The NR male is still in the SC.

1149 – Sinbad is still parked in the same location. Quark goes to Sinbad, solicits and remains with him.

1151 – One of the chicks (my view is too inadequate to ID it) is sighted alone, moving deep into the SC. As the chick reaches the SW section it catches sight of the NR male (who is in the mid-west end area) and immediately dives and isn't resighted at this time (the chick most likely went into the reeds in the SW corner to hide there). It wasn't clear as to whether the NR male saw the chick—regardless, he didn't go after it.

1153 – The NR male has drifted over to the SW corner in the area where the chick had dove and parks there. The chick still hasn't been resighted.

1158 – Remy, Sinbad, Quark and Sparky are sighted near the west shore reeds in the north half of the lagoon. The NR male is still in the SC. He is parked in the mid-SC area and is looking about (as though on "sentry duty"), mainly to the east (probably particularly for members of the SR pair). End observation period.

*Comments:* It appears possible that something may have occurred which could potentially have dire consequences for some or all of the chicks. From the NR male's behavior today the indications are that the SR pair may possibly have lost the SC to the NR pair; that they may have been ousted from the SC by the NR pair who may possibly have annexed it to their territory.

The SC has always been the core area for the SR family and their main activity area. If they have indeed lost the SC it then remains to be seen what impact this would have on the remaining portion of the chick-rearing endeavor. Losing the SC would leave the pair with the West Lagoon, the west half of the inter-lagoon channel, the south ribbon of the river and perhaps some of the river east of the SC as the remaining parts of their territory. Hopefully, this area contains enough resources to support the still growing chicks through the remaining period of their drive toward achieving independence. If not, they would need to come up with an option(s) for attaining sufficient habitat/territory.

Providing a large measure of hope and reassurance here is what overwhelmingly stands as the most likely explanation for the NR male's presence in the SC. The odds are extremely high that this may be a simple case in which, with Sinbad and Remy being a long distance away in the West Lagoon, the NR male took advantage of the opportunity to do some foraging in the SC during their absence. Perhaps a bit at odds with this is the fact that at ~ 1245 (during my continued observations of the family after the formal observation period had concluded) I sighted the NR male in the north section of his territory and within ~ 15-20 minutes he made his way back into the SC (the SR pair still had not returned to the SC). Of course, here again with Sinbad and Remy still in the West Lagoon it may have been just another case of him taking advantage of their temporary absence. Obviously, the key point here is: Why was the SR pair absent from the SC? Was this just down to normal ranging/foraging on their part (as the odds strongly favor—the entire SR family are now routinely seen out of the SC and in the West Lagoon) or had they been ousted by the NR pair in the period since yesterday's observations concluded? A further consideration here is that on a number of occasions in the past in this general area I have seen one or more grebes "sneak" into a part of a pair's territory and forage there while the owners of the territory were in a distant part of their territory. Hopefully, such was the case this time.

Though the odds strongly favor otherwise, it is interesting to speculate on what possible options the SR pair would have if they have indeed lost the SC to the NR pair (and the remainder of their territory is not sufficient for rearing the chicks). One obvious option would be to retake the SC. However, considering that they already have lost the entire northern half of their territory to the NR pair and now (in this possible scenario) the SC, it appears that they probably aren't capable of doing so (though having chicks to feed might possibly give them extra drive and incentive).

The only other possible option would be to expand their territory (or at least their range). The possibilities here are limited. They could attempt to reclaim some or all of the former northern half of their territory but this seems that it would be as unlikely as retaking the SC for the same reasons.

Another possibility would be to annex the narrow channel that flows around the north and east sides of the island that forms the SE bank of the river and the north bank of the south ribbon. However, it's questionable as to whether this would be a sufficient addition. Also, its narrowness and shallowness make it more hazardous, especially for chicks.

A final possibility would be to expand their territory/range to include the east half of the inter-lagoon channel and the north end of the East Lagoon. This area was heavily used by both of the

East Lagoon broods and appears to be quite productive in regard to available prey. Of course, the major obstacle here is that these areas are part of the East Lagoon pair's territory. If it came to this, something that perhaps would work in the SR pair's favor is that it appears that all of the East Lagoon second brood chicks are now independent.  Being that the East Lagoon pair then no longer needs to forage for and feed these chicks perhaps they no longer feel so keen on defending all of their territory. Maybe this might make them a little more lax in this regard and allow the SR family to "borrow" some or all of this part of their territory or allow them to use it sometimes (unless the pair is still keen on defending it for the use of their newly independent chicks).

Relating to the above possibility is a point that I have noted over these past two breeding seasons regarding the hatching times of the broods in the adjacent territories in this area (specifically, the East and West lagoons and the south river). The hatchings of the pairs' clutches have not been synchronous but instead have been staggered. Thus, the nesting periods and various chick rearing/developmental stages at a particular point in time differ between the pairs of the adjacent territories. Such differences allow for the possibility noted above. For example, during the period when Sinbad and Remy were nesting, the East Lagoon pair had chicks who were in their mobile stage at that time and made some use of the West Lagoon while rearing their brood (the pair brought their chicks into the West Lagoon and foraged for and fed them there). [Remy and Sinbad also did some personal foraging in the West Lagoon during that period but with no chicks to feed at that time they didn't really defend or patrol it (nor did they use it much then; recall that they didn't fully annex the West Lagoon as a part of their territory until later).] Such a thing wouldn't be possible for the East Lagoon pair to do right now when Remy and Sinbad are feeding their mobile chicks in the West Lagoon because they would be excluded and/or it would generate a potentially dangerous conflict. One is thus left to wonder if the neighboring pairs monitor each other and deliberately seek to create these staggered hatchings between the pairs in order to set up a situation that can generate this possibility for their respective broods.

With the possible loss of the SC it's fortunate that the SR chicks are currently at a very mobile stage of their development in case the situation calls for wider ranging by the parents and chicks to compensate for the loss. As noted in the 8/28 notes the chicks are progressing well in their march toward achieving independence but all of them (including Sparky and Spot) are still dependent on the parents. Though all of the chicks engage in some independent foraging, none of them are as yet capable of securing their daily food requirements on their own (i.e., they still require their parents' help in the form of feedings). In addition, though Sparky and Spot don't appear to be that far away from doing so, none of the chicks have fledged (using the most minimal definition of fledging here—none of the chicks are capable of taking even short flights; this assessment is based on observations of flight feather growth).

Spot has put on quite a growth spurt over the past few weeks and has nearly caught up to Sparky in regard to size. Spot also does a considerable amount of independent foraging and probably now ranks second highest in the brood's dominance hierarchy. Early on, Spot had been close to Cobra in size but has now clearly surpassed Cobra.

Another possible PB grebe pair was briefly sighted in the far north end of the wide section of the river (the part under and just beyond the Highway 101 overpasses). This small section is either part of the NR pair's territory or borders its north boundary. Is it possible that this pair is pressuring the NR pair and that this could be in some way related to the NR male's presence in the SC? (I didn't get a very good look at these other two grebes so at this time of year it's also possible that they could be two associating newly independent chicks who have recently changed from their juvenile plumage to the winter plumage of an adult.)

The NR male's presence in the SC today is very likely just a case of him taking advantage of the SR pair's temporary absence to do some foraging there. Nevertheless, despite this strong probability I still feel anxious about this situation because the SR pair hadn't yet returned to the SC by the end of observations (this includes approx. 2 hours of continued observations after the formal observation period had concluded). Thus, I don't yet know for certain where the situation stands and whether I will find the SR family back in the SC when I return tomorrow. I expect that I will but I can't help but worry about them in the meantime.

**<u>9/1/17</u>:** Time: 1100-1210.

1100 – The observations begin with a big sigh of relief as the family is sighted back in the SC. Sparky, Spot, Quark and one other chick are all competing over a large crayfish in the NW corner of the SC with Sinbad overseeing the feeding. The intense competition rages for ~ 3 minutes (Sinbad leaves them to it and resumes foraging after a few minutes) with the crayfish changing hands several times and various chicks managing to tear off morsels. Finally, Sparky controls and swallows the large remainder of the carcass (time: 1103).

1106 – Sparky and Spot are parked and preening themselves in the NW section of the SC. None of the others are seen at this time.

1108 – Sparky and Spot are still parked near each other in the NW section. Quark is seen foraging in the NW section. None of the others are seen at this time.

1111 – Phantom is now also seen—he is parked and preening at the SE end of the SC.

1113 – Sinbad is sighted with a medium-sized crayfish at the edge of the reeds just west of the NE point of the SC. Quark is near Sinbad. The chick moves to him, takes and quickly swallows the crayfish. Sinbad then resumes foraging.

1116 – Sinbad is sighted with a large crayfish in the NE section of the SC. Quark and another chick go to him but are slow to make a move at taking the crayfish as Sinbad works it. Phantom (joining the feeding) then rushes in, grabs the crayfish and moves away with it. Neither of the other 2 chicks bother to follow Phantom. Phantom intently works the crayfish for ~ 53 seconds before finally swallowing it. (Shortly after Phantom had taken the crayfish Sinbad rebuffed one of the chicks.) Sinbad resumes foraging.

1122 – Sinbad is now parked and preening at the NE end of the SC. An American coot is foraging at the reeds on the south side of the NE point. Sinbad moves directly at the coot and chases it away. The coot flees, taking a low flight over the surface to the mid-south bank area of the SC. Soon after, Sinbad calls once.

1126 – Phantom, Quark, Spot and Sparky are parked and preening near Sinbad in the NW section of the SC.

1128 – A train noisily passes over the RR bridge. The 4 chicks are spooked by this and move toward the north bank reed beds. After it clears the area they stop and park ~ 6-12 m from the reeds at the mid-north bank. Sinbad remains at his location in the NW section. Quark goes to Sinbad, solicits and remains with him. Sparky soon joins them. Remy and Cobra haven't been sighted yet so I head for the west end of the RR bridge to check for them elsewhere.

1134- Remy is sighted foraging along and within the reeds at the mid-north bank of the south ribbon of the river. She is accompanied by Cobra. Sinbad and the other 4 chicks are parked in the same general area as at 1128. An osprey is perched at the top of one of the two snags that the ospreys usually use as perches when in this area. [These snags are on the east bank of the river, east of (i.e., across the river from) the SC.]

1140 – (I have moved back to my SC observation point on the mid-south bank.) Sinbad and the 4 chicks (all but Cobra) are in the NW section of the SC. Spot has a small fish and is being pursued by Sparky and Phantom as Spot darts about ahead of them (I don't know whether Spot had captured it or had taken it from Sinbad before I arrived and spotted them). The fish is small and could easily be swallowed but instead it changes hands among the 3 chicks a number of times as it is worked by the various chicks for ~ 3 minutes. In the end, it appeared that the fish was just dropped and abandoned (it was as though the chicks had just been playing with or manipulating it; I am also left to wonder if the fish was for some reason unpalatable).

1145 – Sinbad is parked and preening with Phantom in the NW section of the SC. Sinbad calls once and resumes preening. Sparky and Quark are parked apart from each other, also in the NW section. Spot is at the east end of the SC.

1150 – Sinbad is parked with Phantom south of the NE point of the SC. Two other chicks are sighted foraging near the south side of the NE point. The NR male is sighted foraging mid-river in the mid-south half of their territory.

1151 – Sinbad, followed by Phantom and 2 other chicks go into the reeds at the south side of the NE point and go out of my view.

1156 – One chick is briefly sighted moving along the edge of and back into the reeds, west of the NE point.

1200 – A chick emerges from the reeds west of the NE point with a fairly large crayfish (Sinbad is foraging in the reeds in that area so the chick most likely got the crayfish from him). The chick (possibly Phantom) works the crayfish for ~ 40 seconds before swallowing it. Several other chicks are seen at the edge of the reeds (my view is too poor to ID them).

1202 – Sinbad emerges from the reeds at the south side of the NE point and parks just south of it.

1204 – Sinbad moves south along the east end of the SC to the mid-east end, then reverses course and moves back to where he had been and soon begins foraging.

1207 – Spot is sighted with a large crayfish by the reeds west of the NE point (Spot apparently had just taken the crayfish from Sinbad). Spot moves west, keeping ahead of Phantom and one other hotly pursuing chick. Spot works the crayfish while keeping ahead of its 2 sibs, often changing directions in an attempt to elude them. After working the crayfish for ~ 140 seconds under nearly constant pressure from its 2 sibs Spot finally swallows it (time: 1210). The osprey is still perched in the same location as at 1134. End observation period.

*1235 – At ~ 1235 all of the family went into the West Lagoon along the west shore reeds, eventually dispersing from the north end to the south end of the reeds.

*Comments:* It was quite a relief to find the family back in the SC today. The NR male was briefly seen foraging in the mid-section of the south section of the NR pair's territory but wasn't seen anywhere near the SC. It now appears most likely that when he went into the SC yesterday he was just seizing an opportunity to forage there while Sinbad and Remy were absent from and a considerable distance from the SC.

Several of the chicks demonstrated today how proficient they are becoming at working large crayfish by themselves.

**<u>9/2/17</u>:** Time: 1132-1235.

1132 – As observations begin (from the SC), none of the family members are seen at this time. Two family members were seen from a distance in the SW corner as I approached the area but they went into the reeds just before I arrived. None were seen in the West Lagoon as I passed by it on my way to the SC.

1136 – Sinbad emerges from the reeds in the SW corner of the SC with a large crayfish. He is being followed by Quark, Phantom and Cobra. Quark gets the crayfish from Sinbad and works it for ~30 seconds before finally swallowing it. Sinbad had rebuffed one of the chicks who had tried to pursue Quark after Quark had taken control of the crayfish and this caused that chick to cease its pursuit (my view was too inadequate to be able to ID which of the chicks got rebuffed).

1138 – Sparky emerges from the reeds in the SW section of the SC with a large crayfish and is being followed by Spot. Sparky rather casually works the crayfish for ~ 11 seconds and then just drops it (for some reason Sparky doesn't seem very enthusiastic or interested in the crayfish). After Sparky drops the crayfish Spot picks it up and soon swallows it. One of the chicks (it appears to be Phantom) makes its way out from the SW section and goes out of my view as it heads east toward the SE end of the SC.

1142 – Sinbad emerges from the reeds in the SW corner with a large crayfish and is being followed by Sparky. This time Sparky is quite enthusiastic. The chick promptly takes the crayfish from Sinbad and works it for ~ 69 seconds before swallowing it. Sinbad then resumes foraging.

1145 – Cobra and Sparky are parked together ~ 10 m south of the old nest site. Spot is parked and preening ~ 14 m south of the NE point of the SC. None of the others are seen at this time.

1150 – Cobra/Sparky and Spot are still parked in their same locations as at 1145. Sinbad is foraging along and within the reeds west of the NE point.

1154 – Cobra and Sparky have drifted ~ 2 m apart (Cobra is east of Sparky). Sinbad approaches the two chicks from the east with a large crayfish. Cobra (who is nearer to him) spots Sinbad first, rushes to him and takes the crayfish. Cobra works the crayfish for ~ 25 seconds before Sparky joins the feeding and takes the crayfish away from Cobra. Sparky moves away with it and twice as Cobra attempts to pursue Sparky, Cobra is rebuffed by Sinbad and chased away. Sparky works the crayfish for over a minute before finally swallowing it (the feeding lasted ~112 seconds). Sinbad then resumes foraging.

1157 – Sinbad is sighted with a very large crayfish in the NE section of the SC. Cobra is with Sinbad, eagerly looking for a feeding but the crayfish is clearly too large for the chick. Sinbad retains control of the crayfish and after working it for ~ 26 seconds is finally able to swallow it (though he had a difficult time working it down his throat).

1201 – Sinbad moves to the east end of the mid-SC area where he parks and preens. Cobra moves toward Sinbad and solicits him while approaching near. Sinbad is far from receptive to this and charges at Cobra and drives the chick away. Cobra flees from Sinbad and soon parks ~21 m to the NW of him.

1202 – Sinbad is still parked and preening in the same location. He calls once and continues preening.

1207 – Sinbad is still parked in the same location. Sparky moves out to Sinbad and sits by him. Soon after, Sinbad moves a few meters away.

1214 – Spot emerges from the reeds at the mid-north bank of the SC, goes out to Sinbad (still in his same location) and solicits him a number of times but Sinbad is unresponsive to this. Spot then goes back to the reeds west of the NE point (time: 1215). Sparky is still parked in the same general location as at 1207. At this time I leave my SC observation point and head out to the RR bridge to search for Remy (who hasn't been sighted yet today).

1220 – I have returned to my SC observation point—Remy still hasn't been sighted. Sparky is still parked in the same general location. Quark is parked with Sinbad at the east end of the mid-SC area and occasionally solicits him. None of the others are seen at this time.

1225 – Same as at 1220. I again leave my SC observation point to search for Remy. As I'm leaving the SC I sight Cobra emerging from the reeds west of the NE point.

1235 – From the west end of the RR bridge I searched for Remy in the West lagoon and the south ribbon of the river but didn't see her. I have returned to my SC observation point. Sinbad is sighted moving east from the west end with Quark and Cobra following him. They continue toward the SE end of the SC and soon pass from my view (Sinbad appears to be heading for the West Lagoon). Spot and Sparky remain parked in the NW section of the SC. Remy still hasn't been sighted today. End observation period (though I will continue to search for Remy).

*1245 – Remy is finally sighted—she is in the West Lagoon.

*1320 - Remy is now parked with Cobra in the mid-SC area. Sinbad and the other 4 chicks are in the West Lagoon.

*Comments:* Several of the chicks were seen stretching and flapping their wings. Their flight feathers appear to be nearly full grown and perhaps in the cases of Sparky and Spot they are now fully grown or very close to it. If the latter is indeed the case then they may even be capable of taking short flights.

Sinbad was also seen stretching his wings. It was thus noted that the replacement flight feathers for the ones that he molted earlier this season are fully grown or nearly so. It is interesting that Sinbad's habit of rubbing his folded wings together over his back ceased after he finished molting his flight feathers and they began growing back in. It thus is quite possible that this wing rubbing was indeed related to the molting as I had speculated at that time. In fact, Remy has also been observed doing this during her current flight feather molt, though not quite as frequently as Sinbad had done.

As has commonly been observed, Sinbad and several of the chicks were seen regurgitating what look to be crayfish exoskeletons. I have seen the grebes do this numerous times so it is evident that the exoskeletons do not pass completely through the digestive tract. This apparently is a grebe analog to owl pellets.

In a further sign of the increasing maturity of the chicks they displayed several times today that they are no longer apprehensive of or intimidated by the presence and proximity of birds such as American coots and gulls. At one point today Cobra was right in the midst of 3 coots and simply continued about her business and basically ignored them (and vice versa). Similarly, while in the West Lagoon (during my continued observations after the formal observation period had ended) Sparky, Spot and Quark spent some time right next to and sometimes in the midst of a group of western gulls in the mid-lagoon area. They exhibited no apprehension and (as with Cobra and the coots) just went about their business essentially ignoring the gulls (and vice versa, though a few of the gulls did casually gaze at the chicks). Sinbad didn't come out to "rescue" or retrieve them but it wasn't clear if he was aware of the situation.

The chicks are progressing well toward achieving independence (and Sparky, as well as Spot now, still appear more advanced in this respect) and are not that far away from doing so. While they are still dependent on the parents and spend time accompanying and taking food from them, they also spend a considerable amount of time foraging independently (though I still haven't been fortunate enough to see any of them capture prey there seems little doubt that they do). As was noted in my observations of the West Lagoon brood last year, because PB grebes are seldom seen in flight the odds are that the chicks will probably be capable of at least short flights before I actually get to see them fly.

**<u>9/3/17</u>:** Time: 1109-1210.

1109 – As observations begin, Remy is sighted (from the RR bridge) with Quark and Sparky just south of the RR bridge in the north end of the West Lagoon. Remy has a medium-sized fish and is being hotly pursued by Quark (closest to her) and Sparky. It appears that Remy doesn't want to give up the fish (i.e., that she wants to eat it herself) and is dodging attempts by Quark to snatch it from her. Quark is finally able to grab onto it but Remy doesn't release her hold on the fish. A mighty tug-of-war then ensues between these two which, after ~ 7 seconds of strenuous effort, Remy wins. However, soon after Remy wrests back complete control of the fish from Quark, Sparky rushes forward and grabs hold of the fish. Remy again maintains her hold on the fish and another tug-of-war ensues. Sparky eventually wins this and quickly swallows the fish. As the two chicks look to regroup around Remy, she charges at and rebuffs one of them (my view was too poor to be certain of which one but it probably was Sparky), driving the chick away from her immediate proximity (time: 1112). Remy, and then Quark and Sparky also, move north under the bridge to the mid-river area east of the SC. After a short while all three make their way into the SC. At this time I move to my SC observation point on the mid-south bank (time: 1115).

1120 – Remy is now parked and preening herself with Cobra (who intermittently solicits her) in the mid-SC area.

1122 – Sinbad is sighted with Quark, Spot and a third chick in the SW corner of the SC. Sinbad has a large crayfish which Spot soon takes and is then pursued by Quark (the third chick goes back to the reeds). Spot keeps on the move ahead of Quark while working the crayfish and after ~ 17 seconds is able to swallow it. Sinbad (who had been shadowing and overseeing the feeding) then resumes foraging. Spot and Quark go into the reeds at the west end.

1125 – Remy moves to the reeds at the west end of the SC and begins foraging there. Cobra accompanies her and waits by the edge of the reeds, watching to see if she catches any prey.

1127 – Sinbad surfaces from a foraging dive near the reeds north of the nest with a medium-sized crayfish. Sparky rushes to Sinbad, takes the crayfish and swallows it after having worked it for ~ 15 seconds. As Sparky circles back toward Sinbad and begins soliciting him while approaching near, Sinbad charges and pecks at the chick. Sparky immediately dives and flees from Sinbad, resurfacing soon after in the mid-SC area. Sparky then goes back to the reeds in the NW section and forages there. Sinbad has resumed foraging.

1130 – Remy is now parked and preening in the mid-SC area. Sparky is by the old nest site peering into the reeds (Sinbad is probably foraging there). None of the others are seen at this time. There have been no confirmed sightings of Phantom yet today.

1135 – Remy is still parked and preening in the mid-SC area. Quark is sighted foraging in the NW section of the SC. At this time I leave my SC observation point and head for the west end of the RR bridge to check the West Lagoon and south ribbon of the river for Phantom.

1139 – I have returned to my SC observation point (no family members were seen in the West Lagoon). Phantom is finally sighted—the chick is parked in the mid-SC area.

1144 – Remy is sighted with a medium-sized fish ~ 30 m from (north of) my observation point on the mid-south bank. Cobra is with her, eager for the fish. Remy works the fish and doesn't allow Cobra to take it until after ~ 10 seconds. Cobra then works it a further ~ 17 seconds before swallowing it. Both then go to the mid-SC area.

1146 – Remy and all 5 chicks are now spread through the mid-SC area.

1150 – Separately, Remy, Cobra and Spot have moved to the SE end of the SC and have passed beyond my view. Quark is with Sinbad at the mid-area of the east end of the SC. Phantom and Sparky are parked ~ 12 m apart in the NW section.

1152 – Sinbad is still parked with Quark and is preening himself. They have drifted near Phantom and Sparky in the NW section (Phantom and Sparky are now only a few meters apart). Spot and Cobra are in the SE end of the SC and soon move east out of my view. Remy isn't seen at this time.

1156 – Sinbad goes to the reeds at the south side of the NE point of the SC, followed and solicited by Quark and Phantom, and soon begins foraging there. All three of them go into the reeds, followed shortly after by Sparky (all go out of my view).

1158 – Cobra is sighted moving north along the east end of the SC from its SE section. The chick proceeds to the edge of the reeds at the south side of the NE point where Sinbad and the three chicks had gone into the reeds. Cobra parks at the edge of the reeds and peers into them, apparently either watching or looking for Sinbad and the others (Cobra is obviously on the lookout for a feeding).

1200 – Cobra is still parked in the same location, peering into the reeds. None of the other family members are seen at this time.

1203 – Sinbad is sighted with a very large crayfish by the reeds west of the NE point. Spot (who had also made its way to the group that was monitoring Sinbad's foraging) is the only chick to emerge from the reeds and goes to Sinbad. Upon reaching Sinbad, Spot promptly takes the crayfish and begins working it, moving and keeping ahead of Sinbad who shadows Spot to oversee the feeding. Spot has a challenging time working this very large crayfish and is sometimes assisted by Sinbad (whose assistance Spot very willingly accepts). In these latter cases of assistance Sinbad picked up the crayfish after Spot had dropped it on the water while working it and Sinbad then worked it a bit before handing it back to Spot. After the crayfish had been worked a lengthy ~ 140 seconds by the two of them Spot was finally able to swallow it (time: 1206). Sinbad and Spot then park and preen ~ 12 m apart in the NW section of the SC.

1210 – Remy has moved back into the SC and is now parked and preening in the mid-SC area. Spot is still parked and preening in the NW section. Sinbad is foraging in the west end of the SC. Phantom is parked by the reeds there, watching and waiting for Sinbad to capture prey for a feeding. The others aren't seen at this time. End observation period.

*About 50 minutes after the formal observation period concluded, Remy and the accompanying Cobra were seen in the West Lagoon by themselves. Sinbad and the other 4 chicks were in the SC.

*Comments:* The chicks are quite large now and are about at the point where, when seen at a distance it is getting difficult to differentiate parent from chick by size alone. In fact, the two largest chicks (Sparky and Spot) are now pretty similar in size to Remy.

Among the chicks Spot's growth spurt has continued. It appears that Spot may even have slightly surpassed Sparky in size, which would make Spot the largest of the brood. These two are still noticeably the largest of the chicks. However, as they have neared their full size their growth rate has slowed, allowing the other three chicks to close the gap. Quark (previously the smallest chick) has done well and caught up to Phantom and Cobra. These three are now pretty similar in size.

To reiterate the observation noted during my case study of the West Lagoon brood last year, it is obvious how vital the dense reed habitat is for the grebes here. It is within the emergent reeds that they nest and shelter their young. The reeds are the place where both chicks and adults hide when faced with various threats (actual or potential). The reeds are the place where the grebes routinely rest and seek cover. And while they also forage in open water, it is along the edges of and within the reeds that they do the majority of their foraging (at least, this is what I have observed for the SR and West Lagoon pairs while rearing their broods and also for their chicks). The SR pair did much more open water foraging during the period before the chicks hatched than they do now. This is reflected in the general types of prey that are fed to the chicks, with crayfish comprising the great majority of this. [Note though, that the SR pair has fed notably more fish to their chicks (though not so much since 8/18; see below) than the West Lagoon pair did, the latter almost exclusively feeding crayfish to their brood. However, a point that is worthy of consideration here is that the great majority of the observed fish feedings by the SR pair occurred during the first three weeks or so after hatching. I didn't have the opportunity to observe this earliest period for the West Lagoon brood so I don't know whether or not the West Lagoon pair also fed more fish during this early stage.] It is likely of significance in regard to prey type that crayfish are probably much easier to catch than fish. This is an important consideration when trying to feed a large brood of 5 chicks. Unfortunately, I have no information regarding the relative abundance of these two general prey types which can also be an important factor in prey selection; nor do I have any information regarding their relative nutritional aspects.

Today, feedings of fish (i.e., in which a chick actually ate the fish being proffered) were observed during the course of a formal observation period for the first time since 8/17. There were a few feedings in the interim during which a parent attempted to feed a fish but in each of these cases the fish was too large for the chicks to swallow. Also, there is the curious case from 9/1 during which a small fish was manipulated for a lengthy period of approximately 3 minutes by the three chicks involved but was ultimately abandoned by them.

**9/4/17:** Time: 1043-1156.

1043 – As observations begin, Sinbad is sighted (from the west end of the RR bridge) ~ 12 m from the west shore reeds in the mid-section of the West Lagoon. A paddleboard(er) is at the SE end of the river but soon moves out of the area. None of the family is sighted in the SC at this time but if any are there they have probably been spooked by the paddleboarder and are hiding in the reeds, out of sight.

1044 – (I am now at my SC observation point on the mid-south bank.) Phantom is sighted foraging in the mid-SC area.

1047 – Spot is sighted foraging along the reeds in the mid-south bank area of the SC.

1048 – Quark is sighted foraging at the reeds by the old nest site.

1049 – Remy is sighted foraging in the mid-SC area. Spot moves out to the mid-SC and parks in the area where she is foraging.

1051 – Sinbad enters the SC from the SE (returning from the West Lagoon) and moves into the mid-SC area with a small fish. Spot catches sight of Sinbad, rushes to him, takes and quickly swallows the fish. Sinbad then goes to Remy (now parked in the mid-SC) and they meet and greet while voicing extended, continuous chattering vocalizations. Remy remains parked there and Quark goes to her, solicits and remains with her. Sinbad circles through the SW corner and continues east off the SW bank where he deliberately zeroes in on an American coot who is in the SW section. Sinbad charges at and briefly chases the coot who flees from him. The coot flies low over the surface out to the mid-river area east of the SC. Sinbad then parks and preens in the SE section of the SC. Sparky and Phantom go to Sinbad, solicit and remain with him.

1055 – Sinbad is still parked in the same location, now only with Phantom. Sparky and Spot are foraging just east of them in the SE end of the SC. Remy is still parked and preening herself with Quark in the mid-SC area.

1100 – Sinbad is still parked and preening himself with Phantom in the same location (Phantom intermittently solicits him). Spot is now parked ~ 9 m to the east of them. Remy is still parked and preening herself with Quark in the mid-SC area (Quark is intermittently soliciting her). The other two chicks aren't seen at this time (Cobra hasn't been seen yet today).

1102 – Phantom leaves Sinbad and moves to the mid-area at the east end of the SC and parks there.

1103 – Remy moves toward Sinbad. They briefly meet and greet and Remy continues moving toward the SE end of the SC, soon passing beyond my view. Sinbad remains parked and preening in his same location in the SE section. Quark and Spot remain in the mid-SC (Quark didn't follow Remy when she moved away). Phantom goes to Sinbad, solicits him briefly and then goes to the mid-south area of the SC.

1106 – Spot and Sparky are now parked together and preening themselves in the mid-south area. Quark goes to Sinbad, solicits and remains with him.

1108 – Remy comes into the SE section of the SC where she parks and preens. Quark leaves Sinbad and goes to Remy. Quark solicits Remy upon arrival and remains with her. Sinbad moves out to the mid-river area just east of the SC and parks there. Spot moves toward Remy/Quark and parks ~ 4 m to the west of them. Sparky heads E-SE through the SE section of the SC and passes beyond my view.

1111 – Sinbad moves to the SE and passes from my view. Spot goes to Remy/Quark and sits with them. Cobra still hasn't been sighted yet today.

1115 – Remy is still parked and preening herself with Quark in the same location. Spot is now parked ~ 2 m east of them. None of the others are seen at this time. At this point I leave my SC observation point and head for the west end of the RR bridge to check the West Lagoon and south ribbon of the river.

1118 – (From the west end of the RR bridge.) Sinbad is parked with Phantom in the mid-river area, east of the SC.

1120 – Sparky is sighted swallowing a small or medium sized prey item that it captured on its own near the east bank of the river, across from (east of) the SC. This is the first confirmed observation of a chick catching prey. Shortly after, Sinbad moves into the area and charges toward and drives Sparky to the mid-river area (there were at least 4 coots by Sparky at the time and it wasn't clear as to whether Sinbad chased Sparky away because he was concerned about the chick being so close to the coots or was just rebuffing Sparky—it seems more likely to be the former because I haven't observed either of the parents move so far just to rebuff a chick). Sinbad then moves back into the mid-river area. Phantom and Spot are in his general area.

1125 – Sinbad moves to the area near the east bank of the river, across from the SC and begins foraging there. Phantom has accompanied him and sits in that general area watching to see if Sinbad catches any prey for a feeding. Spot and Sparky are in the mid-river area east of the SC. Remy is parked with Quark in the NW section of the SC. Cobra still hasn't been sighted.

1126 – Spot and Sparky go to the area near the east bank of the river where Sinbad and Phantom have gone (obviously, also to monitor Sinbad for a possible feeding).

1127 – Sparky moves back to the mid-river area, east of the SC.

1129 – Remy (still with Quark, now in the NE section of the SC) dives and begins foraging along the east end of the SC (in open water). Sparky has moved near the reeds of the east bank at the SE end of the river.

1132 – Sinbad goes from the east bank of the river to the reeds along the south side of the NE point of the SC and begins foraging there. One chick (it appears to be Phantom) follows him there.

1134 – Remy is sighted just north of the RR bridge in mid-river with a medium-sized fish. Cobra (who is near her location) rushes to Remy, takes and quickly swallows the fish. Remy then resumes foraging. Quark and Cobra accompany her as she does.

1136 – Remy moves south under the RR bridge into the mid-water area at the north end of the West Lagoon. Quark, Cobra and Spot follow her. Sinbad, Sparky and Phantom are in the SC.

1140 – Spot moves near the west shore reeds in the north half of the lagoon and begins foraging.

1142 – Remy moves to the west shore reeds in the north half of the lagoon and begins foraging along the edge of and within the reeds. Quark and Cobra accompany her. Sparky is parked in the mid-SC area. Sinbad and Phantom aren't seen at this time.

1145 – The family is still split: Remy, Spot, Cobra and Quark in the West Lagoon, and Sinbad, Sparky and Phantom in the SC. At this time I move back to my SC observation point.

1147 – (I am now back at my SC observation point on the mid-south bank.) Sinbad moves from the north bank area of the SC to the mid-area. Sparky moves toward Sinbad. As the chick nears him, Sinbad charges at and pecks once at Sparky. Sparky immediately dives to elude him. Sinbad then parks at the east end of the SC and preens. Sparky parks and preens ~ 7 m to the SE of him. Phantom parks and preens ~ 1.5 m east of Sinbad.

1150 – Sinbad, Phantom and Sparky are still parked and preening in their same locations. The others are still in the West Lagoon.

1156 – Sinbad and (following him at a distance) Sparky go to the West Lagoon. Remy, Spot, Quark and Cobra are still there. Phantom remains in the SC, parked and resting in its NE section. Thus, all but Phantom are now in the West Lagoon. End observation period.

*At 1221 Phantom finally went over and joined the rest of the family in the West Lagoon.

*Comments:* The chicks continue to exhibit increasingly more independence in their ranging and foraging. Nevertheless, they clearly are still dependent on the parents. They still seek them out and look to them for guidance. Also, they still solicit the parents for food, spend time accompanying them and continue to take food from Remy and Sinbad. However, they are doing all of the above to a lesser degree now and all of the chicks spend considerable time ranging and foraging independently, both alone and in company with other sibs. It appears that the chicks (especially Sparky and Spot, and probably Cobra) aren't all that far away from achieving independence.

Once again, several of the chicks were observed briefly flapping their wings. The flight feathers on those chicks appear to now be either fully grown or nearly so. It is quite possible that at least some of the chicks are now capable of taking short flights just over the surface. Sinbad was also seen stretching his wings today and the replacement flight feathers from his earlier molt also appear to be fully grown or nearly so. With Remy currently in the molting period for her flight feathers she is now the one whose wing margins are looking a bit ragged.

For some reason Sinbad seems to have a “thing” regarding American coots. For a second time he was observed chasing a coot out of the SC. The coot had done nothing untoward; it was merely going about its own business and paying no attention to the grebes (as I have always observed to be the case with coots). In addition to the above, he apparently chased Sparky away from being in close proximity to a group of coots. On a past occasion he retrieved several chicks who were in proximity to a group of coots by the east bank of the river and led the chicks back to the SC. (Over the past weeks the coots have become numerous and common in the area as the parent coots have begun ranging more widely with their large, newly fledged chicks. Being in proximity to coots at times during the course of a day is now commonplace for the grebes.) Given the innocuous relationship between coots and grebes Sinbad’s behavior and attitude toward them is curious. However, the pair has shown itself to generally be rather sensitive regarding the presence of other birds in the general vicinity of their chicks. Besides the coots the pair has been observed reacting at least mildly to (including alarm calls) the presence of ospreys (numerous times), turkey vultures, great-blue herons, double-crested cormorants, western gulls, mallards, belted kingfishers and even a swooping Forster’s tern. This is in sharp contrast to the West Lagoon pair whom I observed last year. That pair wasn’t observed to react to any other bird species despite the common presence in the lagoon of species such as the coot, osprey, double-crested cormorant, mallard and snowy egret.

As for how the SR chicks relate with the coots, they have come to realize that they are no threat whatsoever. They display no apprehension toward them and in typical PB grebe manner (with the exception of Sinbad!) now just basically ignore them as they go about their business. This relationship was particularly well illustrated in an observation that occurred today during my continued observations after the formal observation period had concluded. Four coots were closely grouped together in the mid-river area east of the SC as they foraged in a modest algal mat. Phantom had been working its way toward them while foraging independently. When Phantom reached them the chick just continued its foraging within this algal mat right in the midst of the coots. Phantom even dove and resurfaced twice smack in the middle of these 4 closely spaced coots and neither Phantom nor the coots paid any heed to each other as each just went about their foraging.

While the SR chicks have come to the realization of the nonthreatening nature of the coots, such is not the case regarding double-crested cormorants. For some reason they generally are still leery of and react apprehensively to the proximity of these cormorants. Here again, this contrasts with what I observed last year with the West Lagoon family: the West lagoon chicks displayed no apprehension toward and paid no attention to the cormorants (and vice versa).

I finally had a confirmed sighting of a chick (Sparky) catching a prey item on its own.

One of the first brood and two of the second brood East Lagoon chicks were sighted foraging independently in the south half of the East Lagoon. All of the second brood chicks are now independent. Despite being independent, some chicks from both broods continue to reside in their natal range.

**9/5/17:** Time: 1103-1205.

1103 – As observations begin, all of the family members are seen in the SC and the mid-river area east of the SC: six are spread from the mid-SC area to the mid-river area and one is in the SW corner of the SC.

1107 – Quark is in the mid-river area east of the SC near 3 American coots. Quark begins heading west and suddenly engages in a short (~ 4 m) flight just above the surface of the water. The chick continues making its way west (in the usual fashion—paddling with its feet) into the SC and ~ 15 seconds later takes a slightly longer flight (~ 15 m) over the surface. Quark then continues paddling along to the reeds in the SW corner where it passes out of my view. For some reason the 3 coots followed Quark into the SW corner (there were no interactions between Quark and the coots).

1113 – Phantom and Spot are parked in the NE section of the SC. None of the others are seen at this time (they are probably in the reeds in various sections of the SC resting or foraging).

1115 – Cobra is sighted foraging at the edge of the reeds by the old nest site. Spot, and shortly after, Phantom move to that area and also begin foraging along the edge of and within the reeds.

1117 – Remy is sighted foraging along the reeds near the old nest site working her way west. A chick goes to Remy as she moves along and solicits her. Remy soon dives and the chick follows her under moments later. Neither is resighted at this time.

1120 – Spot is sighted heading west along the reeds of the north bank of the SC into the NW section. Remy is in the area and sights Spot moving in. From a fair distance away (~ 5-6 m) Remy suddenly charges and pecks once at Spot, who immediately dives to escape her wrath. A chick is sighted in the SW corner with a crayfish (it appears to be Quark) that it had probably taken from Remy (I didn't see how the chick obtained the crayfish so it's possible that it may have captured the crayfish on its own but given Remy's proximity it is most likely that it got the crayfish from her). Within seconds after being sighted, the chick swallows the crayfish.

1124 – Remy is sighted with a medium-sized crayfish by the reeds west of the NE point of the SC. Phantom is nearby and rushes to her, takes the crayfish and ~ 8 seconds later swallows it. Remy then resumes foraging.

1129 – Remy is again sighted by the reeds west of the NE point. She has a small prey item that I'm unable to ID (my view is too poor). A chick rushes to her (again my view is too inadequate to ID the chick), takes and quickly swallows it, then moves back into the reeds. Remy resumes foraging.

1135 – Quark, Phantom and Spot are now parked in the mid-SC area.

1137 – Remy is sighted with a medium-sized crayfish near the reeds west of the NE point. Phantom (who is nearby) rushes to Remy, takes and quickly swallows it.

1138 – Remy parks and preens ~ 35 m south of the old nest site. Quark goes to Remy, solicits and remains with her (the chick continues to intermittently solicit her). Cobra is parked and preening in the NE section of the SC and Phantom is foraging in this area.

1143 – Sinbad enters the SC from the SE, trailed at a distance by Sparky. Sinbad heads for the mid-SC and is met en route and solicited by Phantom. Sinbad parks and preens in the mid-SC; Phantom remains with him.

1145 – Sinbad goes to the west end of the SC and begins foraging. Phantom follows him there and watches to see if he catches any prey for a feeding. Remy has begun foraging near the reeds in the NE section.

1150 – Phantom is still accompanying Sinbad as he forages in the west end of the SC. The other 4 chicks are in or near the reeds by the old nest site. Remy isn't seen at this time.

1153 – Sinbad moves from the west end to the mid-SC area followed by Phantom and dives there. Two other chicks are already in the mid-SC area near a group of 6 coots. When Sinbad resurfaces he parks and preens there. Phantom briefly solicits him and then parks nearby. At this point all 5 chicks are either parked or foraging in the mid-SC area.

1157 – Same situation as at 1153: Sinbad and all 5 chicks are still in the mid-SC area.

1159 – Quark goes to Sinbad (still parked and preening in the mid-SC), solicits him for several seconds and then dives. Soon after, Sinbad calls once and then continues preening.

1200 – Quark goes to Sinbad, solicits him again and then sits with him.

1201 – Remy is sighted with Spot and Phantom near the reeds just west of the NE point of the SC. Remy has a medium-sized crayfish. Spot barges past Phantom, grabs the crayfish from Remy and moves away with it. As Spot is moving away, Remy charges and pecks at the chick from behind (as though she is upset because she had wanted to give the crayfish to Phantom). Spot immediately dives with the crayfish to elude Remy. Spot resurfaces soon after and swallows the crayfish. At this point Sinbad moves toward Remy. Remy spots Sinbad approaching and heads toward him. They meet and greet en route (time: 1202). Sinbad then begins foraging in the NE section of the SC. Remy moves to the mid-SC where she parks and preens.

1205 – Spot moves from the mid-SC area to the reeds in the SW section. Cobra is foraging in the NE section. Quark goes to Remy (still parked and preening in the mid-SC), solicits her briefly (Remy is unresponsive to this) and then goes to the NW corner. Sinbad, Sparky and Phantom aren't seen at this time. End observation period.

*At 1307 Cobra was seen taking a short flight (~ 3-4 m) just above the surface of the water in the NE section of the SC. Cobra is the second chick that has been observed taking a short flight today.

*Comments:* Today was a very special day for the chicks because it marked the achievement of a major milestone in their development. Confirmation was obtained today of what I have suspected based on my observations of their flight feather growth: that the chicks are now capable of at least short flights. Thus, by the most minimal definition of the term, the chicks have fledged. Both Quark and Cobra were observed taking short flights just above the surface of the water. Sparky and Spot are larger than Quark and Cobra (and their growth and development have been ahead of the other three chicks), and Phantom is similar in size and physical development to Quark and Cobra. It thus seems a virtual certainty that, though these other three chicks haven't yet been observed flying, they are also capable of at least short flights. These first observed flights have occurred on the 42nd day after the last egg hatched (considering 7/26 as the day of last hatching and counting it as day one).

As has become common, the parents continued to be quite active in their weaning behavior. In now typical fashion, a number of parental rebuffs of chicks were observed today, both during the formal observation period and in my continued observations afterward. Interestingly (and as would be expected), based on obvious qualitative observations the two largest chicks (Sparky and Spot) are the targets of parental rebuffs notably more often than the three smaller chicks (though all 5 of the chicks receive rebuffs from the parents). In addition, Quark, Phantom and Cobra are much more often allowed to sit with and intermittently solicit a parent without getting rebuffed. Despite the rebuffs, all 5 of the chicks still spend time accompanying and taking food from the parents.

Today's flights were just another sign of how close the chicks are to achieving independence. When they do so (hopefully all of them will continue to hang in there, never a guarantee in the wild) it will be interesting to see how long each of the chicks will hang around in their natal range (and/or how long the parents will allow them to), especially considering that the West Lagoon is open to them. In this regard, it is interesting to note here that some of the chicks from the East Lagoon first brood are still residing in their natal range as are chicks from the second brood (all are independent).

The SC was occupied by a number of American coots throughout the majority of the observation period. The chicks essentially ignored them (and vice versa) and this time, so did Sinbad.

**9/6/17:** Time: 1042-1143; 1150-1200.

1042 – As observations begin, Quark, Phantom, Spot and Sparky are sighted (from the RR bridge) ~ 15-17 m from the west shore reeds in the north half of the West Lagoon. None of the others are seen in the lagoon.

1043 – (From my SC observation point on the mid-south bank) On a check from within the SC, none of the family members are seen there.

1050 – Still no family members have been seen in the SC. At this time I leave the SC and head for my West Lagoon observation point on the mid-east shore. On a quick check of the lagoon from the west end of the RR bridge while en route, Sinbad and Remy are seen in the north half of the lagoon near the west shore reeds, though they are not in proximity.

1058 – I arrive at my West Lagoon observation point. Three chicks are sighted in the mid-water area at the north end of the lagoon (my view is too inadequate to ID them). Remy is in mid-river just north of the RR bridge. It appears that Remy and the 3 chicks are making their way to the SC.

1100 – Remy and the 3 chicks have gone into the SC.

1102 – One chick (it appears to be Phantom) is sighted by the reeds north of the snags slowly moving north, peering into the reeds while progressing (apparently searching for other family members and particularly the parents).

1107 – This chick (Phantom?) has now moved to the north end of the lagoon. A second chick (it appears to be Quark) is there ~ 8 m out from the west shore reeds and is peering toward them (this chick also appears to be looking for the parents or other family members). Phantom (?) reverses course and slowly moves south along the reeds, still searching. (Note: The light conditions currently are such that it's difficult to make out the identifying details on the chicks.)

1109 – Quark (?) moves to the west shore reeds at the north end of the lagoon and begins foraging along the edge of and within the reeds. Phantom (?) goes to the same area and dives into the reeds. Both are now out of my view.

1112 – Remy enters the north end of the lagoon with at least 2 chicks following her.

1114 – Remy parks ~ 13 m from the west shore reeds at the north end of the lagoon. One chick (Quark?) goes to Remy, solicits and remains near her. At least 2 other chicks are foraging at the reeds to the south of Remy.

1117 – The chick near Remy (Quark?) approaches close to her while soliciting and Remy suddenly charges at the chick. The chick immediately turns away and dives to elude her.

1119 – One chick (it appears to be Cobra) moves out to the mid-water area at the north end of the lagoon, dives and begins foraging there. Sinbad hasn't been seen since 1050.

1120 – At this time I leave my West Lagoon observation point and head to the SC to check on possible family members and activities there.

1127 – On my way to the SC I stop for a brief check from the west end of the RR bridge. Sparky is seen parked and preening in the mid-water area in the north half of the lagoon.

1130 – (Still observing from the west end of the RR bridge.) Cobra comes out from the west shore reeds and parks and preens in the mid-water area ~ 8 m south of Sparky.

1133 – I arrive at my SC observation point on the mid-south bank. Remy is now back in the SC. She is parked with Quark in the mid-SC area. None of the others are seen at this time.

1134 – Remy moves to the reeds at the west end of the SC and begins foraging. Quark accompanies her there.

1137-1143: Quark is sighted with a very large crayfish in the SW end of the SC; several other chicks are also there (I didn't see how Quark acquired the crayfish but the chick almost certainly got it from Remy whose foraging Quark had been monitoring). Quark keeps on the move ahead of several pursuing sibs. Quark ends up moving all the way out to the NE section while attempting to lose its pursuing sibs (who either cease pursuit or drop back at a distance) in order to attain some clear space where it can work the crayfish alone. [Quark apparently is cognizant that effectively working this very large crayfish will require dropping it onto the surface where the chick can then rip at and tear chunks from it in order to reduce its bulk for swallowing. In addition, when biting down on the carcass and giving it a vigorous shake (a method used to create breaks in the exoskeleton and tear off pieces) it will be flung onto the water. In general, Quark is apparently aware that when the carcass is on the water (instead of in its firm control while holding it in its beak) it can easily be stolen. Thus, Quark would like to be away from any competing sibs.] Sparky (who already happened to be out in the NE section) catches sight of Quark with the crayfish when Quark moves into the area (Phantom is also now in this section, having followed Quark at a distance). Sparky begins moving toward Quark who then doubles back to the west with Sparky and Phantom in pursuit. As Quark tries to work the crayfish on the move, some scraps occasionally detach or fall away and are snapped up in Quark's wake by the pursuing sibs. Quark continues on into the NW corner where, after having had the crayfish for over 2 minutes, Sparky takes it away from Quark. Sparky then works the crayfish on the move while keeping ahead of various pursuing sibs (including Quark). While near the reeds east of the nest Sparky drops the carcass onto the surface briefly to work it. Just as Sparky begins to pick it up Phantom rushes in and tugs it away from Sparky. Phantom now tries to work the crayfish while being hotly pursued by various sibs and moves into the SW corner. Eventually, Sparky is able to take the crayfish back from Phantom and after a remarkable ~ 5 minutes 44 seconds of some intense competition Sparky finally swallows the crayfish (time: 1143). Observations are temporarily paused at this time. [*Note: The pause was to allow myself the time to fully record this epic feeding episode in my notes.*]

1150 – Observations resumed. Remy is sighted in the NW section of the SC with a medium-sized fish. A nearby chick (my view is too poor to ID it; Cobra is also in the area) rushes to Remy, grabs and quickly swallows the fish. As the chick moves past Remy immediately after eating the fish, Remy suddenly charges and pecks at it. The chick promptly dives to elude Remy. Sinbad is finally resighted—he is parked with Phantom in the SE section of the SC. Phantom intermittently solicits him.

1152 – Sinbad moves E-SE through the SE section with Phantom following and they soon pass beyond my view.

1157 – Sparky and Spot are parked and preening themselves in the west end of the mid-SC area. None of the others are seen in the SC at this time.

1159 – Quark is sighted parked and preening in mid-river, just east of the SC. Spot moves out near Quark. Sparky remains parked in the mid-SC area.

1200 – Quark is still parked and preening in mid-river, just east of the SC. Spot has parked ~ 6 m to east of Quark. Sparky has moved out to the area just west of Quark at the east end of the SC. None of the others are seen at this time. End observation period.

*Just after the formal observation period concluded (at ~ 1212) Remy was observed feeding a medium-sized fish to Phantom in the mid-river area just north of the RR bridge.

*Comments:* The chicks were particularly mobile today (both during the formal observation period and during continued observations thereafter). They were literally all over the place within their natal range (= the SR pair's territory). For example, most of them went from the West Lagoon to the SC, back to the West Lagoon and then back again to the SC. Though they spent time accompanying and taking food from the parents they also did a considerable amount of independent ranging and foraging (including in open water). The distances that they traveled (either alone or with one or more sibs) were impressive. This was but another indication that the chicks are rapidly approaching independence.

The feeding event that occurred during the period from 1137 to 1143 exemplified the point that the chicks will now take and are able to work essentially any size crayfish that the parents catch (but sometimes they are reluctant to do so), though it takes them a rather long time to work a particularly large crayfish (there are, of course, upper limits to what the chicks—and even the parents—can or will tackle). During feedings they typically now take even large crayfish from the parent before the parent has worked the crayfish much, if at all.

A reminder that after the formal observation periods have concluded I nearly always spend additional time (sometimes several hours) observing the SR grebe family. After briefly leaving the area today at the conclusion of the period I returned to the SC and watched the family for about 55 additional minutes. During this time all of the family except Cobra stayed within the SC and the mid-river area to the east of it. Cobra spent this entire period off by itself, mainly in the West Lagoon. Since the time when the chicks first began to do some independent ranging and foraging Cobra has been one of the most independent of the brood. Perhaps due to this, Cobra has been fairly adept at isolating itself with a parent in a situation in which its sibs are quite distant (though such distant isolation is by no means a frequent occurrence). Cobra has a particular affinity with Remy in this regard. It will be interesting to see just when Cobra achieves its independence in relation to its sibs (providing, of course, that Cobra continues to survive—Cobra appears to be doing quite well at this time as do all of the other chicks).

**9/7/17:** Time: 1058-1214.

1058 – As observations begin (in the SC), none of the family members are seen in the SC and surrounding area.

1104 – Still no sightings of any of the family in the SC. At this time I leave the SC and head for my observation point in the West Lagoon to check for them there.

1109 – While passing around the SW end of the West Lagoon en route to my observation point I sight Sparky in the SW section of the lagoon adjacent to the south end of the west shore reed beds. Sparky is observed taking two short flights in this area, approx. one minute apart. The first flight carried for a distance of ~ 9 m and the second was ~ 15 m. In both of the flights Sparky flew just above the surface of the water.

1116 – I arrive at my West Lagoon observation point on the mid-east shore. Sparky is sighted foraging ~ 18 m out from my observation point. After several dives Sparky is no longer resighted at this time. None of the others are seen.

1118 – Sinbad is sighted paddling north near the reeds north of the snags. He veers slightly NE to the mid-water area as he continues heading north through the north end of the lagoon. Three chicks (my view is too inadequate to ID them) are in the mid-water area at the north end of the lagoon. They spot Sinbad moving north and follow him at a distance. Sinbad passes under the RR bridge and parks in the mid-river area just north of the bridge (time: 1120).

1121 – Sinbad (still parked north of the bridge) calls once. Soon after, he makes his way west into the SC. The 3 chicks also separately go into the SC.

1122 – A chick is sighted parked near the west shore reeds just north of the snags. It soon begins moving north along the reeds. The chick eventually stops ~ 8 m from the reeds at the north end of the lagoon and preens (time: 1125).

1130 – The chick (it now appears that it is Sparky) is still parked and preening in the same location. No other family members are seen in the lagoon at this time.

1135 – The chick (Sparky?) moves to the mid-water area at the north end of the lagoon where it parks and preens. At this time I leave my east shore observation point and head for the SC as there are no other family members seen in the lagoon (and I know that at least 4 have gone into the SC).

1143 – I arrive at my SC observation point on the mid-south bank and sight Phantom parked in the mid-area along the east end of the SC. None of the others are seen at this time.

1147 – Phantom is still parked in the same location. Other family members have now become visible. Sparky is parked and preening in the NW section of the SC. Sinbad is parked and preening in the mid-SC area. Remy is parked in the NW section. The others aren't seen at this time.

1151 – Same as at 1147.

1152 – Remy goes to the mid-west end of the SC near the reeds and begins foraging. Sinbad moves west through the mid-SC to the mid-west end. A chick briefly solicits him there and then goes into the reeds. Sparky moves to the mid-SC area and parks and preens.

1153 – Quark comes out from the reeds near the old nest site with a long, discolored reed stem and carries it all the way to the reeds east of the nest. Quark then dives into the reeds (still holding the reed stem) and is lost from view at this time. Sparky is now parked in the NW section of the SC.

1158 – Sinbad moves out near the NE point of the SC from the mid-SC, calls once, lingers there briefly, then moves farther back into the SC.

1159 – A chick is sighted in the NW corner just as it swallows a crayfish. Remy is nearby so it is most likely that the chick had obtained the crayfish from her. Remy then moves east toward Sinbad who is moving west along the reeds of the north bank in the NE section. They meet and greet en route. Remy then parks and preens in the NW section and Sinbad moves to the mid-SC area where he parks and preens. Sparky and Spot are also parked (apart from each other) in the mid-SC area.

1201 – Sparky approaches Sinbad and solicits him. Sinbad suddenly charges at Sparky as the chick nears him. Sparky immediately turns away and dives to elude Sinbad. Sinbad then remains parked in the mid-SC and resumes preening. Sparky parks well away from Sinbad (~ 35 m to the north of him) and preens.

1205 – Sinbad moves to the SE section of the SC where he parks and preens. Remy is still parked in the NW section. Phantom emerges from the reeds at the west end of the SC, goes near Remy and solicits her from a short distance away (Phantom is perhaps leery of getting rebuffed by her). Phantom then goes to the mid-SC area, lingers there for a while and then goes to the reeds in the SW section. Sparky and Spot have gone to the reeds at the west end.

1208 – Spot moves east from the SW corner through the southern half of the SC out into the mid-river area. The chick continues moving east and soon passes beyond my view. Remy moves out near the reeds just west of the NE point and parks there. While parked, she faces toward and peers into the reeds in that area, apparently searching for or watching someone. None of the others are seen at this time.

1214 – Remy moves out to the mid-river area east of the SC and begins foraging. None of the other family members are seen at this time. There have been no confirmed sightings of Cobra up to this point today. End observation period.

*After the formal observation period concluded I searched for Cobra and (as usual) continued to observe the family. I finally sighted Cobra at 1308—Cobra was alone in the north half of the West Lagoon foraging. The rest of the family was in the SC.

*Comments:* Another one of the chicks (Sparky) was observed taking short flights today, thus adding further confirmation that (by the most minimal definition of the term) the chicks have fledged. One of the flights was for a respectable distance of ~ 15 m. Sparky's flights were clearly of a practice or exploratory nature as there were no other noticeable reasons for the chick to have taken flight at the time.

The observation of Quark noted at 1153 in which the chick was seen carrying a long reed stem was quite intriguing. Quark's act of carrying a reed stem from the old nest site area to the new nest area (i.e., the nest utilized for this brood) is strikingly reminiscent of Sinbad's actions during the nest building and nesting periods. Was Quark practicing, "playing" at, or otherwise compelled to mimic nest building/maintenance behavior? Unfortunately, because of the denseness of the reeds I wasn't able to see what Quark did with the reed stem after diving into the reeds by the nest. In fact, with the nest being tucked within the reeds my view of it is so poor that I have no idea as to what condition it currently is in.

For a second consecutive day during observations Cobra was off independently ranging and foraging by itself while the rest of the family was in the SC or just east of it. Cobra has looked rather independent these past two days. Besides Cobra's independent ranging and foraging this chick was not observed accompanying, seeking out, soliciting or taking food from a parent these past two days. Is Cobra on the verge of becoming independent? It will be interesting to observe Cobra over the coming days to see just what this chick's status is.

Only two feedings were observed today: one during the formal observation period and one after it. On the other hand, a number of parental rebuffs of chicks were seen. Of special note, Sparky was the recipient of several particularly focused rebuffs by Sinbad. Though they are commonly seen on a daily basis, the parental rebuffs are sporadic: most of the time when chicks have solicited or remained close to and accompanied a parent they haven't been rebuffed. However, lately the frequency of parental rebuffs appears to have increased, especially toward certain chicks (namely,

Sparky and Spot, the two largest). All of the chicks receive rebuffs but both parents (especially lately) have been noticeably more lenient toward Quark and Phantom in this regard. Perhaps this is because they are smaller and slightly behind (developmentally) Sparky and Spot. As of late, when a chick is observed repeatedly soliciting a parent from up close (including beak to beak solicitations) and/or remaining close alongside a parent for a considerable time it invariably is Quark or Phantom. [Though I wasn't able to observe the hatching order of the brood (recall that hatching was staggered over a 5-7 day period) I have always suspected that Phantom and Quark were the last of the brood to hatch, with Quark being the last. I have also suspected that Sparky and Spot or Cobra were the first two to hatch with Sparky being the first and the hatchings of Cobra and Spot being closely spaced to each other]

Interestingly, most of the parental rebuffs have occurred when the chick is soliciting the parent. Whether or not valid, this makes sense from a weaning perspective as it is thus analogous to mammalian weaning behavior (i.e., weaning rebuffs by mammalian mothers to offspring who are attempting to nurse). Rebuffing a chick while it is soliciting can thus send the very specific (food related) message that the chick needs to do more independent foraging for its daily food requirements. In this light it then appears that the parental rebuffs serve as an inducement (or a kick in the butt) to the chicks to do more independent foraging and to become increasingly less dependent on the parents. Lately, instances are occurring in which the parent rebuffs a chick right after it has taken a proffered food item from the parent. Here again, the rebuffs are associated with feedings by the parents. As with the parental rebuffs in general, Sparky and Spot have been the main targets in these rebuffs. The indication from these cases is that the parents are beginning to become more specific (at least, sometimes) about which of the chicks they want their feedings to go to—in these instances the parents appeared to be upset after Sparky or Spot had taken the proffered prey item, apparently because the parent had intended to give it to one of the other three chicks. The behavior of the parents toward Sparky and Spot leaves little doubt that they recognize that these two chicks are ahead of the others in the rearing process. Accordingly, they are thus providing greater encouragement and inducement to these two chicks (in the form of less tolerance and a higher frequency of rebuffs) to become increasingly more independent. The overall effect in regard to the parents' task of rearing this brood is that the more independent and capable these two are in providing for themselves, the less effort the parents need to invest toward them and the more parental effort they can then devote toward successfully rearing the other three (and thus increasing the odds of success for these three).

In general, it appears that the chicks are now close to the point where they will be able to survive on their own.

Neither of the NR pair has been seen in or even near the SC since the day (8/31) when the NR male spent so much time in there while Sinbad and Remy were in the West Lagoon. Lately I have had some simultaneous sightings of the NR pair. Is this an indication that (if they have been nesting) their nesting period has concluded? (If they do have chicks then they will soon become visible when they become mobile.) Or is it an indication of a failed nesting attempt?

**9/8/17:** Time: 1032-1135.

1032 – As observations begin, Remy is parked with Phantom at the east end of the SC. None of the other family members are seen at this time.

1034 – Remy moves to the mid-SC area and parks there. Phantom accompanies her. Quark, Spot and Sparky are now parked and spread out from each other within the mid-SC and NW section.

1036 – Remy moves to the NW section and parks there: Phantom doesn't follow her but soon moves out of the mid-SC and goes out of my view. Sparky and Spot remain in the mid-SC area, Quark remains in the NW section.

1037 – Spot moves to the area near the reeds east of the nest. Remy is still parked in the same location. Quark is parked ~ 10 m north of her in the NW section. Sparky is still parked in the mid-SC area.

1040 – Remy, Quark and Sparky are still parked in their same locations. None of the others are seen at this time.

1045 – Cobra is sighted foraging in the SW section of the SC, working along to the east. Cobra soon ceases foraging and moves to the mid-SC area where the chick parks and preens. Quark is still parked in the NW section. Sparky is still parked in the mid-SC area. Remy has begun foraging in the NW section.

1047 – Remy is sighted in the NW section with a medium-sized crayfish. Quark is with Remy, eagerly getting ready to take it from her. However, Cobra (who has rushed to her from a fair distance away) darts in, muscles Quark aside, takes the crayfish and swallows it ~ 11 seconds later. Remy then resumes foraging. Cobra, Spot and Quark park in the mid-north section of the SC. Sparky is now parked in the western end of the mid-SC area.

1051 – Cobra, Spot, Quark and Sparky are still parked in their same locations. Remy is still foraging. Sinbad hasn't been seen yet today.

1052 – Remy is sighted with a large crayfish by the reeds just west of the NE point of the SC. Spot is with Remy and promptly takes the crayfish and begins working it. Sparky (who had moved out to that area) rushes to Spot, manages to grab a large scrap from the crayfish while Spot is working it and quickly swallows this. Spot moves about to keep clear of any competition (though Spot is no longer being contested) and after ~ 23 seconds swallows the crayfish (time: 1053). Remy resumes foraging.

1055 – Remy and all of the chicks who are present in the SC have been spooked by a paddleboarder (who has parked at the SE end of the river, east bank) and are hiding in the reeds around the SC. The person who is on the paddleboard then walks onto the bank in that area.

1100 – Remy and the chicks are still hidden in the reeds. The paddleboard is still in the same location but the person using it isn't in view (he is still somewhere on the land in that area).

1101 – Spot ventures out near the edge of the reeds by the old nest sight. Remy goes out to the mid-SC area and begins foraging (the paddleboard person is still somewhere on the east bank, out of view).

1102 – Spot goes back into the reeds, west of the NE point. Remy parks and preens in the west end of the mid-SC area for ~ 1 minute and then resumes foraging.

1104 – A chick is sighted briefly as it dives into the reeds at the south side of the NE point with what appeared to be a small crayfish. If this was indeed a crayfish then the chick had captured it on its own as neither Remy nor Sinbad are in its vicinity.

1106 – Spot moves out into the mid-SC area. A group of 6-7 American coots arrives at the reeds along the south side of the NE point where several of the chicks had gone into the reeds. The coots begin foraging along the edge of and within the reeds.

1107 – Spot dives and begins foraging in the mid-SC area.

1110 – Remy and the chicks have again gone into hiding in the reeds, spooked there by the paddleboarder who is back on his board and is now heading north in the river, east of the SC.

1113 – The paddleboarder enters the channel on the east bank of the river (located across the river from the NE point of the SC) that flows around the island that forms the east bank of the river between there and the SE end of the river. He continues on into the channel and moves out of the area.

1114 – Spot goes from the reeds in the SW corner of the SC to the mid-SC area, cruising along slowly while preening.

1115 – Spot parks in the mid-north section of the SC. None of the others are seen at this time. The coots are still foraging in and along the edge of the reeds in the NE section.

1120 – Sparky emerges from the reeds west of the NE point (paying no heed to the coots and vice versa), moves out near Spot (still parked in the mid-north section), dives once, then goes to the NW section, parks there and preens. Spot then moves to the mid-SC area, parks and preens. None of the others are seen at this time.

1125 – Sparky is still parked and preening in the same location. Spot has drifted to the east end of the SC and is now parked there.

1127 – Spot moves west past Sparky to the mid-west end of the SC and parks there ~ 9 m west of Sparky.

1129 – Quark is sighted foraging along the reeds in the NE section of the SC. The chick works its way west to the NW section. Spot sights Remy foraging in the NW corner and moves to that area.

1131 – Remy moves out ~ 8 m from the reeds in the NW section and parks there. Quark, Sparky and Spot are also now parked in the NW section (a bit spread out from each other) ~ 10-12 m south of Remy.

1132 – Remy dives and begins foraging near the reeds by the old nest site. Quark goes to that area and watches to see if she catches any prey for a feeding. Soon after, Spot also goes there and monitors Remy's foraging.

1135 – Sparky parks ~ 3 m south of the old nest site. Remy is still foraging. She has worked her way east accompanied by Quark, and is foraging in the reeds at the south side of the NE point. Spot is foraging just south of the NE point. Sinbad still hasn't been seen yet today. End observation period.

*A number of noteworthy things were observed during my continued observations of the family after the formal observation period concluded:

*1140 – Cobra was sighted moving from the reeds at the NE point of the SC alone to the mid-river area and foraged there. Remy, Spot, Sparky and Quark were all in the SW corner at the time.

*1143 – Spot attempted to follow and solicit Remy in the SW section of the SC and was soundly rebuffed by Remy three consecutive times. Spot dove to escape Remy after the first rebuff. When Spot resurfaced shortly after, Remy again charged at Spot who again immediately dove to escape her wrath. Upon resurfacing, Spot was charged at by Remy for a third time. Quark attempted to go to Remy and was also rebuffed. At 1145 Remy parked in the NW section of the SC but this time allowed Quark to stay with and repeatedly solicit her.

*1152 – Sinbad was finally sighted as he entered into the SC and called once soon after (it is most likely that he had been in the West Lagoon).

*1203 – Cobra parked with Remy and solicited her often and Remy was tolerant of this.

*1205 – Phantom parked with Sinbad and solicited him often and Sinbad tolerated this. While these two were together, Quark again went to Remy, remained with and frequently solicited her (which Remy again allowed).

*Comments:* The chicks continue on their quest to achieve independence and it appears that they may be able to do so sometime within the next few weeks. It seems the chicks have now reached the point where if something happened to the parents, the youngsters would probably be able to survive on their own (this seems a certainty for Sparky, Spot and probably Cobra). Nevertheless, all of the chicks still spend time accompanying the parents and they all still solicit and take food from them (several feedings were observed today). A second instance of a chick probably having caught a prey item on its own was seen today.

Unlike the previous two days, Cobra spent most of the observation period in the general area of Remy and the other chicks. Cobra was observed taking food from Remy and sitting with and repeatedly soliciting her. However, Cobra also spent time off by itself well distant from Remy and the other family members.

It was again seen today that Remy and Sinbad still allow the three smaller chicks (Cobra, Phantom and Quark) to remain close alongside and to repeatedly solicit them from up close (though even these three aren't successful at doing so on every attempt and sometimes also get rebuffed). Lately I haven't had any sightings of Sparky or Spot sitting alone with a parent, though I have seen them attempt to do so and get rebuffed and driven away (yet further inducement for these two largest chicks of the brood to do more independent foraging). However, they do continue to solicit the parents, though typically from a short distance away rather than up close beak to beak solicitations which often elicit parental rebuffs when attempted.

A fledged, independent chick (possibly from the East Lagoon first brood) was seen foraging in the north half of the NR pair's territory. Shortly after, it was spotted by the NR male and aggressively chased away. A second fledged, independent chick was sighted in the north section of the river north of the Highway 101 overpass bridges. These two chicks were probably keeping company (I wasn't able to ID their genders).

**<u>9/9/17</u>:** Time: 1055-1201.

1055 – As observations begin, Quark is sighted (from the west end of the RR bridge) ~ 17 m out from the west shore reeds in the north half of the West Lagoon. Sinbad is parked in the mid-SC area. Remy is parked in the mid-area along the east end of the SC. None of the others are seen at this time.

1058 – (From my SC observation point on the mid-south bank.) Sinbad is parked and preening with Phantom at the west end of the mid-SC area. Remy is still parked and preening at the mid-area along the east end of the SC.

1100 – Quark enters the SC (having come from the West Lagoon) and moves north along its east end (passing right by Remy) to the reeds on the north bank, just west of the NE point. Phantom is now parked ~ 7 m north of Sinbad in the NW section.

1103 – Quark moves out to the mid-river area east of the SC. Remy begins foraging at the east end of the SC. Spot is sighted parked just south of the reeds on the south side of the NE point.

1104 – Sinbad moves out to the SE section of the SC, calling once en route. As he makes his way there, Quark goes to Sinbad, solicits and follows him closely, repeatedly soliciting him as they move along. The pair continue east through the SE section and pass beyond my view.

1107 – Spot is sighted foraging near the old nest sight. None of the others are seen at this time.

1109 – Sparky is sighted moving to the SE end of the SC where the chick parks and preens. Approx. 45 seconds later Sparky begins moving SE and passes beyond my view.

1113 – Spot is parked in the mid-north section of the SC. Cobra emerges from the reeds along the north bank in the NE section and begins foraging in the mid-SC area. Sparky is now parked at the SE end of the SC.

1115 – Quark moves to the east end of the SC, parks there and preens. Cobra continues foraging and works its way to the SW corner. Sparky begins moving E-SE through the SE section and passes from my view. Neither Sinbad nor Remy are seen at this time.

1120 – Phantom moves to the mid-SC area, parks there and preens. Spot is now parked at the east end of the mid-SC area. None of the others are seen at this time. At this point I head for the west end of the RR bridge to check for the other family members from that vantage point.

1125-1130: (From the west end of the RR bridge.) Sinbad is sighted with a very large crayfish near the reeds along the east bank of the river at its SE end. Quark is with Sinbad and eagerly waits as he begins working it. Sinbad works it only briefly before handing it to Quark. Quark appears to be overwhelmed by this very large crayfish and it is clear that the chick would have preferred for Sinbad to have worked it much more. Quark works it only for a few seconds before just dropping it with the apparent intent of wanting Sinbad to take it and work it further. The carcass begins to sink and Sinbad then retrieves the crayfish, works it briefly and again hands it to Quark. This scenario repeats itself a number of times as Quark is either unable or (for some reason, perhaps intimidated by its large size) unwilling to work the crayfish and is instead wanting Sinbad to work it and prepare it to a point much farther along. After ~ 2 minutes of this Sinbad apparently loses patience with Quark. He picks up the crayfish and purposefully carries it across the river and into the SC with the obvious intent of feeding it to another chick. Quark eagerly follows at his heels. As Sinbad enters into the eastern end of the mid-SC area Phantom (who is still parked in the mid-SC) spots him with the crayfish and rushes to him. As the feeding then proceeds, the crayfish changes hands several times between Phantom and Quark as, first one tries to work the carcass and drops it and then the other picks it up and attempts to work it. Phantom finally takes firm control of the crayfish and finishes working the carcass to the point where Phantom is able to swallow it. This epic feeding lasted a lengthy ~ 4 minutes 5 seconds.

1133 – (I have moved back to my SC observation point.) Quark, Remy, Phantom and Spot are now parked in a line, each separated by ~ 4-5 m, in the NW section of the SC.

1142 – Quark, Remy, Phantom and Spot are still parked in the same general area in the NW section. A group of 6 American coots moves into the mid-SC area.

1144 – Sinbad is sighted with a large crayfish in the NW corner of the SC. Quark (who is nearest to him) rushes to Sinbad and this time doesn't hesitate to take the crayfish (which isn't near as large as the previous one). Quark then works it for ~ 27 seconds before swallowing it.

1146 – Quark, Remy, Spot and Phantom are parked in the NW section of the SC. One of the coots has moved to the reeds along the south bank in the SW section. Sinbad spots the coot there, moves to the SW section, charges at and briefly chases the coot. The coot immediately flees from Sinbad, taking flight to the mid-SC and soon leaves the SC. Sinbad then moves to the mid-SC area where he parks and preens.

1148 – Remy, Spot and Quark move to the reeds. Phantom remains parked in the NW section. Sinbad is still parked and preening in the mid-SC. Sparky and Cobra are not seen at this time.

1151 – Quark is sighted with a large crayfish by the reeds west of the NE point (I didn't see how Quark acquired the crayfish—it is most likely that Quark obtained it from Remy who is probably foraging out of my view in the reeds there). Quark works the crayfish for ~ 33 seconds before swallowing it and then parks and preens in the NW section.

1152 – Sinbad moves to the area ~ 4 m east of Quark and parks there. Phantom moves to the reeds in the SW corner.

1156 – Quark (now resting) is still parked in the NW section of the SC. Sinbad is parked ~ 10 m east of Quark. None of the others are seen at this time.

1158 – Remy is sighted with a large crayfish just south of the old nest site. Quark spots Remy and rushes to her, looking to receive the crayfish from her. However, Remy retains control of it and after working the crayfish for ~ 11 seconds she swallows it. Sinbad passes by both of them as he goes near the reeds at the south side of the NE point and begins foraging. Remy parks and preens in the NW section as does Quark.

1200 – Remy and Quark are still parked and preening near each other in the NW section of the SC. From the reeds at the NE point, Sinbad moves south along the east end of the SC, calling once as he proceeds. He soon stops at the SE end where he parks and preens. Only these three are seen at this time (time: 1201). End observation period.

*Comments:* Though the chicks still spent time accompanying the parents, soliciting and taking food from them, they are looking increasingly more independent. This was especially so for Sparky today.

After leaving the SC at 1115 this chick was off by itself for the rest of the observation period as well as the hour following it (I continued to observe the family for an additional hour after the formal observation period).

Phantom, Quark and Cobra were again allowed to remain close to and repeatedly solicit a parent without being rebuffed.

The feeding that took place starting at 1125 illustrated that there is a size limit beyond which the chicks can handle and that the size limit varies among the chicks. Though Quark wasn't able to handle the very large crayfish involved in this feeding, Phantom was eventually able to swallow it.

Sinbad once again chased a coot out of the SC. As has always been the case, the coot was just going about its business at the time and was no threat whatsoever to the chicks.

It has been a common occurrence to see a preening parent pluck a small feather from itself and hand it to an accompanying chick who then swallows it (this behavior continues, even at the current advanced developmental stage of the chicks). Chicks also commonly will go after, pick up and swallow small feathers that happen to detach while a parent is preening. In addition, for some time now the chicks have been observed to pluck and swallow one of their own small feathers while preening.

The parents, chicks and PB grebes in general preen themselves often. This is just as one would expect for a diving bird that essentially is always in the water. The only times that I have ever seen the parents and chicks out of the water is when they were on the nest and during the small number of brief flights that I have observed.

**<u>9/10/17</u>:** Time: 1004-1105.

1004 – As observations begin, Sinbad is moving through the mid-SC area followed by Phantom. Spot is in the mid-south section. Remy is at the east end of the mid-SC area. Sparky is foraging in the mid-south section. Quark and Cobra aren't seen at this time.

1006 – Sinbad circles the SC from the SW section across the west end to the NW section and then across to the SE section, followed all the way by Phantom. Sinbad (accompanied by Phantom) continues east through the SE end and the pair passes beyond my view.

1010 – Remy is now foraging in the NW corner and Sparky is foraging in the NW section.

1012 – Remy parks and preens at the west end of the mid-SC area. Sparky also ceases foraging and parks in the mid-SC area. Quark is sighted—the chick goes to Remy, solicits her repeatedly for ~ 25 seconds and then goes to the reeds along the mid-north bank.

1014 – Remy is still parked and preening in the same location. Cobra is sighted—the chick goes to Remy and solicits her. Cobra then remains with Remy, soliciting her often. None of the others are seen at this time.

1019 – Spot moves to the NE section of the SC from the mid-river area, east of the SC. Spot then veers south into the mid-SC area and parks and preens there.

1021 – Sparky is sighted slowly moving south along the east end of the SC. Quark is now foraging in the mid-SC. Remy is parked with Cobra in the NW section.

1022 – Quark parks in the mid-SC ~ 3 m from Spot.

1023 – Sinbad is sighted entering the SC from the SE followed at a distance by Phantom. Sparky catches sight of Sinbad and also follows him at a distance. Sinbad moves to the mid-SC area where he parks and preens. Phantom also parks in the mid-SC, ~ 27 m west of Sinbad. Sparky goes to the reeds in the SW corner. Cobra is now parked alone in the NW corner (Remy has begun foraging).

1028 – Remy emerges from the NW corner of the SC, moves east along the north bank and begins foraging near the old nest site. Cobra accompanies her and watches to see if she captures any prey for a feeding. Two other chicks also go to that area.

1030 – Sinbad moves to the west end of the mid-SC area where he parks and preens. Quark goes to Sinbad, solicits and remains with him. Spot moves to the reeds in the mid-west end. Phantom parks in the NW section.

1031 – Cobra is now foraging near the old nest site.

1033 – Remy is sighted with a large crayfish just east of the old nest site. Cobra is with Remy, eagerly looking to receive the crayfish from her. However, Remy retains firm control of the crayfish and swallows it herself after working it for ~ 17 seconds. Sparky and Phantom had spotted her with the crayfish and headed for her but she swallowed it before they arrived. Remy rebuffs two of the chicks as they move closer to her and then parks and preens in the NW section (my view was too poor to ID which of the chicks got rebuffed). Cobra accompanies Remy and intermittently solicits her.

1036 – Sinbad moves to the mid-area at the east end of the SC accompanied by Quark, who intermittently solicits him. Cobra is still parked with Remy. Sparky is parked and preening in the mid-SC area. Spot is parked in the NE section.

1038 – Sinbad is still parked at the SE end of the SC with Quark. Quark leaves Sinbad and goes to the reeds west of the NE point.

1040 – Sinbad moves SE from the SE end of the SC and soon passes beyond my view.

1045 – Cobra is still parked with Remy in the NW section. Sparky, Phantom and Spot are now parked in proximity to each other in the mid-SC. Sinbad and Quark aren't seen at this time.

1047 – Sinbad is sighted engaged in a territorial standoff with the NR male at the territorial boundary on the east side of the NE point of the SC. The confrontation is similar to the previous standoffs that had no fighting, consisting solely of posturing/displaying. The NR female is well away from the standoff—she is in the mid-river area in the mid-southern half of their territory.

1050 – The standoff continues (still no fighting). Cobra is still with Remy in the NW section. Phantom, Spot and Sparky are still in the mid-SC area and have drifted apart.

1051 – Sinbad and the NR male suddenly end the standoff, each just moving away from each other and diving shortly afterward. Sinbad moves to the NE section of the SC. The NR male slowly makes his way north, farther inside the NR pair's territory.

1054 – Remy moves toward Sinbad who is just south of the NE point and they meet and greet there. Sinbad remains parked there. Remy begins foraging near the reeds at the south side of the NE point. Spot is now parked at the east end of the mid-SC area. Sparky is parked in the NW section. The others aren't seen at this time.

1057 – Sinbad moves to the mid-SC area where he parks and preens. Quark goes to Sinbad, solicits and remains with him.

1059 – Cobra emerges from the reeds in the SW section and moves to the mid-SC area where the chick parks and preens. Phantom also moves to the mid-SC and parks there.

1100 – Spot moves out to the mid-river area east of the SC and passes out of my view. Phantom, Cobra and Sparky are parked and preening in the mid-SC area (Cobra and Sparky are near each other). Quark is still with Sinbad and intermittently solicits him.

1105 – A train noisily passes over the RR bridge and this spooks Cobra, Phantom and Sparky a bit and they move toward the reeds along the north bank of the SC. Sinbad remains parked in the mid-SC area and Quark remains with him (Quark continues to intermittently solicit him). Remy and Spot aren't seen at this time. End observation period.

*The following are some noteworthy events that occurred during my continued observations of the family after the formal observation period had concluded:

*1110 – Sparky, Spot and Cobra are in the mid-river area east of the SC. As they begin heading west toward the SC, Sparky takes a short ~ 17 m flight just above the surface to the east end of the SC. Seconds later Cobra also takes a short flight of ~ 10 m to the east end of the SC.

*1118 – I catch sight of the end portion of a breast to breast physical confrontation between two chicks near the reeds in the mid-west end of the SC. This fight (?) was similar to the ones observed between Sinbad and the NR male in which they "stood" straight up as tall as they could, wings flapping and legs churning at the surface to maintain this tall, upright position, bumping breasts and pushing against each other in this manner. It wasn't clear as to whether this was an actual fight between these two chicks or perhaps just a mock battle. Unfortunately, the pair dove out of sight immediately afterward before I had the chance to ID them.

*1124 – Sinbad is sighted feeding a small fish to Phantom.

*1125 – Sinbad is sighted with a large crayfish; Quark is with him, eagerly looking to receive it. However, Sparky (who is nearby) rushes in, takes the crayfish from Sinbad and quickly swallows it.

*Comments:* For the first time since the chicks became mobile no feedings were observed during the formal observation period. However, I saw 5 feedings (including the two noted above) during my period of continued observations after the formal period had concluded. Though they forage independently, all of the chicks also still seek to take food from the parents and still receive feedings from them. Cobra, Quark and Phantom were again allowed to closely accompany the parents and repeatedly solicit them from up close without getting rebuffed (though at times they did get rebuffed).

Another territorial standoff between Sinbad and the NR male was seen today. It was basically similar to all of the others that have been observed in which no physical fighting took place, consisting solely of posturing/displaying at the territorial boundary. It took place at the most recently established boundary at the east side of the NE point of the SC (the boundary runs from there on a line east across the river to the east bank). As usual, the standoff ended abruptly, with the two males suddenly just moving away from each other and going their separate ways into their respective territories.

While Sinbad and Remy have become habituated to the very noisy passing of trains across the RR bridge (though they do sometimes react to them) the chicks are still somewhat spooked by this. If in open water when a train is passing (as was the case today) they typically will either dive and/or retreat toward the reeds (though usually not in a panicked manner). If they haven't reached the reeds by the time the train has passed through they may cease their retreat and resume what they had been doing or engage in a new activity (in the latter case, especially if their activity had been disrupted by the appearance of the train).

When at a distance I am now unable to distinguish the parents from any of the chicks by size alone. The colors of the body plumage on the chicks are now essentially similar to those of the adults and have been for some time. However, the chicks still retain the patterns of black markings/stripes/spots on a white background on the sides of the neck and head that are characteristic of PB grebe chicks at this stage of their development. In addition, their beaks have not yet achieved the size, proportions and color of the beak of an adult.

The fight-like interaction between two of the chicks that was seen at 1118 was quite intriguing. I didn't catch sight of it until its last few seconds nor was I afforded ample opportunity to ID the two chicks. It wasn't clear as to whether it was an actual fight based on aggression or dominance, or perhaps just a mock battle. I am thus left to wonder if PB grebe chicks engage in mock or play battles as do the young of various other species. On the other hand, recall that an intense fight was observed between Phantom and Quark on 8/15 that Sinbad felt compelled to break up.

Several more short flights by chicks were seen today. These flights were taken by Sparky and Cobra. The two chicks were together (with Spot) in the mid-river area east of the SC and had begun heading back to the SC. Sparky took the first flight. Cobra appeared to be spurred on by this and followed Sparky with a short flight seconds later. Here again, there was nothing noticeable that might have caused them to take flight so these most likely were practice, exploratory or developmental flights.

**9/11/17:** Time: 1044-1150.

1044 – As observations begin, Sinbad is parked with Quark (who occasionally solicits him) in the mid-area of the east end of the SC. Remy, Spot and Sparky are parked separately a little to the west of them. Cobra and Phantom aren't seen at this time.

1045 – Spot moves away and is lost from view.

1046 – Quark leaves Sinbad (who is still parked at the east end of the SC) and moves east to the mid-river area east of the SC. Soon after, Remy also moves there as does Sparky.

1047 – Sinbad (still parked in the same location) calls once. Sparky circles back from mid-river to the east end of the SC. As Sparky does so, the chick deliberately chooses a route that allows it to briefly solicit Sinbad, but from a distance (Sparky keeps some distance from Sinbad while soliciting; the chick is obviously leery of getting rebuffed by Sinbad if it were to solicit from close in). After soliciting Sinbad while passing by him, Sparky moves into the mid-SC area.

1049 – Phantom is sighted with a medium-sized fish in the mid-river area, east of the NE point of the SC (I didn't see how Phantom had acquired the fish so I don't know whether the chick had obtained it from Remy or had captured the fish on its own though it most likely was obtained from Remy). After working the fish for ~ 44 seconds, Phantom swallows it. Sparky has again gone to the mid-river area, east of the SC.

1051 – Quark goes to Sinbad (who is still parked at the east end of the SC), solicits and remains with him.

1053 – Sinbad moves to the reeds at the south side of the NE point of the SC and begins foraging. Quark accompanies him and keeps watch to see if he catches any prey for a feeding.

1055 – Sparky is sighted moving away from the reeds at the NE point with a large crayfish which it probably had obtained from Sinbad (I didn't actually see how Sparky had acquired it but Sinbad is foraging in that area so it is most likely that the chick got it from him). Sparky continues on the move ahead of a pursuing sib (further support that Sparky had obtained the crayfish in a feeding by Sinbad) into the mid-SC, working the crayfish while on the go. After working the crayfish for ~ 30 seconds, Sparky swallows it.

1057 – Remy is sighted with a medium-sized crayfish at the SE end of the SC. Spot (who is nearby) rushes to Remy, takes and swallows the crayfish ~ 8 seconds later.

1102 – Quark and Sparky are parked and preening themselves in the mid-SC area. Cobra (seen for the first time today) and Spot are parked and preening ~ 4 m apart in the mid-area along the east end of the SC. Phantom is foraging in the NE section.

1105 – Quark, Phantom and Sparky are now parked and preening in proximity to each other in the mid-SC area. Spot and Cobra are still parked in their same locations, now ~ 1.5 m apart. Remy and Sinbad aren't seen at this time.

1107 – Remy emerges from the reeds in the NW corner of the SC with a medium-sized crayfish. Quark (who is nearest to her) rushes to Remy, takes the crayfish and quickly swallows it. Remy resumes foraging in the reeds along the west end. Phantom remains parked in the mid-SC area. Quark and Cobra move to the west end to the area where Remy is foraging.

1112 – Sinbad enters the SC from the east, Spot is moving well ahead and to the NW of him. Sinbad moves along into the mid-SC. As he does, Remy emerges from the reeds at the west end and heads directly toward him. While voicing extended continuous chattering vocalizations (which become more intense as they near) they briefly meet and greet and proceed in opposite directions. Remy continues heading east. Sinbad goes to the west end of the mid-SC area where he parks and preens. Quark and Phantom follow Sinbad and both solicit him. Phantom then moves away while Quark remains with Sinbad.

1115 – Quark is still parked with Sinbad in the same location. Cobra is sighted moving to the NW section where the chick begins foraging. Spot is foraging in the SW section and Sparky in the mid-SC.

1118 – Sparky parks and preens in the mid-SC area.

1120 – Quark is still parked with Sinbad at the west end of the mid-SC area. Cobra comes out of the NW corner and moves toward Sinbad. As Cobra approaches to within ~ 7 m of Sinbad and is poised to solicit him, Sinbad begins slowly moving toward Cobra as though poised to rebuff the chick. Obviously leery of being rebuffed, Cobra veers SE and stays well ahead of Sinbad who threateningly follows (Quark in turn, follows behind Sinbad). Clearly feeling pressured by Sinbad, Cobra finally dives and moves away while underwater. Sinbad pulls up and soon parks and preens at the east end of the mid-SC area. Quark has followed Sinbad and remains with him. Cobra continues moving SE through the SE section and soon passes from my view. At about this time, Spot moves east along the south end from the SW section and also proceeds through the SE section and beyond my view.

1126 – Quark is still parked with Sinbad, now at the SE end of the SC. Sparky is parked and preening at the NE end. Phantom moves to the NE end and parks and preens ~ 8 m west of Sparky.

1130 – Quark is still with Sinbad in the same location. Phantom and Sparky are also still parked in their same locations.

1132 – Phantom moves out to the mid-river area east of the NE point and begins foraging. Quark is still parked with Sinbad in the same location and occasionally solicits him.

1135 – Quark leaves Sinbad and moves to the NW section of the SC where the chick parks and preens.

1138 – Quark (still in the same location) dives and begins foraging in the NW section. Sinbad has drifted to the east end of the mid-SC area and calls once.

1140 – Spot enters the SC from the SE section. As Spot passes Sinbad's location the chick solicits him on the move from a distance of ~ 5 m (Spot is clearly leery of drawing a rebuff from Sinbad and thus solicited him from a distance rather than from close in). Spot continues on to the area just south of the NE point.

1143 – Sparky is sighted just east of the NE point. The large chick suddenly takes fight, flying for ~ 22 m just above the surface of the water into the NE section of the SC.

1144 – Spot moves to the east end of the SC and again solicits Sinbad on the move from a distance while passing by. This time Spot ventured a little closer, ~ 3 m away, but clearly was still leery of being rebuffed by Sinbad. Spot continues east through the east end of the SC and passes beyond my view.

1145 – Cobra enters the SC from the east and passes near Sinbad while proceeding east. While passing Sinbad, Cobra briefly solicits him from a short distance away and appears wary of being rebuffed while doing so. Cobra then continues on to the SW section.

1147 – Cobra and Sparky are foraging near the reeds in the SW section and work their way to the mid-SC area.

1150 – Cobra has ceased foraging and is parked and preening in the mid-SC area. Sinbad is still parked at the east end of the SC; Sparky is parked ~ 19 m to the west of him. Phantom is moving into the NE section. Remy, Quark and Spot aren't seen at this time. End observation period.

*Shortly after the formal observation period concluded Sinbad went to the West Lagoon and was eventually followed there by all of the chicks. Remy remained in the SC.

*Comments:* The chicks continue to close in on achieving independence though they still spend time looking to their parents for feedings and still take some food from them (in addition to what they may secure on their own). The great progress the chicks have made toward independence is exemplified by the manner in which they now routinely go out into wide open water to forage, travel and even park.

All of the chicks still solicit the parents but Sparky and Spot are now only observed to do so from a short or moderate distance away. This is because when these two largest chicks of the brood attempt to do so from up close they invariably get rebuffed by the parents. It has been a while since I last saw either of them solicit a parent from up close or remain close to a parent without getting rebuffed. It is clear that Sparky and Spot are aware of this reality as evidenced by the fact that they now essentially refrain from soliciting the parents from up close. In addition, it is quite obvious from the often wary manner in which they solicit that they are leery of the possibility of getting rebuffed by the parents, even when they solicit them from a distance.

In contrast, Quark and Phantom typically (but not always) are still allowed to solicit both parents repeatedly from up close as well as being allowed to sit close by and closely accompany them. The situation for Cobra is split: Remy typically allows Cobra to do this (in fact, Cobra appears to have a particular affinity with Remy) while Sinbad's behavior toward Cobra in this regard is variable (hence the reason why Cobra was wary of a possible rebuff from Sinbad when the chick solicited him at 1145 today).

It is quite likely that the difference between the way Sparky and Spot are treated in regard to the rebuffs as opposed to the other three chicks is related to the more advanced state of these two larger chicks. It appears that in this case the weaning process has merely been adjusted in accordance with their more advanced state. In general, the parents clearly are applying their weaning behavior on an individual basis to the chicks according to where each chick is at along the developmental progression.

Sparky was observed taking another short flight today. This flight of ~ 22 m was the longest one that I have seen any of the chicks take up to this point.

While in the West Lagoon after the formal observation period had concluded, Sinbad was once again observed aggressively chasing an American coot away. This was an extended chase because after the initial one the coot fled to the edge of the west shore reeds in the general area where some of the chicks were. Sinbad then charged and chased it a second time, driving it out toward mid-water and out of the area. As usual, the coot had done nothing of a threatening nature to warrant this; it had just been going about its own business.

Over the past several weeks Phantom has put on a growth surge and has noticeably closed its size gap to Sparky and Spot.

**9/12/17:** Time: 1031-1135.

1031 – As observations begin, Quark is parked with Sinbad in the mid-SC area and Phantom, Sparky and Spot are parked in proximity to each other in the mid-north section. Remy and Cobra aren't seen at this time.

1035 – A double-crested cormorant swims between Phantom, Sparky and Spot. The chicks are a bit apprehensive of the cormorant and scatter a short distance apart (their apprehension is fairly moderate—they don't panic or flee).

1036 – Sinbad, accompanied by Quark, moves near the reeds at the west end of the SC and parks there. Phantom, Spot and Sparky resettle, parking and preening in the mid-SC area.

1037 – Remy is sighted as she moves to the west end of the mid-SC area.

1039 – Cobra moves in from the east end of the SC and heads toward Remy. Upon reaching Remy, Cobra solicits her and remains close. Spot and Sparky are parked (separately) in the north half of the SC.

1043 – Sinbad is sighted near the reeds in the SW corner of the SC with Phantom, who has just taken a fairly large fish from him. Approx. 5 seconds later Phantom swallows the fish. Sinbad resumes foraging. Remy moves to the reeds near the old nest site (accompanied by Cobra) and begins foraging. Cobra stays in the area where Remy is foraging and watches to see if she captures any prey for a feeding. Phantom moves to the mid-SC area and parks and preens, apparently satiated for the moment after its large meal.

1045 – Spot moves to the SW corner, then circles around to the reeds by the old nest site (probably having sighted Remy foraging there). Phantom and Sparky are parked apart in the mid-north section of the SC and soon drift into proximity to each other.

1048 – Sinbad moves from the SW corner to a spot ~ 18 m from the reeds at the mid-west end of the SC where he parks and preens.

1050 – Remy moves east through the east end of the SC and continues on into the mid-river area east of the SC, soon passing beyond my view. Spot follows her at a distance and also passes from my view. On a quick check from the west end of the RR bridge it was observed that they went into the south ribbon of the river and foraged there.

1055 – Sparky and Cobra are now foraging in the NE section of the SC. Sinbad and Phantom are still parked in their same locations.

1056 – Quark goes to Sinbad, solicits him upon arrival and remains with him. Soon after, Sinbad goes near the reeds at the mid-west end and parks there. Quark follows and parks with him.

1100 – Sinbad is still parked in the same location with Quark now parked ~ 2 m to the east of him. Phantom is still parked and preening, now in the NW section. Cobra is parked in the mid-SC area. Quark moves close to Sinbad and repeatedly solicits him.

1101 – Sinbad dives into the reeds at the mid-west end and begins foraging. Quark remains in the vicinity and monitors his foraging.

1105 – Sinbad is still foraging in the reeds in the mid-west end; Quark is still monitoring his foraging. Phantom has drifted into the NW corner. Cobra is still parked in the mid-SC.

1106 – Sinbad is sighted with a large crayfish by the reeds at the mid-west end. Phantom (who is now the nearest chick to him) rushes toward Sinbad as do 2 other chicks who are farther away. Phantom easily reaches Sinbad first, takes the crayfish and swallows it ~ 11 seconds later. Remy enters the SC from the SE end, trailed from a distance by Spot. As Remy slows while moving through the east end, Spot passes and solicits her from a distance. Remy immediately reacts to this by charging at Spot who promptly responds by diving to elude Remy.

1107 – Remy moves to the mid-SC area where she parks and preens. Cobra spots Remy and goes to her. Cobra solicits Remy upon arrival and remains with her. Sinbad moves out to the SE end of the SC and continues on through it. He soon passes beyond my view as does Quark, who is following him. Spot is also following at a distance but stops at the east end of the SC.

1108 – Quark moves back into the east end of the SC and parks and preens along its mid-area.

1114 – Remy is now foraging in the NW corner of the SC, accompanied by Cobra. Quark is still parked and preening at the east end of the SC. Phantom is parked and resting in the mid-SC area, well west of Quark. Sparky is parked and resting in the NW section.

1116 – Sparky moves to the NW corner near Cobra in the area where Remy is foraging and monitors her foraging along with Cobra.

1120 – Remy (still foraging) works her way into the reeds at the mid-west end. Phantom is still parked in the mid-SC. Cobra and Sparky are in the NW section, still watching for Remy.

1124 – Cobra moves near the old nest site but soon sees Remy emerge from the reeds at the mid-west end and move to the mid-SC area (Cobra apparently had lost track of Remy's location while she was in the reeds). Cobra then rushes to Remy, solicits her upon arrival and remains with her. Remy then parks with Cobra and preens herself.

1126 – Sparky emerges from the reeds at the mid-west end and moves out near the reeds at the south side of the NE point. Spot is parked there, resting at the edge of the reeds near a stranger PB grebe (this grebe appears to be a fledged, independent and fully grown chick who has recently lost its juvenile markings and is assuming adult winter colors). Phantom soon joins them there. The stranger grebe briefly charges at one of the chicks but all remain in that area. Other than that one brief charge the three chicks and the stranger grebe appear to get along fine with no further aggression or interactions.

1130 – Spot, Sparky and Phantom have begun foraging at the south side of the NE point. The stranger grebe is no longer in sight (I wasn't able to see whether it just went deeper into the reeds or moved away). After the brief charge by the stranger grebe neither the chicks nor the stranger appeared to be concerned about each other. There was no more aggression and no further interactions between them as they just seemed to tolerate each other's presence.

1132 – Remy moves to the reeds in the SW section accompanied by Cobra.

1134 – Sinbad enters the SC from the SE end with a large crayfish and heads toward the mid-SC area. Quark is parked there and spots Sinbad from a distance and rushes toward him. When Quark meets Sinbad en route the chick looks to receive the crayfish from him. However, Sinbad doesn't work the crayfish or hand it to Quark but instead just continues on with Quark swarming around him. Sinbad veers toward the reeds at the mid-south bank of the SC (this is the area where Remy and Cobra had gone). As he nears the reeds he stops and begins working the crayfish with Quark eagerly watching. After only ~ 5 seconds Sinbad hands the crayfish to Quark, who immediately begins working it. As Quark continues to work the crayfish the chick is shadowed by Sinbad (who is overseeing the feeding). They soon drift to the reeds and pass out of my view. Remy then emerges from the reeds, passing Sinbad and Quark as they entered them, and heads toward the east end of the SC. The rest of the feeding took place out of my view (Cobra had probably been in the reeds there with Remy and thus may have competed with Quark for the crayfish). End observation period (time: 1135).

*Comments:* Today was generally similar to yesterday as the chicks work through their final stage in their progression toward independence. Besides their considerable independent foraging and ranging, the chicks continue to spend time accompanying the parents, soliciting and looking to them for feedings. Remy continued to allow Cobra, Quark and Phantom to closely accompany and solicit her, and Sinbad continued to allow Quark and Phantom to do so.

The observation involving the three chicks and the stranger PB grebe was particularly intriguing because it involved chicks from different parents. The stranger appeared to be an independent, grown chick from this season that had just transitioned to adult winter (plumage) colors. Other than the one brief charge by the stranger that was seen, these youngsters appeared to be tolerant of each other's presence; Spot was even lying at rest near the stranger. Fortunately for the stranger, neither Sinbad nor Remy were in a position where they could see it at the time. Of course, it's quite possible that this stranger is one of the East Lagoon first brood chicks.

**9/13/17:** Time: 1032-1145.

1032 – As observations begin, a paddleboard is in the mid-river area east of the SC, heading south. Sinbad is parked near the reeds at the south side of the NE point of the SC, intently peering at the paddleboarder. None of the other family members are seen at this time (if any of the others are in the SC they are probably hiding in the reeds due to the presence of the paddleboarder).

1034 – The paddleboarder moves to the SE end of the river and moves out of the area along the south ribbon of the river. Soon after, Sinbad moves to the NW section of the SC where he parks and preens.

1035 – The NR male is sighted in the mid-river area just north of the territorial boundary as he calls once.

1037 – Spot comes out from the reeds and approaches to within ~ 11 m of Sinbad, apparently with the intention of soliciting him at a distance. However, Sinbad suddenly charges at Spot causing the chick to dive in order to elude him. Sinbad then parks again in the NW corner. Spot goes to the mid-north section of the SC and parks there. Sparky, Phantom and Quark emerge from the reeds and park and preen themselves in the NW section'

1040 – Sinbad is now foraging in the reeds at the NW end of the SC. Quark and Sparky go near the reeds where he is foraging and watch to see if he catches any prey for a feeding. Phantom and Spot are still parked in the NW section.

1044 – Cobra is sighted in the mid-north section of the SC where the chick is parked and preening.

1046 – Phantom and Spot are still parked in the NW section. Sparky is still by the reeds at the mid-west end of the SC monitoring Sinbad's foraging. Remy, Quark (probably in the reeds near Sinbad) and Cobra aren't seen at this time.

1047 – Cobra is sighted moving into the reeds in the mid-west end of the SC where Sinbad (not seen) is probably foraging.

1050 – Sinbad emerges from the reeds at the mid-west end of the SC with a large crayfish. Quark is with him and quickly takes the crayfish and keeps on the move ahead of the hotly pursuing Phantom (who was nearby and came in to join the feeding). Quark soon dives with the crayfish in an attempt to elude Phantom and Phantom dives after Quark. When they resurface Phantom is right next to Quark and grabs onto the crayfish with its beak. Quark retains its hold and a tug-of-war ensues between them. The larger Phantom is able to wrest the crayfish from Quark and the roles are then reversed with Phantom keeping on the move ahead of the pursuing Quark. Phantom works the crayfish on the go and eventually swallows it (the feeding lasted ~ 28 seconds). Sinbad resumes foraging.

1055 – Sparky is by the reeds at the mid-west end of the SC monitoring Sinbad's foraging. Sinbad emerges from the reeds there with a large crayfish. Sparky rushes to him, takes the crayfish and, with Cobra (who was nearby) in pursuit, works it on the move for ~ 27 seconds before swallowing it. Sinbad moves to the SW corner and resumes foraging there. Cobra accompanies him and monitors his foraging from near the edge of the reeds. Sparky parks in the SW section and preens.

1057 – The NR male is sighted foraging just east of the NE point of the SC at the territorial boundary.

1059 – Cobra moves out to the west end of the mid-SC area and parks there. Phantom moves to the SW section and parks there. Sparky is still parked and preening in the SW section. Sinbad isn't seen but is probably still foraging in the reeds in the SW corner.

1102 – Cobra moves out to the mid-river area east of the SC and continues east, passing out of my view. A double-crested cormorant who is foraging in the SW corner of the SC passes between Phantom and Sparky who are still parked in the SW section. Sparky stays put and pays little attention to the cormorant. However, Phantom appears to be a bit apprehensive of it and glides away to the NW section, then returns after the cormorant has cleared the area. The cormorant had just ignored the chicks as it continued on with its foraging.

1104 – Sparky and Phantom are parked near the reeds in the mid-west end of the SC. As a train noisily passes over the RR bridge the two chicks are made uneasy by this and move to the north toward the reeds there. After the train clears the area they move back to where they had been. Remy still hasn't been sighted yet today.

1106 – The NR male is again sighted foraging east of the NE point of the SC at the territorial boundary.

1107 – Sinbad emerges from the reeds at the mid-west end of the SC and parks and preens near Sparky and Phantom.

1110 – Phantom and Sparky are now parked ~ 3 m apart in the NW section of the SC. Sinbad moves farther east into the mid-SC area and parks and preens there.

1112 – Spot moves out near Sinbad, keeping to the north of him, and solicits Sinbad from a distance. Spot is then immediately charged at by Sinbad. Spot promptly turns away and dives to elude him. Sinbad then parks again in the mid-SC area and preens. Spot eventually parks ~ 8 m north of Sinbad.

1115 – Sinbad is still parked and preening in the mid-SC area. Phantom, Spot and Sparky are well strung out, parked in a line to the north of him. Phantom and Sparky are resting.

1116 – Spot moves to the area just south of the NE point and begins foraging. The NR male is no longer seen in that general area.

1117 – Sinbad moves to the reeds at the mid-west end of the SC, dives into them and begins foraging. Phantom follows closely behind him, parks just outside the reeds and watches to see if he catches any prey for a feeding. He is soon joined by Sparky.

1120 – Cobra moves to the mid-area at the east end of the SC from the east, parks there and preens.

1121 – Spot joins Phantom and Sparky briefly as they continue to monitor Sinbad's foraging, then moves to the SW section and parks there.

1124 – Sinbad is now foraging along the north bank in the NW corner (as he foraged inside the reeds the chicks apparently lost track of him). He is spotted first by Phantom, then by Sparky and Spot and in turn all three chicks move to that area and monitor his foraging from outside the reeds.

1126 – Sinbad moves from the NW corner to the mid-area at the east end of the SC (calling once en route) where he parks and preens. Cobra (still parked at the east end) spots him moving into the area, solicits him briefly from a distance, then moves to the NW section and parks there. Phantom, Sparky and Spot have begun foraging in the NW corner.

1128 – As a train noisily passes over the RR bridge Cobra (who is made uneasy by this) slides into the NW corner.

1130 – From the NW corner Cobra moves diagonally across the length of the SC to the SE section. The chick continues east through this area and passes beyond my view. Sinbad is still parked and preening in the mid-area at the east end.

1131 – Sinbad begins moving south along the east end of the SC, calling once as he moves along. Spot follows at a distance but soon moves ahead of him, keeping at a distance and anticipating Sinbad's course. Both continue SE and soon pass beyond my view.

1134 – Phantom goes to the west end of the mid-SC area, parks there and preens. None of the others are seen at this time. Remy still hasn't been seen yet and Quark hasn't been seen again since early in the observation period (both may be in the West Lagoon).

1136 – Sinbad enters the SC from the SE section and goes to the mid-SC area where he parks and preens.

1137 – Phantom moves near Sinbad and solicits him from close by (but not beak to beak), then parks ~ 1-1.5 m away from him (Phantom appears to be somewhat leery of a possible rebuff as the chick seems reluctant to go right up next to him as it typically does).

1143 – Sinbad and Phantom are still parked in the same location; Phantom is ~ 1.5-2 m from Sinbad.

1144 – Spot emerges from the reeds in the SW section of the SC, moves out to the mid-SC area and parks and preens there. Phantom moves close to Sinbad and solicits him (still not beak to beak) and remains from ~ .5-1.5 m away, still appearing a touch leery of a possible rebuff.

1145 – Spot is still parked and preening in the same location. Phantom is parked ~ 1.5 m from Sinbad. None of the others are seen at this time and Remy has yet to be seen today. End observation period.

*The following are some noteworthy events that occurred during my continued observations of the family after the formal observation period had concluded:

*1159 – While searching for Remy after the formal observation period had concluded I finally sighted her NE of the snags in the West Lagoon. Shortly after, I sighted Quark with her (intermittently soliciting Remy) ~ 9 m from the west shore reeds in the mid-north section of the lagoon.

*1215 – Sparky was observed to have captured a medium-sized fish on its own. Sparky also worked the fish on its own, swallowing it after doing so for ~ 62 seconds.

*About an hour after the formal observation period had concluded Cobra was seen with Remy (the pair was parked together ~ 8 m from the west shore reeds in the north half of the West Lagoon). As Remy preened, Cobra repeatedly solicited her. Soon after I had sighted them, Sinbad went near the pair and rebuffed Cobra away from Remy a number of times, the subsequent rebuffs coming when Cobra tried to make its way right back to Remy after being rebuffed. When Sinbad finally moved away from the area Cobra wasted little time in returning to and repeatedly soliciting Remy, who continued to preen herself as she had been doing throughout this incident.

*Comments:* Sparky was observed to have caught a medium-sized fish in open water in the mid-SC area today. Considering the difficulty level of capturing a fish, this was a significant display of how far this chick's skills and abilities have progressed. Sparky has always been a little ahead of its sibs right from the start and it seems a certainty that (though Sparky still seeks and takes some food from the parents) it is now capable of surviving on its own. The others may also be capable of this (notably Spot and Cobra) but there seems less certainty in their cases. Cobra seems to have regressed a bit lately in that it is spending more time accompanying parents on foraging forays and in keeping close company with Remy.

There was a possible indication today that Sinbad's weaning behavior toward Phantom may be becoming more strict. Prior to today, despite the occasional rebuff, Phantom has generally had no qualms about going right in close to Sinbad when soliciting or wanting to sit with and accompany him. However, in the segment at the end of the observation period (beginning at 1137) Phantom appeared reluctant to do so and somewhat leery of a possible rebuff. Phantom did no close-in beak to beak soliciting and generally kept from about half to 1 ½ meters or more away when sitting with him rather than right alongside as Phantom typically has done. It thus seems likely that Phantom has recently received a significant rebuffing from Sinbad (or several of them) that has made a notable impression in the chick's mind, making Phantom somewhat wary about approaching very close to him.

**<u>9/14/17</u>:** Time: 1017-1120.

1017 – As observations begin, Sinbad is parked and preening in the mid-SC area. Phantom is parked near Sinbad ~ 1.5 m east of him. Spot is parked and resting in the NE section of the SC. The other family members aren't seen at this time.

1021 – Same as at 1017.

1022 – Phantom moves in close to Sinbad and solicits him briefly (but not in the beak to beak manner), then goes near the reeds at the north bank in the NW corner. Soon after, the chick dives into the reeds there.

1025 – Sinbad is still parked and preening in the mid-SC area. Spot is still parked and resting and has drifted to the NE end of the SC. The others aren't seen at this time.

1026 – Sinbad moves out near the reeds at the south side of the NE point of the SC. Phantom spots him moving there, comes out from the NW corner and heads toward Sinbad. Sinbad lingers at the edge of the reeds briefly, then dives into them and begins foraging. Phantom soon arrives there and goes into the reeds after Sinbad. Just before Sinbad went into the reeds he met Remy who was coming out of them. Remy then parked and preened in the NE section.

1030 – Sinbad and Phantom are still in the reeds along the south side of the NE point. Remy is still parked and preening in the NE section. Spot moves over near the reeds at the south side of the NE point where Phantom has just taken a large crayfish from Sinbad. Spot goes near Phantom but doesn't bother to pursue or challenge Phantom for the crayfish. Phantom works the crayfish for ~ 34 seconds before swallowing it. Sinbad resumes foraging.

1033 – Quark enters the SC from the SE end and heads toward Remy. However, Quark dives well before reaching her and then goes to the NW corner after resurfacing.

1035 – Phantom goes to the mid-north section of the SC and parks and preens there. Cobra is sighted moving to Remy (who is still parked in the NE section). Upon reaching Remy, Cobra briefly solicits her from up close and then goes to the south side of the NE point.

1037 – Sinbad goes from the NE point to the SE section where he parks and preens. Cobra moves back to Remy, solicits and remains with her. Phantom is still parked in the mid-north section.

1042 – Quark is sighted with a large crayfish that the chick caught on its own in the SW corner. Quark works the crayfish for ~ 36 seconds before swallowing it, then moves into the mid-SC area and parks and preens.

1044 – Cobra is still parked with Remy in the same location. Sinbad is still parked and preening in the SE section of the SC. Phantom is parked and resting in the NE section. Spot is parked and preening in the mid-north section. Quark is parked and preening ~ 15 m west of Spot. Sparky hasn't been seen yet today.

1047 – Quark goes to Sinbad (who is still parked in the SE section), solicits him repeatedly and remains with him.

1050 – Quark is still with Sinbad, often soliciting him. They are now in the mid-SC area. Cobra is still with Remy (they have drifted to the mid-area along the east end of the SC). Spot and Phantom are parked ~ 2.5 m apart, also in the mid-area at the east end of the SC but well away from Remy and Cobra.

1055 – Cobra is still parked with Remy. The pair has drifted to the mid-river area east of the SC (Cobra is intermittently soliciting Remy). Quark is still with Sinbad in the mid-SC area (Quark intermittently solicits him). Spot and Phantom are parked ~ 4 m apart at the mid-area along the east end of the SC.

1058 – Remy goes near the east bank of the river, across from the SC, dives and begins foraging. Cobra closely accompanies her.

1059 – Remy is sighted with a large crayfish near the east bank of the river, east of the SC. Cobra (who is still with her) soon takes the crayfish and works it for ~ 20 seconds before swallowing it. Remy then resumes foraging.

1100 – Quark has left Sinbad (who is still parked in the mid-SC) and isn't seen at this time. Phantom is now parked in the mid-river area east of the SC. Spot is parked west of Phantom at the far east end of the SC. Remy is foraging near the east bank of the river, east of the SC, still accompanied by Cobra. The pair soon passes beyond my view.

1102 – Phantom and Spot move together to the mid-river area east of the NE point of the SC and begin foraging (this area is at the territorial boundary with the NR pair, neither of whom is seen in the area). The pair soon work their way east near the mouth of the channel on the east bank that flows around the island that forms the east bank at the SE end of the river. Sinbad moves near the reeds at the south side of the NE point. He slowly moves along the edge of the reeds while peering into them (apparently looking for someone). He eventually moves into the reeds, going out of my view as he disappears into them.

1108 – Cobra enters the SC from the mid-river area and goes near the reeds at the south side of the NE point and begins foraging.

1110 – Sinbad emerges from the reeds at the south side of the NE point and moves to the mid-SC area (calling once en route) where he parks and preens. Cobra moves to the mid-area along the east end of the SC and parks there.

1112 – Quark moves to the east end of the SC from mid-river.

1114 – Remy enters the SC from the SE with a large crayfish. Cobra (still parked at the mid-east end) is closer to Remy than Quark, who is a little farther to the NE. Both race to Remy. Cobra (taking advantage of its closer proximity) wins the race and quickly takes the crayfish from Remy. Quark initially pursues Cobra but then follows Remy as she continues moving west. Cobra works the crayfish for ~ 17 seconds before swallowing it.

1116 – Quark goes to Sinbad (who is still parked and preening in the mid-SC area). The chick repeatedly solicits Sinbad and remains with him. Cobra parks and preens at the SE end of the SC.

1118 – Quark is still parked with Sinbad in the mid-SC area. Remy emerges from the reeds at the west end with a large crayfish and heads toward Quark/Sinbad. Quark spots Remy approaching, rushes to her, takes the crayfish and swallows it after working it for ~ 18 seconds. Cobra has moved to the NE section and is parked and preening there.

1120 – Sinbad moves out to the SE section of the SC and continues on through it, soon passing beyond my view. Quark initially follows him but then turns back, goes to the mid-north section of the SC and parks ~ 7 m to the east of Remy (who is now parked and preening in this area). Cobra begins foraging along the reeds of the north bank in the NE section, working in a westward direction. The others aren't seen at this time. Sparky hasn't been seen yet today. End observation period.

*1216 – Sparky is finally sighted—the chick is entering the SC from the SE. Sparky continues moving west along the southern half of the SC. Remy is in the SW section with several of the other chicks. When Sparky nears Remy's area Remy suddenly and strongly rebuffs the chick, charging and pouncing at Sparky. Sparky initially dives to elude Remy who then dives after the chick. The two resurface apart from each other but Remy immediately charges and pounces at Sparky a second time. The chick again dives to escape Remy who follows Sparky down. The action then moves into the SW corner out of my view. Shortly after, a grebe is sighted flying out of the SW corner heading east (the grebe is flying just above the surface). The grebe takes a fairly long flight of 45-50 m into the SE section, dives immediately upon landing and isn't resighted. This all happened very quickly and I didn't get a good enough view to identify the flying grebe. However, given the circumstances there seems little doubt that it was Sparky. Regardless, Sparky didn't flee out of the SC for any length of time. At ~ 1230 Sparky was sighted emerging from the reeds in the SW corner and going to the mid-SC area where the chick parked and preened (Sparky's feathers were quite soaked suggesting that Sparky had been foraging in the SW corner—note that after entering the SC at 1216 Sparky had been headed for the SW corner before being intercepted and driven away by Remy).

*Right after the incident with Sparky, Quark also received a fairly strong rebuff from Remy. However, this didn't appear to have any lasting effect on Quark. Not long after this, Remy parked in the mid-SC area and with no hesitation and no hint of apprehension Quark went to Remy and closely and repeatedly solicited her (including beak to beak solicitations) and remained close alongside her. For Remy's part, she was completely tolerant of this.

*About 40 minutes after having been soundly rebuffed by Remy, Sparky received a rather curious feeding from her. At the time, Remy was foraging in the reeds in the mid-north bank area of the SC. Sparky was busy preening a little to the west of her. When Remy emerged from the reeds with a medium-sized crayfish Sparky quickly spotted and moved to her. Remy routinely handed the crayfish off to Sparky but as the chick was moving away from her with it, Remy suddenly charged at Sparky from behind. Sparky instantly fled with the crayfish and when Remy closed and pounced, Sparky dove to elude her. Remy immediately dove after Sparky in hot pursuit. When they resurfaced shortly after, Remy was only ~ 1 m behind and still in "pursuit mode." Sparky (still holding onto the crayfish) attempted to flee but Remy was instantly on the chick's heels and pounced a second time. Sparky again dove with Remy diving right behind in pursuit. This time Sparky stayed under for a longer time and moved a much farther distance while underwater. When the two resurfaced Sparky was well away from Remy and quickly swallowed the crayfish. Upon resurfacing, Remy initially looked about intently for Sparky but quickly calmed and ceased her pursuit.

*Comments:* Another chick was observed catching a prey item on its own today. On this occasion Quark captured and consumed a large crayfish. This occurred at 1042 and yet at 1047 (a mere 5 minutes later) Quark went to Sinbad, repeatedly solicited him and remained at his side.

All 5 of the chicks were observed to get at least one feeding from a parent today (this includes my observations during and after the formal observation period).

Underscoring Sparky's burgeoning independence, the chick spent the entire formal observation period off by itself. Judging by the increasingly antagonistic behavior of the parents toward Sparky, this aspect of the chick's development has not been lost on them.

Phantom again was reluctant to go right up close to Sinbad when soliciting and sitting with him. As was observed yesterday, Phantom seemed a bit leery of a possible rebuff. The chick again went near but not up close to Sinbad when soliciting him (no beak to beak soliciting) and stayed near (~ .5-1.5 m or more away) but not right next to him when sitting with or accompanying him. It thus appears that Sinbad's behavior toward Phantom has become somewhat stricter in regard to weaning and inducing independence.

On the other hand, Phantom still has no qualms whatsoever about going right up close to Remy when soliciting and sitting with her. This contrast between the tolerances of the two parents underscores the obvious point that they are individuals. It thus isn't surprising that their individual behavior toward the chicks may vary somewhat. In regard to weaning/independence inducing behavior Remy clearly has been the more lenient of the two. She currently still allows Quark, Cobra and Phantom to solicit her from up close (including beak to beak soliciting) and to stay close alongside when sitting with and accompanying her. At this point, Sinbad only allows Quark to do so (and as noted above, he allows the near presence of Phantom). Neither allows this of Sparky and Spot (in fact, it has been a while since either of these two chicks were observed engaging in close-up beak to beak solicitations or sitting close alongside a parent). From a sociobiological perspective it's interesting to contemplate whether this situation of the female (the mother) being the more lenient is the general pattern among this species (i.e., in consideration of the fact that, having been the one who produced the eggs, the mother has a greater investment in the brood).

**9/15/17:** Time: 1005-1125.

1005 – As observations begin, Sinbad is sighted (from the west end of the RR bridge) parked and preening in the mid-water area in the north half of the West Lagoon. None of the other family members are seen at this time.

1006 – Quark emerges from the west shore reeds in the north half of the lagoon and moves toward Sinbad. Quark solicits Sinbad briefly from nearby while passing and continues across the lagoon to the area near the east shore just south of the mouth of the inter-lagoon channel.

1008 – Cobra emerges from the west shore reeds in the same general area where Quark had come out. Cobra moves toward Sinbad and solicits him from a distance. Sinbad lightly rebuffs Cobra with a very short, distant charge (he never got close to Cobra) but it was enough to cause Cobra to turn away and dive. Cobra then eventually moves out to the mid-water area at the north end of the lagoon where the chick parks and preens.

1010 – Sinbad is still parked in the same general location.

1016 – Quark goes to Sinbad (still parked in the same general location), solicits and remains with him for ~ 1.5 minutes, then moves away. At this time I leave the west end of the RR bridge and head for my SC observation point on the mid-south bank.

1018 – I arrive at my SC observation point. None of the family is seen at this time.

1024 – Sparky is sighted entering the SC from the SE section. The chick proceeds to the reeds in the SW section and begins foraging there (Sparky then quickly goes out of my view).

1029 – None of the family members are seen in the SC at this time (though Sparky is probably still in the reeds in the SW section). The others are probably in the West Lagoon so I leave the SC and head for my east shore observation point in the lagoon.

1031 – A quick check from the west end of the RR bridge en route to my West Lagoon observation point: Remy is sighted near the west shore reeds in the north half of the lagoon with a few chicks nearby. A grown, independent stranger PB grebe chick is sighted foraging in mid-river, just north of the RR bridge. Another stranger PB grebe (either an adult female or another grown, mature chick) is foraging ~ 8 m from the east bank of the river across from (east of) the SC.

1041 – I arrive at my east shore observation point in the West Lagoon. Sinbad is parked ~ 6 m from the west shore reeds in the north half of the lagoon. Phantom and Sparky (now in the West Lagoon) are parked near each other ~ 4 m from the reeds north of the snags.

1045 – Quark is sighted next to Sinbad with a large crayfish (Quark has probably just taken it from Sinbad)—they are by the reeds just north of the snags. Quark soon swallows the crayfish. Just as Quark is doing so a chick (probably Sparky) approaches the area and is rebuffed and driven away from the immediate area by Sinbad. Sinbad then resumes foraging.

1050-1054: Sinbad is sighted with a large crayfish by the reeds just north of the snags (Quark is with him and several other chicks are nearby). Quark quickly takes the crayfish and heads east toward open water while being pursued by Phantom and Sparky. An intense, long distance pursuit then ensues with Phantom intensely pressuring Quark and Sparky following more at a distance. Quark cuts a serpentine route going out to the mid-water area and then heading north in a desperate attempt to keep ahead of the pursuing sibs (notably Phantom). On several occasions Phantom dives and then resurfaces next to Quark, causing Quark to abruptly change course. Quark eventually turns west in the north half of the lagoon and heads toward the west shore reeds. By the time Quark reaches the reeds the harried chick has gained some separation from Phantom (and Sparky). Quark then dives into the reeds and isn't resighted at this time. Phantom stops by the edge of the reeds but doesn't follow Quark into them (Sparky had given up its pursuit by this time). The feeding had already lasted ~ 3 minutes 18 seconds by the time that Quark went into the reeds. This was easily the longest distance that I have seen covered in a competitive feeding pursuit. (Though I couldn't see into the reeds to verify it, with

Phantom and Sparky having abandoned their pursuit it is a virtual certainty that Quark was finally able to work and swallow the crayfish that it had fought so hard to retain.)

1057 – One chick is sighted parked and preening near the west shore reeds at the north end of the lagoon (my view is too poor to ID the chick). None of the others are seen at this time.

1059 – The chick sighted at 1057 dives and begins foraging.

1101 – Two chicks are sighted foraging at the north end of the lagoon, one near the reeds and one in the mid-water area.

1103 – Sinbad is sighted in the mid-water area at the north end of the lagoon. He makes his way north under the RR bridge (calling once en route) with one chick following at a distance.

1104 – Sinbad begins foraging in the mid-river area just north of the bridge. He soon surfaces from a foraging dive with a very large crayfish. The chick that had accompanied Sinbad rushes to him. Sinbad works the crayfish and eventually hands it off to the chick. Sinbad then heads west into the SC and soon passes from my view. Following him at a distance while working the crayfish, the chick also passes from my view. The feeding has lasted ~ 75 seconds up to this point.

1108 – None of the family members are seen in the lagoon at this time. At this point I head out for the SC to continue observations there.

1112 – While rounding the south end of the lagoon I sight Quark and Spot near the west shore reeds in the north half of the lagoon.

1120 – I arrive at my SC observation point on the mid-south bank. Sparky is sighted moving west from the east end of the SC. The chick soon stops and parks in the mid-SC area. Sinbad is in the west end of the SC near the reeds east of the nest. He soon moves to the mid-SC area where he parks and preens. As he does so, Sparky moves a little farther east in the mid-SC. Sinbad did not threaten or appear to do anything noticeable to drive Sparky away; his mere proximity (though he wasn't close) appeared to be enough to cause Sparky to move a little farther away.

1123 – Spot enters the SC from the SE section and goes to the mid-SC area where the chick parks and preens ~ 5 m to the north of Sparky.

1125 – Spot moves to the area near the reeds at the south side of the NE point of the SC and begins foraging. Sinbad and Sparky are still parked in their same locations. None of the other family members are seen at this time. End observation period.

*Comments:* Most of the family spent the majority of their time during the observation period in the West Lagoon. Sparky moved back and forth between the lagoon and the SC several times.

The feeding that took place at 1050 involved what was by far the longest distance pursuit that I have observed during a competitive feeding. Phantom's intense pursuit of Quark was quite remarkable. Despite this, Quark managed to hold onto the crayfish.

**9/16/17:** Time: 1132-1238.

1132 – As observations begin, Sinbad is parked and preening in the mid-SC area. Sparky enters the SC from the SE and goes to the mid-SC and parks and preens ~ 5 m north of Sinbad. None of the others are seen at this time.

1137 – Sinbad is still parked in the same location. Sparky goes near the reeds along the south side of the NE point of the SC and begins foraging. One other chick is briefly sighted foraging there.

1139 – Remy is sighted foraging by the reeds near the old nest site. Phantom is sighted moving in her direction. As Phantom nears Remy she suddenly charges and pounces at the chick. Phantom instantly turns away and dives to elude her and then moves to the reeds in the NW section. Remy goes to the mid-SC area where she parks and preens.

1142 – Sparky moves to the mid-SC area and forages there. Sinbad and Remy are still parked (well apart) in the mid-SC area.

1145 – Same as at 1142.

1146 – Sparky ceases foraging and parks and preens in the mid-SC area. Remy has drifted into the NE section, south of the NE point. Sinbad has drifted to the mid-west end of the SC.

1148 – Phantom emerges from the reeds in the NW corner and goes to Sinbad (still parked in the mid-west end). Phantom repeatedly solicits him from close by (but not right up next to him and Phantom does no beak to beak soliciting) and remains near him.

1149 – Remy moves out near the reeds at the south side of the NE point and begins foraging.

1150 – Sinbad (Phantom is still parked near him) moves to the mid-SC area where he parks and preens. Phantom remains in his location and doesn't follow him. Sparky is still parked and preening in the mid-SC area.

1153 – Sparky moves out to the mid-river area east of the NE point of the SC and begins foraging. Phantom moves back near Sinbad (still parked in the mid-SC area). Sinbad soon moves out to the mid-area along the east end of the SC with Phantom following, usually from .5-2 m behind. Sinbad parks there and Phantom remains near him, sometimes getting close beside him. Spot, Cobra and Quark haven't been seen yet today.

1155 – A person is heard calling loudly from nearby and this spooks Phantom and causes the chick to leave Sinbad (Phantom soon goes out of my view). Sinbad remains parked in the mid-area along the east end of the SC. None of the other family members are seen at this time.

1157 – From mid-river, Sparky goes near the reeds at the NE point of the SC and forages there.

1159 – Remy goes to a point ~ 20 m south of the old nest site where she parks and preens.

1200 – Sinbad (now parked at the NE end of the SC) dives and begins foraging.

1201 – Sinbad moves west along the north bank of the SC. Sparky follows him at a distance. While following, Sparky solicits Sinbad from a distance. Sinbad responds by charging at Sparky and driving the chick away. Sinbad then proceeds to the NW section and parks there. Sparky parks and preens in the mid-north section.

1203 – Sparky heads east, passing ~ 6 m south of Remy (now parked in the mid-SC). While moving east, Sparky spots Phantom entering the east end of the SC from mid-river. Phantom soon parks at the east end. Sparky moves toward Phantom but continues past Phantom and goes just east of the SC. Sparky lingers there briefly, then moves back into the SC. Remy has moved from her location and is no longer seen.

1206 – Sinbad is still parked in the NW section. Phantom goes near Sinbad, solicits him from fairly close by (but not right next to him) and remains with him, generally keeping from ~ .5-1.5 m from him. Remy is now parked south of the NE point. Sparky is parked in the mid-SC area.

1211 – Quark is sighted—the chick enters the SC from the SE end and parks in the mid-SC area ~ 1.5 m from Sparky.

1215 – Phantom leaves Sinbad (they are still parked in the NW section) and moves to the NW corner. Phantom lingers there briefly and then goes back near Sinbad.

1216 – Sinbad goes near the reeds at the mid-west end of the SC and Phantom follows. Quark goes to Sinbad and briefly solicits him. Sinbad drifts a little farther north along the west end and parks there. Quark goes to the reeds at the north bank in the NW corner and begins foraging. Phantom goes to the mid-SC area and parks there.

1220 – Quark moves out to the SE section of the SC, continues east and passes beyond my view. Sinbad moves to the reeds by the nest and begins foraging. Phantom comes over near the edge of the reeds there and monitors Sinbad's foraging, watching to see if he catches any prey for a feeding.

1222 – Cobra is sighted in the mid-river area east of the SC moving east toward the east bank and soon passes from my view.

1225 – Sinbad and Remy are now in the mid-north section of the SC. A chick moves toward Remy (my view is too inadequate to ID the chick) but she charges at it and forces the chick to dive and flee.

1227 – Sinbad moves to the NW section where he parks and preens. Quark goes to Sinbad, repeatedly solicits him from up close (including beak to beak solicitation) and sits close to him.

1231 – Remy is sighted with a very large crayfish near the reeds in the NW corner and begins working it. It appears as though she is going to keep it for herself. However, Phantom spots Remy from nearby, rushes to her and snatches the crayfish from right out of her beak. Phantom immediately begins moving away with it. Remy appears angered by this. She charges at Phantom from behind and chases after the chick briefly as it flees from her with the crayfish. Remy soon ceases her chase and lets Phantom go but Sparky then begins pursuing Phantom as Phantom works the crayfish on the go. Sparky continues the pursuit until Phantom finally swallows the crayfish ~ 91 seconds after having snatched it from Remy.

1235 – Sinbad (with Quark following closely along and repeatedly soliciting him) moves out to the mid-river area east of the NE point and parks there with Quark.

1238 – Cobra moves from the mid-river area to the SE end of the SC, parks there and preens. Sinbad, accompanied closely by Quark, moves to the reeds at the south side of the NE point and begins foraging. Quark monitors his foraging. Phantom also goes to that area and monitors Sinbad's foraging. Sparky is in the mid-river area, east of the SC. Remy isn't seen at this time. Spot hasn't been seen during the observation period. End observation period.

*The following are some noteworthy events that occurred during my continued observations of the family after the formal observation period had concluded:

*1244 – Quark receives a feeding of a large crayfish from Sinbad at the south side of the NE point of the SC.

*1339 – Cobra took a long flight of ~ 80-90 m from the mid-river area east of the SC to the SW corner of the SC. Cobra began the flight flying just above (but clear of) the surface of the water but about halfway through the flight Cobra notably gained altitude up to a height of about 1.5-2 m (and probably a little higher). After landing, Cobra very purposefully paddled back out to mid-river, east of the SC and took off on a similar long flight of ~ 90-100 m into the SW corner of the SC. Once again, about midway into the flight Cobra gained altitude to a height of ~ 1.5-2 m. After the second flight Cobra paddled all the way out near the east bank of the river across from the NE point of the SC and began foraging in that area.

*Spot was possibly sighted at ~ 1430 near the west shore reeds in the north half of the West Lagoon. However, the sighting was too distant and brief to be 100% certain that it was Spot.

*Comments:* From a qualitative perspective today's observations seemed to have a different "feel." Though some feedings were seen after the formal observation period concluded (and one during the period which seemed more like a theft), the parents never really seemed to get into "chick feeding" mode. While the chicks still look some to the parents for feedings, at this point it appears that all of them may be capable of surviving independently. And judging by the long, beautiful flights taken by Cobra today, they are probably all capable of leaving their natal range should they choose to (or perhaps get excluded by the parents?). An interesting point regarding Cobra's long flights is that when I first saw the three remaining chicks from the West Lagoon brood last year take similar flights, they were gone from the area the next day (however, it is perhaps significant that that was much later in the season, mid-October). It will be interesting to see how much longer the various chicks of the SR brood remain in their natal area, especially considering that some of the now independent East Lagoon first and second brood chicks have remained in and around their natal area up to this point and that the season still has at least a month or so remaining.

The manner in which Cobra set up for its second flight was rather interesting. After having flown from mid-river to the SW section of the SC, Cobra could simply have flown back out to mid-river for a second flight. Instead, Cobra paddled all the way back out to mid-river and again took off from there and flew to the SW section. Cobra's behavior left no doubt that it had deliberately gone out there specifically for the purpose of taking that second flight. One is thus compelled to wonder why Cobra chose to expend all that energy paddling back out there and doing the flight in this manner? Did Cobra feel more secure flying into the SC, the place where the chicks feel most secure and know best? Did it have anything to do with the flight conditions (though there wasn't much of a breeze at the time)?

There were no confirmed sightings of Spot today. I had a possible sighting of Spot in the West Lagoon but I was too distant and the sighting too brief to be certain that it was Spot.

Quark was still allowed to closely solicit (including beak to beak solicitations) and remain close to both Sinbad and Remy today and Cobra was allowed to do so with Remy.

A third brood of the season for the East Lagoon pair was sighted in the East Lagoon. Three young, probably recently mobile chicks were seen by and within the reeds at the east shore in the mid-section of the lagoon. This is the same area where I saw the chicks from the first two East Lagoon broods when they were at a similar stage. Unfortunately, there is much of the East Lagoon that I'm not able to view so I was never able to see where the pair had their nest(s). It would have been interesting to see if the pair used the same or different nest(s) for each of the broods. I suspect that producing three broods in a season probably isn't particularly common for this species (from what I've observed over the past two years I would guess that one or two is probably the norm).

**9/17/17:** Time: 1144-1245.

1144 – As observations begin, Cobra is parked with Remy in the mid-river area just east of the SC. Sinbad is sighted with a large crayfish near the reeds by the old nest site. Sparky (who is nearby) rushes to Sinbad who hands the crayfish off to the chick. However, as Sparky begins moving away with it, Sinbad charges the chick from behind and causes Sparky to dive with the crayfish and flee. Sinbad dives after Sparky and when they resurface he continues his pursuit of the chick. Sinbad causes Sparky to dive a second time with the crayfish but the chick is able to swallow the crayfish soon after resurfacing (this feeding lasted ~ 19 seconds).

1145 – Sinbad then moves to the mid-SC area where he parks and preens. Sparky also parks and preens in the mid-SC, ~ 15 m NW of Sinbad. Phantom is parked and preening in the mid-north section of the SC. Cobra is parked with Remy at the SE end of the SC. Quark is foraging in the NW corner. Spot isn't seen at this time.

1148 – Sinbad moves east to the SE section of the SC and continues on through it, passing beyond my view.

1150 – Sparky moves to the SE section of the SC and parks there. Phantom moves to the mid-SC area and parks and preens. Cobra is still with Remy in the same location. Quark has moved across the length of the SC and through the SE section, passing beyond my view. Spot still hasn't been seen.

1154 – Remy moves to the mid-area along the east end of the SC with Cobra closely accompanying her. Phantom parks near the pair. Sparky is still parked and preening, now in the mid-SC area.

1157 – Phantom goes out to the mid-river area east of the SE section of the SC and moves beyond my view. Cobra is still with Remy in the same location. Sparky is now parked in the SE section of the SC.

1200 – Sparky moves out to mid-river just east of the SC. Remy, with Cobra, moves a little farther west inside the SC from the east end. Spot is sighted foraging near the east bank of the river, across from the SC.

1202 – A train noisily passes over the RR bridge. Remy and Cobra are made a bit uneasy by this and begin moving north toward the reeds along the north bank of the SC. The train passes before they get near the reeds and Remy then goes back to the mid-SC area with Cobra following closely. Sparky moves across the river and begins foraging near the east bank, east of the SC.

1205 – Spot is sighted with a large crayfish that it has captured on its own by the east bank of the river, east of the SC. Spot crosses the river nonstop, working the crayfish while moving along. The chick goes all the way to the mid-SC before finally swallowing the crayfish after having worked it for ~ 106 seconds. Spot then parks in the mid-SC area. Remy, closely accompanied by Cobra, moves east through the SE section of the SC and passes beyond my view.

1209 – Sparky enters the SC from the SE end with a crayfish and is being hotly pursued by Remy. Remy puts on a burst of speed and pounces at Sparky. Sparky instantly dives to elude Remy but she dives right after the chick. When they resurface Remy briefly continues her pursuit, then pulls up and parks at the SE end of the SC where she is joined by Cobra. Sparky continues moving away while working the crayfish and soon swallows it. Sparky then parks in the SE section of the SC and preens. [I didn't see the beginning of this incident but it appears that the following is what most likely had happened: Remy had caught this crayfish and probably had intended it for Cobra but Sparky moved in and took it from her instead (or perhaps stole it from Cobra). This then apparently raised the ire of Remy and she rebuffed and aggressively chased Sparky.]

1213 – Remy, closely accompanied by Cobra, heads toward the mid-north bank of the SC. On the way, Remy passes near Spot (who is still parked in the mid-SC area) and suddenly charges at Spot, causing the chick to dive and flee. Remy then continues to the reeds along the mid-north bank and begins foraging there. Cobra closely follows her there and monitors her foraging from the edge of the reeds.

1220 – Spot has resettled in the mid-SC. Sparky has moved through the SE section of the SC heading east and passes beyond my view. Remy is foraging in the reeds along the north bank of the SC in the NW section; Cobra is monitoring her foraging. A stranger PB grebe (probably a grown, independent fledgling) is sighted lying at rest within an extensive algal mat in mid-river along the territorial boundary between the SR and NR pairs.

1225 – Spot moves east through the mid-river area east of the SC. While doing so, Spot and a second grown stranger PB grebe fledgling pass near each other with neither paying any attention to the other. Remy is now sighted foraging in the mid-area along the east end of the SC where she soon ceases her foraging and parks and preens. This time she is alone (i.e., Cobra isn't with her).

1228 – One chick is briefly sighted foraging near the east bank of the river, across from the SC (my view is too poor to ID the chick).

1229 – Cobra goes to Remy (who is still parked at the east end of the SC), remains with and repeatedly solicits her.

1235 – Cobra is still with Remy and intermittently solicits her (the pair has drifted into the mid-SC area). Remy soon goes to the reeds just west of the old nest site, dives into the reeds and begins foraging. Cobra accompanies Remy and monitors her foraging. None of the other family members are seen at this time (Sinbad and Quark haven't been seen for a while so it is quite probable that they are in the West Lagoon).

1239 – Remy has worked her way into the reeds in the NW corner of the SC. Cobra continues to accompany her and monitor her foraging. At one point when Cobra popped out from the reeds it appeared that the chick had just finished swallowing a prey item.

1240 – Sinbad enters the SC from the SE end and heads west to the mid-SC area (calling once en route) where he parks and preens.

1243 – Sinbad moves to the NE section of the SC and parks and preens there.

1244 – Sinbad goes near the reeds at the south side of the NE point of the SC and begins foraging.

1245 – Sinbad is still foraging along the reeds at the south side of the NE point. Remy is still foraging among the reeds in the NW corner. Cobra is still accompanying her and monitoring her foraging. None of the other family members are seen at this time. End observation period.

*The following occurred during my continued observations of the family after the formal observation period had concluded:

*1250 – Remy is sighted with a medium-sized fish near the reeds east of the nest. Cobra rushes to Remy, looking to receive the fish from her. However, Remy keeps the fish for herself and works it for ~ 70 seconds before swallowing it. As she finishes swallowing the fish, Sinbad emerges from the reeds nearby with an extremely large crayfish. Cobra leaves Remy and goes to Sinbad, looking for a feeding from him instead. Remarkably, what then ensues is a crayfish feeding the likes of which I haven't seen among this brood since the chicks were much younger (i.e., since the time when the chicks became capable of working large crayfish on their own). Specifically, this was the type of feeding in which the parent repeatedly tears off scraps from the crayfish and hands them off to the chick(s), thus reducing the bulk of the carcass to a point where it is small enough for the chick to swallow. Sinbad worked the crayfish in this manner, tearing off scraps and handing them to Cobra (Sinbad also ate some of the scraps himself). What differed in this feeding compared to when the chicks were younger is that Cobra also worked the crayfish with Sinbad and a number of times also dove after the carcass when it sank and tore off a scrap while it was still underwater. However, initially Cobra was reluctant to take or work the crayfish, apparently put off or intimidated by its large size. Cobra didn't really join in until Sinbad had worked the crayfish a bit. After the crayfish had been worked for over 2 minutes the carcass was reduced to the point where Cobra was able to swallow it. After this feeding Cobra again went to Remy (who was parked and preening in the mid-SC area) and stayed with her.

*1259 – Sparky is sighted with a medium-sized crayfish that the chick has captured on its own. Sparky works the crayfish for ~ 30 seconds before swallowing it.

*Comments:* Sinbad's noteworthy feeding of Cobra at 1250 was somewhat surprising. This is because for some time now he hasn't shown such patience when a chick has been reluctant to take an extra-large crayfish from him. In such cases he has allowed the chick several chances to take it but if the chick remains reluctant he typically has either taken it to another chick or eaten the crayfish himself. This feeding was also notable for its retrograde aspects.

Lately Cobra has spent more time closely accompanying a parent than any of the other chicks. In Cobra's case this time has all been spent with Remy since Sinbad doesn't allow Cobra to closely accompany him. (Second in this regard is Quark who is allowed by both Remy and Sinbad to closely accompany them.) On the surface, this seems as though it may be a bit of a regression for Cobra but I observed something in my observations of the West Lagoon brood last year that perhaps is similar. In that case two of the chicks also began spending more time accompanying the parents in the period just before they became fully independent and left their natal area. It was as though they were seeking to take advantage of a good thing for a final time before its imminent end.

Sparky and Spot both were seen capturing a crayfish on their own today which they then worked and swallowed.

What appeared to be two East Lagoon second brood and three East Lagoon first brood chicks (all now independent) were seen together in the reeds in the SW corner of the East Lagoon.

**9/18/17:** Time: 1104-1210.

1104 – As observations begin, Sparky is parked in the mid-river area just north of the RR bridge. Sinbad is parked and preening in the mid-section along the east end of the SC. Quark is with Remy in the NW section of the SC and Phantom is parked near them. Spot moves east through the SE section and passes beyond my view. Cobra isn't seen at this time.

1107 – Quark leaves Remy and goes to the SE section of the SC.

1110 – Remy moves toward the NE section of the SC and several chicks follow at a distance. One chick (possibly Spot) approaches and solicits her from a distance. Remy instantly responds by charging and pouncing at the chick. The chick immediately turns away and dives to elude Remy who then dives after the chick. They resurface near each other and Remy again charges at the chick. The chick responds as before by diving to elude her but Remy follows the chick under. When the two resurface Remy charges at the chick yet again and this scenario repeats itself for a third time. Upon resurfacing, Remy finally ceases her aggressive chasing of the chick. Sinbad is still parked at the east end of the SC.

1112 – Remy moves toward mid-river, east of the SC, closely accompanied by Quark. The pair continues east toward the east bank and passes beyond my view. Sparky moves into the mid-SC area.

1115 – Phantom goes near Sinbad and briefly solicits him, then moves away. Sparky moves to the mid-river area, east of the SC, continues east and passes beyond my view.

1118 – Sinbad has drifted to the mid-SC area. None of the other family members are seen at this time.

1123 – I make a quick check of the West Lagoon from the west end of the RR bridge: none of the family members are seen there at this time (though it is possible that some could be out of sight in the reeds).

1125 – (Back at my SC observation point on the mid-south bank.) Sinbad is still parked in the mid-SC area.

1126 – Sinbad (still parked in the mid-SC) dives and begins foraging.

1130 – Sinbad is foraging along the reeds in the NW section of the SC. None of the others are seen at this time.

1132 – Sinbad emerges from the NW corner and moves to the mid-SC area where he parks and preens.

1134 – Phantom enters the SC from the SE end and heads toward Sinbad. Phantom passes near Sinbad and briefly solicits him, then parks and preens in the NW section, well away from Sinbad. Phantom is now in the midst of 6-7 American coots—neither Phantom nor the coots pay any attention to each other. Cobra hasn't been seen yet today.

1139 – Phantom goes to Sinbad and solicits him while approaching. Sinbad instantly responds by charging at and driving Phantom away. Shortly after, Quark enters the SC from the SE end, goes directly to Sinbad, solicits him from up close and remains with him. Sinbad is completely tolerant of this, despite having just rebuffed Phantom.

1141 – Sinbad, accompanied by Quark, heads into the SE section. Phantom follows fairly close to Sinbad. The trio continue east through the SE section and pass beyond my view.

1144 – Sparky is sighted with a large crayfish east of the SE end of the SC. The chick is on the move being pursued by Phantom and soon swallows the crayfish. After swallowing it, Sparky twice charges and pecks at Phantom, driving Phantom away. (I didn't see how Sparky had acquired the crayfish. However, given the pursuit by Phantom it seems most likely that Sparky had obtained the crayfish in a feeding by one of the parents.)

1145 – Remy is sighted with a large crayfish east of the east end of the SC. Sparky is near Remy and rushes in, snatches the crayfish from her and moves away with it. Remy immediately charges after Sparky and a fairly long chase ensues with Remy aggressively pursuing the chick. During the chase Sparky dives several times (while continuing to hold onto the crayfish) in an attempt to escape but Remy dives at the chick's heels and continues her pursuit. After ~ 17 seconds Remy finally ceases her chase and lets Sparky get away. Sparky continues on into the SC where the chick works and eventually swallows the crayfish (time: 1146). Remy parks and preens in the mid-area along the east end of the SC. Sparky moves to the mid-SC area where the chick parks and preens.

1150 – Sparky and Remy are still parked in their same locations. None of the others are seen at this time.

1155 – Sparky is still parked in the mid-SC area. Remy moves north to the NE third along the east end of the SC and parks there. Cobra still hasn't been sighted today.

1158 – Sparky is still parked in the mid-SC area. A great-blue heron flies low from the SW corner and heads E-NE over the SC. The heron passes low almost right over Sparky but the chick shows no alarm and just rather casually watches the heron pass by.

1159 – Spot moves into the SC from the SE and goes to the mid-SC area where the chick parks and preens. Sparky has drifted out to the mid-area along the east end of the SC. Remy is still parked in the NE end of the SC.

1204 – Sparky moves near Spot in the mid-SC area and parks ~ 5 m away from Spot.

1205 – A train very loudly passes over the RR bridge and this spooks the grebes a bit. Sparky and Spot retreat to the area near the reeds along the mid-north bank of the SC and Remy goes near the reeds at the south side of the NE point. After the train passes, Sparky moves toward Remy. As Sparky approaches Remy she charges at the chick causing Sparky to dive and flee from her. Remy then begins foraging among the reeds at the south side of the NE point. Sparky moves ~ 8 m south of that area and parks there and apparently monitors Remy's foraging from that location. Spot has moved back to the mid-SC area and resumes preening.

1210 – Remy is still foraging and has worked her way west. She is now foraging among the reeds along the mid-north bank of the SC. Sparky is at the edge of the reeds in that area monitoring her foraging. Spot is still parked in the mid-SC area. None of the other family members are seen at this time. Cobra has not been sighted yet today. End observation period.

*The following are some noteworthy events that occurred during my continued observations of the family after the formal observation period had concluded:

*1213 – Sinbad enters the SC from the SE with Quark following a fair distance behind. Sinbad moves toward Spot (still parked in the mid-SC) and as he draws near he charges at and chases Spot, driving the chick out to the NE section. Sinbad then goes to the west end of the SC. Quark follows him there.

*1217 – Sinbad is sighted by the reeds east of the nest with a medium-sized fish which he hands off to Quark who then swallows it.

*1223 – Sinbad is sighted with a very large fish by the reeds at the mid-west end of the SC. Quark and Spot are swarming around him as he begins working the fish and even briefly attempts to swallow it himself. As Sinbad works the fish it changes hands a number of times between the three of them with both Spot and Quark attempting and failing to swallow it. Finally, after ~ 2.5 minutes, Spot (shadowed by Sinbad and Quark) manages (with great effort) to swallow the fish.

*1257 – Cobra is finally sighted—the chick is in the east side of the West Lagoon near the mouth of the inter-lagoon channel. Cobra soon leaves the lagoon, moves under the RR bridge and goes to Remy, who is parked in the mid-area along the east end of the SC. Upon reaching Remy, Cobra repeatedly solicits her from up close (including beak to beak solicitations) and remains close alongside her.

*Comments:* The parents are becoming increasingly brusque with the chicks. In general, their rebuffs are becoming more frequent and aggressive and oftentimes more extended and persistent (for example, some rather long chases have been observed lately). When chicks solicit a parent there is an increasing likelihood the chick will get rebuffed. Lately, sometimes a parent will even charge at and chase a chick who is just lying at rest (as occurred in the 1213 incident involving Sinbad and Spot). In addition, sometimes during a feeding a parent will chase a chick after it has just handed a prey item to it. All of this is especially so regarding Sparky and Spot. There are still a few exceptions to this general trend and these involve the two smallest chicks, Cobra and Quark (see below, especially regarding Cobra's current relative size). Both Sinbad and Remy still allow Quark to solicit them from up close and to closely sit with and accompany them. Remy, but not Sinbad, also allows Cobra to do so. Despite this, Quark and Cobra do sometimes get rebuffed by the respective tolerant parent(s).

Quark appears to have surpassed Cobra in size; thus Cobra is now the smallest of the brood (Spot notably increased its size gap over Cobra some time ago and Phantom surpassed Cobra in size a bit later). However, neither is runty and they appear to be well-developed with good weight and in good health. Even at their current advanced stage of development it is difficult to determine the gender of the chicks with certainty, especially since their beaks have yet to achieve fully adult size and proportions. Nevertheless, at this point I am certain about the gender of two of the chicks: Cobra is definitely a female and Phantom is definitely a male. Quark appears to be a male and with the chicks now approaching their full adult size it then is not surprising that Quark (and Phantom) should surpass Cobra (in all of the adult pairs that I have seen in this area the male is larger than its female partner).

The black juvenile markings (stripes, spots, blotches) on the side of the head and neck of the chicks have mostly turned brown and it appears that the fading process is beginning. When these markings fade away I will no longer be able to identify the individual chicks.

**9/19/17:** Time: 1054-1210.

1054 – As observations begin, Cobra is sighted (from the west end of the RR bridge) near the reeds along the east bank at the SE end of the river, just north of the RR bridge. Sinbad, Spot, Sparky, Quark and Phantom are sighted in the north half of the West Lagoon. Remy isn't seen at this time.

1055 – Sinbad moves to the east side of the lagoon and goes just inside the mouth of the inter-lagoon channel. Spot, Sparky and Phantom follow him there. Cobra goes into the West Lagoon.

1057 – Remy is sighted with Quark close beside her in the mid-water area in the mid-north half of the lagoon.

1100 – I prepare to head to my West Lagoon observation point on the mid-east shore of the lagoon but pause when I sight a chick (it appears to be Spot) with a crayfish that it has taken from Sinbad inside the mouth of the inter-lagoon channel. This chick is pursued by two others but soon swallows the crayfish.

1105 – I go to the area at the base of the west side of the RR bridge to see if I will be able to have a better view of Sinbad and the accompanying chicks from there. At least 3 and possibly 4 of the chicks are sighted inside the mouth of the inter-lagoon channel where Sinbad is foraging.

1108 – Sinbad is sighted inside the mouth of the inter-lagoon channel where he has just handed a crayfish to a chick (my view is too inadequate to ID the chick). The chick works the crayfish and soon swallows it. Sinbad then resumes foraging. (My view of Sinbad and the chicks is poor as they constantly pass in and out of sight.)

1112 – Sinbad is sighted with a very large crayfish (still just inside the mouth of the channel). Three chicks hover around him, looking for a feeding. However, Sinbad keeps the crayfish for himself. He works the crayfish for over 2 minutes before (with much effort) he is finally able to swallow it.

1115 – Spot comes out from the channel and moves west ~ 10 m into the lagoon where the chick parks and preens.

1117 – Sinbad moves out of the channel and parks in the mid-water area west of the channel. Remy emerges from the west shore reeds in the north half of the lagoon and moves out near Sinbad. At this point I head for my West Lagoon observation point on the mid-east shore of the lagoon.

1127 – I arrive at my east shore observation point. Sinbad is parked and preening in the mid-water area in the north half of the lagoon. Remy is foraging near and within the reeds in the area north of the snags. None of the chicks are seen at this time.

1133 – Sinbad slowly moves north and parks ~20 m out from the west shore reeds at the north end of the lagoon.

1136 – Sinbad (still in the same location) dives and isn't resighted at this time. None of the others are currently seen.

1139 – Sinbad is resighted—he is now north of the RR bridge in the mid-river area east of the SC. He soon heads west and goes into the SC.

1140 – None of the family members are currently visible in the lagoon (though Remy and at least some of the chicks are probably just out of view in the reeds). At this time I leave my West Lagoon observation point and head for my SC observation point.

1141 – As I begin leaving, a chick is sighted moving west across the north half of the lagoon from the east shore area (probably from the mouth of the inter-lagoon channel). The chick dives in the mid-water area and begins foraging.

1147 – While rounding the south end of the lagoon I sight Remy and 3 chicks in open water in the north half of the lagoon (none of them are together).

1157 – On a quick check from the west end of the RR bridge en route to the SC, Remy and 3 chicks are sighted in the north half of the West Lagoon. I then continue on and arrive at my SC observation point. Sinbad is parked at the NE end of the SC. None of the others are seen in the SC.

1201 – Sinbad drifts into the mid-SC area.

1202 – Sinbad (still in the mid-SC) dives and begins foraging.

1204 – Sinbad emerges from the NW corner and moves to the mid-SC area where he parks and looks about.

1210 – Sinbad is still parked in the same location and looking about. He appears to be on "sentry duty" in the SC. None of the other family members are seen in the SC at this time. End observation period.

*The following occurred during my continued observations of the family after the formal observation period had concluded:

*1235 – Cobra is sighted with Remy in the mid-water area in the north half of the West Lagoon. Cobra is close beside Remy and intermittently solicits her.

*1246 – (At this time only Sinbad, Phantom and Sparky are in the SC.) Phantom is monitoring Sinbad's foraging from near the reeds, east of the nest (Sinbad is foraging in the reeds in the nest area). Sinbad emerges from the reeds with a large crayfish. Phantom rushes to Sinbad who soon hands the crayfish to Phantom. The chick then works the crayfish for ~ 46 seconds before swallowing it. (Sparky is not near them at this time—the chick is in the reeds in the SW section.)

*1307 – Sinbad (now in the West Lagoon) is parked ~ 25 m from the west shore reeds in the north half of the lagoon. Quark goes to Sinbad from the north end of the lagoon and repeatedly solicits him from close up (including beak to beak solicitations). Quark then remains close beside him. Cobra is closely accompanying Remy, who is foraging among the reeds along the north bank of the south ribbon of the river.

*Comments:* For nearly all of the observation period the chicks were too distant or otherwise too poorly seen to identify (when they could be seen at all). However, all 5 of them were seen. The parents continue to conduct feedings for the chicks despite the fact that the chicks spend so much time foraging on their own. For their part, the chicks still spend some time looking to the parents for feedings and all continue to take some food from them. At this point there seems little doubt that all of the chicks are capable of surviving on their own (including the two smallest, Quark and Cobra—though they are now nearly fully grown) and that they are capable of flying away from the natal area.

Sinbad and Remy still allow Quark to solicit them from close up (including beak to beak solicitations) and to closely accompany and sit beside them. Remy also still allows Cobra to do so.

The affinity among the chicks persists. As expected, it has evolved as the chicks have developed and isn't as strong as it was when they were young and tended to stay together as a group. All of the chicks now spend some time off by themselves. However, they also spend time ranging, foraging and resting in the reeds in company with one or more of their sibs. In general, the chicks are still gregarious and still tend to gravitate toward one another, just not nearly to the extent that they did when they were young (and of course, the parents still serve as a focal point around whom the chicks congregate, especially in regard to feedings). A further indication of this loosened affinity is reflected in the fact that the times when all 5 of the chicks are seen gathered together in proximity are far less frequent now. Observations of the East Lagoon chicks indicate that, at least for some of the chicks, this affinity will possibly persist into the period after the chicks become fully independent. I have regularly seen two of the now fully independent East Lagoon first brood chicks keeping company. I have also had a few sightings of some East Lagoon first brood chicks keeping company with some now also fully independent East Lagoon second brood chicks.

All three of the young East Lagoon third brood chicks were seen resting in the reeds with one of the parents along the east shore in the southern half of the East Lagoon.

A person conducting depth soundings on the river from the RR bridge related to me that he recorded a depth of 15 feet in the mid-river area off the north side of the bridge.

**9/20/17:** Time: 1040-1145.

1040 – As observations begin, Sparky and Quark are parked (separately) in the mid-SC area. Cobra is parked in the mid-area along the east end of the SC. Phantom is in the SW section. Sinbad, Remy and Spot aren't seen at this time.

1044 – Quark moves to the SW corner.

1045 – Sinbad emerges from the SW corner of the SC and moves out to the mid-area at the east end of the SC where he parks and preens. He is closely accompanied by Quark who remains with Sinbad and solicits him often. Cobra has drifted to the mid-river area, east of the SC. Sparky is still parked in the mid-SC. Phantom moves to the NE section of the SC where the chick parks and preens.

1047 – Cobra goes to the reeds at the NE point of the SC and begins foraging. Sinbad is still parked in the same location with Quark, who is repeatedly soliciting him. Remy and Spot haven't been seen yet.

1049 – The NR male is parked near the reeds just north of the NE point of the SC (and thus, just north of the territorial boundary).

1052 – Quark leaves Sinbad and goes near the reeds at the mid-west end of the SC. The chick lingers there briefly and then moves well east across the length of the SC, continuing on through the mid-river area and passing beyond my view.

1055 – Phantom moves near Sinbad and solicits him from a short distance away. Soon after, Sinbad moves to the west end of the SC near the reeds. Phantom follows him from a short distance behind.

1058 – Sinbad moves to the NW section of the SC and 4 of the chicks (all but Cobra; Spot has now been sighted) make their way there also. Sinbad rebuffs one of the chicks on his way there.

1100 – Remy enters the SC from the east with a large crayfish and heads toward the NW section (the 4 chicks are still there monitoring Sinbad's foraging). Quark spots Remy first and rushes to her, uncontested by any sibs (Sinbad has just captured a crayfish and the other three chicks are swarming around him). Quark takes the crayfish and works it for ~ 49 seconds before swallowing it (my focus was on Quark's feeding so I missed the outcome of Sinbad's feeding).

1103 – Remy parks and preens in the NE section of the SC. Cobra goes to Remy from mid-river, solicits and remains with her (the NR male is no longer in this general area and hasn't been seen again since around 1049). Sinbad is foraging in the NW section with at least 2 chicks monitoring him.

1110 – Sinbad has worked his way to the reeds west of the NE point of the SC. He is sighted by the reeds there with a large crayfish. Quark is near Sinbad and goes to him. Quark takes the crayfish and begins working it on the move while trying to keep ahead of two hotly pursuing sibs (whom Quark had beaten in a race to Sinbad due to his closer proximity). Quark continues to work the crayfish while evading the sibs and after ~ 36 seconds of this ducks into the reeds and out of my view. The two pursuing chicks and Sinbad (who has been shadowing this trio while overseeing the feeding) follow Quark into the reeds. They are not resighted at this time so I don't know the outcome of this feeding. Cobra is still with Remy (intermittently soliciting her) in the NE section of the SC.

1115 – Cobra is still with Remy in the same location. Phantom parks and preens ~ 5 m east of them. Sparky, Quark and Spot are by the reeds at the south side of the NE point monitoring Sinbad, who is foraging in the reeds there.

1117 – Sinbad emerges from the reeds at the NE point and moves toward mid-river, followed closely by Quark.

1118 – Sinbad parks and preens in the mid-river area just east of the mid-east end of the SC. Quark is no longer with him.

1120 – Cobra is still with Remy—they have drifted out to mid-river, just east of the SC. Sinbad is still parked and preening in the same location as at 1118.

1121 – Sinbad moves to Remy and Cobra (still together) and charges briefly at Cobra to chase the chick away from Remy. Cobra flees to mid-river to the east. Sinbad then moves ~ 15 m to the west of Remy where he parks and preens. Cobra then returns to Remy and stays with her. Soon after, Quark goes to and remains with Sinbad, intermittently soliciting him. Spot is parked at the NE end of the SC ~ 12 m from Sinbad and Quark.

1125 – Cobra is still with Remy—the pair is now in the mid-river area east of the NE point of the SC. Sinbad moves out to the mid-river area. Quark accompanies him and Spot follows from a distance. Sparky is parked and preening in the mid-area along the east end of the SC.

1130 – Cobra is still with Remy and Quark is still with Sinbad in their same respective locations. Spot is still parked in the general vicinity of Sinbad and Quark. Sparky is still parked and preening at the east end of the SC.

1134 – Sinbad moves to the reeds at the south side of the NE point of the SC accompanied by Quark. Spot also follows Sinbad but at a distance and also solicits him briefly from a distance as Sinbad moves along. Sinbad dives into the reeds there and begins foraging. Quark and Spot wait outside the reeds and monitor his foraging. Phantom is sighted foraging in the reeds west of the NE point. Remy and Cobra had begun moving west toward the SC but are no longer seen.

1137 – Phantom is sighted with a large crayfish by the reeds at the south side of the NE point just before the chick swallows it (Sinbad is nearby so it is most likely that Phantom obtained the crayfish from him). Quark and Sparky are also in that area. Sinbad resumes foraging in the reeds and Quark and Sparky resume monitoring his foraging from just outside the reeds.

1140 – Two chicks are briefly sighted in the reeds west of the NE point. None of the other family members are seen at this time.

1144 – Remy emerges from the SW section of the SC followed closely by Cobra. The pair move east through the SE section and continue east into the mid-river area, soon passing beyond my view.

1145 – Quark and Phantom move from the NE section of the SC to the reeds near the nest. None of the other family members are seen at this time. End observation period.

*Comments:* For some reason all of the chicks seemed particularly keen on looking to the parents for feedings today, both during and in my continued observations after the formal observation period.

Sinbad was again observed chasing Cobra away from close proximity with Remy. However, as soon as Sinbad moved away from their area Cobra returned to Remy and remained with her. Though Sinbad chased Cobra away from Remy he still allowed Quark to solicit him from up close and to remain close beside him.

The NR pair still hasn't been seen with any chicks or even carrying nest material. It's late in the season so if they aren't seen with chicks soon then it appears they will go through the season without having produced any chicks.

A few more instances of aggressive, specifically directed chick to chick competitive interactions were observed today during my continued observations after the formal observation period had concluded (notably, Sparky briefly charging and pecking at Spot in an attempt to exclude Spot from the vicinity of the foraging Sinbad). Overall, it has been somewhat surprising how few of these interactions have been seen, especially compared with all that I observed among the West Lagoon brood last year and even in casual observations of the first two East Lagoon broods this season. Since becoming fully mobile the feeding competition among the SR brood has consisted mostly of: spotting and reaching a provisioning (prey bearing) parent first; barging in and muscling sibs aside in order to receive or snatch the prey item from the parent; and chasing after and attempting to steal the prey item from the sib who possesses it (and in the latter's case, attempting to hold onto it).

**9/21/17:** Time: 1020-1123.

1020 – As observations begin (from the west end of the RR bridge), Remy is parked at the west end of the mid-SC area. None of the other family members are seen in the SC. Two of the chicks are sighted near the west shore reeds in the mid-north area of the West Lagoon.

1025 – (From my SC observation point on the mid-south bank.) Remy is still parked in the same location. None of the others are seen in the SC at this time,

1033 – Remy (now preening) is still parked in the same location. None of the others are seen in the SC so I head for my West Lagoon observation point.

1041 – I arrive at my West Lagoon observation point on the mid-east shore. Quark, Spot and Phantom are sighted in the area east of the snags. None of the others are seen at this time.

1045 – Sinbad is sighted—he is parked in the mid-water area in the north half of the lagoon. Quark is with him and Sparky, Spot and Phantom are in the general vicinity around him. Quark is intermittently soliciting Sinbad (including beak to beak solicitations).

1047 – One of the chicks (probably Spot) moves east toward the east shore and passes beyond my view. One of the other chicks (not Quark) begins foraging in the mid-water area there.

1050 – Quark is still parked with Sinbad in the mid-water area in the north half of the lagoon. One other chick is parked in their general area but isn't close to them.

1052 – Remy is sighted moving south under the RR bridge into the north end of the West Lagoon with a large crayfish. Three chicks are in the general area (Quark is closest to her) and all rush toward Remy. Quark (taking advantage of its closer proximity) wins the race, takes the crayfish from Remy and keeps on the move ahead of 2 and then just one pursuing sib as the other abandons its pursuit. Quark finally swallows the crayfish after having worked it for ~ 75 seconds.

1054 – Sinbad has left the West Lagoon and gone into the SC. Remy and 3 chicks are spread out near the west shore reeds in the north half of the lagoon. There have been no confirmed sightings of Cobra at this point.

1057 – Remy is still parked near the west shore reeds in the north half of the lagoon. Quark goes to Remy, solicits and remains with her.

1100 – Quark is still with Remy near the west shore reeds in the north half of the lagoon. Spot moves across the lagoon from the east side to the west side and parks and preens ~ 5 m from the reeds in the north half.

1101 – Remy moves to the reeds near the snags and begins foraging. Quark accompanies her and monitors her foraging from just outside the reeds.

1104 – Spot moves near Quark and also monitors Remy's foraging. Sparky and Phantom soon join them and all 4 chicks monitor Remy's foraging.

1108 – Remy is now foraging near the snags. The 4 chicks are a bit strung out along the reeds but all are still monitoring her foraging (a few of the chicks have probably lost sight of just where she has worked her way to).

1109 – Remy is sighted near the snags and one of the chicks (it appears to be Phantom) is seen by her just as it finishes swallowing a prey item. Right after the chick swallows the prey item, Remy charges and pounces on the chick, causing it to dive and flee.

1110 – Remy comes out from the reeds just north of the snags. Two chicks are sitting together just north of her. Remy charges at each one in turn, causing them to dive and flee.

1111 – Remy resumes foraging in the reeds by the snags. Two chicks (they appear to be Sparky and Spot) monitor her progress.

1115 – Remy surfaces from a foraging dive by the reeds near the snags with a medium-sized crayfish. Sparky rushes to her and Remy hands it off to the chick. Sparky immediately begins moving away with the crayfish but while doing so, Remy charges at Sparky from behind. Sparky instantly dives with the crayfish to elude her. Remy dives after the chick and when they resurface Sparky quickly swallows the crayfish. Remy then resumes foraging in the reeds near the snags. Phantom comes over and monitors her foraging from outside the reeds.

1119 – Remy surfaces from a foraging dive (unsuccessful) near the snags. Phantom has been monitoring her foraging and is near her. Remy suddenly charges at Phantom causing the chick to dive and flee. Remy then resumes foraging and Phantom continues monitoring her but from a little farther away. One other chick is also monitoring her from a distance.

1122 – Remy surfaces from a foraging dive near the snags with a large crayfish. Phantom rushes to her, looking for Remy to hand over the crayfish. However, Remy retains possession of the crayfish and begins working it. As Remy works it, Phantom attempts to snatch it away several times but Remy is able to thwart these by quickly turning away. Finally, after having worked the crayfish for ~ 30 seconds Remy hands it off to Phantom. Phantom then works the crayfish on the go while keeping ahead of a pursuing sib who has joined the fray; Remy closely shadows the pair as she oversees the feeding. At this point a train noisily passes over the RR bridge (time: 1123). This appears to spook Phantom who then ducks into the reeds with the crayfish. The pursuing chick and Remy follow at Phantom's heels and all go out of my view inside the reeds. They are not resighted in the immediate aftermath (I thus don't know the ultimate outcome of this feeding). None of the other family members are seen at this time. Cobra hasn't been seen yet today (i.e., no confirmed or even probable sightings). End observation period.

*Comments:* The parents continue with their increasingly brusque rebuffs of chicks, including to chicks to whom they have just handed a prey item. Despite this, they continue to conduct feedings for the chicks. The chicks again seemed particularly keen on looking to the parents for feedings today (though they also foraged independently).

Cobra wasn't sighted today. The chicks are at the stage now where a non-sighting could be due to the chick having moved away from its natal range. Cobra has gone unsighted in a previous observation period so I'll just have to wait and see if she is sighted tomorrow before speculating any further.

**9/22/17:** Time: 1043-1148.

1043 – As observations begin, Sinbad is sighted moving SE through the SE end of the SC and he soon passes beyond my view. None of the other family members are seen at this time.

1045 – Sparky emerges from the reeds at the NE point of the SC and moves south along the east end of the SC. Sparky continues south toward the SE point and passes beyond my view. None of the others are sighted at this time.

1050 – None of the family members are seen in the SC at this time and it appears that Sinbad and Sparky have probably gone to the West Lagoon (they were headed in that direction when sighted). I thus leave my SC observation point and head for the west end of the RR bridge to make a brief check of the West Lagoon.

1055 – From the west end of the RR bridge Spot is sighted in the mid-water area of the north half of the West Lagoon moving east. Quark and Phantom are near the west shore reeds in the north half of the lagoon. At this time I head for my West Lagoon observation point on the mid-east shore of the lagoon.

1104 – I arrive at my east shore observation point. Spot is sighted in the mid-water area in the north half of the lagoon. None of the others are sighted at this time.

1106 – Sinbad is sighted parked ~ 20 m from the west shore reeds in the north half of the lagoon. Spot moves near Sinbad and parks ~ 3 m away from him.

1110 – Sinbad and Spot are still parked in their same locations.

1112 – Spot moves to within ~ .5 m to 2 m of Sinbad and repeatedly solicits him. Spot appears leery of being rebuffed by Sinbad (the chick looks to be poised to flee should this become necessary) but contrary to his recent behavior toward Spot he allows the chick to solicit him from close by (though Spot doesn't attempt any beak to beak solicitations). Spot then remains fairly close to Sinbad (from ~ 1-3 m away). Sinbad soon heads NE across the lagoon and Spot follows fairly closely. From the east half of the north end of the lagoon Sinbad passes under the RR bridge and then heads NW toward the NE section of the SC. Spot follows at a distance but moves closer to Sinbad as he nears the reeds west of the NE point (time: 1115). When they reach the edge of the reeds Sinbad dives into them and begins foraging. Spot monitors his foraging from just outside the reeds. As Sinbad works his way west they pass from my distant view.

1120 – Two chicks are sighted near and within the west shore reeds in the north half of the lagoon.

1122 – One chick is sighted in mid-water just south of the RR bridge. It passes under the bridge and heads for the SC. A second chick is sighted following it at a distance. Both make their way into the SC and pass from my view.

1125 – None of the family members are seen in the West Lagoon at this time. At this point I leave this observation point and head for the SC. Remy hasn't been seen yet today nor have there been any confirmed sightings of Cobra.

1134 – I arrive at my SC observation point on the mid-south bank. Quark is with Remy ~ 6 m south of the old nest site (Quark is intermittently soliciting her). Sparky is parked in the NE section to the east of them. Sinbad is parked in the mid-area along the east end of the SC.

1137 – One chick moves near Sinbad, solicits him briefly and then moves away.

1138 – Remy, accompanied by Quark, moves toward the mid-SC. Sparky follows along ~ 3 m ahead of her. Remy suddenly charges briefly at Sparky, causing the chick to dive and flee. Remy then parks in the mid-SC area. Quark parks with Remy and intermittently solicits her.

1142 – Remy goes to the SW corner of the SC followed closely by Quark. Remy then dives into the reeds there and begins foraging. Quark monitors her foraging from outside the reeds. Sparky moves to that area and also monitors Remy. Sinbad is now parked ~ 25 m south of the NE point.

1144 – Spot moves from mid-river and goes to Sinbad (still parked south of the NE point) and repeatedly solicits him from fairly close-up (from ~ .5-1.5 m). Spot then remains near Sinbad, soliciting him often.

1147 – Remy emerges from the reeds at the mid-west end of the SC with a large crayfish. Sparky and Phantom are by the reeds in the SW section (Phantom is much closer to her than Sparky). They both spot Remy with the crayfish and race to her. With less distance to cover, Phantom easily wins the race, takes the crayfish from Remy and quickly swallows it. Remy then moves to the reeds in the NW corner and resumes her foraging there. Sparky and Phantom go to that area and monitor her foraging from outside the reeds (time: 1148). Spot soon joins them. There still have been no confirmed or even possible sightings of Cobra. End observation period.

*1235 – Spot is sighted in mid-river, north of the RR bridge working and then swallowing a crayfish that the chick captured on its own.

*Comments:* It appears likely that Cobra has become the first of the brood to leave their natal range. For a second consecutive day Cobra was not seen. It is still too soon to say with certainty that Cobra is no longer in the area. Though I searched for Cobra after the formal observation periods yesterday and today it is still possible that the chick could have been foraging or resting in the reeds out of my view when I checked the various areas. However, if Cobra isn't seen for a third consecutive day tomorrow then the odds will very much favor the probability that Cobra is no longer active (for whatever reason) in the natal range (i.e., the SR pair's territory).

Though Sinbad continued his brusque rebuffing of chicks today, at times he was notably more lenient than he has been lately. During the formal observation period he allowed Spot to solicit him from close by (though Spot didn't attempt any beak to beak solicitations) and to remain near and follow him closely. During my continued observations after the formal observation period he also allowed this of Phantom.

The juvenile markings on the side of the neck and head of the chicks continue to turn brown and fade. This is making it more difficult to identify the individual chicks, especially when at a distance. As previously noted, when these markings are gone I will no longer be able to distinguish the individual chicks.

All three of the young East Lagoon third brood chicks were seen today at the edge of and within the reeds along the east shore in the middle section of the East Lagoon. During the brief time that I observed them several feedings took place and the female of the pair remained with them while the male was out foraging.

**9/23/17:** Time: 1120-1230.

1120 – As observations begin, Sparky, Quark and Spot are seen (from the west end of the RR bridge) in the NW section of the West Lagoon. (From my SC observation point) None of the family members are seen in the SC at this time.

1123 – Remy is sighted parked and preening ~ 5 m south of the NE point of the SC.

1125 – Remy (still in the same location) dives and isn't resighted at this time.

1127 – Sparky is sighted moving west along the south bank of the SC. When Sparky reaches about mid-south bank the chick reverses course and works its way east back into the SE section and soon passes beyond my view.

1131 – Remy is resighted as she parks and preens ~ 13 m south of the NE point of the SC. None of the others are seen at this time.

1135 – Remy is still parked and preening in the same location. At this time I leave the SC and head for my West Lagoon observation point to check for family members there.

1137 – (Brief check from the west end of the RR bridge while en route to my West Lagoon observation point.) Sparky is foraging near the west shore of the lagoon just south of the RR bridge. Sinbad is sighted with Phantom near the reeds north of the snags.

1145 – I arrive at my West Lagoon observation point on the mid-east shore. Sparky is sighted near the reeds south of the snags looking toward the reeds as though monitoring someone in the reeds (perhaps Sinbad is foraging there). None of the others are seen at this time.

1148 – Sparky is still in the same general area and is obviously monitoring someone who is in the reeds (Sinbad is probably foraging there but isn't seen). Phantom joins Sparky. The two of them move a little farther to the south, dive and aren't resighted at this time (they probably went into the reeds).

1150 – Sinbad emerges from the reeds south of the snags, moves out to the mid-water area, dives and isn't resighted at this time.

1153 – Spot is sighted near the west shore just south of the RR bridge.

1155 – Sparky is sighted parked and preening in the mid-water area in the north half of the lagoon. Sparky soon dives and begins foraging.

1157 – Sparky is still foraging in the mid-water area in the mid-section of the lagoon.

1158 – One other chick is sighted foraging to the north of Sparky.

1200 – Sparky moves near the snags and parks there, looking toward the reeds.

1202 – Phantom and Sparky park and preen near each other ~ 6 m from the reeds, north of the snags. Sparky soon goes near the reeds and parks there.

1205 – A train noisily passes across the RR bridge. This spooks the chicks a bit and the three chicks who have currently been sighted (Sparky, Phantom and Spot) all retreat to the south. By the time that the train passes they have reached the area east of the snags. Sparky and Phantom then turn and head north through the mid-water area. Spot remains east of the snags and parks and preens there.

1208 – Sparky continues north and passes under the RR bridge. Sinbad emerges from the west shore reeds in the south half of the lagoon and heads north, calling once as he moves along through the mid-water area. Spot goes near Sinbad and follows him at a distance. Phantom doubles back from the north half of the lagoon, moves close to Sinbad, solicits him (but doesn't engage in any beak to beak solicitations) and closely follows him north. Spot continues to follow at a distance. Sparky is no longer seen (perhaps having gone into the SC). Cobra hasn't been seen yet today.

1210 – As Sinbad approaches the RR bridge from the south he spots a chick (probably Sparky) ahead, just north of the bridge. Sinbad suddenly charges at the chick causing it to take a short, low flight over the surface while fleeing from him. Sinbad then makes his way to the SC. Phantom and Spot lag well behind but follow him there.

1216 – Sinbad, Phantom and Spot have passed from my view after entering the SC. None of the family members are currently seen in the lagoon so I head for my observation point in the SC.

1225 – I arrive at my SC observation point on the mid-south bank. Sparky is parked and preening in the mid-north section of the SC. Sinbad is parked and preening in the mid-area along the east end of the SC. None of the others are seen at this time.

1227 – Quark emerges from the reeds in the SW corner and parks ~ 5 m from the reeds. The chick then gazes back toward the reeds as though monitoring someone who is in there (Remy is probably foraging in the reeds there). Sparky begins moving toward that area.

1229 – Remy emerges from the reeds in the SW corner with a small crayfish. Quark rushes to Remy, takes the crayfish from her and quickly swallows it.

1230 – Remy moves to the mid-west end of the SC ~ 7 m from the reeds and parks there. Quark moves to the NW section where the chick parks and preens. Sinbad is still parked in the mid-area along the east end of the SC. None of the other family members are seen at this time. End observation period. For a third consecutive observation period Cobra has not been seen.

*1251 – Quark goes to Remy (who is parked in the mid-SC area), solicits her from up close and remains close alongside her.

*After the formal observation period ended I searched for Cobra, including in the surrounding areas, and didn't see her.

*Comments:* For a third consecutive day Cobra wasn't seen. It now appears highly likely that Cobra has left the natal range and is fully independent. Of course, because Cobra wasn't actually observed leaving (nor was she seen when I checked the surrounding areas for her) it can't be ruled out that she was a victim of predation or otherwise died. However, given the following circumstances the odds greatly favor the conclusion that Cobra has left the natal range and is now fully independent. Right from the start Cobra has always been one of the most independent of the brood. As of the first day that Cobra went missing (9/21) all of the chicks had almost certainly been capable of surviving on their own (and for a number of days already). In addition, all of the chicks had fledged and no doubt had the ability to fly away from their natal range at that time. Of particular relevance here is that Cobra (along with Sparky) was observed taking flights more often than any of the other chicks and it was Cobra who took the beautiful back to back very long flights on 9/16 (Cobra is the only one of the brood thus far who has been observed taking such long flights, however I suspect that others, and perhaps all of them, have done so but I just wasn't present to witness them—pointing out the obvious here, I have only observationally sampled a fraction of each of their lives). Along with the above, Cobra's development had progressed very well (as has been the case for all of the chicks) and she appeared to be in good health with good weight. Finally, quite simply, the chicks are now at the stage when they can fully be expected to leave their natal range. Considering 7/26 to be the last day of hatching for the brood (and counting it as day 1) and 9/21 as the day she left then Cobra achieved full independence on the 58th day after the last egg in the clutch hatched.

All four of the remaining chicks appear well capable of leaving the natal range and becoming fully independent and have appeared so for a number of days (Sparky appeared to be capable of doing so noticeably before the others with Spot following not too long after). In considering this situation, on the surface there seems no pressing need for the chicks to leave yet and it may even be somewhat advantageous to remain a bit longer (note that at least some of the East Lagoon first and second brood chicks have continued to reside in and near their natal area after becoming independent). For one thing, Remy and Sinbad aren't going to be rearing another clutch this season so there's no need for them to clear the chicks out for that reason. The SR pair's current territory has proven itself well capable of supporting the 7 of them (and they now appear to be down to 6) so there is no need for the chicks to leave on that account. Staying longer will allow the chicks to be a little more developed and gain a little more experience at being independent before striking out on their own or in company with another sib(s). It also means that the chicks will have their parents as a safety net for a bit longer (though at some point the parents will cease doing any feedings). The West Lagoon brood from last season didn't leave their natal area until mid-October so it will be interesting to see how much longer these 4 chicks will remain (or perhaps, how much longer before they are driven out by the parents?). It will also be interesting to see if Cobra returns, and if so, after a period of absence would Cobra then be treated as an intruder?

Since the time that the East Lagoon first brood chicks became independent I have regularly seen fledged, independent chicks in this general area. Lately (as one might expect because it's late in the season and many chicks have become independent) I have seen quite a few more. They sometimes encroach into the core sections of the three territories here (getting aggressively driven away by the owners when sighted by them) but mainly stick to more peripheral and lesser used sections of these territories such as: the southern end of the West Lagoon; the inter-lagoon channel; the south ribbon of the river; the eastern third of the river (including along the east bank) in the section east of the SC; the southernmost third of the East Lagoon; and the small section that lies to the immediate north of the NR pair's territory. Over the past week I have regularly seen from 4 to 6 of these independent chicks by and within the reeds in the SW corner of the East Lagoon. Today I didn't see any of them there and I also saw less of them among the other usual areas. Is this an indication that a number of these independent chicks have left the area or did they just happen to be hidden within the reeds? Obviously, further observations over the coming days will clarify this situation.

Among other things, it seems possible that such local assemblages of independent chicks may serve as a means by which chicks can find prospective mates.

Though Sinbad continues to deal out some strong rebuffs to chicks, he again was more lenient than he has been toward Phantom and Spot. He once again sometimes allowed them to solicit him from close-in (though no beak to beak solicitations were attempted by these chicks) and to sit by and accompany him in close proximity. Both Sinbad and Remy continue to allow Quark to regularly solicit, sit by and accompany them from close-up (including allowing beak to beak solicitations).

Only one feeding was observed during the formal observation period today and just a few during my continued observations of the family afterward.

**9/24/17:** Time: 1054-1155.

1054 – As observations begin (from the west end of the RR bridge), Phantom is with Sinbad near the reeds north of the snags in the West Lagoon. Phantom is remaining close to Sinbad. In the SC, Quark is with Remy at its NE end (Quark is soliciting Remy often). None of the other family members are seen at this time.

1100 – (From my SC observation point on the mid-south bank.) Quark is still with Remy in the same location. Remy soon goes to the reeds at the south side of the NE point of the SC and begins foraging. Quark accompanies Remy and monitors her foraging from outside the reeds.

1101 – Quark goes into the reeds where Remy is foraging and is lost from view (as is Remy). None of the others are seen in the SC at this time.

1104 – A stranger PB grebe who appears to be a mature independent chick (it no longer has its juvenile markings) enters the SC from mid-river and (unseen by Remy) goes into the reeds in the SW corner.

1106 – Remy and Quark are sighted together ~ 7 m from the reeds in the NE section of the SC. Both are preening (I didn't see whether or not a feeding had occurred in conjunction with their having moved there from the reeds)

1112 – Quark is still with Remy in the same location. A train noisily passes over the RR bridge. This spooks the pair a bit and they slowly move toward the reeds. The train passes through before they reach the reeds and they stop ~ 4 m south of the old nest site.

1115 – Remy goes to the edge of the reeds in the mid-west end of the SC, dives into them and begins foraging. Quark accompanies her to the edge of the reeds and monitors her foraging from out there. None of the others are seen at this time.

1118 – Remy emerges from the reeds by the nest with what is either a small fish or a small crayfish and moves toward Quark. Quark spots Remy approaching, rushes to her and takes and quickly swallows the prey item. Remy then goes to the reeds in the NW corner and resumes foraging there. Quark follows Remy and monitors her foraging from outside the reeds.

1121 – Remy emerges from the reeds in the NW corner with a medium-sized crayfish. Quark goes to Remy, takes the crayfish and works it briefly before dropping it onto the water. Remy retrieves the crayfish, works it a little more and then hands it off to Quark. This scenario repeats itself a number of times with Quark also sometimes diving to retrieve the crayfish. In the end, it appeared that Remy was the one who eventually swallowed the crayfish but my view was too poor to be certain of this. I wasn't able to determine why swallowing this particular medium-sized crayfish was so problematic for Quark. Quark typically is able to quickly swallow small to medium-sized crayfish.

1124 – Remy emerges from the reeds at the mid-west end of the SC with a large crayfish. Quark rushes to Remy as she begins to work it. Shortly after, Remy hands the crayfish off to Quark who begins working it and soon detaches a large chunk from the tail. As Quark swallows this, Remy picks up the carcass and works it for ~ 25 seconds before swallowing it herself. While Remy is busy working the carcass down her throat, Quark picks up a large claw from the water that had become detached and swallows it. Remy then moves to the mid-SC area where she parks and preens. Quark initially forages briefly in the NW corner (unsuccessfully) and then goes to Remy and remains with her (time: 1127).

1130 – Quark is still with Remy in the mid-SC area. Spot, Sparky and Cobra have not been sighted yet today.

1134 – Remy moves out to the mid-area along the east end of the SC where she parks and preens. Quark doesn't follow her this time and instead remains parked in the mid-SC area. Sparky is sighted moving into the SC from mid-river. The chick proceeds to the mid-SC area and parks ~ 4 m from Quark.

1137 – Remy goes to the reeds in the SW corner. Quark sees Remy moving there and follows after her. Sparky also soon moves to that area. All three of them are now out of my view in the reeds.

1143 – A stranger mature, independent PB grebe chick (it no longer has its juvenile colors) is sighted moving toward the SC from the mid-river area. Just as the stranger chick reaches the east end of the SC, Sinbad also happens to be making his way to the SC from the SE (probably returning from the West Lagoon). The stranger chick soon spots Sinbad and instantly engages in a sinking dive in an apparent attempt to elude detection by Sinbad. However, Sinbad did spot the chick from a distance. He then moves toward the direction where the stranger went under and dives after it. Sinbad soon resurfaces near where the stranger had dove, parks there and intently looks about. The stranger wasn't resighted after it dove (it probably had fled to the east while underwater). Sinbad lingers there for a while and then heads into the SC.

1146 – Remy is now parked and preening in the mid-SC area. Sparky is also in the mid-SC area, not close to Remy but in her general vicinity. Remy suddenly charges at Sparky, causing the chick to dive and flee from her. Sparky soon resurfaces in the mid-south section of the SC. Remy again parks in the mid-SC area and preens herself.

1147 – Sinbad is sighted moving toward Remy (who is still parked in the mid-SC) from the west end. They greet briefly when Sinbad reaches her and then he continues on to the mid-area along the east end of the SC where he parks and preens.

1148 – Sparky moves east through the SE section of the SC and soon passes beyond my view.

1151 – Remy and Sinbad are still in their same respective locations. None of the chicks are seen at this time.

1155 – Remy is still parked in the mid-SC area. Sinbad is now parked ~ 6 m south of the NE point of the SC. None of the chicks are seen at this time. Spot and Cobra have not been sighted yet today. End observation period.

*The following occurred during my continued observations of the family after the formal observation period had concluded:

*1157 – Sparky enters the SC from the SE section and makes its way toward the NW corner. Sinbad (parked just south of the NE point) spots Sparky moving there and instantly follows at a distance. Sparky becomes aware that Sinbad is trailing from a distance and when the chick reaches the reeds at the NW corner (it had been moving at a steady but not rapid pace) it dives into them, apparently to hide. Sinbad reaches the edge of the reeds there soon after and also dives into them, apparently to search for Sparky. Sinbad soon emerges from the reeds and parks in the NW section. Approx. 5 minutes later Sparky is sighted parked in the mid-north section, well away from Sinbad (the chick apparently having done a proficient job at eluding him).

*1200 – Quark emerges from the reeds in the SW section and goes to Sinbad (who is parked in the NW section). Quark briefly solicits Sinbad from up close. When Sinbad doesn't respond Quark leaves him and moves out near the old nest site.

*1202 – A mature stranger PB grebe chick (probably the one that went into the SW corner at 1104 and had been hiding in the reeds) is sighted rapidly fleeing out of the SW corner heading east. The stranger chick is being aggressively pursued at a distance by Sinbad. The panicked chick takes flight well ahead of Sinbad who then also takes flight in hot pursuit. The chick speeds ahead in a low, skipping flight over the surface out of the SC and lands in the mid-river area east of the SC. Upon landing the stranger chick immediately dives. Sinbad has rapidly pursued the stranger, also in a low, skipping flight over the surface. He lands a little behind where the chick had gone under and instantly dives and continues his pursuit of the intruder underwater. At this point they pass beyond my view, Sinbad having chased the intruder well out of the SC.

*1221 – Phantom is sighted at the east end of the SC closely soliciting and then closely following Sinbad.

*Comments:* It is possible that a second chick from the brood has "flown the coop." Spot was not seen today (including during my search for the chick after the formal observation period had concluded), thus raising the possibility that this chick has also left the natal range. As always, any serious speculation regarding Spot must await further observations over the next several days.

Cobra was not seen for a fourth consecutive day. It now appears a virtual certainty that Cobra has left the natal range and is now fully independent.

Both Sinbad and Remy were particularly brusque toward Sparky today. One is thus left to wonder if Sparky will also be leaving the natal range over the next day or two. Regardless, Sparky's case is interesting to contemplate. Right from the start Sparky has been a little ahead of its sibs and appeared to be the first of the brood who became capable of surviving on its own (I suspect that Sparky probably was the first of the brood to hatch). Yet Sparky has chosen to remain in the natal range up to this point. Is this perhaps due to any or all of the aspects discussed in yesterday's notes (i.e., regarding why a newly independent chick might want to remain a little longer in its natal range)? And/or the fact that, being the dominant chick, Sparky has remained in order to take advantage of its ability to best its sibs during feedings (in fact, during my continued observations after the formal observation period had concluded Sparky was observed charging, pecking at and driving Phantom away from close proximity to where Remy was foraging at the time)?

Sinbad appears to have reverted back to his previous permissive behavior toward Phantom in regard to weaning behavior. For a third consecutive day he has allowed Phantom to solicit him from up close and to sit close beside and closely accompany him. He also still allows this of Quark (who spent the majority of the observation period beside and accompanying Remy).

On two occasions today Sinbad aggressively chased mature, independent stranger chicks out of the SC. A fair number of such chicks continue to be seen in this general area (today this included at least three of the mature, independent chicks whom I had regularly seen in and around the reeds in the SW corner of the East Lagoon—recall that I had not seen any of them there yesterday and wondered whether they had left the area). In general, these mature, independent chicks are quite proficient at avoiding confrontations with the various adult territory owners. When they do get attacked or chased they just take it in stride and continue with their usually successful avoidance tactics.

**<u>9/25/17</u>:** Time: 1059-1207.

1059 – As observations begin, Quark and Phantom are parked ~ 4 m apart in the NW section of the SC. None of the other family members are seen at this time.

1103 – Sinbad emerges from the NW corner of the SC and moves east along the north bank. Quark and Phantom follow him at a modest distance. At about midway along the north bank Sinbad turns and moves toward one of the trailing chicks (I'm not able to determine which one) and causes it to dive and move away. Sinbad then continues east, going to the mid area along the east end of the SC where he parks and preens (alone). Phantom parks ~ 5 m south of the old nest site. Quark is no longer seen.

1110 – Sinbad is still parked in the same location. Phantom is now parked in the mid-SC area. Remy emerges from the reeds in the SW section of the SC, moves into the mid-SC area and parks and preens. None of the others are seen at this time.

1114 – Sinbad has drifted to just south of the south side of the NE point of the SC. Phantom goes to Sinbad and repeatedly solicits him from up close (including beak to beak solicitations) and follows him closely as he moves ~ 7 m farther south and parks there. Phantom lingers by Sinbad but soon parks ~ 3 m to the south of him. Remy moves into the reeds at the mid-west end of the SC and begins foraging.

1115 – Sparky is sighted moving south along the east end of the SC from just east of the NE point. Upon reaching the mid-area along the east end Sparky dives and begins foraging.

1117 – Quark emerges from the reeds west of the NE point and moves toward Sinbad (who is parked in the NE end of the SC) and solicits him from ~ 4 m away. Quark lingers there for about a minute and then goes to Remy (now parked in the NW section). Quark solicits her from up close and remains close alongside her. Phantom has begun foraging in the NE section.

1121 – Sinbad moves south along the east end of the SC toward the SE point and passes beyond my view. He is soon resighted as he doubles back and trails (at a distance) Sparky who has just entered the SC from its SE end. It initially appears as though Sinbad is going to trail and eventually charge at Sparky but instead he soon reverses course and moves east through the SE section and again passes beyond my view. Sparky parks and preens in the mid-SC area (time: 1123). Quark is still with Remy in the NW section.

1130 – Sparky moves east through the SE section of the SC and passes from my view. Remy is still parked with Quark—they have drifted to the mid-north section of the SC. Phantom is parked and preening ~ 2-3 m to the west of them. Spot has not been seen yet today.

1134 – Remy goes into the reeds at the south side of the NE point and begins foraging. Quark accompanies Remy and initially monitors her foraging from outside the reeds before going in after her. Phantom moves to ~ 14 m south of that area and parks there but doesn't monitor Remy's foraging. Sparky enters the SC from the SE, goes to the mid-SC area and parks and preens.

1135 – Phantom (still parked in the same location) spreads and briefly flaps his wings in several separate bursts of flapping.

1137 – Phantom moves all the way out to the east half of the river, east of the NE point of the SC. Something in Phantom's manner during this move indicates that a long flight is imminent (on several previous occasions—Cobra this season and West Lagoon chicks last year—I have observed chicks move in this manner prior to taking a long flight). Upon reaching a point in the east half of the river Phantom stops, turns toward the SC and launches on a long (~ 100 m), low, skipping flight above the surface into the SC, landing in the mid-SC area. Sparky is in the vicinity where Phantom lands and briefly rushes and pecks once at Phantom who flees from Sparky's short, brief attack. As was the case with Cobra on 9/16, it is quickly evident that Phantom is going to engage in back to back flights. After landing (and fleeing from Sparky's charge) Phantom immediately begins making his way straight back out to the east half of the river, east of the NE point (time: 1142). As with the first flight, Phantom stops, turns toward the SC and launches into a second long (~ 70-75 m), low, skipping flight above the surface and lands in the area south of the NE point.

1143 – Sinbad enters the SC from the SE with a large crayfish and heads toward the mid-SC area. Sparky is still parked there and spots Sinbad approaching from a distance. Sparky then rushes to Sinbad who hands the crayfish off to Sparky. The chick works the crayfish for ~ 17 seconds before swallowing it. Sinbad then goes to the mid-area along the east end of the SC where he parks and preens.

1147 – Quark is still with Remy; the pair is now parked in the mid-north section of the SC.

1148 – Remy, accompanied closely by Quark, goes out toward Sinbad (still at the east end of the SC) but stops and reverses course when reaching ~ 4 m from him. Remy then goes into the reeds just west of the old nest site and probably begins foraging there (she has moved out of my view). Quark follows Remy and initially monitors her from the edge of the reeds before going in after her (both are now out of my view).

1151 – Sinbad has drifted out to mid-river, just east of the SC.

1153 – Phantom parks and preens ~ 7 m south of the old nest site. Remy has worked her way to the NW corner and sits outside the reeds there, intently staring into them as though she has possibly spotted an intruder. She soon goes into the reeds where she passes out of my view.

1156 – Remy is back in view in the NW corner, again intently staring into the reeds from just outside them.

1157 – Sparky enters the SC from the SE and moves to the mid-north section.

1159 – Remy comes out from the NW corner and moves east along the north section of the SC (I have no clue as to what it was that had captured Remy's attention in the NW corner); Sparky solicits her from a distance as she passes by. Sinbad has moved into the NW corner and Quark goes to and solicits him. Quark then closely accompanies Sinbad as he moves south along the reeds at the west end and goes into the reeds north of the nest. Quark follows him into the reeds.

1200 – Remy moves out to mid-river, east of the SC and passes beyond my view. Sparky is parked in the NW section of the SC.

1204 – Phantom moves near Sparky and parks ~ 1.5-2 m away.

1205 – Phantom and Sparky are parked near each other in the NW corner peering toward the reeds (Sinbad isn't seen but he is probably foraging there so the two chicks are probably monitoring him).

1207 – Remy enters the SC from the east with a large crayfish and heads toward the NW corner. Sparky (who is closer to the approaching Remy) and Phantom spot her from a distance and rush toward her. With less distance to cover Sparky easily wins the race and meets Remy en route in the mid-SC. Remy hands the crayfish off to Sparky but within a second of doing so she suddenly charges and pounces at Sparky from behind as the chick is moving away. Sparky quickly swallows the crayfish while fleeing and dives in an attempt to escape Remy but she dives after the chick and pursues Sparky underwater. Sparky resurfaces just south of the NE point and Remy resurfaces right behind the chick. Remy then immediately pounces at Sparky again. Sparky promptly dives and flees with Remy following at the chick's heels. This time Sparky moves a considerable distance while underwater and resurfaces all the way over by the east side of the NE point of the SC. Remy apparently ceased her chase while the pair was underwater and resurfaces south of the NE point, well away from Sparky. Spot still hasn't been seen yet today. End observation period.

*The following are some noteworthy events that occurred during my continued observations of the family after the formal observation period had concluded:

*1215 – Sinbad is sighted with a medium-sized crayfish in the mid-SC area. Sparky (the only chick in the general area) goes to Sinbad. Sinbad holds out the crayfish for Sparky to take. However, when the chick grabs hold of it Sinbad apparently decides that he doesn't want Sparky to be the one who gets the crayfish and retains his hold on it. A tug of war then ensues and Sparky is able to yank the crayfish away from Sinbad and begins moving away with it. Sinbad immediately charges and pounces at Sparky, who instantly dives and flees with the crayfish. Upon resurfacing, Sparky is able to quickly swallow the crayfish before Sinbad again charges and pounces at the chick. Sparky promptly dives and flees. Sinbad dives after Sparky but ceases his chase when he resurfaces.

*1230 – Spot is finally sighted—the chick is in the SC and eventually parks and preens in the mid-SC area.

*1320 – Spot is observed with a large crayfish that the chick has captured on its own in the mid-river area north of the RR bridge which it then swallows.

*Comments:* For a second consecutive day Spot wasn't seen during the formal observation period. However, Spot was finally sighted in the SC during my continued observations of the family after the period had concluded. Over the past few days Spot has spent considerable time away by itself.

Cobra was not seen for a fifth consecutive day. It thus appears certain that Cobra has left the natal range and is now fully independent.

Along with Phantom's long flights and Spot capturing a large crayfish by itself the chicks again demonstrated today that they are capable of being fully independent.

As with Cobra on 9/16, Phantom chose to fly into the SC for both of his long flights. Also as with Cobra, Phantom's choice to do so on the second flight was clearly quite deliberate as he purposefully went back out to the east half of the river specifically so that he could take his flight into the SC. Unlike Cobra's flights Phantom stayed low and didn't gain altitude during the flights.

It has become rather common for both Sinbad and Remy to charge, pounce at and chase Sparky and Spot immediately after handing a prey item off to them (though this obviously wasn't observed over the past two days regarding Spot). However, they don't do so on every such occasion. Sometimes while a parent is looking to hand the prey item off to Quark or Phantom, Sparky or Spot will barge in and steal it from the parent and this always elicits a rebuff from the parent. Some of these chases (such as the one at 1207 involving Remy and Sparky) are becoming quite extended.

All three of the third brood East Lagoon chicks were seen at the edge of and within the reeds at one of their two most used feeding stations along the east shore in the mid-section of the East Lagoon. Two feedings were seen. One involved a crayfish and the other involved a medium-sized fish that was fairly large for such small chicks but the largest chick was able to swallow it. The sight of these small chicks serves as a ready reminder of just how far the SR chicks have come along.

**9/26/17:** Time: 1027-1130.

1027 – As observations begin, Quark is foraging in the south ribbon of the river. Remy is parked in the NE section of the SC. Sparky is parked in the mid-SC area. None of the other family members are seen at this time.

1030 – Sparky moves to the SW section and begins foraging. Remy moves to a point ~ .5 m south of the old nest site and parks there.

1032 – Remy moves near the reeds just west of the old nest site, dives into the reeds and begins foraging. Sparky is foraging along the mid-south bank of the SC.

1035 – Sparky is still foraging along the reeds in the mid-south bank area. None of the others are seen at this time.

1040 – Remy emerges from the reeds west of the old nest site and goes into the NE section where she parks and preens. None of the others are seen at this time.

1045 – I make a quick check of the West Lagoon from the west end of the RR bridge: none of the family members are seen there at this time (though it is, of course, possible that some of them could be hidden within the reeds). Remy is still parked in the NE section of the SC.

1046 – (Back at my SC observation point.) Remy goes near the reeds west of the NE point of the SC and dives into them. Sparky emerges from the reeds in the SW corner, goes to the mid-SC area and parks and preens there.

1047 – Remy is sighted with a large crayfish near the old nest site which she begins working. Sparky (still in the mid-SC) spots Remy with the crayfish and heads toward her. However, instead of rushing to Remy and looking to receive or take the crayfish from her as Sparky typically does, the chick just pauses by her ~ 1.5-2 m away and peers at Remy. Sparky is clearly interested in the crayfish but also is clearly reluctant to go in close to Remy to solicit or attempt to receive it from her. Remy swallows the crayfish after working it for ~ 23 seconds, having shown no intent whatsoever toward giving it to Sparky. Sparky then heads toward the mid-SC area and Remy intently follows the chick at a modest distance, her attention clearly focused on Sparky. While following Sparky, Remy suddenly hurries her pace a bit and moves more quickly toward the chick. Though Remy didn't engage in a full charge, this was enough to cause Sparky to dive and flee (Remy never got close to Sparky but her intent was obvious). Sparky then parks in the mid-south section and Remy parks and preens in the mid-SC area (time: 1048).

1054 – Remy and Sparky are still parked in their same respective locations.

1058 – Sparky heads east along the south bank of the SC, moves through the SE section and passes beyond my view. Remy is still parked and preening in the mid-SC area. Sinbad, Phantom and Spot haven't been seen yet today.

1101 – I make another quick trip to the west end of the RR bridge to check for the others (only Remy is currently seen in the SC). Sparky is sighted in the mid-river area just north of the bridge with a large crayfish that the chick captured on its own. Sparky swallows the crayfish after working it for ~ 11 seconds. Sinbad is sighted moving south along the west shore in the north half of the West Lagoon. Phantom emerges from near the mouth of the inter-lagoon channel and heads across the lagoon toward Sinbad. Upon reaching Sinbad, Phantom repeatedly solicits him from up close and follows him closely. As Sinbad, accompanied by Phantom, nears the area north of the snags two other chicks (they appear to be Quark and Spot but are too distant for me to be 100% certain of this) come out from the reeds and go to the general area around him.

1106 – (Back at my SC observation point.) Sparky enters the SC from the SE and moves west along the south bank. While doing so Sparky gives a brief, fairly casual chase after a water strider bug. The water strider flees from the chick along the surface but Sparky soon ceases the chase and continues west into the SW section. Remy is now parked in the NE section.

1111 – Remy goes to the west end of the SC and begins foraging. Sparky moves out to the NE end of the SC. The chick lingers there briefly, then slowly makes its way west along the north bank.

1114 – Remy emerges from the reeds at the mid-west end of the SC with a medium-sized prey item (I can't see it well enough to ID the captured prey). Sparky (now near the old nest site) spots Remy from a distance and moves toward her. Remy also moves toward Sparky. It initially appears as though she might feed the prey item to the chick but she swallows it well before Sparky nears her. Then, with Sparky a modest distance away while still approaching, Remy charges toward the chick causing Sparky to turn away, dive and flee. Both are soon lost from view.

1120 – None of the family members are seen in the SC at this time.

1123 – Same as at 1120 (no family members seen). The NR female is foraging near the reeds just north of the NE point of the SC and is working her way north.

1125 – Sparky is parked near the reeds of the north bank in the NW section of the SC. The chick is peering intently toward the reeds there (perhaps Remy, not seen, is foraging in the reeds there and Sparky is monitoring her?).

1130 – Sparky is the same as noted at 1125. None of the others are seen at this time. End observation period.

*1140 – Sinbad is sighted foraging in the reeds near the snags in the West Lagoon. Spot, Phantom and Quark are near the snags monitoring his foraging.

*Comments:* No feedings were observed today either during or in my continued observations after the formal observation period. However, I spent essentially the entire observation period in the SC where only Remy and Sparky ventured during this time. Nevertheless, it was interesting to note that although Remy did do some foraging, she never entered into her "chick feeding" mode with Sparky being the only chick present in the SC. Her more brusque behavior toward Sparky aside, it is perhaps of interest to note that on numerous occasions when Remy was foraging a considerable distance away from any chicks, she has carried prey items that she has captured over that long distance to the chicks. Remy made no such efforts during today's observations. In the two instances in which Remy was observed with a prey item she ate them herself and displayed no intent of wanting to give them to

Sparky, despite her having obviously seen Sparky in the general vicinity. As for Sparky, the chick clearly was quite leery of being rebuffed by Remy on these occasions. Sparky typically rushes up to Remy when she has a prey item and looks to receive or even just take it from her. In both instances today Sparky was obviously reluctant to do so and in fact, did not. It would have been interesting to see if Remy would have foraged for Quark and/or Phantom if they had come into the SC later in the observation period.

When I sighted Quark, Phantom and Spot in the West Lagoon all three were looking to Sinbad for feedings. Though I didn't see any feedings during the short period that I observed them, the odds are great that Sinbad probably did administer at least some feedings to them. It is most likely that had he not been conducting any feedings at least one or all of the chicks would have made their way over to the SC to see if Remy would do so (recall that none of these three chicks were seen in the SC during my observations).

Sparky was again observed to have captured a crayfish on its own.

Sparky's interest in and brief chase of a water strider bug at 1106 is rather intriguing (fortunately, this incident took place very near my observation point on the south bank of the SC where I was close enough to be able to see the diminutive water strider). Recall from my notes on 8/11 that the chicks (who were then quite small) were observed to peck and dab at the surface of the water as though perhaps capturing and eating insects (such as water striders) or other small invertebrates. This was speculative on my part because these instances were much too distant for me to be able to discern details of such small magnitude (i.e., I was much too far away to be able to see small insects and other small invertebrates). Thus, I couldn't see exactly what the chicks were doing. Sparky's interest and brief (though fairly casual) chase of the water strider lends some credence to this possibility or at the very least raises the question of whether they feed on these insects. One is then left to speculate on whether the diet of PB grebe chicks differs somewhat from that of adults (as occurs in various other species) with the chicks supplementing their feedings from the parents by eating small insects and/or other small invertebrates (prey that they generally won't bother with as adults)?

**9/27/17:** Time: 1045-1153.

1045 – As observations begin, Quark is parked in the mid-north section of the SC, Sparky is parked in the SE end of the SC and Phantom is parked in the mid-river area, east of the SC. Sinbad, Remy and Spot aren't seen at this time.

1048 – Sparky begins foraging by the SE point of the SC. Phantom moves south toward the West Lagoon and passes beyond my view.

1055 – Quark is now parked in the mid-SC area. Sparky (having moved deeper into the SC) is sighted moving east along the south bank in the SW section of the SC. Sparky continues on through the SE section and passes beyond my view.

1100 – Quark is still parked (and now preening) in the mid-SC area. Sparky is foraging along the mid-east end of the SC.

1105 – A mature, fully independent stranger PB grebe chick (it no longer has its juvenile colors and markings) is sighted foraging in the SE section of the SC working its way west, deeper into the SC. Quark (still parked) has drifted to the NE end of the SC.

1106 – The stranger chick has worked its way to the reeds along the mid-south bank of the SC.

1108 – Quark moves diagonally across the length of the SC (SW from the NE end) and goes into the reeds in the SW corner.

1109 – The stranger chick moves east out of the SC into the mid-river area and continues east, soon passing beyond my view.

1114 – Sparky is sighted in the SE end of the SC with a medium-sized crayfish that the chick captured on its own. Sparky works the crayfish for ~ 11 seconds before swallowing it. The chick then goes to the mid-SC area and parks and preens. The NR male is parked near the reeds ~ 20 m north of the NE point of the SC.

1120 – Sparky is still parked in the mid-SC area. None of the other family members are seen at this time. Remy, Sinbad and Spot have not been seen yet today.

1124 – I make a brief check of the West Lagoon from the west end of the RR bridge: Spot is sighted ~ 25 m from the west shore reeds in the north half of the West Lagoon. Spot solicits Sinbad from long distance as he emerges from the reeds to the west of Spot. Remy is sighted in the south half of the lagoon near the south end of the extensive west shore reeds.

1126 – (I have returned to my SC observation point.) Sparky is still parked and has drifted to the mid-east end of the SC.

1128 – Quark emerges from the reeds in the SW corner and moves to the mid-SC area where the chick parks and preens.

1133 – Quark is still parked and preening in the mid-SC area. Sparky is still parked in the mid-east end of the SC.

1134 – Sparky begins foraging in the mid-river area, east of the SC. Quark is still parked and preening in the mid-SC area. None of the others are seen at this time.

1137 – Quark is still parked in the mid-SC area. None of the others are seen at this time. I leave the SC and head for my West Lagoon observation point.

1139 – En route to my West Lagoon observation point I make a quick check from the west end of the RR bridge: Sinbad is parked and preening in the mid-water area in the north half of the lagoon. Remy is sighted near the west shore reeds in the north half of the lagoon.

1148 – I arrive at my West Lagoon observation point on the mid-east shore. Phantom and Spot are sighted east of the snags. Initially, both are monitoring Remy (she is foraging in the reeds by the snags). Then Phantom moves out to ~ 18 m from the reeds and parks and preens while Spot moves to ~ 3-4 m from the reeds and monitors Remy from there.

1151 – Remy emerges from the reeds south of the snags. Spot moves toward Remy but with the chick still a fair distance away while approaching, Remy charges toward Spot and causes the chick to turn around and flee. Remy then parks and preens east of the snags. Spot moves to a point ~ 8 m from the west shore reeds in the north half of the lagoon (well north of Remy) and parks and preens there.

1153 – Sinbad is sighted moving south off the west shore toward Remy. Phantom spots Sinbad and moves to him. Upon reaching Sinbad, Phantom solicits him from up close and accompanies him closely as he continues on toward Remy. As Sinbad approaches Remy they engage in extended, chattering vocalizations. They meet and greet while Remy remains parked. Sinbad then continues south, accompanied by Phantom. Sinbad soon goes into the reeds south of the snags and begins foraging. Phantom monitors his foraging from just outside the reeds. End observation period.

*1212 – Spot receives a feeding from Remy (a large crayfish) just north of the snags in the West Lagoon.

*1221 – Phantom receives a feeding from Remy (a large crayfish) near the west shore reeds in the north half of the West Lagoon. Moments after handing the crayfish off to Phantom, Remy charges at the chick, causing Phantom to dive with the crayfish and flee from her. Upon resurfacing, Phantom is quickly able to swallow the crayfish. Remy then makes one more brief charge at Phantom (who flees a short distance away) and ceases her chase.

*Comments:* Today's observations were fairly quiet and uneventful. No feedings were observed during the formal observation period. However, two feedings (one each to Spot and Phantom from Remy) were seen during my continued observations after the period had concluded.

Sparky and Quark spent the entire observation period by themselves in the SC and the river to the east of the SC. Sinbad and Remy spent the entire period away from these areas (they were probably in the West Lagoon for the entire period).

Sparky was again observed to have captured a crayfish on its own.

Sinbad still allowed Phantom to solicit him from up close and to closely accompany him, though he was observed mildly backing Phantom away twice.

As previously noted, though the chicks appear capable of surviving on their own there seems no pressing need for them to leave their natal range at this time and there are some advantages in staying. Obviously, at some point the chicks will have to get started on their individual lives as Cobra has already done. In the meantime, it will be interesting to see if any or all of the remaining four chicks stay until the end of the season or beyond (or if the parents will even allow this).

Interestingly, over the past weeks as the chicks appear to have become capable of surviving on their own, the frequency of Sinbad's calling has quite noticeably decreased. Of further note, only one territorial standoff (seen on 9/10) between Sinbad and the NR male has been observed since 8/28. Of course, it is certainly possible (and likely probable) that others have occurred since then but during times when I wasn't around to observe them. Regardless, even if others have occurred, the fact that I have observed only one since that time strongly indicates that the frequency of standoffs has also notably decreased—note specifically here that seven territorial standoffs were observed in the period from 7/29 (when the first one was observed) to 8/28 versus only one since then. (The time spent conducting daily observations of the SR family both during and after this period has been comparable.) Despite the above, the movements and ranging by the grebes indicate that the territorial boundary has remained in the same location where the last territorial standoff was observed.

**<u>9/28/17</u>:** Time: 1050-1158.

1050 – As observations begin, Sinbad is sighted (from the west end of the RR bridge) near the snags in the West Lagoon and Phantom is in the NE section of the SC. The other family members aren't seen at this time.

1053 – (From my SC observation point on the mid-south bank.) Phantom moves to the SE section of the SC and begins foraging.

1056 – On a brief check of the West Lagoon from the west end of the RR bridge Quark is sighted with Remy in the mid-water area at the north end of the lagoon.

1101 – (From the west end of the RR bridge.) Remy is now foraging in the reeds along the north bank of the south ribbon of the river. Quark is accompanying Remy and monitors her foraging from outside the reeds. Remy emerges from the reeds with a medium-sized crayfish. Quark goes to her, takes the crayfish and quickly swallows it. Remy then resumes foraging in the reeds. Quark is now in the midst of a group of American coots (probably the parents and their fledglings) and one of them (a parent) pecks Quark lightly on the rump once to get the chick to move aside so that it can pass by (Quark is inadvertently blocking the coot's way). Quark then casually slides over ~ .5 m and continues to monitor Remy's foraging as the coot passes by. The coots just continue on their way.

1105 – (Back at my SC observation point.) Phantom is parked and preening in the NW section of the SC. None of the others are seen at this time.

1110 – Phantom moves out to the SE section of the SC and continues on through it, soon passing beyond my view.

1114 – Sinbad enters the SC from the SE, followed at a distance by Phantom. Sinbad pauses briefly while moving through the SE end and Phantom moves in close to him. Sinbad then goes to the mid-SC area where he parks and preens. Phantom accompanies Sinbad and remains close beside him.

1118 – Sinbad begins moving west, solicited closely by Phantom (including beak to beak solicitations) as he moves along. Sinbad makes his way to the edge of the reeds in the mid-west end of the SC, dives into them and begins foraging. Phantom monitors his foraging from just outside the reeds.

1123 – Phantom is still monitoring Sinbad's foraging from outside the reeds in the same general location as at 1118. Sparky and Spot haven't been seen yet today.

1128 – Phantom is still monitoring Sinbad's foraging from the same general location.

1130 – Phantom shifts his monitoring location a little to the south to the area east of the nest (outside of the reeds).

1131 – Phantom goes into the reeds just north of the nest and moves out of my view.

1132 – Sinbad emerges from the reeds in the SW corner followed closely by Phantom. Phantom soon dives while still in the SW corner and isn't resighted at this time. Sinbad continues on to the north section and stops ~ 17 m south of the old nest site where he parks and preens. Shortly after, Phantom is briefly sighted foraging in the NW section.

1136 – Remy enters the SC from the SE and moves toward the NE point of the SC. As she does, Phantom approaches and solicits her from a distance. Remy soon turns and heads west through the north section. Phantom anticipates her course and moves ahead of her at a distance. As they near the NW corner Remy charges ahead at Phantom who flees a short distance, circling around behind Remy and keeping at a distance. Remy then goes into the reeds and begins foraging; Phantom monitors her foraging from outside of the reeds. Sinbad is no longer seen at this time.

1142 – Phantom has moved out to the mid-SC area. Remy emerges from the reeds at the mid-west end, parks ~ 14 m from the reeds and preens.

1144 – Phantom moves out to the area just east of the NE point of the SC (Sinbad may be foraging in the reeds in that area) and parks there.

1145 – Sinbad emerges from the reeds just west of the NE point with a large crayfish. Phantom spots Sinbad and rushes to him. Sinbad hands the crayfish off to Phantom who works it for ~ 50 seconds before swallowing it (time: 1146). Sinbad resumes foraging in the reeds east of the old nest site and Phantom monitors him from outside the reeds there.

1151-1158: Quark enters the SC from the SE and heads toward the area south of the old nest site where Phantom is monitoring Sinbad (still foraging in the reeds by the old nest site). Sinbad emerges from the reeds east of the old nest site with a particularly large crayfish. Phantom rushes to Sinbad and comes away with a large scrap from the crayfish (probably a claw) and moves ~ 1 m away to swallow it. As Phantom does so, Quark reaches Sinbad, takes the crayfish and moves south with it. Phantom initially pursues Quark but quickly ceases as Quark dives with the crayfish and resurfaces well away from Phantom in the mid-SC area. Quark then begins a remarkably long and intense working of the crayfish, moving into the mid-area along the east end of the SC while so engaged. The chick repeatedly tears off scraps (some of these being fairly large) from the crayfish. While doing so, the carcass often is flicked or laid onto the surface of the water where it sinks, thus requiring Quark to dive completely under the surface numerous times to retrieve it. In this manner Quark works this very large crayfish for a notably long period of ~ 6 minutes 17 seconds before finally finishing the last of it. Quark then goes to Remy (who is parked in the mid-SC area), solicits her briefly and remains close beside her. Sinbad is also now parked in the mid-SC area, farther away to the east of them. Phantom is foraging near the reeds in the NE section. Sparky and Spot haven't been seen yet today. End observation period.

*1219 – Sparky is finally sighted—the chick is just inside the west end of the inter-lagoon channel (i.e., just south of the NE edge of the West Lagoon). Sparky soon enters into the NE section of the West Lagoon and heads north, eventually passing under the RR bridge and turning toward the SC. By 1223 Sparky has entered into the SE end of the SC. However, soon after Sparky enters the SC the chick is spotted from a distance by Sinbad who then begins moving directly toward Sparky. While still a fair distance away, Sinbad aggressively charges toward Sparky. The chick immediately flees from the fast approaching Sinbad, initially by paddling furiously with its feet but then launching into a short (~ 8 m), skipping flight over the surface that carries Sparky deeper into the SC. Upon landing, Sparky instantly dives and isn't resighted at this time. From a distance behind, Sinbad dives after Sparky but soon resurfaces. Sinbad then cruises around the SE section of the SC and the mid-river area to the east of it searching for Sparky. Later, at ~ 1227 Sparky was sighted in the west end of the SC. At 1230 Sparky exited the SC through the SE end and went to the mid-river area to the east.

*Comments:* Neither Sparky nor Spot were seen during the formal observation period though Sparky was finally sighted during a search after the period had concluded. Spot remains unsighted, again raising the possibility that this chick has left the natal range. As always, any serious consideration regarding Spot's situation must await further observations tomorrow and on the subsequent days. Spot also recently went unseen on 9/24 but was sighted the next day.

Phantom (from Sinbad) and Quark (from Remy and Sinbad) both received feedings from the parents today. Quark's handling of that very large crayfish at the end of the observation period was quite impressive.

Sinbad's ready charge and pursuit of Sparky upon sighting the chick as well as his extended search for Sparky after the chick had dove and gone out of sight while fleeing, compels one to wonder if he is going to mount a concerted effort to attempt to exclude/drive out Sparky from the territory.

For the first time I observed a direct interaction between an American coot and one of the grebe family (i.e., other than Sinbad's anomalous chases of coots). While monitoring Remy's foraging in the south ribbon of the river (at 1101) Quark was in the midst of a foraging group of coots (probably parents and their fledged chicks). While looking to move by Quark in this rather cramped segment of this narrow ribbon of the river, one of the adults gave Quark a light peck on the rump, apparently to get Quark to move aside so that it could pass. Quark casually slid over a half meter (remaining focused on Remy), the coot passed by, Quark continued monitoring Remy and that's all there was to the interaction. Basically, it was analogous to a person tapping someone on the shoulder who was inadvertently blocking the way. The general area here is currently teeming with coots (mainly parents and their fledglings) and they are now regularly in proximity with the grebes and have been for some time since the coot chicks began roaming about.

All three of the East Lagoon third brood chicks were seen today. They were in and along the edge of the section of the reeds along the east shore in the mid-segment of the East Lagoon that is one of their two favorite feeding stations. The chicks are gradually becoming increasingly more mobile. One of them was sighted in open water near the west shore of the lagoon yesterday.

**9/29/17:** Time: 1052-1200.

1052 – As observations begin, Sinbad is seen chasing Sparky out of the NE section of the SC into the mid-river area. Sparky dives while fleeing from Sinbad and isn't resighted at this time. Phantom is parked and preening in the mid-north section of the SC. Remy is parked and preening ~ 6 m west of Phantom. Quark and Spot aren't seen at this time.

1056 – Quark emerges from the reeds in the SW corner and goes to Remy (still parked and preening in the mid-north section, now only ~ 2 m from Phantom). Quark solicits Remy upon reaching her and remains beside her.

1059 – Sinbad is sighted moving to the mid-area along the east end of the SC where he parks and preens.

1102 – Remy, closely accompanied by Quark, moves to the edge of the reeds at the mid-west end of the SC. She pauses there briefly, then dives into the reeds and begins foraging. Quark monitors her foraging from just outside the reeds. Phantom slowly moves toward that area, apparently with the intention of also monitoring Remy. Sinbad moves to the area just south of the NE point of the SC where he parks and preens.

1104 – Quark dives into the reeds in the area where Remy is foraging. Phantom is still monitoring her from outside the reeds.

1105 – Remy emerges from the reeds at the mid-west end of the SC with a large crayfish and Phantom rushes to her (Quark is still in the reeds). Remy hands the crayfish to Phantom who begins moving away with it. However, a moment later Remy charges at Phantom from behind. Phantom dives with the crayfish to flee from Remy, who dives after him. They are not resighted at this time—they probably went into the reeds where they were out of my view. Sinbad is no longer in sight.

1106 – Quark has moved back out to the edge of the reeds at the mid-west end of the SC and is parked there, peering into the reeds.

1108 – Remy comes out from the reeds in the NW corner. Quark spots Remy and goes to her. Remy then goes back into the reeds in the NW corner and Quark follows her in. None of the others are seen at this time.

1111 – Quark is now monitoring Remy's foraging from outside of the reeds in the NW corner.

1113 – Phantom is sighted moving into the mid-SC area from the east where he soon parks and preens.

1115 – Quark goes back into the reeds in the NW end of the SC. Phantom is still parked and preening in the mid-SC area.

1117 – Sinbad is sighted moving to the mid-east end of the SC where he parks and preens.

1120 – Sinbad is still parked and preening at the mid-east end of the SC. Phantom is still parked in the mid-SC area.

1121 – Quark emerges from the reeds near the nest and moves out to the mid-SC area. Moments later Remy emerges from the reeds at the mid-west end of the SC. Quark spots Remy and goes to her, solicits her from up close upon arrival and remains with her.

1125 – Quark is still with Remy and intermittently solicits her. Sinbad is still parked and preening in the mid-area along the east end of the SC. Phantom is still parked (and now resting) and has drifted into the NE section. Spot hasn't been seen yet today.

1127 – Sinbad goes near the reeds at the south side of the NE point of the SC and parks there.

1129 – Phantom moves to Sinbad and initially solicits him from a short distance away but then goes near Sinbad and remains with him. Soon after, Sinbad moves west along the north bank to the NW corner, accompanied closely by Phantom, and begins foraging. Phantom monitors Sinbad's foraging and is soon joined by Quark.

1131 – Quark comes out from the NW corner with a large crayfish that the chick had received from Sinbad and moves to the mid-SC, pursued by Phantom. Quark works the crayfish for ~ 109 seconds with Phantom all over Quark, attempting to steal it. Phantom was able to pluck some morsels and pick scraps from the water as Quark worked the crayfish but Quark managed to hold onto and eventually swallow it. Sinbad has resumed foraging.

1134 – Sinbad emerges from the NW corner with a large crayfish and heads toward the mid-SC area. Phantom spots Sinbad and rushes to him. Sinbad hands the crayfish off to Phantom who then begins working it. At this time Quark approaches them and Sinbad suddenly charges at and chases Quark, who dives and flees. Phantom then works the crayfish in peace for ~ 35 seconds before swallowing it.

1140 – Sinbad is now parked and preening in the mid-area along the east end of the SC. Phantom is near the reeds east of the nest, peering into the reeds (Remy isn't seen but may be foraging in the reeds by the nest).

1145 – Sinbad has moved a short distance north and is parked in the NE end of the SC. Phantom is now parked in the mid-north section of the SC. Sparky is briefly sighted moving west (into the SC) along the south bank in the SE section. While moving along, Sparky is clearly being stealthy and apparently attempting to avoid detection by the parents as the chick mainly moves underwater, raising only its head barely above the surface when it resurfaces. After this initial sighting, Sparky isn't seen again at this time.

1147 – Remy emerges from the reeds in the SW corner and moves out ~ 21 m from the reeds at the mid-west end of the SC where she parks and preens.

1148 – Phantom goes to Sinbad (who is still parked at the NE end of the SC) and solicits him from up close. Sinbad soon responds by moving west along the north bank (accompanied by Phantom) to the NW corner where he begins foraging. Phantom monitors his foraging.

1151 – Sinbad is sighted with a large crayfish in the NW corner. Phantom goes to Sinbad, takes the crayfish and begins working it. The chick initially breaks off a few appendages, letting the carcass fall onto water and sink while swallowing them (Sinbad isn't overseeing the feeding and has already resumed foraging). Phantom then makes numerous dives to retrieve the carcass but apparently is unable to relocate it as he is never seen resurfacing with it. Phantom gives up the search after ~ 1.5 minutes.

1153 – At the time when Phantom gives up the search Sinbad is nearby with another crayfish that he has just captured. Phantom heads toward Sinbad but Sinbad retains this one and swallows it himself.

1154 – Remy is now parked and preening in the mid-north section of the SC. Sinbad moves out to the mid-area along the east end of the SC where he parks and preens. Phantom is foraging in the mid-SC. The others aren't seen at this time.

1157 – Remy is still parked and preening in the mid-north section of the SC. Phantom moves toward Remy, solicits her briefly from a short distance and soon parks ~ 4-5 m away from her. Shortly after, Remy goes out near the reeds west of the NE point and parks there. Quark emerges from the reeds in the SW corner, moves to the mid-west end of the SC and parks there. Sinbad is still parked at the mid-east end of the SC.

1200 – Sinbad, Remy, Phantom and Quark are all still in their same general locations as at 1157. A large group of American coots (over two dozen) suddenly begins entering the SC from mid-river, east of the SC. Most notably, a number of them engage in a kind of "group flight," flying into the SC from mid-river in low, often skipping flights over the surface (these appear to be mainly fledglings who perhaps, as a group, are taking practice, exploratory or conditioning flights). The grebes appear to be somewhat put-off by this sudden, rather chaotic activity all around them. With coots flying by and landing all around Phantom, he partially submerges to the point where only his head and neck are above the surface. Remy goes into the reeds just west of the NE point. Sinbad makes his way from the east end of the SC to the SW corner. Quark is no longer in sight (probably having gone into the reeds at the west end). Sparky hasn't been sighted since 1145. Spot hasn't been seen yet today. End observation period.

*1212 – Sinbad is sighted with a large crayfish in the NW corner of the SC. Quark (who is nearby) rushes to Sinbad who hands the crayfish off to the chick. Moments later (as Quark is moving away with the crayfish), Sinbad suddenly charges and pounces at Quark from behind. Quark instantly dives with the crayfish in an attempt to elude Sinbad but Sinbad dives right after Quark. They apparently go into the reeds as they are not resighted at this time.

*Comments:* For a second consecutive day Spot was not seen (note that I also conducted a search for Spot within the SR territory and through this general area after the formal observation period concluded). This considerably increases the probability that Spot has left the natal area and has become fully independent. If so, then Spot would be the second chick from the brood to have done so, with Cobra having been the first. It is, of course, also possible that Spot may have been a victim of predation or otherwise died. However, given that Spot appeared to be in good health with good weight, was well capable of surviving on its own and was at the stage when it could be expected to leave, it is most likely that Spot has just "flown the coop" (i.e., providing that Spot has actually left the natal range). Any firm conclusion regarding this matter must await further observations over the next few days. Regardless of whether or not Spot has actually left the natal range, the significant point here is that for two consecutive days Spot has not been seen associating with or receiving feedings from the parents. This indicates that Spot is living independently.

For all practical purposes and considerations, the brood is now down to just two chicks (Phantom and Quark) who are being attended to by the parents and still receive some feedings from them (i.e., feedings that are willingly proffered and intended by the parents and not sneakily snatched or stolen from them, as Sparky in particular has become quite adept at doing). Phantom and Quark again were the only chicks to receive feedings from the parents (Remy and Sinbad both administered some feedings). Sparky has not been observed getting a feeding from a parent since 9/25. The clear indication here is that, though Sparky continues to reside in the natal range this chick has been living independently (and has essentially been doing so for some time now). One has to wonder how much longer Sparky will remain in the natal range, especially given that the parents (particularly Sinbad) have become increasingly intolerant of Sparky's presence (and Sparky is quite aware of this, as reflected in the manner in which Sparky was observed trying to sneak undetected into the SC today in the 1145 sighting).

Over the past few days Sinbad's behavior toward Quark has begun to change. He is becoming increasingly brusque with Quark and is rebuffing Quark more often. Today Sinbad even charged and chased Quark moments after handing a crayfish off to the chick. At this point, each of the "remaining" chicks has a parent to which it can still solicit from close up and closely accompany and sit alongside without being rebuffed very often: Quark is still allowed to do so with Remy, and Phantom with Sinbad.

**9/30/17:** Time: 1121-1224.

1121 – As observations begin, none of the family members are seen (this includes in the West Lagoon).

1127 – Remy is sighted with a large crayfish near the reeds in the NW section of the SC; Quark is with her. Remy works the crayfish briefly before handing it off to Quark. Quark then works the crayfish for ~ 64 seconds before swallowing it (Remy oversaw the feeding as Quark worked the crayfish). Remy then resumes foraging in the NW section and Quark monitors her. None of the others are seen at this time.

1131 – Remy emerges from the reeds in the NW section with a small prey item (my view is too inadequate to ID it). Quark goes to Remy, takes the prey item and quickly swallows it. Remy then resumes foraging in the reeds there and Quark monitors her from the edge of the reeds.

1135 – Remy emerges from the reeds in the NW section and moves out to the mid-SC area where she parks and preens. Quark goes to Remy, solicits her briefly and then settles next to her.

1138 – Remy goes near the reeds in the SW section and parks there. Quark accompanies Remy, settles close beside her and intermittently solicits Remy.

1142 – Remy goes to the edge of the reeds east of the nest, accompanied closely by Quark. Remy then dives into the reeds and begins foraging while Quark monitors her from just outside the reeds. None of the others are seen at this time.

1143 – Quark goes into the reeds after Remy. Both are now out of my view and none of the family is seen at this time.

1144 – Remy emerges from the reeds in the SW corner with a large crayfish. She looks to give it to Quark but can't find the chick because Quark is in the reeds out of view. Remy then begins moving toward the reeds just as Quark comes out from them. Quark goes to Remy, takes the crayfish from her and begins working it. Remy assists Quark and both have turns working the crayfish. Both also detach and eat large chunks from the carcass which they finish after working it for ~ 64 seconds (time: 1145). After finishing the crayfish Remy remains parked near the reeds in the SW corner. Quark remains in her vicinity but dives a number of times in the area of the feeding as though searching for more scraps from the crayfish that they had just worked on. Sinbad, Phantom, Sparky and Spot haven't been seen yet today.

1150 – Remy (still parked in the same location) dives into the reeds in the SW corner. Quark has been active in the SW corner in Remy's vicinity, often diving (though it doesn't appear as though the chick has been foraging). After yet another dive Quark resurfaces with a stick (~ 5-6 inches long) in its beak. Quark holds onto it briefly, then drops the stick and lets it sink. The chick then dives again and retrieves it. Upon resurfacing, Quark holds it momentarily, then releases it and lets it sink. This time Quark doesn't retrieve it and soon parks ~ 4-5 m east of the nest.

1154 – Remy emerges from the reeds at the mid-west end of the SC with a large crayfish. Quark goes to Remy who hands it off to the chick and a rather unorthodox feeding then ensues in which the two basically share the crayfish (somewhat similar to the 1144 feeding but much more extreme regarding the latter). Each takes shifting turns working the crayfish and retrieving it when it sinks, and tearing off various sized morsels and larger chunks. Interestingly, Remy several times tears off scraps and hands these to Quark (as she used to do during feedings when the chicks were younger), who then swallows them. After ~ 67 seconds they finally finish eating the crayfish (time: 1155). Overall, it appeared that they essentially had shared the crayfish with each having consumed a roughly similar proportion of it.

1200 – Quark is with Remy—they are parked together ~ 15 m east of the nest. At this point, Remy and Quark are the only ones who have been seen today.

1203 – Remy is parked with Quark in the same location. At this time I leave the SC and go to the west end of the RR bridge to check the West Lagoon for the others. Sparky is sighted foraging in the mid-river area east of the SE point of the SC, just north of the bridge.

1207 – Phantom is sighted with Sinbad near the reeds north of the snags in the West Lagoon.

1209 – (I'm now back at my SC observation point on the mid-south bank.) Sparky enters the SC from the SE end and moves west along the south bank to the SW corner. Quark is seen diving in that general area but Remy isn't seen. Sparky soon goes into the reeds in the SW corner and goes out of my view.

1213 – Sparky is briefly sighted foraging in the reeds in the SW corner. None of the others are seen at this time.

1214 – Sparky suddenly comes bursting out of the SW corner while carrying a large crayfish and being aggressively chased by Remy. Sparky promptly breaks into a short, low, skipping flight of ~ 5 m while continuing to flee from Remy with the crayfish. Sparky dives immediately upon landing and Remy (from a modest distance behind) dives after the chick. They are apart in the mid-SC area when they resurface and Remy wastes no time in again charging at Sparky. Sparky dives to flee from Remy who again dives after the chick. They are still separated when they resurface and Remy charges at Sparky a third time, causing the chick to once again dive to flee from her. This time Sparky travels a considerable distance while underwater and resurfaces in the mid-south section, well away from Remy, and continues SE (Remy ceases her pursuit after resurfacing and not finding Sparky nearby). Sparky now begins working the crayfish and after doing so for ~ 34 seconds is able to swallow it. Quark (who apparently had been on the scene when this "feeding" episode began) had followed Sparky at a distance but made no serious attempts to move in close and try to steal the crayfish. [Sparky had obtained the crayfish while out of my view in the reeds. Thus, I don't know whether Sparky had captured the crayfish on its own and was subsequently spotted and chased by Remy or had taken (or stolen) it from her. Given the circumstances and actions of this episode (including the involvement of Quark and the striking tenacity of Remy's pursuit) the overwhelmingly most likely scenario is that Sparky had cheekily snatched the crayfish from Remy who probably had intended to give it to Quark.]

1220 – Sparky has parked (and is preening) ~ 9 m south of the NE point of the SC. Sinbad, followed by Phantom, has moved into the SC from the SE and is now parked with Phantom in the mid-SC area. Remy is parked in the SW section.

1224 – Sinbad, closely accompanied by Phantom, moves to the mid-west end of the SC. While en route he passes by and briefly greets Remy (still parked in the SW section). Sinbad then parks in the

NW section; Phantom remains close beside him. Sparky is still parked just south of the NE point. Quark is foraging in the mid-north section. Spot still hasn't been seen today. End observation period.

*1227 – Remy is sighted near the NE point of the SC aggressively chasing Sparky out of the SC. Sparky dives while fleeing from Remy and isn't resighted at this time.

*1234 – Phantom receives a crayfish feeding from Sinbad in the NW corner of the SC.

*1245 – Sinbad is sighted with a very large fish near the reeds in the NW section of the SC. Phantom rushes to him and is soon joined by Sparky. The two chicks swarm around Sinbad but he doesn't appear that he is ready (or willing?) to give it up (or perhaps, just not to Sparky). Sinbad keeps on the move ahead of the hotly pursuing chicks (with Sparky being particularly aggressive in its attempt to grab the fish). Just as Sinbad is about to dive with the fish Sparky grabs hold of it also and these two dive side by side, each with a hold on the fish. When they resurface Sparky has the fish (apparently having wrested it away from Sinbad while they were underwater) and Sinbad pursues Sparky over a long distance (from the NW section to the SW section and then to the SE end of the SC), charging, pouncing at, and diving after Sparky (who dives a number of times while trying to elude Sinbad). However, Sparky keeps on the move ahead of Sinbad and manages to hold onto the fish. Finally, after about 3 minutes of this intense pursuit, Sparky is able (with much difficulty) to swallow the fish while at the east end of the SC. At this point Sinbad ceases his pursuit. [It is most likely that Sinbad had either wanted to keep this fish for himself or give it to Phantom after he had worked it a bit.]

*Comments:* For a third consecutive day Spot was not seen (this includes a search of the SR territory and this general area after the observation period had concluded). It now appears highly probable that Spot has left the natal range. If Spot has indeed left and is considered to have done so on the first of these three consecutive days (9/28), then this would mean that Spot left the natal area 65 days after the last egg of the clutch hatched.

Sparky continues to reside in the natal range though this chick clearly is now only a peripheral member of the family. The parents have little tolerance for Sparky's presence. Nevertheless, Sparky proved today that it can still manage to obtain a prey item from a parent. However, it appears that Sparky must now do so by sneaking or barging in and snatching the prey from an unwilling parent (as at the 1245 incident when Sparky wrested that large fish from Sinbad, and probably also in the 1214 incident—given the circumstances and actions in that incident, it is extremely likely that Sparky had snatched that crayfish from Remy).

In contrast to Sparky, both Phantom and Quark willingly received feedings from the parents. Two of Remy's feedings to Quark were notable in that the pair essentially shared a crayfish. In the second of these (at 1154) Remy even several times tore a morsel from the carcass and handed it to Quark as she routinely used to do during feedings when the chicks were younger.

Sinbad still allows Phantom to solicit him from up close and to closely accompany and sit close alongside him, and Remy still allows this of Quark.

**<u>10/1/17</u>:** Time: 1110-1210.

1110 – As observations begin (from the west end of the RR bridge), none of the family members are seen in the SC but Sinbad is sighted parked in the mid-water area in the north end of the West Lagoon.

1115 – (From my SC observation point on the mid-south bank.) None of the family members are seen at this time.

1120 – Brief check from the west end of the RR bridge: Sinbad is parked with Phantom in the mid-water area in the north half of the West Lagoon.

1124 – (Back at my SC observation point on the mid-south bank.) None of the family members are seen in the SC. At this time I head for my West Lagoon observation point on the mid-east shore of the lagoon.

1130 – I arrive at my West Lagoon observation point. Sparky is briefly sighted in the south half of the lagoon.

1131 – Sinbad is sighted with a large crayfish and is moving SW through the center area in the mid-section of the lagoon (he appears to be looking for Phantom). Phantom suddenly surfaces near Sinbad and he hands the crayfish off to the chick. Phantom works the crayfish for ~ 52 seconds before swallowing it (Sinbad oversaw the entire feeding). Sinbad then goes to the west shore reeds north of the snags, dives into them and begins foraging. Phantom follows Sinbad and monitors his foraging from outside the reeds.

1136 – Phantom is still monitoring Sinbad from outside the reeds just north of the snags. None of the others are seen at this time.

1139 – Sinbad emerges from the reeds just north of the snags with a large crayfish. Phantom goes to Sinbad, takes the crayfish and works it (with some assistance from Sinbad) for ~ 74 seconds before swallowing it. Sinbad then resumes foraging in the reeds north of the snags. Phantom engages in some foraging dives in the area just to the east of the reeds there.

1143 – Quark is sighted with a large crayfish (which Quark probably took from Remy, who is nearby) east of the snags. Quark works it for ~53 seconds before swallowing it. Remy then resumes foraging in the area east of the snags and Quark parks there.

1145 – Remy is sighted chasing and pouncing at Phantom in the area east of the snags. Soon after, she parks and preens ~ 3 m from the reeds, north of the snags. Phantom parks and preens ~3-4 m to the north of her.

1149 – Quark, Remy and Phantom are still parked in their same respective locations.

1151 – Quark goes to Remy (still parked and preening near the reeds north of the snags) and parks alongside her.

1153 – Phantom begins foraging along the edge of the reeds north of the snags near Remy and Quark. Sinbad is sighted in the mid-water area in the north half of the lagoon.

1155 – Quark is still parked with Remy in the same location as at 1151. Phantom is still foraging along the reeds north of the snags. Sinbad is still parked in the same area as at 1153. Sparky hasn't been seen again since 1130. Spot hasn't been seen yet today.

1200 – Quark is now parked ~ 3 m from Remy. Phantom begins moving north off the west shore reeds. Sinbad goes to the west shore reeds in the north end of the lagoon. He lingers there briefly, then dives into the reeds and begins foraging.

1202 – Remy is now parked ~ 3 m from the west shore reeds in the north half of the lagoon. None of the others are seen at this time.

1205 – Remy has moved (probably into the reeds) and is no longer seen. Phantom parks near the west shore reeds in the mid-north half of the lagoon. None of the others are seen at this time.

1207 – Sinbad emerges from the west shore reeds in the mid-north half of the lagoon near Phantom with a large crayfish. Phantom goes to Sinbad, receives the crayfish from him and begins working it as Sinbad looks on. As Phantom works the crayfish, Quark (who is nearby) moves toward him. However, Sinbad doesn't allow Quark to get to Phantom. As Quark nears them Sinbad suddenly charges, pounces at and chases Quark away from Phantom. Phantom then works the crayfish unchallenged for ~ 56 seconds before swallowing it. Sinbad then parks and preens ~ 12 m from the reeds there. Phantom goes into the reeds and apparently begins foraging. Quark eventually goes to the area by the snags.

1210 – Sinbad is still parked and preening near the west shore reeds in the mid-north half of the lagoon. Phantom is still foraging in the reeds in that area. Quark is by the snags. Remy isn't seen at this time. Sparky hasn't been seen since 1130. Spot hasn't been seen yet today. End observation period.

*Comments:* For a fourth consecutive day Spot wasn't seen. It now appears a virtual certainty that Spot has left the natal range making Spot the second chick from the brood to have done so.

The family is now at a point where only two of the chicks (Phantom and Quark) remain in its inner circle, with Sparky being peripheral and Cobra and Spot having left the natal range. Two distinct dyads have formed among the parents and the two remaining "inner circle" chicks: Remy and Quark; Sinbad and Phantom. Remy's attentions are mainly focused on Quark and Sinbad's mainly on Phantom. Thus, Remy now allows only Quark to solicit her from up close (including beak to beak solicitations) and to closely accompany and sit close beside her; Sinbad allows this only of Phantom. In addition, each parent was observed driving their non-focal chick away from competing with their focal chick during a feeding situation. Phantom and Quark are well aware of this situation and readily go to their respective parent while being cautious and a bit leery about approaching the other parent. Of course, aside from the time that they spend with the parents, Phantom and Quark also range and forage independently.

Sinbad continues to make territorial calls but as previously noted, he has been doing so notably less often.

All three of the East Lagoon third brood chicks were again seen today. At one point the three chicks ventured out into open water in the mid-segment of the East Lagoon. There is noticeable size disparity among them, mainly with one of them appearing to be notably ahead of the other two in both size and behavior.

**<u>10/2/17</u>:** Time: 1103-1205.

1103 – As observations begin, Remy is parked in the west end of the SC ~ 4 m east of the nest. Four mallards and a large group of American coots are around her. None of the other family members are seen at this time.

1105 – The four mallards are close together near Remy. Remy dives and comes up under the mallards (a deliberate attempt to chase them away) and causes all four to take flight together and flee out of the SC. They stay together and land by the reeds along the west shore of the river, well north of the NE point of the SC. Remy then parks and preens ~ 21 m from the reeds in the mid-west end of the SC. (The mallards had just been sitting and were doing nothing of note that would have warranted their having been chased away.)

1110 – Remy is still parked at the mid-west end of the SC. She moves near the reeds east of the nest, then slowly moves north along the reeds, peering into them as she moves along (as if looking for someone). Remy soon dives into the reeds along the north half of the west end and goes out of my view.

1114 – Sparky enters the SC from the SE, moves west along the south bank to the SW corner and parks there and gazes toward the reeds.

1117 – Sparky moves into the reeds in the SW corner of the SC and goes out of my view. None of the others are seen at this time.

1118 – Quark is sighted foraging in the SW section of the SC in open water. Quark soon moves into the mid-north section and parks and preens there.

1123 – Remy emerges from the reeds in the NW end of the SC and moves toward the mid-SC area. Quark (still parked in the mid-north section) spots Remy, moves to and solicits her and remains with Remy as she parks and preens in the mid-north section.

1125 – Remy goes near the reeds west of the old nest site, dives into them and begins foraging. Quark follows Remy and monitors her foraging from the edge of the reeds.

1130 – Sparky emerges from the reeds in the SW corner and moves to a point ~ 4 m from the reeds in the mid-west end, parks there and preens. Quark is still monitoring Remy (she is out of my view in the reeds) from just east of the old nest site at the edge of the reeds.

1132 – Sparky goes to the edge of the reeds at the north bank in the NW section of the SC.

1133 – Remy emerges from the reeds just west of the old nest site with a large crayfish. Remy initially looks for Quark, who is now in the reeds and apparently had lost track of Remy. It's clear that she wants to give the crayfish to Quark but the chick is still in the reeds and out of view. Remy soon gives up on her search for Quark and swallows the crayfish herself. She then parks and preens near the reeds, west of the NE point. Sinbad and Phantom haven't been seen yet today.

1137 – Sparky moves from the edge of the reeds in the NW section and heads east along the north bank toward Remy's location. However, Sparky (clearly having spotted Remy in the distance) dives well before reaching her and isn't resighted at this time. In the meantime, Remy moves from her location and is lost from sight.

1142 – Sparky bursts from the reeds at the mid-south bank of the SC with a large crayfish and is being hotly pursued by Quark. While trying to shake Quark after exiting the reeds Sparky breaks into a short (~ 4 m), skipping flight over the surface and lands in the mid-SC area. Quark continues its hot pursuit as Sparky turns east and moves into the SE section while keeping ahead of Quark. After ~ 40 seconds Quark finally ceases its pursuit and goes back to the reeds in the SW section. Sparky, now able to work the crayfish in peace, continues slowly moving east while working it, eventually ending up in the mid-river area, east of the SC. Approx. 97 seconds after first being sighted with the crayfish, Sparky finally swallows it. [Sparky had acquired the crayfish while in the reeds out of my view so I didn't see how the chick had obtained it. There are three main possibilities: Sparky could have taken it from Remy (who hadn't been seen since 1137 and thus could have been foraging in the reeds there for Quark); Sparky could have captured it on its own (with Quark being nearby and then attempting to steal it); or Sparky could have stolen it from Quark after Quark had captured it or taken it from Remy.

Given past history, the first or third possibilities seem most likely, with Remy being involved (Quark's return afterward to the location where the incident had begun suggests that Remy probably was foraging in the reeds there and Quark had been accompanying her).]

1145 – Sparky is now parked and preening in the mid-area along the east end of the SC.

1148 – Remy is sighted—she is parked and preening in the NE section of the SC. (Has she been in this area since last being sighted at 1137? If so, then Sparky obviously couldn't have gotten the crayfish from her in the 1142 incident. On the other hand, the grebes can and do readily move back and forth across the SC and can do so in a short space of time. And again, it is somewhat suspicious that after ceasing its pursuit of Sparky, Quark went back to the area where Sparky had obtained the crayfish. Was this because Remy was foraging there?) Sparky is now foraging in the mid-river area, just east of the SC.

1150 – Quark goes to Remy (who is now parked and preening in the mid-SC area), solicits and remains close alongside her.

1152 – Sinbad is sighted at the SE end of the SC chasing Sparky north along its east end. Sparky is well ahead of Sinbad, indicating that Sparky had probably spotted Sinbad approaching while he was still fairly distant. Sparky dives and soon resurfaces in the mid-area along the east end of the SC. Sinbad again charges at Sparky from a distance. Sparky once again dives while fleeing, this time going a long distance while underwater, and is able to elude him. Sinbad rushes to the NE point of the SC where he parks and intently peers toward the reeds there and to the east (he obviously has lost track of Sparky and is searching for the chick). In the meantime, I sight Sparky who is by the old nest site moving west along the north bank, clearly unseen by Sinbad. Sparky continues moving west along the reeds of the north bank, turning south through the NW corner and continuing south along the west end into the SW corner where Sparky finally parks.

1158 – Quark is still parked with Remy in the mid-SC area. Sinbad is parked and preening just east of the NE point. Sparky is now parked in the NW corner. Phantom still hasn't been seen today.

1205 – Quark is still parked with Remy in the mid-SC area and intermittently solicits her. Sinbad is parked at the NE end of the SC. Sparky is lying at rest just inside the reeds in the NW corner. Phantom and Spot haven't been seen yet today. End observation period.

*1209 – Phantom is finally sighted. The chick is foraging along the east bank of the river, across from (east of) the SC. Phantom soon crosses the river, enters the SC, goes all the way to the SW end and parks and preens there.

*1211 – Phantom moves north along the reeds at the west end of the SC from the SW corner. As Phantom moves into the NW corner he passes near Sparky who is still lying at rest just inside the edge of the reeds there. Curiously, Phantom stops by Sparky and briefly solicits his sib from a short distance away, just outside the reeds (did Phantom actually mean to solicit his sib for food or did Phantom initially think that it was Sinbad?). Phantom then moves into the edge of the reeds and sits close alongside Sparky for ~ 25 seconds before spotting Sinbad (who is now foraging in the reeds along the north bank in the NW section). Phantom then moves out and goes into the reeds to Sinbad, solicits him and remains there with him as he continues to forage.

*1212 – Sinbad resurfaces from a foraging dive next to Phantom with a large crayfish (they are still just inside the reeds in the NW section along the north bank). Sinbad hands the crayfish off to Phantom who works it for ~ 15 seconds before swallowing it.

*Comments:* It has now been five consecutive days since Spot was last seen. It thus appears certain that Spot has left the natal range (providing of course, that Spot hasn't died).

In the incident at 1105 Remy's reason for chasing the mallards away was not apparent. The mallards were doing nothing of note. They were just sitting there and are of no threat to the grebes. Remy was seen driving mallards out of the territory on a previous occasion (on 7/7) so perhaps for some reason (as with Sinbad regarding American coots) she just has a "thing" regarding mallards? At the same time that Remy was driving the mallards away she was surrounded by over a dozen coots but paid no heed to them.

Though now only a peripheral member of the family, Sparky continues to reside in the natal area. For whatever reason, Sparky has appeared quite intent on remaining here, despite the parents' notable intolerance of the chick's presence. In general, Sparky just takes the parental rebuffs and aggressive chases in stride, almost "matter of fact-like." Sparky is always wary of being rebuffed or chased when sighted by or in the vicinity of a parent but isn't putt-off by this. In this respect, Sparky's general manner toward these is analogous to that of ravens or magpies by a carcass that is being fed upon by a large predator or scavenger: they don't feel really threatened when backed away or chased from the carcass and continue to hang around it. Similarly, Sparky seems to regard the parental rebuffs and chases more as just a scolding rather than a real threat, the chick thus not seriously fearing them and brushing them off as not being serious enough to make it want to leave what it apparently feels is still a good situation (i.e., remaining in an area—the natal range—that it knows so well (and where it is the dominant chick) with the added bonus of sometimes still being able to cheekily snatch a free meal from the parents, as likely was the case in the 1142 incident). Regardless, one would expect that Sparky won't be hanging around for much longer.

The two parent-chick dyads (Remy and Quark; Sinbad and Phantom) continued on as usual today.

All three chicks from the East Lagoon third brood were seen today. The chicks are now fully mobile and are accompanying the parents on foraging forays.

**10/3/17:** Time: 1047-1155.

1047 – As observations begin, Quark is parked with Remy in the mid-SC area and Phantom is parked with Sinbad ~ 4 m south of the old nest site. Sparky isn't seen at this time.

1050 – Same as at 1047.

1052 – Sinbad goes into the reeds along the north bank of the SC in the NW section. Phantom accompanies Sinbad and monitors his foraging from outside the reeds. Quark is still with Remy in the mid-SC area.

1054 – Sinbad emerges from the reeds in the NW corner with a large crayfish. Phantom rushes to Sinbad, takes the crayfish and works it for ~ 16 seconds before swallowing it. Sinbad then resumes foraging in the reeds and Phantom monitors him from outside the reeds.

1056 – Remy goes near the reeds east of the nest with Quark following along. She then dives into the reeds and begins foraging. Quark monitors her foraging from outside the reeds.

1058 – Remy emerges from the reeds by the nest, goes into the reeds along the mid-north bank and resumes her foraging there. Quark initially doesn't follow her but then moves to the mid-SC area, parks and monitors Remy's foraging from there. Phantom moves south along the reeds at the west end of the SC, parks in the mid-west section and peers into the reeds (he is probably monitoring Sinbad who isn't seen at this time but is probably foraging in the reeds there).

1103 – Phantom moves a little farther south along the reeds at the west end and parks east of the nest. Quark soon moves to that general area also, dives there and isn't resighted at this time.

1105 – Sinbad exits the reeds in the SW section with a large crayfish. Phantom goes to Sinbad, takes the crayfish from him and swallows it ~ 5 seconds later. Sinbad then resumes foraging in the reeds in the SW section. Phantom moves to the reeds to monitor him and passes out of my view. Sparky hasn't been seen yet today.

1108 – Sinbad emerges from the reeds in the SW section and goes to the mid-SC area where he parks and preens. Phantom (not seen) doesn't accompany him this time.

1110 – Remy is sighted as she parks and preens in the mid-SC area. Sinbad slowly moves to the SE end of the SC where he parks and preens. Phantom is foraging in open water in the SW section.

1114 – Phantom moves out to Sinbad (still parked and preening at the SE end of the SC), solicits him briefly from up close, then dives and isn't resighted at this time.

1115 – Phantom is resighted as he parks ~ 1.5-2 m from Remy, who is still parked and preening in the mid-SC area.

1117 – Quark emerges from the reeds in the SW corner and goes out near Remy but not to her.

1118 – Remy moves farther south in the mid-SC area and parks there. Sinbad moves to the mid-SC area. Phantom goes to Sinbad while he is moving there and briefly follows him before diving (Phantom isn't resighted at this time).

1119 – Quark moves out to the mid-river area east of the SE point of the SC and parks there.

1122 – Quark moves back into the SC, goes to Remy (still parked in the mid-SC area), solicits and remains close alongside her.

1123 – Sinbad is now moving south along the east end of the SC trailed a short distance behind by Phantom. They continue south and soon pass beyond my view.

1130 – Quark is still with Remy in the mid-SC area and intermittently solicits her. None of the others are seen at this time.

1135 – Quark leaves Remy, goes into the reeds at the NW end of the SC and begins foraging.

1140 – Remy is still parked and preening in the mid-SC area. None of the others are seen at this time.

1142 – Quark emerges from the reeds in the NW corner, goes out to Remy, solicits her repeatedly (including beak to beak solicitations) and remains close alongside her.

1146 – Remy goes to the NW corner, dives into the reeds there and begins foraging. Quark closely accompanies her there and monitors her foraging from outside the reeds.

1149 – Quark dives into the reeds in the NW corner where Remy has been foraging and goes out of my view.

1155 – Remy ceases foraging, exits the reeds and parks and preens in the NW corner. Quark (unseen) is probably still in the reeds there. None of the others are seen at this time. Sparky has not been seen yet today. End observation period.

*Comments:* It was a quiet, fairly uneventful observation period. Sparky wasn't seen today (including during a search for the chick during my continued observations of the family after the formal observation period concluded). Has Sparky finally decided to leave the natal range? As always, serious consideration of this question must await further observations over the upcoming days.

The two parent-chick dyads of Remy-Quark and Sinbad-Phantom continued on as usual today.

The NR pair still hasn't been seen with any chicks. It's late in the season so at this point it is probably safe to conclude that they didn't produce any chicks this season. There was a period earlier in the season when there was an alternating pattern of sightings of the NR male and female which suggested that they may have been incubating a clutch. If so, then the clutch apparently failed.

**10/4/17:** Time: 1050-1200.

1050 – As observations begin (from the west end of the RR bridge), Sinbad is parked and preening in the mid-area along the east end of the SC. Phantom is parked and preening near the east bank at the SE end of the river. None of the others are seen at this time.

1055 – (From my SC observation point on the mid-south bank.) Sinbad moves to the mid-SC area where he parks and preens. None of the others are seen at this time (Phantom may still be at his 1050 location but if so he can't be seen from my current observation point).

1057 – Three stranger PB grebes are sighted parked in an extensive algal mat in the mid-river area north of the NE point of the SC. They appear to be mature, independent chicks who have transitioned from their juvenile plumage. Sinbad moves to the SW section of the SC and begins foraging. He soon goes into the reeds and out of my view.

1100 – None of the family members are seen at this time.

1105 – Sparky is sighted moving north along the east end of the SC from the SE. Sparky has spotted and is moving toward what appears to be a mature stranger grebe chick that has transitioned from its juvenile plumage. This stranger chick is by the NE point of the SC and may be one of the mature chicks who were sighted at 1057. The stranger chick enters the SC, moves west along the north bank and disappears into the NW corner. Sparky follows the stranger (at a considerable distance) to the NW corner, parks outside of the reeds there and peers toward where the stranger had gone (time: 1106).

1107 – Sinbad suddenly appears and aggressively chases Sparky out of the NW corner. While fleeing from Sinbad, Sparky takes a short (~ 3 m), skipping flight over the surface into the SW section. Upon landing, Sparky dives and isn't resighted at this time. Sinbad lingers in the SW section briefly, looking around for Sparky (who is probably hiding in the reeds there). He then heads east through the SE section and soon passes beyond my view (he is probably heading for the West Lagoon). Sinbad called once as he moved through the SW section.

1115 – Sparky emerges from the reeds in the SW corner, moves east through the SE section of the SC and passes beyond my view.

1117 – (I go to the west end of the RR bridge to check for family members from there.) Sparky is sighted foraging in the mid-river area east of the SE point of the SC, just north of the bridge. Sparky soon resurfaces from a foraging dive with a large crayfish. The chick works the crayfish for ~ 11 seconds before swallowing it. Sparky then resumes foraging. Quark hasn't been seen yet today.

1122 – Sparky (still foraging in the same location) resurfaces from a foraging dive with another large crayfish. Sparky works the crayfish for ~ 17 seconds before swallowing it and then resumes foraging.

1125 – (I have moved back to my SC observation point.) None of the family members are seen at this time.

1133 – Still no family members seen in the SC so I head for my West Lagoon observation point on the mid-east shore of the lagoon.

1146 – I arrive at my West Lagoon observation point. Remy is sighted parked and preening near the west shore reeds in the south half of the lagoon. Phantom is parked ~ 4 m from the reeds, south of the snags.

1148 – Sinbad is sighted foraging near the west shore reeds in the south half of the lagoon. Phantom moves toward his general area. Remy is still parked in the same location as at 1146.

1150 – Sinbad parks briefly in the mid-water area, east of the south end of the west shore reeds. Phantom moves in to ~ 4 m of him. Sinbad then dives and isn't resighted at this time.

1152 – Phantom is now foraging in the mid-water area in the south half of the lagoon. Sinbad is foraging in the southern third of the lagoon. Remy is still parked in the same general location as at 1146. Quark still hasn't been seen yet today.

1157 – Sinbad parks and preens ~ 7 m from the west shore reeds in the south half of the lagoon. Remy is parked east of the snags and Phantom is parked ~ 6-7 m to the north of her.

1200 – Remy begins foraging by the reeds just south of the snags. Sinbad is still parked and preening in the same location as at 1157. Phantom goes to Sinbad, solicits him from up close and remains close alongside him. Quark still hasn't been seen yet today. End observation period.

*Comments:* Quark wasn't seen today, including during my continued observations of the family (and a search for Quark) after the formal observation period concluded. Considering how regularly Quark has been with and around Remy is it of particular significance that Quark wasn't seen near her today? Once again, any serious consideration about whether Quark has left the natal area must await further observations over the upcoming days.

Sparky wasn't seen yesterday but was sighted a number of times today. At this point it appears that Sparky is being chased away essentially every time that the chick is sighted by Sinbad and Remy when they are near enough to do so. Sinbad was observed aggressively chasing Sparky during the observation period and Remy was seen chasing Sparky out of the SC on two separate occasions during my continued observations of the family after the formal period had concluded. Sparky demonstrated just how capable it is of surviving on its own when the chick was observed capturing two large crayfish within a 5 minute period (1117-1122) in the mid-river area just north of the RR bridge (I have seen Sparky capture crayfish in this particular area on several previous occasions so it appears to be a favored foraging location for the chick).

The juvenile markings on the chicks continue to fade (they are particularly faded on Sparky). The chicks are thus becoming increasingly more difficult to identify, especially from a long distance away and when they are wet.

No feedings were observed today and neither parent ever appeared to be in "chick feeding mode."

**10/5/17:** Time: 1045-1145.

1045 – As observations begin, Sparky is parked in the mid-river area, east of the SE point of the SC. A stranger PB grebe chick is parked ~ 2.5 m from Sparky (it is a mature chick that has transitioned from its juvenile plumage). No interactions between them are noted. Sinbad is parked in the NW section of the SC. Remy, Phantom and Quark are not seen at this time.

1048 – Phantom is sighted moving north along the reeds at the west end of the SC and eventually parks ~ 2 m away from Sinbad (Phantom doesn't go up close to Sinbad as he has typically done).

1049 – Sinbad goes into the reeds north of the nest and begins foraging. Phantom follows and monitors Sinbad's foraging from outside the reeds at the mid-west end of the SC.

1052 – Sinbad emerges from the reeds and moves toward Phantom. Phantom moves away and Sinbad follows at a distance. Phantom then moves to the reeds in the NW section and dives into them, apparently to get away from Sinbad (Sinbad was moving at a normal pace toward Phantom but with obvious intent). Sinbad (who never got close to Phantom) then goes to the mid-SC area where he parks and preens.

1054 – Phantom exits the reeds in the NW corner, parks there and looks toward Sinbad (still in the mid-SC area). Sinbad soon dives and isn't resighted at this time.

1055 – Phantom moves out to the mid-river area east of the SC and begins foraging (he is soon lost from my view). Remy and Quark haven't been seen yet today.

1100 – None of the family members are seen at this time.

1104 – Remy is sighted foraging by the reeds at the south side of the NE point of the SC. She soon dives into the reeds and goes out of my view.

1107 – Remy is briefly resighted foraging at the edge of and within the reeds just west of the old nest site.

1109 – Remy parks and preens ~ 3 m from the reeds in the NW section of the SC. None of the others are seen at this time.

1112 – Sinbad enters the SC from the SE trailed at a distance by Phantom. Sinbad makes his way to the NW end, meeting and greeting with Remy as he passes by her on the way. Sinbad dives into the reeds there and begins foraging. Phantom follows Sinbad but appears to be a bit leery of him and keeps at a distance. Phantom then monitors Sinbad's foraging from outside the reeds. Remy remains parked in the same location as at 1109.

1118 – Phantom is still monitoring Sinbad (who is in the reeds but not seen) from just outside the reeds at the mid-west end of the SC. Remy is still parked and preening but has drifted to the NE section.

1121 – Remy goes to the NW corner by the reeds there. She is soon lost from view.

1122 – Sinbad is now foraging in the NW corner. Phantom goes near Sinbad and solicits him from a short distance away (getting as close as ~ .5 m) but doesn't go right up close to him as Phantom usually does. Sinbad soon moves toward the mid-SC area. Phantom follows at a distance but backs away when Sinbad turns and moves toward him. Sinbad then goes to the mid-SC area where he parks and preens. Phantom goes to the NW corner and passes from my view.

1126 – Phantom is sighted moving east through the SE section of the SC and soon passes beyond my view. Sinbad is still parked and preening in the mid-SC area.

1130 – Sinbad is now parked in the NE section of the SC. None of the others are seen at this time.

1135 – Sinbad moves to the mid-SC area where he parks and preens.

1138 – Remy emerges from the reeds at the mid-west end of the SC, moves out to the area just west of the NE Point and goes into the reeds there. Sinbad moves out near the reeds at the south side of the NE point. He dives there and isn't resighted at this time.

1143 – Remy emerges from the reeds west of the NE point, moves out ~ 10 m from the reeds and parks and preens.

1145 – Remy is still parked and preening but has drifted to ~ 35 m south of the NE point. None of the others are seen at this time. Quark hasn't been seen yet today. End observation period.

*1201 – Phantom approaches Sinbad (who is parked in the NW section of the SC). As Phantom nears within ~ 3 m of Sinbad he suddenly charges and pounces at Phantom, causing Phantom to turn away and dive while fleeing. Thereafter, Phantom stays clear of Sinbad.

*Comments:* As noted below, today was a very significant day for Remy and Sinbad.

Quark wasn't seen for a second consecutive day. This considerably increases the probability that Quark has left the natal range. If this proves to be the case then Quark will be the third chick from the brood to have left the natal range. As always, any serious considerations regarding Quark's status must await further observations over the upcoming days.

Though independent, Sparky continues to reside in the natal range.

For a second consecutive day no feedings were observed. Only one chick (Phantom) even looked to the parents for a feeding. Phantom was observed soliciting Sinbad a few times and spent some time monitoring him when he was foraging. However, this was all in vain as neither Sinbad or Remy ever appeared to be in "chick feeding mode" and instead clearly looked to be just foraging for themselves when they engaged in this activity (this also appeared to be the case yesterday).

Along with the above it is also noteworthy that Sinbad's behavior toward Phantom was drastically different today. He didn't allow Phantom (or Phantom was too reluctant) to solicit him from up close or to closely follow or sit close to him. To the contrary, on several occasions he intently moved toward Phantom in order to turn the chick away and even aggressively charged and chased Phantom away. Sinbad's new attitude toward Phantom was also reflected in Phantom's behavior. The chick clearly was wary of a possible rebuff when approaching and soliciting Sinbad and deliberately kept its distance. It appears that (though Phantom has almost certainly been capable of leaving the natal range and surviving on his own for quite a while now) Sinbad has finally decided to leave the chick to its own means.

Given all of the above and the fact that the parents apparently are no longer attending to any of the chicks it can now be stated that the chick rearing period has officially come to an end! This has occurred on the 72nd day since the last chick hatched out, though all of the chicks appeared to be capable of leaving the natal range and surviving independently quite a bit sooner (for example, note that Cobra left after day 58).

The black throat patch on Sinbad and Remy that is part of the breeding season plumage has begun to dissipate lately, especially on Remy.

All three of the East Lagoon third brood chicks were seen. All were sighted in open water in the East Lagoon.

**10/6/17:** Time: 1035-1137.

1035 – As observations begin, Sparky is parked near the reeds at the SE end of the river, just north of the RR bridge. Remy is parked just east of the SE end of the SC. None of the other family members are seen at this time.

1037 – Remy enters the SC along the SE section and forages there. Phantom is sighted near the reeds east of the nest but soon goes out of view.

1043 – Remy goes to the mid-SC area where she parks and preens.

1045 – Remy is still parked and preening in the mid-SC area. None of the others are seen at this time.

1047 – Sinbad is sighted foraging in the NW corner of the SC. Remy (still parked in the mid-SC) dives and isn't resighted at this time.

1052 – Sinbad is now foraging along the reeds west of the old nest site. Phantom emerges from the reeds in the SW corner, moves to the mid-SC area, dives there and isn't resighted at this time.

1054 – Sparky enters the SC from the east and goes all the way to the SW corner where the chick soon goes out of view.

1058 – Remy emerges from the reeds west of the NE point of the SC, parks ~ 4 m from the reeds and preens.

1059 – Remy moves south along the east end of the SC. She continues SE into the mid-river area east of the SE point and passes beyond my view.

1103 – What appears to be a mature PB grebe chick (a stranger who has transitioned from its juvenile plumage) is sighted parked in the mid-river area north of the NE point of the SC.

1105 – Sparky exits the reeds in the SW corner, moves to an area ~ 3 m from the reeds at the mid-west end and parks and preens. Quark hasn't been seen yet today.

1110 – Sparky is still parked and preening but has drifted into the NW section.

1112 – Sparky moves into the mid-SC area, parks and lies at rest. None of the others are seen at this time.

1115 – Sparky is still parked and resting but has drifted to the mid-north section.

1117 – A stranger mature PB grebe chick (it has transitioned from its juvenile plumage) enters the SC from the SE end and makes its way west along the south bank into the SW corner where it goes out of view.

1120 – The stranger chick seen at 1117 exits the SW corner, moves east along the south bank and continues on through the SE section, soon passing beyond my view. Sparky sights the stranger from a distance and moves toward its direction (time: 1122). Sparky stops when reaching the SE section, lingers there for ~ 45 seconds, then continues SE and soon passes beyond my view.

1125 – None of the family members are seen at this time.

1126 – On a quick check from the west end of the RR bridge Sparky is seen ~ 4m from the reeds at the SE end of the river, across from the SE point of the SC.

1132 – (I'm back at my SC observation point on the mid-south bank.) Two unidentified chicks enter the SC from the SE end and go west to the area near the shore at the mid-south bank. One of these is a mature stranger chick who has transitioned from its juvenile plumage (i.e., its juvenile markings are completely gone). This stranger chick begins foraging near the shore in this area and the other chick remains nearby and watches it. The second chick now appears to be Phantom but my view is too inadequate to be certain (the continued fading of the juvenile markings on the SR chicks is making identification increasingly more difficult).

1135 – The two chicks are still at the mid-south bank. The stranger chick is foraging, the one that appears to be Phantom remains nearby.

1137 – A chick that appears to be Sparky is sighted parked and preening in the mid-north section of the SC. The stranger chick (the one first noted at 1132) heads east and moves out of the SC. The chick that I can now definitely identify as Phantom goes to the mid-SC area and parks there. Remy and Sinbad aren't seen at this time. Quark hasn't been seen yet today. End observation period.

*Comments:* For a third consecutive day Quark was not seen. It now appears fairly certain that Quark has left the natal range. If not seen tomorrow the odds for this will be extremely strong.

The general observations today served to underscore the conclusion noted yesterday that the chick rearing period has ended. For a third consecutive day no feedings were observed. Furthermore, no chicks were seen accompanying or looking to the parents for a feeding. Even more telling is that no parent-chick interactions were seen at all.

The juvenile markings on the chicks have nearly faded away making it very difficult to identify the individual chicks. It appears that anytime from this point these markings will fade completely and I will then no longer be able to identify them. The problem of chick identification is made even more difficult by the fact that there are also a number of mature stranger chicks in the area (and some of these are occasionally seen inside the SR pair's territory, as was evident today).

**10/7/17:** Time: 1025-1135.

1025 – As observations begin (from the west end of the RR bridge), Remy is parked near the reeds just north of the snags in the West Lagoon and Sinbad is parked ~ 6 m to the east of her. None of the chicks are seen at this time.

1030 – (From my SC observation point on the mid-south bank.) None of the family members are seen at this time.

1035 – A stranger PB grebe is sighted foraging in the NE section of the SC (the grebe is a male and is either an adult in winter plumage or a mature chick that has transitioned from its juvenile plumage—his beak appears that it may be just shy of fully adult proportions).

1040 – The stranger grebe is still foraging in the NE section.

1045 – The stranger grebe is still foraging in the NE section.

1050 – The stranger grebe is still foraging in the NE section. None of the family members are seen at this time.

1053 – Quark (?) emerges from the reeds in the SW corner, moves to a point ~ 5 m from the reeds at the mid-west end of the SC and parks and preens (Quark had not been seen for three consecutive days prior to this sighting). The stranger grebe is now parked in the mid-north section of the SC. *[See the post observation comments regarding an interaction between Quark and an American coot that took place while Quark was parked here.]*

1055 – Quark (still parked in the same general location) goes into the reeds at the mid-west end of the SC and begins foraging.

1056 – The stranger grebe moves to the NW section, begins foraging there and soon works his way into the mid-SC area.

1059 – The stranger grebe is still foraging and has worked his way to the area along the mid-south bank.

1102 – The stranger grebe works his way east through the SE section into the mid-river area east of the SC and passes beyond my view.

1105 – Quark exits the reeds in the SW corner and moves near the reeds north of the nest. Quark stops and preens there for about one minute and then goes back into the reeds in the SW corner (and out of my view).

1110 – None of the family members are seen at this time. I leave the SC and head for my West Lagoon observation point on the mid-east shore of the lagoon to check for family members there (i.e., other than Quark).

1118 – I arrive at my West Lagoon observation point. None of the family members are seen at this time.

1121 – Remy is sighted ~ 7 m east of the snags where she is parked and preening.

1122 – Sinbad is sighted parked ~ 4 m from the reeds in the northern third of the lagoon.

1125 – Remy and Sinbad are still parked and preening in their same respective locations. None of the chicks are seen at this time.

1127 – A train noisily passes over the RR bridge. This makes Sinbad a little uneasy and he moves close to the west shore reeds in the mid-north half of the lagoon.

1128 – Remy moves near the reeds ~ 22 m north of the snags and parks there. Sinbad (still in the same location) dives and isn't resighted at this time (he has probably begun foraging).

1135 – Remy is still parked and preening in the same location as at 1128. Sinbad isn't seen at this time. Sparky and Phantom haven't been seen yet today. End observation period.

*1137 – Phantom is finally sighted—he is foraging in open water in the eastern half of the mid-section of the West Lagoon. Phantom resurfaces from a foraging dive with a large crayfish, quickly swallows it and resumes foraging.

*1138 – Phantom is still foraging in the eastern half of the middle third of the lagoon. While Phantom is on the surface Sinbad suddenly surfaces near Phantom and charges and pounces at the chick. Phantom immediately dives to flee from Sinbad, who promptly dives after him. Phantom resurfaces a fair distance away from Sinbad but Sinbad quickly relocates Phantom and again charges after the chick. Phantom again dives in an attempt to elude Sinbad and this scenario of Sinbad charging and chasing Phantom and Phantom diving to elude him repeats itself a number of times as Sinbad continues to pursue the chick for several minutes. In his flight from Sinbad, Phantom eventually goes to the edge of the west shore reeds well south of the snags, dives and isn't immediately resighted. Sinbad quickly makes his way to the edge of the reeds there and intently peers into them, apparently thinking that Phantom had gone into the reeds when he dove. However, it is soon apparent that Phantom has finally slipped away from Sinbad when the chick is sighted moving north along the reeds north of the snags. Phantom continues north and soon stops and parks near the reeds in the mid-north half of the lagoon. Sinbad continues to search for Phantom by the reeds in the south half of the lagoon where Phantom had dived before slipping away. He eventually ceases his searching, moves out ~ 8 m from the reeds and parks.

*1145 – Remy is parked and preening ~ 6 m north of Phantom (who is still parked near the west shore reeds in the mid-north half of the lagoon). Remy focuses on her preening and no interactions occur between the two. Remy soon begins foraging in the reeds there and Phantom appears to be monitoring her from a distance (~ 4 m out from the reeds) but also looks about, apparently keeping a lookout for Sinbad (who is now parked in the mid-water area in the south half of the lagoon). Remy clearly is just foraging for herself and no feedings occur. Phantom soon goes out to the eastern half of the north section of the lagoon and begins foraging there.

*Comments:* Once again, no feedings were observed and no chicks were seen soliciting or accompanying the parents. In fact, during the formal observation period no parent-chick interactions were seen. It wasn't until after the period had concluded that a parent-chick interaction was observed (and this involved Sinbad attempting to chase Phantom out of the area). For a period of approx. 2 minutes Phantom just happened to be in the general area where Remy was foraging in the reeds and it appeared that the chick monitored her briefly from a distance from outside the reeds. However, it was clear that Remy was just foraging for herself and there was no connection between the two. Basically, today's observations again just served to reinforce the fact that the chick rearing period for this brood has ended. Sinbad's extended chase of Phantom underscored the point (especially to Phantom) that the parents will no longer be attending to the chicks; that their "chickdom" has come to an end.

Quark was almost certainly seen in the SC today (note that the juvenile markings on this chick were quite faded; thus, I wasn't quite 100% certain that it was Quark but I have little doubt that it was). This marks the first time in four days that Quark was (likely) seen. Unfortunately, there is no way of knowing for certain whether Quark had been in the natal range all along but just went unseen during those three consecutive days (which is certainly probable given that the chicks are now independent) or had temporarily left the natal range and has now returned. I searched the surrounding areas for Quark on each of those three days but there are numerous portions of these areas that are hidden from view (including sizeable sections of the East Lagoon and the inter-lagoon channel). It is also quite possible that Quark just happened to be hidden in the reeds in a particular area during the period that I searched it (and this includes within the natal range).

There were no confirmed sightings of Sparky today. As usual, any serious considerations about whether or not Sparky has left the natal range must await further observations over the upcoming days (providing that Sparky's juvenile markings haven't faded to the point where I can no longer ID the chick!).

In regard to the brood as a whole, at this point it appears pretty certain that Cobra and Spot have left the natal range and Phantom (definitely) and Quark (almost certainly) are living independently within it. Because thus far it has only been a single day that Sparky hasn't been seen it is as yet unknown whether or not this chick has left the natal range.

For a second time an American coot was seen giving a light peck to the rump of one of the chicks. The general situation was the same as during the first incident (which occurred on 9/28): the chick was inadvertently blocking the coot's path. In this case, the chick that almost certainly was Quark was parked in the SW section of the SC in the path of a coot that was approaching from Quark's rear. When the coot reached Quark it paused at the chick's rear and gave a light peck to Quark's rump. Quark then quite casually slid over about a half meter and the coot just as casually went by and continued on its way. Here again, this incident was basically analogous to a situation in which a person taps someone on the shoulder who had inadvertently been blocking the way. Ironically, Quark was the chick who was involved in both of these incidents.

**<u>10/8/17</u>:** Time: 1040-1154.

1040 – As observations begin, none of the family members are seen at this time. A mature PB grebe chick that has lost its juvenile markings is foraging by the mid-south bank of the SC (this is definitely a stranger chick).

1045 – Still no family members have been seen. The stranger chick has moved away and is no longer seen.

1048 – I make a quick check from the west end of the RR bridge: Phantom is sighted moving north under the bridge and appears to be heading for the SC. Sinbad and Remy are parked in the north half of the West Lagoon. A mature stranger PB grebe chick is sighted by the east bank at the SE end of the river.

1055 – (Back at my SC observation point on the mid-south bank.) None of the family members are seen at this time.

1058 – A mature stranger PB grebe chick goes to the reeds at the south side of the NE point of the SC and begins foraging.

1100 – None of the family members are seen so I head for my West Lagoon observation point on the mid-east shore of the lagoon.

1109 – I arrive at my West Lagoon observation point. Sinbad and Remy are sighted near the west shore reeds in the northern third of the lagoon. Remy remains parked and preening while Sinbad goes into the reeds and begins foraging. None of the chicks are seen at this time.

1113 – Sinbad emerges from the reeds north of the snags and moves out to the mid-water area east of the snags where he parks and preens. Remy is still parked and preening in the same general area as at 1109.

1118 – Sinbad has drifted into the east half of the middle third of the lagoon and passes from my view. Remy has drifted to the mid-water area in the northern third of the lagoon.

1123 – Sinbad moves across the lagoon from the east shore area in the middle third of the lagoon to ~ 6 m from the reeds south of the snags where he parks and preens. Remy is still parked and preening at the north end of the lagoon.

1127 – Remy (still parked in the mid-water area at the north end of the lagoon) dives and isn't resighted at this time. Sinbad is still parked and preening in the same location as at 1123.

1130 – A paddleboarder moves south under the RR bridge and into the West Lagoon. Sinbad (still parked in his same location) quickly spots him, flees into the west shore reeds and hides there (he has moved out of my view).

1132 – The paddleboarder has progressed deep inside the lagoon. Sinbad remains hidden in the reeds. Remy hasn't been resighted since 1127. With no chance of seeing the grebes here while the paddleboarder is present I decide to head for my SC observation point. Sparky and Quark haven't been seen yet today.

1140 – I arrive at my SC observation point. Frustratingly, I find that the paddleboarder has moved north and is now in the river east of the SC. None of the family members are seen at this time—if any are present in the SC and surrounding area they would have been spooked into the reeds by the paddleboarder.

1142 – The paddleboarder enters the channel along the east bank of the river and leaves the area.

1145 – Remy has come out of the reeds and is parked and preening near the reeds in the NW corner of the SC.

1147 – Phantom is sighted parked and preening ~ 8 m from the reeds at the mid-west end of the SC. Remy is parked ~ 11 m north of Phantom but takes no apparent notice of the chick.

1151 – Phantom slowly moves over into the NW end of the SC ~ 2-3 m from the reeds. Remy is still parked and preening in the NW section. She appears to be quite focused on her preening and pays no noticeable attention to Phantom. However, Phantom appears to be keeping a wary eye on her (Remy is ~ 22 m away from Phantom).

1154 – Phantom goes into the reeds at the NW end of the SC and begins foraging. Remy is still parked and preening in the same location as at 1151. Sparky and Quark haven't been seen yet today. End observation period.

*1209 – Remy aggressively chases a mature stranger PB grebe chick out of the SC (Remy chased the chick from the SW section of the SC out into the mid-river area east of the NE point of the SC).

*1219 – Remy aggressively chases Quark out of the SC (Quark's juvenile markings are faint but are still just distinguishable enough to allow me to positively ID this chick). Quark had finally been sighted shortly before being chased away.

*Comments:* Today's formal observation period was quiet and (other than yet another intrusion by a paddleboarder) uneventful. However, two notable incidents occurred during my continued observations after the period had concluded. Both of these incidents involved Remy aggressively chasing a chick out of the SC. In the first of these she chased a mature stranger PB grebe chick (it has transitioned from its juvenile plumage) away and in the second she chased Quark out. It was somewhat surprising that she didn't also chase Phantom away during the final segment of the observation period. Though Phantom wasn't close to her, he was in her general area. At that time Remy was very focused on her preening and paid no noticeable attention to Phantom. Nevertheless, there seems little doubt that she was at least aware of Phantom's presence. Yet, for some reason she chose not to charge at and chase Phantom as she typically has been doing. In contrast, Phantom kept an eye on Remy and clearly was wary of a possible charge by her.

The observations today once again underscored the fact that the chick rearing period for this brood has concluded. No feedings were observed, no solicitations by the chicks were seen and no chicks were seen accompanying or looking to a parent for feedings. Most revealing of all is that the only parent-chick interaction observed was Remy's aggressive chase of Quark.

Quark was seen in the SC again today as was Phantom. It is thus known that, though living independently, both continue to reside in the natal range. Sparky wasn't seen for a second consecutive day. This, of course, calls into question whether Sparky has left the natal range. As usual, serious consideration regarding this question must await further observations over the upcoming days.

**10/9/17:** Time: 1052-1155.

1052 – As observations begin, a chick is briefly sighted in the SE section of the SC—I was unable to identify or determine if it was one of the family members before it dived and was lost from view. There are no confirmed sightings of any of the family at this time.

1053 – Remy is sighted entering the SC from the SE. She proceeds into the mid-SC area, dives and isn't resighted at this time.

1054 – Quark is sighted in the SW corner moving north along the reeds at the west end. After passing north of the nest Quark dives into the reeds and goes out of view.

1057 – Remy exits the reeds at the mid-west end of the SC, moves out to the mid-river area east of the SC and passes beyond my view.

1100 – None of the family members are seen at this time.

1101 – Sinbad enters the SC from the SE and goes to the reeds along the north bank in the NW section. Sinbad then dives into the reeds and begins foraging.

1105 – Quark exits the reeds at the mid-west end of the SC, goes to a point ~ 6 m south of the old nest site and parks and preens.

1108 – Sinbad is sighted intently moving east along the south bank of the SC and soon dives. Shortly after, he is suddenly seen chasing a chick out of the SE section. The chick breaks into a short, low flight over the surface while fleeing into the mid-river area east of the SC. (When first sighted prior to this chase Sinbad apparently had been stalking the chick.) I wasn't able to ID the chick—its juvenile markings were extremely faded but not completely, so it is possible that it could have been a chick from the SR brood (but definitely not Quark who is still parked south of the old nest site).

1109 – Quark is still parked in the same location as at 1105.

1113 – Quark goes into the NW corner of the SC and moves out of my view.

1118 – Brief check from the west end of the RR bridge: Sinbad is sighted near the east shore in the middle third of the West Lagoon. Remy is near the reeds north of the snags.

1125 – (Back at my SC observation point on the mid-south bank.) A mature stranger PB grebe chick is sighted entering the SC along the reeds at the south side of the NE point where it dives and isn't resighted at this time.

1130 – The stranger chick is resighted foraging in the NW corner of the SC. Sparky and Phantom haven't been seen yet today.

1135 – None of the family members are seen at this time. I leave the SC and head for my West Lagoon observation point on the mid-east shore of the lagoon.

1144 – I arrive at my West Lagoon observation point. Sinbad is parked near the reeds north of the snags. He is peering into the reeds as though looking for someone. None of the others are seen at this time.

1147 – Sinbad is now parked near the west shore reeds in the mid-north half of the lagoon. He moves east a few meters, calling once as he does.

1148 – Sinbad moves east to the mid-water area in the mid-north half of the lagoon where Remy is now foraging and parks there.

1150 – Remy is still foraging in the mid-water area, now in the middle third of the lagoon. Sinbad goes near the west shore reeds in the mid-north half of the lagoon, parks there (often peering into the reeds) but is soon lost from sight.

1155 – Remy (still foraging) works her way over near the reeds north of the snags and continues her foraging there. Sinbad isn't seen at this time. Sparky and Phantom haven't been seen yet today. End observation period.

*Comments:* Today's observation period was typical of what has been seen in all of the periods since the chick rearing period ended (10/5). No feedings were observed nor were there even any parent-chick interactions other than a possible incident (at 1108) of a parent chasing a chick out of the SC. The parents have finished their work rearing the chicks and are now back to just taking care of themselves. They also continue to patrol their territory (Sinbad and Remy both went back and forth between the West Lagoon and the SC during the observation period) and attempt to exclude chicks (their own and strangers) by aggressively chasing them away (though the chicks have nevertheless been persistent).

Quark was again seen in the SC while Phantom wasn't seen today. Also, for a third consecutive day Sparky wasn't seen, thus increasing the possibility that Sparky has left the natal area. Unfortunately, it is quite likely that I won't be able to determine when (or if?) these three chicks leave the natal area. This is because it has about reached the point now where their juvenile markings are so faint (as they transition from their juvenile plumage) that I will no longer be able to identify the individual chicks or even distinguish them from other mature chicks in the area.

The black throat patch on Remy is fading fairly quickly as is her dark beak ring. Sinbad's black throat patch has diminished a little but is still very prominent as is his dark beak ring.

A brood of 5 young, recently mobile chicks was seen in the northern third of the East Lagoon and the east end of the inter-lagoon channel. The three chicks from the East Lagoon third brood are still being reared in the southern two thirds of the East Lagoon (developmentally, they are well ahead of the chicks in this newly discovered brood). It is thus evident that the East Lagoon is now split into two territories. This was pretty late in the season to have produced a brood, perhaps as late as any pair should dare to produce in this area. It is again interesting to note how widely separated in their developmental stages these two adjacent broods are. One is thus compelled to again consider the question of whether this wide gapping between the broods of adjacent pairs is deliberate (as discussed in the 8/31 notes)?

Given that the chick rearing period has ended for the SR brood and that the chicks are independent, and that identification of the individual chicks very soon won't be possible I will end my schedule of daily observations tomorrow. At that time I will conclude this study and continue with biweekly follow-up observations for a short period.

**10/10/17:** Time: 1055-1159.

1055 – As observations begin (from the west end of the RR bridge), Sinbad is sighted in the mid-water area east of the snags in the West Lagoon. Remy is parked ~ 1.5 m from the reeds in the SW corner of the SC. A mature stranger PB grebe chick is parked in the NW corner.

1059 – (From my SC observation point on the mid-south bank.) Remy (still parked in the same location) dives and isn't resighted at this time.

1104 – A stranger PB grebe (either an adult in non-breeding plumage or a mature chick that has transitioned from its juvenile plumage) enters the SC from the SE end and moves west along the south bank in the SE section. It soon dives and isn't resighted at this time. Remy emerges from the reeds in the SW corner and makes her way out to the SE end. While there she suddenly charges off to the east (promptly breaking into a low, skipping flight over the surface) out into mid-river and rapidly passes beyond my view. The manner in which she did this leaves little doubt that she was chasing a PB grebe away. However, the subject of her chase was out of my view right from the start and thus couldn't be identified (it is quite probable that it was the stranger grebe whom I had just seen entering into the SC).

1110 – One of the SR chicks enters the SC from the east (its juvenile markings are so faded that I can't be certain of its identity though it appears to be Quark). While Quark (?) is still passing west through the east end of the SC Remy suddenly surfaces ~ 1 m behind and aggressively charges at and chases Quark (?). The chick immediately turns east and flees toward mid-river, east of the SC. While fleeing, Quark (?) promptly breaks into a low flight, as does Remy in pursuit. Quark (?) lands in the mid-river area a short distance ahead of where Remy lands. Remy then charges at and chases the chick for a second time. Quark (?) again breaks into a low flight while fleeing (this time heading south) with Remy flying after the chick in pursuit. When they land in the mid-river area east of the SC with Remy a short distance behind, she charges at and chases Quark (?) for a third time. This time Quark (?) initially heads south but in a fairly long, low flight circles west and goes into the SC again. At this point Remy abruptly ceases the chase, landing and remaining in mid-river east of the SC amid several dozen American coots. When Quark (?) lands in the SC (now well away from Remy) the chick paddles at a good pace straight to the reeds at the NW end of the SC, enters into the reeds and goes out of view.

1115 – None of the family members are seen at this time.

1120 – None of the family members are seen at this time.

1123 – Quark (?) is briefly sighted foraging in the NW corner. A stranger PB grebe (it appears to be an adult in non-breeding plumage) is sighted foraging in the SW section of the SC.

1126 – The stranger grebe has worked its way east into the SE section. It continues east and soon passes beyond my view.

1130 – None of the family members are seen in the SC at this time so I head for my West Lagoon observation point on the mid-east shore of the lagoon.

1133 – On a quick check from the west end of the RR bridge while en route to my West Lagoon observation point Remy is sighted moving south under the bridge into the West Lagoon.

1143 – I arrive at my West Lagoon observation point. Sinbad is sighted parked ~ 9 m from the west shore reeds in the mid-north half of the lagoon. Remy isn't seen at this time nor are any of the chicks.

1145 – Remy is sighted foraging near the west shore at the north end of the lagoon, just south of the RR bridge.

1147 – Sinbad (still parked and preening) has drifted to the mid-water area in the mid-north half of the lagoon.

1153 – Sinbad moves west to the reeds by the snags, dives into the reeds and goes out of view.

1155 – None of the family members are seen at this time.

1158 – Remy is sighted parked and preening ~ 10 m from the west shore reeds in the mid-north half of the lagoon (she has probably just concluded a foraging session).

1159 – Remy is parked and preening in the same location. Sinbad exits the reeds by the snags and moves out ~ 35 m east of the snags where he parks and preens. No chicks are seen at this time and probably only Quark has been seen today. End observation period.

*Comments:* Once again, the only parent-chick interaction that was seen involved a parent aggressively chasing a chick away. Only one such incident was seen today and it involved Remy chasing a chick (who appeared to be Quark) out of the SC. It was an extended chase that covered a fair distance. As is typical, the chick was persistent and ultimately Quark (?) ended up in the SC where the chick had wanted to go in the first place.

The juvenile markings on the chicks are now so nearly completely faded that I can no longer identify the individual chicks with certainty. The chick who I saw today probably was Quark but I couldn't be certain of this. The situation has now reached the point where if I should sight a mature chick that is in a similar stage as the SR chicks I won't be able to tell whether it is a SR chick or a stranger. This being the case and given the fact that all of the chicks are now fully independent, **daily observations will cease as of today and this case study has now concluded.** However, I will continue for a short while with biweekly informal follow-up observations.

As this study concludes, the post-chick rearing dispersal story of the individual chicks as of today is as follows. It appears a virtual certainty that Cobra and Spot have left the natal range and that Quark is fully independent but currently still residing in the natal range. The cases for Sparky and Phantom are uncertain at this time. It has now been four consecutive days since there was a confirmed sighting of Sparky and two consecutive days for Phantom. Thus, all that can be stated with certainty regarding these two is that as of 10/6 Sparky was fully independent but still residing in the natal range and that as of 10/8 the same can be stated for Phantom. Regardless of the current uncertainty about whether or not these two have left the natal range (the case for Sparky having done so seems pretty strong but is less so for Phantom), the most significant point to note is that Remy and Sinbad did an amazing job and successfully reared all five of the chicks from their brood to independence.

Over the entirety of this study a tally was kept of the general type of prey item that the parents fed to the chicks in each of the feedings that were observed during the course of each formal observation period (note again that observation periods typically were from one hour to one hour and 15 minutes in length, with some lasting longer). Only feedings in which the chick(s) was actually seen eating the proffered prey item (and in cases in which this wasn't actually seen but there was no doubt that it was eaten by a chick) were included in this tally. Other than one possible exception, the feedings consisted exclusively of either crayfish or fish (in the one possible exception it appeared that the prey item was a dragonfly but my view was too inadequate to be 100% certain of this). Overall, crayfish was the predominant food item that was fed to the chicks. Of the total of 186 feedings that were observed during the formal observation periods, crayfish were fed in 128 (68.8%) of these while only 58 (31.2%) of the feedings consisted of fish. However, this predominance of crayfish feedings wasn't always the case during the course of the chick rearing period. Interestingly, over the first few weeks after hatching, fish were fed overwhelmingly more often than crayfish. During the period through 8/8/17 (recall that hatching began on 7/21 or 7/22 and concluded on 7/26 or 7/27) a total of 54 feedings were observed. Fish feedings comprised a whopping 41 (75.9%) of these with only 13 (24.1%) of them consisting of crayfish. On 8/9/17 the shift toward the overwhelming predominance of crayfish feedings began. Of the 132 feedings observed during the period from 8/9 through the end of the study an overwhelming total of 115 (87.1%) of these consisted of crayfish while fish feedings comprised only 17 (12.9%) of them. It is interesting to note that the beginning of this shift on 8/9 coincided with the day that the chicks were first observed going over to the West Lagoon (this represented a major expansion in the chicks' mobility).

## Post Chick Rearing Follow-Up Notes and Observations

**10/12/17:** Time: 1100-1200.

Only one chick was seen that definitely was one of the SR brood. The juvenile markings on this chick were extremely faded and nearly completely gone. I am fairly certain that this chick was Quark but not 100% so. Quark (?) was sighted in the west end of the SC, sometimes parked near the reeds and sometimes within the reeds foraging or resting. Neither of the parents was seen in the SC today: Sinbad and Remy spent the entire observation period in the West Lagoon. Sinbad was heard calling once.

A few other mature chicks were seen in the SC and surrounding area. Because the juvenile markings have nearly completely faded on the SR chicks (in other words, their transition from their juvenile plumage is nearly complete) and I can barely identify the individual chicks (if at all) it's possible that some of the mature chicks seen today could have been SR chicks. It appears that from this point on I will probably no longer be able to distinguish the SR chicks from other mature stranger chicks.

**10/14/17:** Time: 1115-1203.

The parents spent nearly all of the observation period in the West Lagoon. Remy was seen only in the West Lagoon. Sinbad left the West Lagoon for a short trip to the SC and returned soon afterward.

One definite mature chick and an individual who was either an adult in non-breeding plumage or a mature chick were seen in the west end of the SC (this was at a time when Remy and Sinbad were both in the West Lagoon). The latter seemed too far along to be one of the SR chicks.

The former could possibly have been one but with its juvenile markings being completely gone I couldn't identify it one way or the other (though I suspect that it may have been Quark). Unfortunately, the SR chicks are now at the point where they have transitioned from their juvenile plumage so their juvenile markings are completely gone. Thus, I can no longer identify the individual chicks nor can I even distinguish them from mature stranger chicks. Therefore, from now on when I see a mature chick in the natal range the best that I'll be able to do is to state whether it is possible that the chick could be one of the SR brood (i.e., I won't be able to confirm whether or not it is).

On two separate occasions the two unidentified grebes noted above came together briefly. As they neared each other there was some brief tenseness but this quickly subsided and there was no aggression between them. In fact, they behaved as though there was familiarity between them. One is then left to wonder if perhaps they are sibs (perhaps even SR sibs)? Or maybe a newly forming pair (the one that may have been Quark appeared to be a male and the other was a female)?

**__10/17/17__:** Time: 1112-1200.

The many open water algal mats that have been building up in the river, the SC and the West Lagoon have been cleared away. It is most likely that the strong winds that the area has been experiencing are responsible for clearing them away.

Lately during my observations Remy and Sinbad have been spending the great majority of their time in the West Lagoon (today they spent the entire observation period there).

No mature chicks were seen today in the SR pair's territory or the surrounding areas. As always, further observations over the upcoming days will be required in order to determine if the mature chicks who have been residing in this area have left. Several mature chicks were seen in the south end of the East Lagoon.

The black throat patch on Remy is nearly gone. Her dark bill ring is also fading but is still quite conspicuous. Sinbad's black throat patch and bill ring are still prominent though they have diminished some.

All three of the East Lagoon third brood chicks were seen in the middle segment of the East Lagoon. All appear to be progressing well toward independence.

**__10/20/17__:** Time: 1214-1332.

Sinbad and Remy were both seen today though neither was in view for much of the time during the observation period (Remy wasn't sighted until the end of the period). All sightings of them were in the West Lagoon—neither was seen in the SC or north of the RR bridge.

Only one possible mature chick was seen in the area today. This individual was either a very mature chick or an adult in nonbreeding plumage. It has now reached the point where I can't reliably distinguish a very mature chick from an adult in nonbreeding colors as even the beaks of these mature chicks have achieved adult size, proportions and color or nearly so. The individual noted above was seen in the north half of the SC, notably in the NE section south of the NE point. With the juvenile markings having completely faded (being replaced by adult nonbreeding colors) on the SR brood I can no longer identify the individual chicks or even distinguish them from mature stranger chicks. Thus, realistically all that I can state regarding this individual is that it could be one of the SR chicks but I can't really say one way or the other. At this time it appears that if any of the SR chicks are still residing in the natal range then probably no more than one is doing so. With Remy and Sinbad apparently ranging mainly (but not exclusively) within the West Lagoon lately (at least, this is the indication based on what has been seen during the observation periods), this means that for most of the time other grebes (including any remaining SR chicks) can spend much time in the SC during which there is no threat of being chased out by Sinbad or Remy.

**10/23/17:** Time: 1105-1210; 1225-1235.

As has been the case lately, Sinbad and Remy were seen only in the West Lagoon during today's observations. Remy's black throat patch is nearly completely gone. Her bill ring has also faded considerably but is still noticeable. Sinbad's black throat patch and bill ring have diminished some but are still very prominent.

At least four PB grebes were seen in or near the SR pair's territory. These individuals were either adults with nonbreeding plumage or mature chicks who have developed the plumage and beaks of adults in nonbreeding plumage. At this point such chicks are difficult if not impossible to distinguish visually. Two of these four were seen in the SC. One of the latter had been in the SW section of the SC when the second one entered the SC from the east ~ 25 minutes later. As the second grebe made its way west through the mid-SC, the first one spotted it and intently moved directly toward it in typical PB grebe threat posture: keeping a low, flattened body posture with the head/neck held low and flat with the neck drawn tight and compact to the shoulders and the beak pointing straight forward. There was momentary tenseness as they drew near but this instantly vanished when they met en route and greeted, briefly touching breast to breast. It initially appeared as though the first grebe (the one that was approaching in a threatening manner) might charge and attack the other grebe but no aggression occurred and they behaved as though they recognized and were familiar with each other. [I have seen similar such approaches a number of times involving Remy and Sinbad. In these cases it appeared that the individual who was approaching in the threatening posture was initially unsure as to whether the grebe it had sighted was its mate and was thus poised to drive it away if not (the aggressive threat posture while approaching being meant to intimidate the possible intruder). In each case, the individual ultimately recognized that the targeted grebe was its mate and a typical amicable greeting then ensued. While engaging in parental rebuffs Remy and Sinbad were sometimes seen approaching a chick in this threatening posture. In these instances the incident ended with the parent charging at and chasing the chick.] After this very brief meeting they each went their separate ways within the SC. About 5 minutes later these two had a second, similar coming together with the same individual doing the threatening approach and with the same nonaggressive result. Unfortunately, at this point I could no longer tell if either or both of these two were SR chicks or even be certain as to whether they were mature, transitioned chicks or adults in nonbreeding plumage (though they appear to be mature chicks). Regardless, it was intriguing that when the two met they behaved as though they were familiar with and recognized each other (the absence of aggression despite the threatening posture on one of them as it approached was particularly noteworthy). One of these individuals is a male and the other a female. This raises an obvious question if these two are mature chicks as appears to be the case: Is this a newly formed pair (or one in the process of forming)? And if so, with Sinbad and Remy now seldom if ever returning to the SC are they attempting to stake out a territory here? Adding further intrigue is my suspicion that the male could possibly be Quark.

**10/26/17:** 1050-1150.

Once again, the only confirmed sightings of Remy and Sinbad were in the West Lagoon. At one point during the observation period Remy was seen catching and quickly swallowing a crayfish. Her black throat patch is nearly completely gone with only some very faded dark splotches remaining. Remy's bill ring has also faded considerably but is still conspicuous and noticeable even from a distance. Sinbad's black throat patch and bill ring are still prominent.

During the observation period a pair of PB grebes was seen associating with each other. For the most part, they benefitted from the absence of Remy and Sinbad from the northern half of their territory (i.e., the SC and the section of the river east of the SC). At one point this pair was parked together in the mid-river area just east of the SC (they also went deep into the SC during the observation period). Both individuals were in adult nonbreeding plumage and one was a male and the other a female. If these two were definitely known to be adults it would then be obvious that they are a mated pair. Unfortunately, at this time of year things aren't so simple. The mature chicks from this season have long since transitioned from their juvenile plumage and have progressed to the point where they now sport adult nonbreeding plumage. Thus, in regard to plumage they are now virtually indistinguishable to me from adults who are currently in their nonbreeding plumage. The beaks of the mature chicks had lagged behind in achieving fully adult size, proportions and color and I have mainly been using this feature to distinguish the chicks from the adults. However, the beaks on the very mature chicks have at this stage essentially attained adult size and proportions. I am thus no longer able to reliably use this feature as a means of distinguishing the mature chicks from adults (if I had more experience with this species I would perhaps have discovered other features that I could reliably key on).

In general then, I can no longer reliably distinguish these mature chicks from adults in nonbreeding plumage. Lacking this ability means that there are various plausible possibilities regarding who this pair of associating individuals are that I'm unable to choose between or rule out with certainty. As noted above, if they are adults then their amicable, associative behavior would indicate that they are a mated pair. If they are mature chicks then there are two possibilities that immediately come to mind. They could be a newly formed pair or working through the process of perhaps becoming a pair (note here that I am not familiar with and know nothing about the pair formation process of PB grebes). It is perhaps also possible that they could be familiar members of the same brood who, though independent, are continuing to associate with each other. If the latter was the case then it could even be possible that they are two of the chicks from the SR brood (recall that with their juvenile markings now gone I am no longer able to ID and distinguish the SR chicks). Based on what I have observed I think it is most likely that these two are mature chicks and are a newly formed pair (or are in the process of doing so), and that it is quite likely that they are the same individuals whom I saw in the SC on 10/23.

**10/30/17:** Time: 1100-1150.

As has become typical over the past weeks (and essentially, since the time that the chick rearing period concluded for the SR brood) Remy and Sinbad were seen only in the West Lagoon during observations. Remy's throat patch appears to be completely gone other than some faint blotching. Her bill ring is quite faded but is still visible (though not very clearly) and her white eye rings have narrowed though they are still present. Sinbad's black throat patch and bill ring are still prominent.

At least five other PB grebes were seen in the SR pair's territory and the surrounding area. This included sightings in the SC and the section of the river east of the SC, the NE end of the West Lagoon, and in the NR pair's territory. Among those seen was an associating pair. Though when sighted they were too distant for me to be certain that they were the pair sighted on 10/26, they did appear to be. Once again, at this stage I was unable to reliably determine whether the various individuals who were seen were mature chicks or adults in nonbreeding plumage.

**This concludes regular "official" observations for this study.**

# South River Family Album

**Figure 1.** The arrow points to the location within the reeds along the north bank of the Semicircle where the well-hidden first nest of the South River pair lies. This nest and the clutch of eggs within it was later lost when there was a sudden dramatic rise in the water level that the floating nest was unable to cope with (photo was taken on 6/14/17)

**Figure 2.** The arrow points to the location of the pair's new (second) nest tucked within the reeds along the mid-west bank of the Semicircle. Remy is sitting on the nest incubating the clutch—she and the nest are barely visible through the surrounding reeds (photo was taken on 7/7/17).

**Figure 3.** Remy works on swallowing a large fish that she had captured within the Semicircle. The Semicircle is the core area of the South River pair's territory and, appropriately, is where they nested.

**Figure 4.** An intense territorial challenge by the North River pair erupted on 7/23/17. Two prolonged savage fights were observed on this day when the North River male made deep incursions into the Semicircle and engaged members of the South River pair. Here, in the second of these fights, Sinbad and the North River male are engaged in fierce battle as Remy (left) remains close and offers moral support to Sinbad as well as delivering an occasional firm peck or two at the North River male.

**Figure 5.** Sinbad (nearest) and the North River male engage in the first of the 8 observed territorial standoffs (this one occurred on 7/29/17) along the newly established territorial boundary. The North River female is seen in the distance (only the males participated in the standoffs).

**Figure 6.** The first observed territorial standoff (as depicted in Fig. 5) was interspersed with brief bouts of fighting, the last of which is shown here.

**Figure 7.** Sinbad. In comparing this photo of Sinbad with that of Remy in Fig. 44 the apparent differing gender related features noted in the 6/24/17 notes are readily discernable.

**Figure 8.** Remy (left) and all five of the chicks. The chicks have only recently become mobile. Note the considerable size disparity among the brood. (Photo taken on 8/6/17)

**Figure 9.** Sinbad (nearest) and the North River male engaged in another territorial standoff along the territorial boundary. (Photo taken on 8/6/17)

**Figure 10.** A feeding: Sinbad hands a fish off to Sparky. (Photo taken on 8/7/17)

**Figure 11.** The 5 chicks of the South River family: Back-Sparky; middle row (left to right)-Spot and Cobra; front row (left to right)-Quark and Phantom. (Photos in Figs. 11-25 taken on 8/8/17)

**Figure 12.** Back row (left to right)-Spot and Cobra; front row (left to right)-Quark and Phantom.

**Figure 13.** Left to right: Cobra (swallowing a crayfish she had obtained from a parent), Quark and Sparky.

**Figure 14.** Left to right: Cobra, Spot, Quark, Sparky and Phantom.

**Figure 15.** Left to right: Phantom, Quark and Sparky.

**Figure 16.** Back to front: Quark, Sparky and Phantom.

**Figure 17.** Left to right: Quark, Sparky and Phantom.

**Figure 18.** Quark (left) and Phantom with Remy.

**Figure 19.** Sparky works on swallowing a fish that Sinbad has just handed off to the chick.

**Figure 20.** Sparky (left) and Phantom.

**Figure 21.** Left to right: Sparky, Quark and Spot.

**Figure 22.** Spot.

**Figure 23.** From left to right: Spot, Phantom, Sparky and Quark (facing to the right).

**Figure 24.** From left to right: Spot (nearest), Phantom, Sparky and Quark. Note the differing patterns of the markings on the back of the head and neck of the chicks. This is one of the features that was used to identify the individual chicks.

**Figure 25.** Front-Quark; back (left to right)-Spot and Cobra.

**Figure 26.** Remy with all 5 of the chicks. The close-up photos of the chicks from 8/8 make the chicks appear larger than they actually were at this time. Here, Remy serves as a convenient scale to illustrate how small the chicks still were at this point. Right flank-Cobra; center (front to back)-Sparky, Spot, Phantom, Remy; left flank-Quark. (Photo taken on 8/10/17)

**Figure 27.** Quark soliciting Sinbad for a feeding. (Photo taken on 8/18/17)

**Figure 28.** Quark with Sinbad. (Photo taken on 8/18/17)

**Figure 29.** Sparky regurgitating crayfish exoskeleton. The grebes were commonly seen doing this, indicating that ingested crayfish exoskeletons didn't pass all the way through the digestive tract. (Photo taken on 8/20/17)

**Figure 30.** Sparky. Note that the white markings on the top of the chick's head are fading away (compare this with the photos of Sparky taken on 8/8/17). (Photo taken on 8/20/17)

**Figure 31.** Phantom accompanying Sinbad. (Photo taken on 8/29/17)

**Figure 32.** Cobra (front) ranging in proximity to two American coots. (Photo taken on 9/2/17)

**Figure 33.** Cobra. (Photo taken on 9/2/17)

**Figure 34.** Sparky. (Photo taken on 9/2/17)

**Figure 35.** Sparky (left) and Spot. Note that by this time the white markings on the top of the chicks' heads are gone. (Photo taken on 9/2/17)

**Figure 36.** Spot. A large fish (probably a species of carp) swimming just below the surface to Spot's right is barely visible in the upper portion of the photo. (Photo taken on 9/2/17)

**Figure 37.** An osprey who has perched on a snag on the east shore of the river east of the Semicircle (well out of the picture to the right) has suddenly begun calling loudly and continuously and this has drawn the focused attention of Sinbad and the 4 chicks in his vicinity. Back row (left to right)-Spot, Phantom, Sparky; Front row-Quark (left), Sinbad. (From 9/3/17)

**Figure 38.** Back-Cobra accompanying Remy; Front-Sinbad. (Photo taken on 9/5/17)

**Figure 39.** Sparky attempts to swallow a crayfish that the chick had obtained from a parent but is unable to do so at this point because it is still too large and bulky. Phantom had been pursuing Sparky but briefly ceased before rejoining his pursuit (see Figs. 40 and 41).

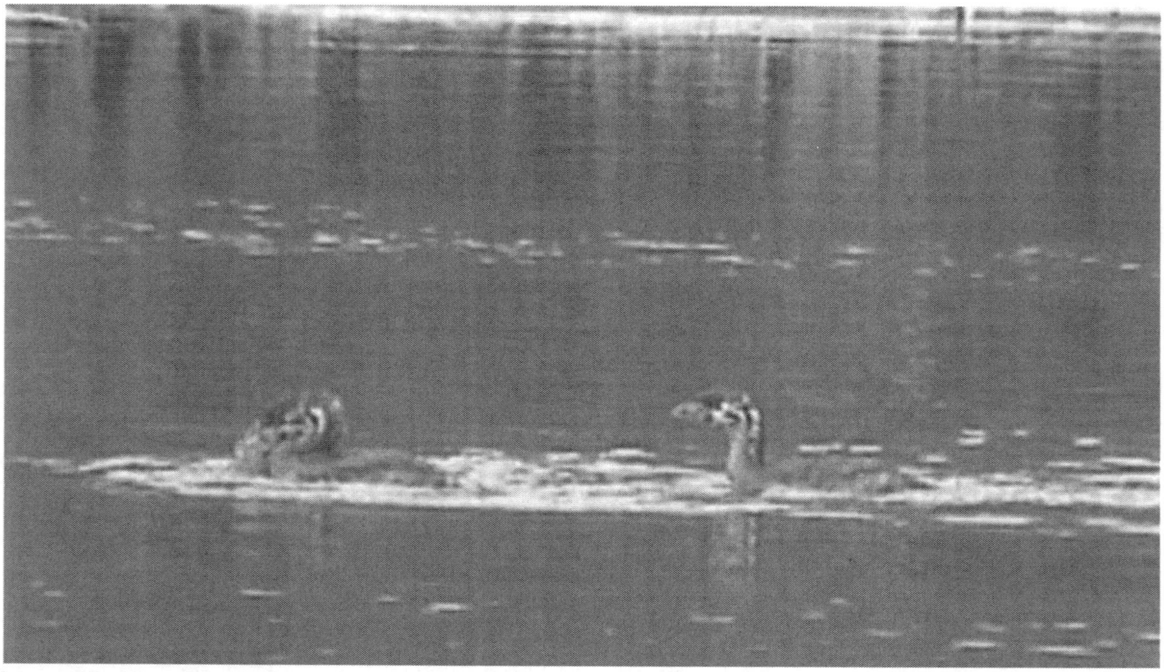

**Figure 40.** (Continued from Fig. 39) Having been unable to swallow the crayfish Sparky continues to work it while again being hotly pursued by Phantom.

**Figure 41.** (Continued from Figs. 39 and 40) Sparky continues to work the crayfish on the move as Phantom continues his pursuit. By continuing to work the crayfish Sparky eventually reduced its bulk to the point where the chick was then able to swallow it. Phantom succeeded in obtaining some scraps from the crayfish while Sparky was working it. (Figs. 39- 41 from 9/6/17)

**Fig. 42.** Sparky stretching out a wing and leg. As can be seen here Sparky's flight feathers are well developed and the chick is almost certainly capable of at least short flights at this point (two of the smaller chicks were seen taking short flights the previous day). In fact, Sparky was observed taking two short flights the following day. (Photo taken on 9/6/17)

**Figure 43.** Cobra (left) and Spot. (Photo taken on 9/8/17)

**Figure 44.** Cobra (right) with Remy. Over the course of the chick rearing period a particular affinity developed between these two and it became common to see them together. (Photo taken on 9/10/17)

**Figure 45.** Quark (left) with Sinbad. (Photo taken on 9/12/17)

**Figure 46.** Sparky regurgitating a clump of crayfish exoskeleton. (Photo taken on 9/12/17)

**Figure 47.** Sparky. By this point the black juvenile markings on the sides of Sparky's face and neck have transitioned to brown (Photo taken on 9/12/17)

**Figure 48.** Quark. (Photo taken on 9/14/17)

**Figure 49.** Sinbad. (Photo taken on 9/14/17)

**Figure 50.** Cobra (left) solicits Remy from up close. In contrast, by this point Sinbad no longer allowed Cobra to solicit him from up close or to closely accompany or sit close by him. Three American coots can be seen in the background. (Photo taken on 9/17/17)

**Figure 51.** Sparky working a crayfish seconds before swallowing it. (Photo taken on 9/17/17)

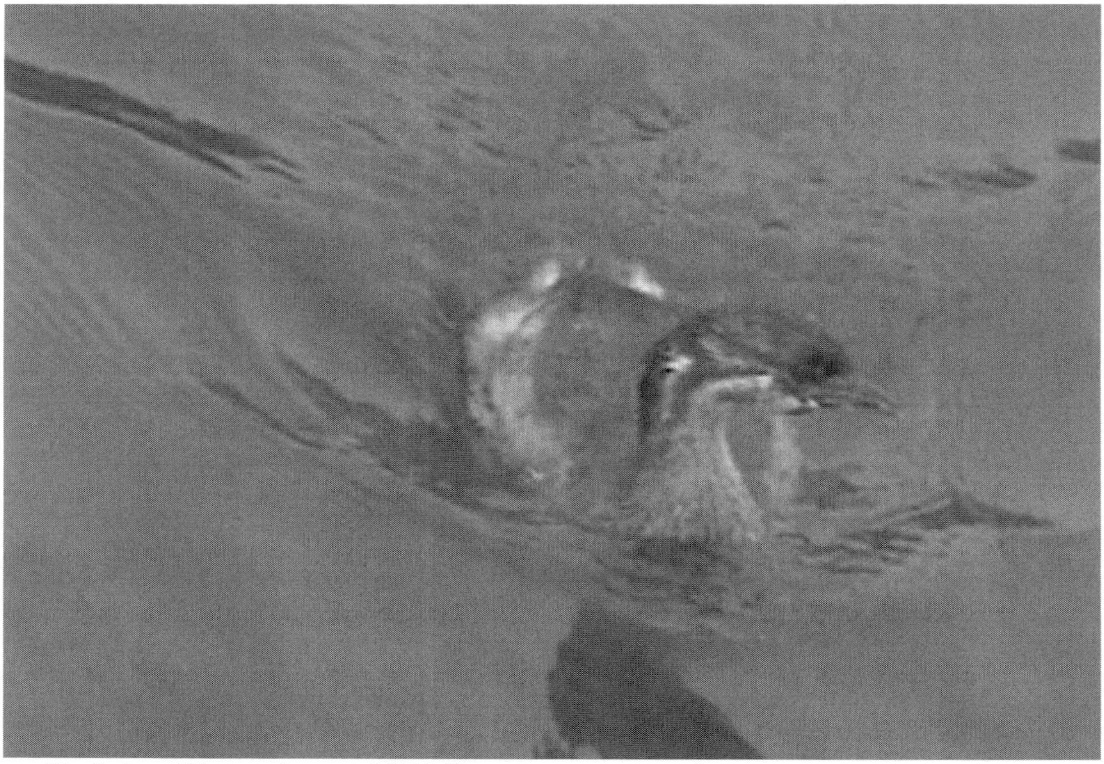

**Figure 52.** Cobra. (Photo taken on 9/18/17)

**Figure 53.** This photo of Cobra was taken on 9/18/17. Several days later (on 9/20 or 9/21) Cobra would begin her fully independent life when she became the first of the brood to leave the natal area for good (she was not seen again thereafter).

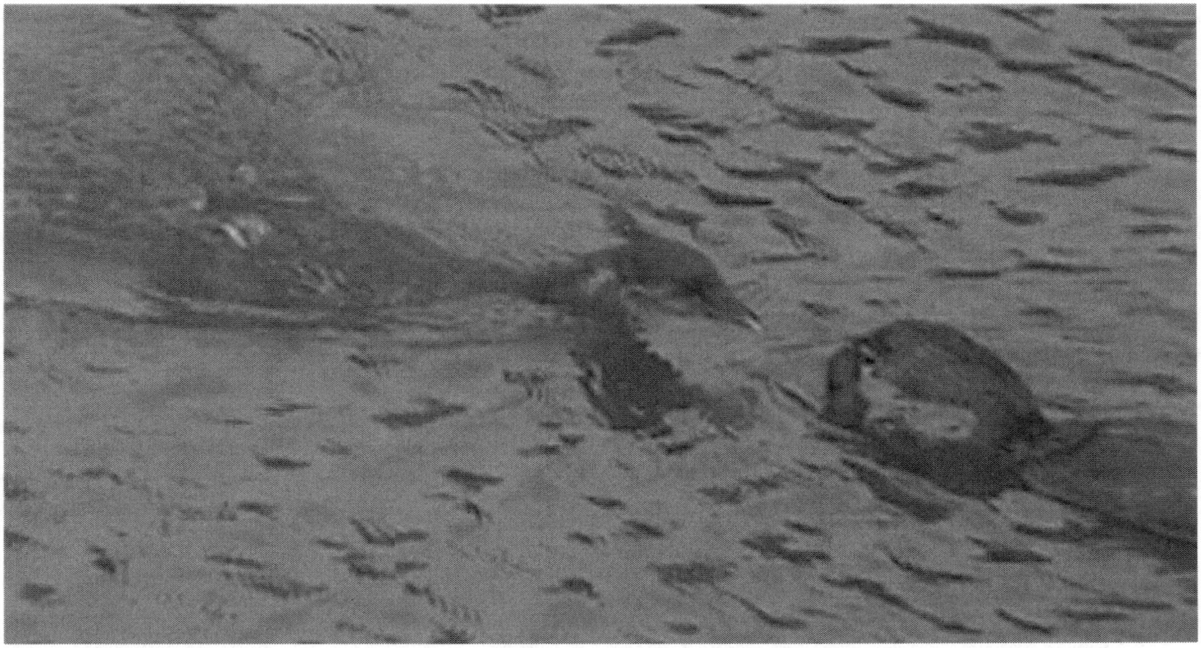

**Figure 54.** Figs. 54-58 comprise a series of photos (from 9/18/17) depicting a feeding in which Sparky took a crayfish from Remy but was then aggressively rebuffed by her in a persistent, extended chase. Here, Sparky rushes to Remy who is offering up the crayfish to the chick.

**Figure 55.** Sparky (left) takes the crayfish from Remy and begins moving away with it.

**Figure 56.** As Sparky moves away with the crayfish Remy suddenly charges at the chick from behind.

**Figure 57.** Remy aggressively pounces on the fleeing Sparky (still holding the crayfish) who is preparing to dive in an attempt to escape Remy's wrath.

**Figure 58.** When Sparky dove (see Fig. 57) Remy dived at the chick's heels and continued her chase underwater. Above: the pair has resurfaced and a persistent Remy continues her aggressive chase. Sparky (still holding onto the crayfish) is about to dive a second time in an attempt to elude her. Remy again dove after the chick but this time ceased her chase after resurfacing. Sparky was then finally able to work and swallow the crayfish.

**Figure 59.** A feeding: Phantom works on a crayfish that the chick has just obtained from Sinbad; Sinbad follows the chick and oversees the feeding. (Photo was taken on 9/20/17)

**Figure 60.** Sparky. (Photo taken on 9/22/17

**Figure 61.** Sparky stretching a wing and leg. Note the well-developed flight feathers. (Photo taken on 9/22/17)

**Figure 62.** On 9/25/17 Phantom was observed taking two long, low back to back flights. Above: Phantom on the first of his long flights.

**Figure 63.** Phantom on the first of his long flights.

**Figure 64.** Phantom on the first of his long flights.

**Figure 65.** Sparky. Note how the dark juvenile markings on the side of the chick's face and neck have begun fading away at this point. (Photos taken on 9/26/17)

**Figure 66.** Quark (left) with Remy. By this point Quark had surpassed Remy in size. (Photo taken on 9/30/17)

# *Chapter 5*

## General Summary for the South River Family Study

The nesting and chick rearing periods for the SR pair and their brood are described in considerable detail in the above field notes and comments. These observations have been presented in chronological order, as they were recorded, so that readers can follow the processes and events as they unfolded (and thus adding a further layer of detail; this also allows the readers to reflect on what has occurred up to each point in time and to speculate on what may yet transpire and in what direction things might be headed as though they themselves were there conducting the ongoing observations). All of this occurred over a lengthy, eventful four month period. In order to pull it all together in a more concise form a summary of these processes will be presented here. The summary will be fairly comprehensive. However, as is the nature of summaries, it generally will not be as detailed as the field notes so the reader is referred to the latter for more detailed accounts of the various topics discussed here. In order to eliminate some of the repetition the reader will specifically be referred to the field notes for various topics. It should be appreciated that due to its nature as a case study comprised of daily, continuous, ad lib observations of this grebe family this study does more than just provide a description of the nesting and chick rearing process. It also serves as a narrative of the lives of the seven individuals who make up this family; in other words, it tells the story of the lives of these seven individuals over the course of this four month period.

In retrospect, Remy and Sinbad did an amazing job in successfully rearing all five of their chicks to independence. In doing so they had to overcome considerable adversity and there was much drama along the way. This included suffering the loss of their first clutch, as well as facing a major territorial challenge at a critical time and having to deal with the resultant crucial loss of the northern half of their territory.

The SR pair was first seen on a nest on 6/9/17. The nest was located within the SC in the reeds at the mid-section of the north bank. At that time I wasn't able to confirm whether or not any eggs were present. When I next checked on the pair on 6/13/17 I was able to confirm that eggs were present and that they were incubating a clutch (daily observations were initiated on this day). At that time three eggs were definitely seen in the nest, probably a fourth and possibly more. (Due to its location in the reeds I was never afforded a clear look at the nest. My view was always partially obscured by the surrounding reed stems and was analogous to looking through a picket fence that was made up of various sized slats.) The nest was a floating structure that had a central bowl section which was raised above the surface. It was constructed of local vegetation, mainly reed material. It was fairly bulky and obviously had been made sufficiently stout to support the weight of the sitting parent while keeping the eggs safely above the surface of the water (in fact, it was well able to simultaneously bear the weight of both parents). The nest was situated within a small clump of emergent reeds that lay just in front of the dense stand of tall reeds that line the north bank of the SC (these reeds—bulrushes and cattails—extend along the west bank of the river in a lush, continuous band from the immediate south side of the Highway 101 overpass bridges all the way to the mid-south bank of the SC). The reeds in this clump were only about half the size (though still fairly tall) of the main reed bed behind it. The gap between this clump and the main bed was barely noticeable, less than one third of a meter. It appeared that the nest was attached to some of the surrounding reeds as a means of keeping it in place. Overall, the nest was well concealed and was quite difficult to see, even with a sitting parent.

In the course of my observations I soon learned the extent of the SR pair's territory. It consisted of the following areas: the segment of river extending from the north side of the RR bridge (i.e., the north end of the West Lagoon) at its southern boundary to a point along the river about halfway between the RR bridge and the Highway 101 bridges at its northern boundary; the entire SC; and the south ribbon of the river. At this time the West Lagoon had not been claimed by any PB grebe pair and was a common area—it was sometimes being utilized by both the East Lagoon pair (notably while rearing their chicks) and the SR pair (mainly by Remy but only occasionally).

The pair's nesting behavior was similar to what will be described later. All appeared to be going well. However, when I arrived to begin my observations on 6/21 I discovered that disaster had struck the nesting pair. In the time between then and the conclusion of my previous day's observations the water level in the river and lagoons had suddenly rose dramatically and the floating nest wasn't able to cope with this situation. The badly compromised nest was still floating within the clump of reeds where it had been constructed but its above-surface section was very low and barely above the surface. Its integrity had been destroyed by the rising water and it had begun breaking apart. It was obvious that the eggs had been fatally inundated by the rising water and none of them were seen in the nest or anywhere else—the eggs had either been swept out of the nest or fallen through it. There seems little doubt that this sudden rise in the water level had been caused by a period of extra-high tides during which the surf flowed all the way over the beach and poured into the lagoons (it had been months since any rain had fallen in the region; also, the water level in the upstream reservoir was too low for the authorities to have even thought of conducting a release).

[This raises another probable factor in regard to the greater suitability of the lagoon for nesting by the grebes during the dry season. With its connection to the ocean being cut off during the dry season (with the exception of uncommon events such as that noted above) the water level in the lagoon essentially remains stable—a situation that otherwise would not exist with an open connection due to the daily fluctuations of the water level that would be created by the regular tidal cycles.]

This was a significant loss for the SR pair considering all that they had already invested toward producing this brood: they had built a nest which had now been destroyed; the eggs they had produced were lost (four eggs had definitely been seen and the clutch probably contained at least one or two more); and they had already worked through what probably had been at least half or nearly so of the incubation period. Yet the SR pair proved themselves to be quite resilient. Undaunted by this loss, the pair wasted little time preparing for a second nesting attempt. On 6/23 they were observed engaged in the construction of a new floating nest. Quite a bit of construction had already been done on this nest indicating that they likely had begun their work on it the day before or even sooner. This new nest was located deep inside the SC within the dense reed beds that lined the mid-section of its western bank. It was well concealed within a clump of emergent reeds that extended slightly out from and was connected to (i.e., part of) the main reed bed. The area immediately outside the reeds that was adjacent to the nest within was surrounded by a fairly extensive floating algal mat. Though it's not known if such was the intended effect, this mat effectively enhanced the concealment of the nest.

When going to the nest both members of the pair habitually dove while outside of the reeds when still a considerable distance away (including when carrying nest material). They then covered the remaining distance to the nest while underwater and resurfaced within the reeds by the nest where they were much more difficult to see. Aside from the fact that it was easier to go under the algal mats that surrounded the approach to the nest area rather than through them, it is quite probable that they did this as a means of trying to keep from revealing the location of the nest (they did this when going to their first nest also). They followed a similar pattern when leaving the nest: they dove in the nest area (within the reeds) and didn't resurface until well away from the nest (outside of the reeds). The pair adhered to this pattern of nest approach and departure throughout the entire time that they constructed and utilized the nest (the latter included the incubation period and the period immediately after hatching during which the brood remained on and near the nest).

Both members of the pair participated in nest construction (the nest work was conducted intermittently). This mainly involved diving to gather the vegetation that was used for nest construction and placing or working this onto the nest (some vegetation was also gathered from the surface of the water). The nest material was gathered from the area around the nest and was also collected from various distances away and then carried to the nest. The nest was constructed mainly, and perhaps solely, of reed material. Most of this was discolored in various shades of brown (i.e., not fresh). However, sometimes one of the pair (Sinbad) was seen collecting a long, fresh, green reed stem (usually from a fair distance away), carrying this to the nest and working the stem into it (perhaps these fresh stems were integral in regard to maintaining the buoyancy of the floating nest?). It quickly became evident that the pair was also utilizing much vegetation from the recently destroyed nest from their first nesting attempt in the construction of the new nest. On numerous occasions the members of the pair (especially Sinbad) were observed specifically going to the vicinity of the old nest site, diving and resurfacing with vegetation that they collected there, and then carrying this to and adding it onto the new nest (Sinbad was sometimes observed making numerous such trips in succession). The pair constructed their new nest in what seemed a remarkably short time. Perhaps utilizing much material from the old nest was instrumental in expediting its construction—this provided a ready source of nest material that they had already amassed and thus saved considerable time and effort from what would otherwise have been required if they had to start from scratch.

The new nest was basically the same as their first nest: it was situated in a similar type location (within the reeds in what appeared to be fairly shallow water but deep enough to dive in); it was a floating nest constructed mainly of reed material; it was fairly bulky with a sufficiently raised central bowl section and attachments to the reeds to hold it in place; and it was built stout enough to easily bear the weight of both parents simultaneously.

As noted above, the nest was constructed in what seemed a remarkably short time. By 6/24 it already appeared ready for use or nearly so. On this day members of the pair were seen sitting well-settled on the nest. Even more telling was that a copulation also took place on the nest, demonstrating that it was well-capable of bearing the combined weight of Sinbad and Remy. Note that this was only three days after their first nest had been destroyed. Regular, daily nest construction and/or maintenance continued on, during and after this day.

Interestingly, even before any eggs were laid both Remy and Sinbad intermittently took stints of various lengths of time lying well-settled on the nest, as though incubating. Nest exchanges even occurred between them. This behavior generally increased as egg laying approached. Four copulations were observed in the days before the first egg(s) was laid. These occurred on 6/24, 6/26, 6/27 and 6/28 (for details regarding these copulations see the field notes for these respective days). All of the copulations took place on the nest.

Up to this point Sinbad had done noticeably more nest construction work and nest sitting than Remy. For her part, Remy spent noticeably more time foraging than her mate. Considering the much greater nutritional demands faced by females because they are the ones that have to produce the eggs, this seems quite logical from a biological perspective.

On 6/29 the first egg(s) of this new clutch was seen in the nest. This was only 8 days after the pair had lost their first clutch (i.e., their first clutch of the season that I was aware of—it is, of course, possible that they could have had another clutch before I began my observations of the SR pair; however, if the latter was the case then it would almost certainly mean that they had lost it as it was early enough in the season that if they had successfully hatched a brood they would have still been in the process of rearing it at that time). This meant that the first egg was laid sometime between the end of my 6/28 observations and the beginning of my 6/29 observations. Thus, laying for this clutch began on either of these days. If serious incubation began on 6/29 then the hatching window for any egg laid on that day opened on 7/21. One egg was definitely seen in the nest but my view was too inadequate to be able to confirm whether or not it was the only one (similar to the situation with the first nest, because this new nest was also tucked within the reeds my view was akin to peering through a picket fence with various-sized gaps between the slats; in addition, my best view was essentially from water level—I didn't have any decent vantage points looking down on the nest that would allow me a better look into the nest bowl). Nevertheless, from what I was able to see I felt pretty certain that only one egg was present at that time.

Though my view was too inadequate to be 100% certain, it appeared that one egg was added to the clutch on each of the next two days, making for a total of at least three eggs as of 7/1. On the subsequent days during the egg laying period my views of the nest were so poor that I had no clue as to how the egg laying was progressing, what the actual clutch size was, or even on what day egg laying had concluded (on some days all that was visible to me was a small patch on one or two eggs). Unfortunately, over the course of the entire incubation period I was never afforded a definitive view of the clutch so I don't know exactly how many eggs it contained. With my limited views, the best that I was able to determine during this period was that four eggs were definitely present and probably one or two more. Ultimately, when hatching concluded it then became obvious that the brood consisted of five chicks but it was never known whether or not the clutch consisted of only five eggs (in other words, did the clutch actually contain one or more additional eggs that were lost or nonviable?).

Another event of note that occurred during the egg laying period was seen on 7/2. On this day a fifth observed copulation took place. As with the previous four copulations that had been observed, this one was also conducted on the nest (see the 7/2 field notes for the details). It was interesting that this copulation took place during the egg laying period, 4-5 days after laying had begun (the other four observed copulations occurred before egg laying had commenced). No more copulations were seen after this one (this, of course, doesn't preclude the possibility that further copulations may have occurred during the extensive periods when the pair wasn't being observed).

During the incubation period the lives of Sinbad and Remy generally revolved around four main activities: incubating the eggs, nest M/E (maintenance/enhancement), foraging, and territorial maintenance.

Incubation was essentially shared equitably among the pair. Both members took their turns incubating the eggs in alternating shifts. In frequency, typically 1-3 nest exchanges occurred within a one hour period (the pair was never observed to have left the eggs uncovered for a lengthy period of time). Throughout the incubation period Remy and Sinbad were quite diligent in regard to alternating incubation shifts though both sometimes took a more extended break from incubation duty (thus leaving their mate with an extended incubation shift). In duration, incubation shifts generally were in the 20-40 minute range (but could be shorter or longer) and were either continuous or interspersed with one or more brief breaks (in the latter case such breaks typically, but not always, were for the purpose of conducting some brief nest M/E work). The longest incubation shift that I observed was a one hour shift taken by Remy (it was actually an unknown length longer being that she was already engaged in it when I first sighted her at the beginning of that particular observation period). On that occasion Sinbad didn't seem to be having much luck foraging and it appeared that he deliberately took a longer break than usual in order to give himself more foraging time.

Nest exchanges were conducted in several ways. In one of the more common manners that the pair exchanged nesting duty the individual who was coming in to relieve its partner went to the side of the nest and waited for its mate to stand and vacate it. The nest sitting individual then either immediately left the nest or waited a variable short time longer before finally walking off. On a few occasions, while sitting in the water at the side of the nest the relieving individual reached out with its beak and gave a gentle poke to its nest-sitting partner as a prod for it to stand and vacate the nest.

In the other most common manner of nest exchange the nest-sitting individual didn't wait for its mate to reach the nest. In such cases the incubating bird caught sight of its approaching partner while it was still a considerable distance away. Then, while its approaching partner was still at a distance, the nest sitting individual stood up, walked off the nest and headed directly for its mate. The two would then meet en route, greet briefly (either by brushing lightly as they passed or just by passing each other in very close proximity) and continue on in opposite directions, with the relieving bird making its way to the nest to assume incubation duty. Oftentimes when nesting duty was exchanged in this manner it appeared obvious that the nest-sitting individual was restless and eager to be relieved (especially while engaged in a long incubation shift), in most cases, probably so that it could do some foraging as well as just being able to stretch and freely move about. In some instances in which the individual was especially eager to be relieved it walked off the nest and sat in the water by its side or moved just outside the edge of the reeds east of the nest and intently peered to the east, apparently looking for its mate. If its mate didn't show up, the individual eventually (and reluctantly) went back onto the nest and resumed incubation. On a few occasions a very restless and eager Sinbad voiced a call while settled on the eggs or in the water by the nest, the call almost certainly directed at Remy (no response was noted from Remy; however, on 6/27 Sinbad gave an especially long, drawn out call while "nest-sitting" in the period just before egg laying began that Remy did respond to—in this case the call precipitated a chain of actions, including a "nest exchange," that ultimately led to a copulation).

Usually, upon reaching the nest or (in the first noted manner of nest exchange) after its mate had left it, the relieving individual either immediately went onto the nest or sat in the water by it for a short period before doing so, or briefly engaged in some nest M/E before climbing onto it. After climbing onto the nest and before settling down onto the clutch, the individual nearly always spent at least some time repositioning/rotating the eggs and/or rearranging the nest material around them.

Having settled on the eggs, the individual then typically engaged in some preening (sometimes the individual engaged in preening right after climbing onto the nest while standing over the eggs prior to settling on them). During the course of an incubation shift it was common for the incubating bird to periodically stand and briefly fuss with the eggs.

When relieved of incubation duty, that individual (and Sinbad in particular) initially often engaged in some nest M/E work (sometimes this was quite brief and other times the individual did so for several minutes or more while its mate settled on the nest and incubated the clutch; also, sometimes after the nest-sitting bird vacated the nest the pair simultaneously engaged in some nest M/E work before the incoming partner settled on the nest to begin its incubation shift). After doing so or immediately after leaving the nest if no nest M/E work was done, the individual usually parked somewhere in the SC and spent time preening. Typically (though not always) the individual who had been relieved used the break to do some foraging. Besides foraging, other activities during incubation breaks mainly included the activities that were noted above: nest M/E work, territorial maintenance, and much preening.

Both members of the pair engaged in nest M/E work though overall, Sinbad did notably more of this than Remy. The pair spent considerable time attending to this task and did so regularly. It appears that these floating nests require much attention and constant upkeep in order to keep them stout and buoyant. While engaging in their nest M/E work during the incubation period, besides collecting nest material from the vicinity of the nest or from varying distances away, the pair (especially Sinbad) continued to make trips to and utilize nest material from the old nest.

Territorial maintenance by the grebes appears to involve calls, evicting intruders who venture into the territory, and maintaining a visible physical presence throughout the territory. Regular foraging through the various sections of the territory plays a role in the latter. In addition, Sinbad and Remy regularly parked themselves out in open, conspicuous places within various parts of the territory where they would be easily visible to other PB grebes. This appears to serve to intimidate and advertise to any other grebe that this territory has been claimed. When parked in this manner they often were observed looking about, apparently (among other things) keeping a lookout for possible intruders. With Remy spending a bit less time incubating and engaging in nest M/E and more time foraging and moving about the territory than Sinbad, she was probably a little more involved in presenting a physical territorial presence than he was during the incubation period. As for the calls, they appear to serve a role in the establishment and maintenance of a territory. Sinbad was the only one of the pair whom I observed calling. In fact, in my more casual observations of the NR and East Lagoon pairs it was only the males of these respective pairs who I observed calling. These observations raise the possibility that it may only be the males who regularly voice these territorial calls.

Probably the most obvious element of territorial establishment and maintenance is physically excluding interlopers from the territory by means of aggression (including chases) and/or displays.

Over the course of this study both Sinbad and Remy were observed chasing intruder PB grebes from the territory and Remy proved herself to be quite feisty in this regard. However, based on my observations of the interactions between the SR and NR pairs, it appears that it is the male of the pair who engages in the major fighting and territorial standoffs in situations involving the pairs of neighboring territories (perhaps the involvement mainly of males in these direct confrontations is a factor in the evolution of the larger size of the males of this species). Eight territorial standoffs between the SR and NR pairs were observed and all eight involved only Sinbad and the NR male. In addition, it was the males who specifically sought to engage in the physical fighting in the territorial conflicts. In the instances in which both of the males and at least one of the females was present during a physical territorial encounter it was the males who did the fighting. For example, in the second of the physical fights that were observed on 7/23 when the NR male made another deep incursion into the SC, both Sinbad and Remy were present together to confront him. Though Remy remained focused on and in close proximity to the confrontation it was Sinbad who engaged the male in this fierce fight. On two occasions early in that confrontation Remy did briefly attempt to physically join Sinbad in the fight but she was quickly overpowered by the NR male (in both instances Sinbad promptly re-engaged the NR male, causing him to switch his focus off of Remy and back to Sinbad; note that Remy is noticeably smaller than both of the males). Thereafter, Remy continued to support Sinbad through the remainder of the fight by following the combatants closely and sometimes taking a firm peck or two at the NR male. Nevertheless, the battle was essentially fought between the two males. On 7/28 Remy was observed in a fierce physical battle with the NR male in the mid-river area east of the SC (Sinbad was not in the area at the time—he was settled on the nest with the chicks). Unfortunately, I didn't see how this altercation began (it was already in progress when first sighted). However, it seems unlikely that Remy had attacked the NR male and had been the one who precipitated it. She was no match for the NR male and badly lost the fight. Rather than engaging the NR male she was trying to escape from him (which she finally succeeded in doing). Considering the latter points, it is most likely that the NR male had made an incursion into the SR pair's territory and jumped an unsuspecting Remy while she was foraging. As for the NR female, she was never observed to have been involved in or joined in on any of the territorial confrontations even though she had been nearby when a few of the confrontations between her mate and Sinbad took place (nor was she even observed making any incursions into the SR pair's territory south of the NE point of the SC). The indication from the observations was that the confrontational aspects of the attempt to take over the SR pair's territory were conducted solely by the NR male with the NR female taking no active part in the confrontations (though she did aid her partner in physically occupying the northern half of the SR pair's territory during the takeover period).

In regard to preening, this is obviously a very important activity for PB grebes in general (specifically, in regard to thermoregulation). All members of the SR family spent much time preening and did so often (only self-preening was observed; no mutual preening was seen). This is just what one would expect from an aquatic bird that spends essentially all of its time on and under the water. In fact, the only times that I ever observed any of the family members out of the water was when they were on the nest or briefly in the air during the few flights that were seen.

Hatching began on 7/22. This was the day that a hatched chick was first seen at the nest. The circumstances were such that it is virtually certain that this chick hatched out on 7/22 and not the previous day (see the field notes and comments for 7/22). These circumstances were as follows: only one chick was seen on this day; Sinbad (who was incubating the clutch at the time) suddenly began behaving as though something was happening with the eggs and the indication was that a hatching was probably in progress; the chick was first sighted shortly after this and less than two minutes after this sighting Sinbad was observed picking up the eggshell from a hatching and removing it from the nest. Counting the day of laying as day one, this first hatching occurred 24 or 25 days after the first egg of the clutch was laid (recall that the first egg was laid on either 6/28 or 6/29).

On the next day (7/23) it was noted that at least two additional chicks had hatched and some unhatched eggs remained in the nest. On the remaining days of the hatching period the views of the nest that I was afforded were too inadequate to be certain of how the hatching was progressing. The only thing that I was certain of was that 7/26 was the last day that an unhatched egg was seen on the nest. This meant that hatching concluded on either 7/26 or 7/27 (i.e., sometime between the end of the 7/26 observation period and the beginning of the 7/27 observations). Of these two days it appears most likely that the last hatching occurred on 7/26. Before observations concluded on that day the parents' behavior at the nest indicated that a pipping or hatching may have been in progress. With hatching having begun on either 7/21 or 7/22 and concluded on either 7/26 or 7/27 then hatching for this clutch was staggered over a period of from 5 to 7 days (it appears almost certain that the first hatching was on 7/22 and it is most likely that the last hatching occurred on 7/26 so it is quite probable that hatching was staggered over this particular 5 day period). [The staggered hatching that was observed for the SR brood essentially confirms what I had surmised regarding the West Lagoon brood that I observed during last year's breeding season (recall that I was not familiar with the nesting habits of this species). The chicks had already hatched (and were quite young) when I first discovered them and I thus noted that the considerable size disparity that existed among the chicks of this brood indicated that hatching had probably been staggered over a period of days).]

A record was kept of how many minutes each member of the pair spent incubating the eggs. In regard to how the pair shared incubation duty I need do no more here than repeat my comments from the 7/28 notes: *"Overall, incubation of the eggs had been equitably shared between Sinbad and Remy, with Sinbad doing a bit more than his mate. Considering 7/26 as the last day of incubation, the pair logged a combined total of 1,560 minutes (26 hours) incubating the eggs during the [formal] observation periods that were conducted over the course of the incubation period (observation periods generally were from 1 hr. to 1hr. 15 mins. in length). Individually, Sinbad incubated for 826 minutes (53% of the total minutes) and Remy did so for 734 minutes (47% of the total minutes).*

*Sinbad thus spent about 6 % more time incubating the eggs than did Remy. Because this is the only PB grebe incubation period that I have ever observed I don't know how representative this is for the species as a whole."* (See table below.)

***No. Minutes Each Member of South River Pair Spent Incubating Their Clutch**

| Parent | Incubation Minutes | % of Total Incubation Time |
|---|---|---|
| Remy (mother) | 734 | 47% |
| Sinbad (father) | 826 | 53% |
| **Total** | **1,560 (26 hours)** | **100%** |

**Includes only the time that the pair spent incubating the clutch during the 'formal' observation periods.*

For what it's worth, the pair incubated the eggs for a combined 79.8 % of the total time that they were observed over the course of the incubation period (i.e., considering only the 'formal' observation periods). Note that this figure basically covers only incubation by the pair during the portion of the day from about mid-morning to mid-afternoon. Therefore, I don't know if incubation by the pair was similar during the portions of the day (and notably during the night) that the pair wasn't observed. Nevertheless, the indications are that, overall, the eggs were probably incubated during the great majority of the time.

No specific record was kept of how much time each of the pair spent engaged in nest M/E activity. Nevertheless, while both shared in the nest M/E work it was extremely obvious that Sinbad did notably more of this than did Remy. The pair conducted nest M/E work throughout the entire incubation period as well as the early post-hatching period during which the nest was still being utilized for the young brood.

It was on the second day of hatching (7/23) that Sinbad and Remy were thrust into what was a potentially devastating situation for them and their brood. This was the day that the NR pair (most notably, the NR male) seriously launched their bid to take over the SR pair's territory. Two fierce, savage fights were observed between members of the two pairs on this day. These occurred at times when the NR male (the NR female was not involved) made two deep incursions into the SC (see the 7/23 field notes and comments for details). The first of these fights was especially alarming because it took place around and sometimes even on the nest. This serious territorial challenge could not have happened at a worse time for the SR pair—a time when they had newly hatched chicks to care for (at least three chicks had hatched by this time) and the unhatched remainder of the clutch that still needed to be incubated. None of the chicks were seen during the period that these fights occurred. They most likely were hiding within the dense reeds. Some chicks and eggs were finally seen after the second fight but because of my limited views of the nest and the fact that I was never afforded a definitive look at the clutch throughout the entire nesting period (and thus never knew exactly how many eggs it contained) I will never know for certain if any chicks and/or eggs were lost in these confrontations.

From this point until this territorial conflict was resolved the situation remained tense. During this time it was clear that Sinbad and Remy were concerned about and preoccupied with the NR pair and they essentially restricted their ranging and activities to the SC (sometimes also briefly going into the section of the river just east of the SC). Undoubtedly mindful of the serious encroachment that the NR male had made into the SC they apparently were very concerned about ensuring that they didn't leave their fragile newly hatched chicks and the remaining unhatched eggs vulnerable to the intruders. In the meantime the NR pair was getting increasingly more comfortable with being in the northern half of the SR pair's territory. The situation at this point presented somewhat of a quandary for Sinbad and Remy. If they wanted to try to hold onto the northern half of their territory they needed to go out and defend it. However, it appeared as though they felt that it was too risky to go out there and attempt to drive the NR pair away since this would entail leaving their chicks and unhatched eggs more vulnerable to these intruders.

Though the situation remained tense, no further physical fighting between the pairs was observed until the 7/28 fight involving Remy and the NR male that was noted above. Recall that the NR male probably initiated this fight: he made an incursion into the section of the SR pair's territory in the river east of the SC and apparently jumped Remy while she was busy foraging.

The territorial conflict carried on for a tense week before apparently finally coming to a resolution on 7/29. On this day Sinbad and the NR male engaged in an intense territorial standoff which was interspersed with brief bouts of fighting along what turned out to be the newly established territorial boundary between the now revised territories of these two pairs (see the 7/29 field notes for details regarding this incident). This new boundary was located on a line across the river just north of the NE point of the SC (note that from the day that the territorial conflict began on 7/23 nether Sinbad or Remy were ever observed venturing north of this boundary line). So, Sinbad and Remy had been able to hold onto the core area of their territory (the SC which, of course, contained their nest) but they lost the entire northern half of it to the NR pair. This was a serious loss for Sinbad and Remy that had potentially grave implications regarding the rearing of their new brood. This is because what remained of their territory (the SC, the section of river east of the SC and the south ribbon of the river) was not sufficient for rearing their large brood of chicks. Fortunately, at that time the West Lagoon was unclaimed by any PB grebe pair and Sinbad and Remy were able to annex this to what remained of their territory. This addition then gave them sufficient habitat in which to rear their brood. I feel that the annexation of the West Lagoon was the crucial factor that ultimately allowed them to successfully rear all 5 of their chicks.

For a period of just over three weeks after the first one was seen, territorial standoffs were regularly observed. Five more were seen during this period (on 8/5, 8/6, 8/8, 8/12 and 8/15; see the field notes for these days for details), all taking place at the territorial boundary. Given the number of times that these were seen during my observational sampling periods (which comprised only a fraction of each day) it seems quite possible that territorial standoffs were a daily occurrence (or nearly so) during this period, perhaps even occurring multiple times during the course of a day. After this period only two more territorial standoffs were observed (on 8/28 and 9/10) indicating that the frequency of standoffs had decreased considerably. It appears likely that in the more intense period after the new territorial boundary had been established between the two pairs it required more frequent maintenance from them (specifically, the males of each pair). Thereafter, the boundary had become solidified and accepted and no longer required frequent maintenance. No more territorial standoffs were observed after 9/10. Actually, the standoff on 8/28 (which was especially long in duration) revealed that the territorial boundary was pushed slightly farther south by the NR pair (i.e., the NR male) to a point nearly adjacent to (but still just north of) the NE point of the SC. However, this readjustment was negligible and had no real effect on the situation: the main point here (and the major aspect regarding the location of the territorial boundary) is that the boundary line remained just north of the NE point of the SC (though the SR pair lost the north half of their territory Sinbad was quite resolute in ensuring that the pair retained the SC and the section of river to the east of it; for the NR male's part, he appeared well-satisfied with the sufficiently large section of territory that he had usurped and at that point probably was not willing (nor was there the need) to risk pushing the situation any further, especially considering how steadfast Sinbad was regarding this section of their territory.

The newly established territorial boundary was well-adhered to by both pairs. As noted above, neither Remy nor Sinbad were observed venturing north of it after it had been established (and actually, not from the time that the territorial conflict began). The NR female was never observed venturing south of the boundary and the NR male was only observed doing so once. On that occasion (on 8/31) he spent at least several hours well inside the SC, often foraging. At the time the entire SR family was in the West Lagoon, well distant from the SC (a SR chick later made its way deep into the SC but hid in the reeds after sighting the NR male; it didn't appear that he had caught sight of the chick). As things turned out, this apparently was simply a case of the NR male taking advantage of the temporary absence of Sinbad and Remy to sneak in some foraging in the SC.

Only two of the territorial standoffs (those observed on 7/29 and 8/12) involved some physical fighting. All of the others consisted solely of stereotypical posturing and displaying.

Ironically, the aftermath of this territorial conflict ultimately ended out with both pairs having sufficient habitat for rearing a brood whereas prior to it only the SR pair had this. Before their territorial challenge, the NR pair had established a territory in the section of river on the immediate north side of the SR pair's territory (in fact, on 7/8 a territorial standoff was even observed between

Sinbad and a member of the NR pair—probably the male—at that far northern territorial boundary where the boundary was located at that time). This territory was much too small for rearing a brood. There seems little doubt that the NR pair's bid to usurp the SR pair's territory (or at least a significant portion of it) was undertaken as an attempt to remedy this situation by expanding their range south (they couldn't expand to the north because their current territory was situated at the northern extent of suitable habitat in this section of the river). By taking over the entire northern half of the SR pair's territory the NR pair succeeded in expanding their territory to the point where it contained sufficient habitat for rearing a brood. When the SR pair then annexed the West Lagoon after their loss they restored their territory to the point where they had sufficient habitat for rearing their brood. Unfortunately for the NR pair, their effort was in vain because they failed to successfully rear a brood this breeding season (they were never seen with any chicks). It isn't known for certain if they even made a nesting attempt. At one point the nature of the sightings of the pair indicated that they may have been incubating a clutch (see the 8/13 field note comments). If so, then their nesting attempt failed. I'm not familiar enough with this species to be able to determine the relative ages of adults but I wonder if perhaps the NR pair was a young, inexperienced pair that was entering their first breeding season together?

Remy and Sinbad continued as usual with their alternating incubation shifts throughout the hatching period. They also continued with their usual activities during their incubation breaks, including nest M/E work—and of course, with hatching having commenced these activities now also included provisioning the hatched chicks. Feedings probably began on the first day of hatching but because of my limited view of the nest I wasn't able to unequivocally confirm any feedings until 7/24. [What was particularly frustrating in regard to verifying the possible feedings that were observed prior to 7/24 is that in each of these instances the possible feeding took place while the chick(s) was on the nest and the possible provisioning parent was on the far side of the nest where my view of its actions was blocked (by the nest as well as by the incubating partner who was sitting on the nest). Thus, I wasn't able to see if feedings actually took place.]

In the immediate days after hatching (that is, as each chick made its appearance into the world during the hatching period) the chicks mainly stayed on the nest (this included climbing onto the parent's back while that parent was sitting on the nest incubating the eggs). Of particular note, when a parent was on the nest the chicks were essentially always on the nest with the parent. The chicks did sometimes go onto the water around the nest. Such instances mainly occurred during nest exchanges between the parents or at other times when the nest was unoccupied (such as when one or both of the parents were engaged in nest M/E work) and when a parent came to the nest area to conduct a feeding (in these latter cases, even if the chicks were on the nest with a parent they might come off of it and go to the feeding parent—at this early point all of the feedings were administered close to the nest or to chicks who were on it).

As hatching proceeded toward its last day the chicks were still almost always on the nest (including being on the back of an incubating parent), though those who had hatched out began increasing the time they spent off the nest but stayed near it and always remained within the reeds. A spot that they particularly frequented when off the nest was the area about 1 m to the north of it where nearly all of the observed feedings were conducted at this time. This spot (referred to in the field notes as the "north feeding station") immediately became the regular feeding station during the period when the chicks were still tied to the nest and its immediate vicinity (it's most likely that Sinbad and Remy predominantly used this specific location because it was the only spot close to the nest within the reeds that had enough space in which to maneuver and conduct feedings). As the number of hatched chicks increased, the parents stepped up their feedings accordingly. In the days after hatching concluded this stepping up of feedings became particularly noticeable as feeding the chicks became a major daily activity for Remy and Sinbad.

The pair continued to be very protective of the newly hatched chicks in the days immediately after hatching had concluded (the ongoing territorial conflict probably making them even more so). Except for some brief periods when both Remy and Sinbad were away foraging (with at least one of them doing so in the general area of the nest on these occasions), they kept to a strategy of keeping one parent at the nest guarding and looking after the chicks while the other foraged (for the chicks and for itself). The nest-sitting parent generally spent its shift lying on the nest with the chicks on its back and/or on the nest around it. As during the incubation period, the pair regularly exchanged chick-sitting duty.

It was first noted on 7/30 that the parents had begun the process of "loosening the apron strings" in regard to the chicks. On this day they continued with their protective pattern of keeping one parent at the nest to guard and look after the chicks while the other was away foraging and engaging in various other activities. However, the chicks were sometimes left by themselves at the nest while both parents simultaneously foraged a fair distance away. This was the first time that I had seen the parents do this since the time when the first chick hatched. This was probably largely a reflection of the fact that the chicks were growing and the demands on the parents to feed them were increasing.

This trend of loosening the apron strings progressed steadily from this point. On the next day (7/31) Remy and Sinbad continued taking turns sitting on the nest guarding and babysitting the chicks but spent less time doing so. The chicks were thus left to spend more time at the nest by themselves while the parents were simultaneously out foraging for them. As usual up to this point, the chicks remained on or near the nest but were noticeably more active and spent more time off the nest. A few were seen diving briefly. Also, for the first time some of the chicks were observed going to the inside edge of the reeds by the nest (while there they often looked east out into open water, probably looking out for parents approaching with food).

Not surprisingly, there was an association between the progression of the parents' loosening of the apron strings and the chicks' progression toward becoming fully mobile. As is already evident here, the chicks did not just suddenly become fully mobile. Instead, their attainment of full mobility was progressively achieved. As noted above, during the hatching period and the days immediately after it concluded the chicks were almost always on the nest but did sometimes go onto the water near the nest. During this period, from the time of hatching the chicks began increasing the time they spent off the nest (most notably, going to the north feeding station) but stayed near the nest when off of it and always remained within the reeds. As the parents increasingly spent more time when both were simultaneously away from the nest foraging for their brood, the chicks increased the time that they spent off the nest. Note again that on 7/31 chicks were seen venturing to the inside edge of the reeds (just east of the nest) for the first time.

The next day (8/1) was a milestone day for the chicks as they were observed venturing outside the reeds for the first time (note that it wasn't until this episode that I was finally able to confirm that the brood consisted of 5 chicks; in addition, it wasn't until this episode that I was able to begin the process of identifying the individual chicks). On this occasion the chicks remained outside the reeds for about a half hour before Sinbad led them back to the nest. The chicks didn't venture far from the reeds and mainly remained within the algal mats that bordered the outside edge of the reeds. Nor did they venture very far from the nest: they remained within its general vicinity, ranging only as far as approximately 20-25 m north of the nest (within the algal mat). They were mostly chaperoned by the parents but were sometimes there by themselves while the parents simultaneously foraged for them. At least 5 feedings were observed during this time; these were the first observed feedings that were conducted outside of the reeds (see the 8/1 field notes). Though sometimes slightly spread out, the chicks generally stuck together as a cohesive group. Considering 7/22 as the day hatching began and 7/26 as the day it concluded, then it wasn't until the tenth day after hatching began (and the sixth day after hatching concluded) that the chicks were first seen venturing outside of the reeds. In other words, the chicks remained tied to (i.e., on or near) the nest until this time. [In an interesting prelude to this episode it appeared that the parents worked together to ensure that all was safe and clear before leading the chicks out from the reeds. Remy went out to the east end of the SC, vigilantly looked about and then voiced a series of single, one note calls. Sinbad (sitting by the side of the nest with the chicks and alertly peering out toward Remy) then appeared to respond to Remy's apparent "all clear" calls by leading the chicks out from the reeds (see the 8/1 field notes). The pair was observed doing the same general thing the next day prior to leading the chicks out from the reeds, this time with their respective roles reversed.]

From this day forward the chicks no longer remained strictly tied to the nest and its immediate vicinity. This also marked the end of the parents' continuous 'close protective phase' of their rearing of this brood (that is, they ceased the continuous, day long alternating chick-sitting shifts at the nest, though they continued to engage in less regular chick-sitting bouts at the nest for a short time longer). On the following day (8/2) Sinbad and Remy spent a combined total of only about 9 minutes sitting on the nest with the chicks during the formal observation period while the chicks took another small step toward becoming mobile. For the majority of that period the chicks were by themselves while the parents were out foraging. Over the first half of the observation period the chicks remained at the nest (on it and in the water near it). About midway through the period the parents led them away from the nest and the chicks then spent the second half of the period outside of the reeds, mainly in the algal mat in the area from the nest north to the NW corner. On this occasion the chicks also spent time in open water beyond the edge of the algal mat. At one point Sinbad led the chicks out approximately 8 m from the edge of the algal mat—up to this time this was the farthest out in open water that the chicks had been observed to venture. During this period all five of the chicks were seen engaging in brief dives.

Over the next few days Remy and Sinbad continued to guide the chicks on their steady progression toward becoming fully mobile. As they had done in the previous few days, on 8/3 the parents led the chicks out into the algal mat and only a short distance beyond, keeping within the west end of the SC in the general area of the nest. On this day the chicks mainly ranged within the algal mat or at the edge of the reeds during the time that they spent outside of the reeds. Once again they also spent much time by themselves at or near the nest while both parents were simultaneously out foraging for them. However, the parents continued to spend some time (in irregular shifts) settled on the nest with the chicks.

On 8/4 the chicks took their biggest step to date toward becoming fully mobile and the parents loosened the apron strings to the greatest extent that they had done up to that point. Of particular note regarding the latter, this was the first observation period during which the parents spent no time sitting on the nest with the chicks. In addition, it was the first observation period in which the chicks spent the majority of the time out away from the nest. The chicks were actually somewhat mobile on this day but restricted their movements to within the SW quadrant of the SC and especially in the SW corner. Of further note, they took most of their feedings away from the nest area and beyond the reeds.

The next day (8/5) marked another milestone for the chicks as they finally became mobile. They ranged through much of the SC (going as far as its east end) and spent considerable time in open water (this included spending time parked in open water). They also moved out on their own to parents who were parked in open water and moved to parents whom they sighted approaching their area with food. The brood thus became mobile on the tenth day after the last egg hatched (considering 7/26 as the day of the last hatching). Though the chicks became mobile, the parents still spent a little time on this day sitting on the nest and babysitting them. In regard to feedings, chick mobility was a welcome happening for the parents. It meant that the parents would now be able to take the chicks to the general areas where they wanted to forage or the chicks could just accompany the parent as it foraged. Overall, this considerably reduced the distances that the parents had to carry their captured prey to the chicks (and thus saved time and energy). This also meant that the parents would be able to expand their foraging range for chick feedings. On the other hand, this mobility carried with it greater risk in the lives of the chicks and made it more difficult for the parents to watch over and protect them.

From this time to the point when they achieved independence the chicks steadily enhanced their mobility (the expanding ranging by the chicks can be followed chronologically in the daily field notes). This was due to factors such as: the enhancement of their physical skills and abilities as their physical development progressed (ultimately including the ability to fly); steadily expanding their ranging, doing so while accompanying parents (especially during initial expansions) and sibs and also independently (note that throughout the entire chick rearing period the chicks were never observed to stray outside of their natal range—in other words, they remained within the SR pair's territory; some of the chicks did finally leave the natal range but not until becoming fully independent); experience and confidence gained through increasing familiarity with their natal range, through their increasing independence and maturity, and through their growing life experience in general.

In the initial days after becoming mobile the chicks restricted their ranging to the SC. However, it wasn't long before the chicks' ranging was expanded in a major way: on 8/9 they were seen in the West Lagoon for the first time. Though the chicks were already in the West Lagoon when first sighted on this occasion, there is little doubt that it was one of the parents who had taken them (i.e., led them) there (there seems zero chance that the chicks would have made this trip on their own at this early stage in their lives). In fact, it was under the security of a leading parent(s) whom they accompanied that the young chicks made their initial trips out of the reeds and to the various reaches of the SC.

When the chicks became mobile the days of the parents spending any time sitting on the nest closely guarding and babysitting them came to a quick end. The last day that a parent was observed doing so was on 8/7. This thus concluded the phase during which the parents utilized the nest in their chick production/rearing endeavor. However, for some time after the nest area continued to serve the chicks as a kind of home base area, a place of relative safety and security that the chicks could retreat to if alarmed or feeling threatened (in fact, for a while after the chicks had become mobile the parents still sometimes led them to the nest area at times when something caused the parent to become concerned about their welfare). [As a precaution, I should qualify the above in regard to how much longer the parents utilized the nest. Because I didn't observe the family at night I don't know whether or not the parents spent the night on the nest with the chicks. If so, then I could not dismiss the possibility that they continued to do so for some period beyond when I ceased seeing them utilize the nest during my daytime observations.]

The young chicks were commonly seen going onto the back of a parent (typically tucking themselves under the parent's folded wings) throughout the hatching period and into the period when they began coming out from the reeds. They did so with both Remy and Sinbad, including during times when the parent was well-settled on the nest either incubating the unhatched eggs or (after hatching had concluded) just babysitting them. There were times when anywhere from just one chick to the entire brood was on the parent's back. The last time that a chick was seen on the back of a parent was on 8/4. Note that this was the day before the chicks became mobile and on the ninth day after the last egg had hatched. [Having had the opportunity to observe the SR brood from hatching through to independence I now have a reasonable idea of just how old the West Lagoon brood was when I first discovered them the previous year. Recall that those chicks were already hatched and mobile when I first saw them and began my daily observations of the family. At that time the chicks were quite young (they still had the plumage and beak colors of young chicks) but were beyond the stage in which they went onto the backs of their parents. In addition, based on my observations of the SR brood, it appears that they had only fairly recently become mobile. I would thus estimate that the chicks were at an age of approximately two and a half to three weeks from when the last egg of their clutch had hatched.]

Upon my first definitive view of the SR brood on 8/1 some notable size disparity was present among the chicks, though not nearly to the extent of what was seen among the West Lagoon brood last year and none of the chicks were runty. There seems little doubt that it was the staggered hatching of the eggs in the clutch that was responsible for producing such notable size disparity among such young chicks (recall that hatching occurred over a 5-7 day period—most likely over the 5 day period from 7/22 to 7/26). The size disparity thus probably results from the fact that as each chick hatches it gets a head start over any unhatched brood mates because it begins receiving feedings while its unhatched sibs are still in their egg. Ideally, this situation would be particularly advantageous for the first chick to hatch because it would have no competitors and thus is guaranteed to be the sole recipient of the food from all of the feedings that occur prior to the subsequent hatchings. All things being equal, a chick will be slightly larger and slightly more advanced developmentally over subsequently hatched sibs and thus gain a competitive advantage (through being more dominant) that it can carry through the chick rearing period. Theoretically, the wider the gap between hatchings, the greater the advantage for the prior hatched chick.

It's unfortunate that due to my limited view of the nest I wasn't able to observe the hatching order of the chicks during the hatching period. Based on the factors noted above (and particularly on size) I suspect that the hatching order for the brood was as follows. I feel quite certain that Sparky was probably the first of the brood to hatch. Sparky was the largest of the chicks and from the start all the way through the chick rearing period was always a bit ahead of its sibs. In addition, Sparky was the most dominant of the brood and (though not the first to leave the natal area) was the first of the chicks to become independent. It isn't clear-cut as to who probably was second and third as it appeared to be close between Cobra and Spot. Recall that only one chick was sighted on the first day of the hatching period (and this very likely was Sparky) but two additional chicks were seen on the second day. The hatching times of these two chicks probably were close, thus conveying little or no advantage in regard to hatching time to the first of these two to hatch. Initially, they were the second and third largest of the brood and were very close in size, with Cobra perhaps being very slightly larger (this would change in the latter portion of the chick rearing period). In addition, they were equal developmentally throughout the entire chick rearing period. They remained a bit behind Sparky but slightly ahead of Phantom and Quark throughout. In regard to the dominance hierarchy among the brood they ranked second and third. Initially it was a very close second and third with Cobra probably just outranking Spot. However, in the latter portion of the chick rearing period Spot became the very clear-cut second ranked chick (the change between these two quite likely was gender related). Because Cobra initially was slightly larger and slightly more dominant over Spot I would lean toward Cobra as being the second of the chicks to hatch with Spot being a close third. The fourth and fifth chicks in the hatching order seem quite certain: Phantom (initially the fourth largest of the brood) probably was the fourth to hatch and Quark (initially the smallest of the brood) was probably the last. Probable support for this may also be evident from the behavior of the parents toward the chicks in that Phantom and Quark were the last chicks of the brood who were allowed to closely solicit and accompany the parents.

As noted above, none of the chicks were runty and this remained the case throughout. In fact, over the course of the chick rearing period all five of the chicks progressed well and always looked healthy and appeared to have good weight. Some changes in the relative sizes of the chicks did occur over the course of the chick rearing period. As the largest chicks approached their adult sizes their rate of growth slowed and this allowed the smaller chicks to close the size gap to them. In the latter segment of the rearing period it appeared that Spot caught up to and slightly surpassed Sparky in size, thus becoming the largest of the brood from that point. Also in this latter segment, both Phantom and Quark surpassed Cobra in size, making Cobra (who earlier had clearly been surpassed by Spot) the smallest of the brood from that point. This almost certainly was gender related. Visually determining gender among the chicks was generally more difficult because the salient physical features used to differentiate the sexes typically aren't so clearly expressed in the chicks. However, in the case of two of the chicks they were: Cobra clearly was a female and Phantom clearly was a male. I wasn't certain about the gender of the other three but they appeared to be males (by the end portion of the chick rearing period I felt certain that they were indeed males). In all of the PB grebe pairs that I have seen the male of the pair has been larger than its mate. From this observation it then appears that on average, male PB grebes are larger than females though this size dimorphism isn't pronounced. Therefore, Cobra's being surpassed in size by those other three sibs as they were reaching their adult sizes was probably just down to a manifestation of the normal size dimorphism between the sexes.

Feeding of the chicks by Remy and Sinbad was basically similar to what I observed with the West Lagoon family during the previous year's breeding season. Both parents shared in this vital task and it was common (i.e., after the chicks became mobile) to see the chicks split between Sinbad and Remy (this includes times when the parents were very distant from each other, with some chicks being with one of the parents in the SC and some with the other parent in the West Lagoon). During the hatching period and the days following it (that is, before the chicks began coming out from the reeds) the chicks were tied to the nest and its immediate area so this is where all of the feedings were conducted (with most of the observed feedings occurring at the north feeding station). This meant that the parents had to carry all of the prey items that they captured for feedings during this period to the nest area. The parents thus often had to travel a considerable distance with their captured prey, sometimes even from the West Lagoon, though they typically foraged for the feedings within the SC and the section of the river to east of it. When the chicks first began emerging from and spending time out of the reeds on 8/1 the first feedings to be conducted beyond the reeds away from the immediate nest area were observed. Soon after (on 8/5), the chicks became mobile. From that time the chicks steadily expanded their ranging (with the parents usually leading their initial expansions), eventually reaching the point where feedings might be conducted from pretty much anywhere within the SR pair's territory. For the feedings the parents did commonly forage in open water but the great majority of their foraging was done along the edges of and within the reeds.

With one possible exception (a possible dragonfly feeding) all of the observed feedings consisted of either a crayfish or a fish (I wasn't able to ID the particular species of each). In every feeding that was observed the parent had only a single prey item to feed (in other words, only one crayfish or one fish). The crayfish feedings were conducted in the same manner as those described for the West Lagoon family (the reader is referred to those descriptions). Also similar to the West Lagoon family situation, when the chicks were younger more than just one chick (i.e., in a group feeding situation) might obtain food during a crayfish feeding. However, later on when the chicks became capable of swallowing large prey items and working a crayfish carcass on their own only one chick typically obtained food from a crayfish feeding (see the last paragraph of the 8/30 notes for a detailed explanation).

Except for one probable exception, the fish that were fed were always swallowed whole. Other than this single probable exception (on 7/28) there was no tearing off of scraps to feed multiple chicks and to reduce the bulk to a size that could be swallowed as was commonly done with large crayfish. The proffered fish had to be of a size that the chick could swallow otherwise the chick simply would not be able to eat it. On occasion a parent was observed offering a fish that was too large for the chick(s). In these instances, after watching the chick's unsuccessful attempts to swallow the fish and apparently realizing that it was too large for the youngster the parent then swallowed the fish itself. In group feeding situations the fish sometimes changed hands (or beaks, to state it more accurately) several times among the chicks with several of them making unsuccessful attempts to swallow it. Though the parents didn't tear scraps from the fish, medium to large-sized fish nevertheless were worked by the parent (and roughly so) before being offered to a chick (or consumed by the parent), sometimes for 1.5 minutes or more. It seems probable that this was done in order to kill or incapacitate the fish prior to its being swallowed. Fish were always swallowed head first (though I couldn't always see if this was the case involving small-sized fish) and crayfish were always swallowed tail first with the tail partially doubled over on itself ventrally.

A summation of the relative amounts of these two general food types that were fed to the chicks by the parents was presented in the field notes for 10/10 and will be repeated here. "*Overall, crayfish was the predominant food item that was fed to the chicks. Of the total of 186 feedings that were observed during the formal observation periods, crayfish were fed in 128 (68.8%) of these while only 58 (31.2%) of the feedings consisted of fish. However, this predominance of crayfish feedings wasn't always the case during the course of the chick rearing period. Interestingly, over the first few weeks after hatching, fish were fed overwhelmingly more often than crayfish. During the period through 8/8/17 (recall that hatching began on 7/21 or 7/22 and concluded on 7/26 or 7/27) a total of 54 feedings were observed. Fish feedings comprised a whopping 41 (75.9%) of these with only 13 (24.1%) of them consisting of crayfish. On 8/9/17 the shift toward the overwhelming predominance of crayfish feedings began. Of the 132 feedings observed during the period from 8/9 through the end of the study an overwhelming total of 115 (87.1%) of these consisted of crayfish while fish feedings comprised only 17 (12.9%) of them. It is interesting to note that the beginning of this shift on 8/9 coincided with the day that the chicks were first observed going over to the West Lagoon.*" (See table below.)

***Proportion of General Prey Types Fed to the Chicks in Observed Feedings**

| | ****Number (and %) of Prey Type Fed** | | |
|---|---|---|---|
| **Time Period** | **Crayfish** | **Fish** | **Total No. Feedings** |
| 7/22/17 - 8/8/17 | 13 (24.1%) | 41 (75.9%) | 54 |
| 8/9/17 – 10/10/17 | 115 (87.1%) | 17 (12.9%) | 132 |
| **Overall Totals** | 128 (68.8%) | 58 (31.2%) | 186 |

**Includes only the feedings that were observed during the 'formal' observation periods.*
***All of the observed feedings that occurred during the 'formal' observation periods consisted only of either fish or crayfish.*

Some notable atypical feedings were observed over the course of the chick rearing period. In two instances one parent assisted its mate with a feeding. Both cases occurred early in the chick rearing period. In each instance the parent who assisted was nearby when its mate was conducting a group feeding and joined in (in both instances the prey item that was being fed was a large crayfish). In the first case, Remy was busy conducting a crayfish feeding with all 5 chicks swarming around her when Sinbad approached the group with a small fish that he had just captured for a feeding. After handing the fish off to one of the chicks Sinbad then joined Remy and assisted her by picking up and retrieving small scraps that detached from the crayfish while she was working it and handing these to the chicks (see the 1209 entry in the field notes for 8/5). In the second case Remy happened to be nearby while Sinbad was conducting a crayfish feeding involving all 5 chicks. Remy joined in the feeding and assisted Sinbad with each then taking turns working the crayfish, tearing morsels from it and handing these to the chicks and each also taking their turns diving after and retrieving the carcass during times when it sank as they were working it (see the 1143 entry in the field notes for 8/9).

In two other feedings of note Remy and Quark actually shared the feeding. Both of these cases occurred on 9/30, each involving a large crayfish. In both instances Remy took the crayfish to Quark for a feeding but each then took their turns working the crayfish and retrieving it during the times that it sank, and tearing off morsels that each consumed (see the 1144 and 1154 entries in the field notes for 9/30).

A third feeding that also involved the sharing of the food item between a parent and a chick (in this instance, Sinbad and Cobra) took place on 9/17. In this feeding Sinbad had captured a very large crayfish which he offered up to Cobra. However, Cobra initially appeared to be a bit intimidated by this extra-large crayfish and wouldn't take it at this time. In what at this advanced stage of the chick's development was a retrograde feeding method, Sinbad then began repeatedly tearing scraps from the crayfish and handing these to Cobra. The initial segment of the feeding proceeded in this manner but after Sinbad had worked and reduced the carcass somewhat Cobra joined him in working the carcass and the two of them then shared it in a manner similar to what was observed between Remy and Quark. When the mass of the large crayfish had been sufficiently reduced by their shared feeding, Cobra assumed control of the carcass and swallowed it (see the "1250" entry in field notes for 9/17).

The chicks solicited the parents for food often and continued to do so even after reaching the stage when it was clear that they could survive independently (given that PB grebe parents don't regurgitate food for the chicks these solicitations more accurately are instances of the chick urging the parent to forage for it). The general circumstances under which the chicks solicited the parents are the same as those noted for the chicks of the West Lagoon family. Similarly, the chicks performed these solicitations in the same ways as described for the West Lagoon chicks (the reader is referred to the notes on the West Lagoon family for the details on these latter two topics). The SR chicks performed both the close-in (beak to beak) solicitations and those given from a short distance away from the parent (the latter becoming increasing more frequent after the parents began rebuffing chicks, with some chicks eventually performing only these due to their apprehension regarding a possible parental rebuff). While soliciting a parent and also commonly while in the vicinity of a parent it appeared that the chicks often uttered strings of continuous low peeps (as they aged they did so progressively less often). I was rarely close enough to actually hear these peeps however I often saw them moving their beaks in a manner which indicated that they were voicing them.

In obtaining food from the parents the chicks employed three main strategies: staying in one specific area and waiting there for the parents to deliver feedings (because they were tied to the nest area during the hatching period and immediately after, this was the initial manner in which the chicks received their feedings—of course, the chicks had no other options at that time but did still make use of this strategy later when they did have choices); moving to the area where a parent was foraging and monitoring its progress (in particular, watching and waiting for it to catch prey items for feedings); accompanying a parent and remaining with that parent even when the parent isn't foraging. These are the same general strategies that were utilized by the West Lagoon chicks. These strategies are discussed in detail in the end segment of the field notes for 8/28 so the reader is here referred to those notes.

As was observed with the West Lagoon chicks, the competition among the SR chicks for feedings was intense. Not surprisingly, dominance was a major factor in this competition with the larger, more dominant chicks typically besting their subordinates during group feeding situations. However, through guile and persistence subordinates were sometimes able to win in these group situations. Competitive advantages conveyed by larger size and being ahead developmentally included: greater strength and more bulk with which to physically dominate, push aside and outmaneuver lesser competitors; the capacity to swallow larger food items and to more quickly and easily swallow large food items in general (this latter ability is particularly helpful in cases in which a chick in possession of a food item is being contested by sibs—the quicker the chick can swallow the food item, the lesser the chance of it being stolen); faster swimming (paddling) speed (this is particularly advantageous when racing other sibs to an approaching, food bearing parent). Considering that size and being slightly ahead developmentally appear to be important factors in regard to dominance, hatching order then is probably highly correlated with dominance order among broods.

The general nature of the feeding competition among the SR chicks and the manner in which it was conducted were basically the same as described for the West Lagoon chicks and will not be repeated here (the reader is referred to the relevant observations and discussions in the field notes for both families). Only a very general summary will be presented here. In the earlier stage of the chick rearing period the competition mainly consisted of muscling aside and trying to outmaneuver rival chicks, attempting to steal a food item that was in a rival's possession (this included pressuring and hotly pursuing that rival; in addition, a number of tug-of-wars were seen), and being vigilant for parents approaching with a food item and outracing rival sibs to that parent (or in general, attempting to be the first to reach the food-bearing parent). Later, the feeding competition ramped-up considerably when the chicks also began engaging in direct aggressive actions taken by more dominant chicks toward subordinate ones—that is, a more dominant chick specifically targeted aggressive actions at a subordinate chick(s) (the first such incident was seen on 8/15). These targeted aggressive actions consisted mainly of charging, pecking at and briefly chasing the subordinate(s) away from the vicinity of a foraging parent or one that was conducting a feeding (the dominant chick might employ one, two or all three of these actions during a particular episode). In doing so, the more dominant chick's intent was to exclude competing sibs from close access to a feeding parent or create a situation in which it would be closest to and able to gain first access to a foraging parent when it captured a prey item. In one particularly notable incident that occurred on the first day that such interactions were observed a fierce fight broke out between the more dominant chick and its targeted subordinate sib. This incident took place at the end of a feeding administered by Sinbad in which he attempted to feed a fish that was too large for the chicks and ended up swallowing it himself. The fight involved the two lowest ranked chicks of the brood: Phantom (ranked fourth) and Quark (ranked fifth and last). During the attempted feeding (at which Sparky was also present) Phantom charged, pecked at and briefly chased Quark from the immediate area. However, undaunted by this Quark soon rejoined the feeding. When the "feeding" concluded, Phantom suddenly attacked Quark. This time Quark staunchly stood up to Phantom and an intense fight ensued between them. The two fought fiercely (with Phantom getting the better of a determined Quark) for approximately 5-6 seconds before Sinbad intervened and broke it up (see the 1118 entry in the field notes for 8/15). Over the course of the chick rearing period this was the only observed fight that resulted from these directed aggressive competitive feeding interactions.

For the SR brood these direct aggressive actions taken by more dominant chicks toward subordinate ones in competitive feeding situations were notably less frequent and played a lesser role compared to what was observed among the West Lagoon chicks during my observations of that family. Since becoming mobile the feeding competition among the SR brood consisted more prominently of spotting and reaching a provisioning (prey bearing) parent first or barging in and muscling competing sibs aside.

Regardless of where they stood in the dominance hierarchy of the brood, through the various strategies and tactics they employed all of the chicks succeeded in obtaining a healthy share of the feedings provided by the parents (and to the great credit of Remy and Sinbad, they did a fantastic job in ensuring that this was the case). This of course included Quark. Though Quark was the smallest and lowest ranking of the sibs, this chick was no pushover. Quark was a feisty, dogged competitor who readily mixed it up with its sibs during feedings. Despite typically being bested by its brood mates during group feedings, Quark still occasionally managed to come away with and swallow the prey item that was being proffered by the parent. Among Quark's key qualities regarding feedings was being proficient at vigilantly watching for parents approaching at a distance with a prey item and being the first to spot the parent. This then allowed Quark to gain a head start in the race to the parent (reaching the parent first meant that Quark was able to take the prey item before having to compete with any sibs for it or otherwise have the first shot at doing so).

There were some intriguing observations which suggested the possibility that when very young (from the hatching period early into the period after the chicks became mobile) the chicks may have supplemented the parental feedings with a different kind of surface foraging of their own. Specifically, chicks were observed to peck and dab at the surface of the water as though perhaps capturing and eating insects (such as water striders) or other small invertebrates. Unfortunately, in these instances I was always too distant to be able to discern items as small as insects and other small invertebrates so I was never able to confirm whether or not this was the case. Possible support for this was seen in a later close up observation in which Sparky displayed interest in and briefly chased a water strider bug (see the 1106 entry in the field notes for 9/26). Repeating the relevant comment from the 9/26 notes: *"One is then left to speculate on whether the diet of PB grebe chicks differs somewhat from that of adults (as occurs in various other species) with the chicks supplementing their feedings from the parents by eating small insects and/or other small invertebrates (prey that they generally won't bother with as adults)?"*

Food wasn't the only thing that the parents fed to the chicks. On numerous occasions a parent was observed plucking a small feather from its breast or flank and handing this to a chick who then immediately swallowed it (chicks were also seen picking up and swallowing small feathers from the surface of the water that detached from a parent while that parent was preening). I observed this same behavior in the West Lagoon family during my 2016 study. It appears that these ingested feathers may play a role in the digestive process or with the movement of food through the digestive tract. Probably related to the latter possibilities is that both the parents and the chicks were commonly seen regurgitating the exoskeletons of crayfish. It is thus evident that the crayfish exoskeletons do not pass completely through the digestive tract. This apparently is a grebe analog to owl pellets.

Each chick's course to achieving independence can be followed chronologically in detail in the daily notes so only some of the basic aspects of this process will be noted below.

For the chicks, the process of working their way toward independence was similar to what was observed with the West Lagoon chicks. They followed the same general pattern of spending increasingly more time ranging and foraging independently. Though the chicks followed this same general pattern there was variability among the brood regarding how much of this each did during the course of the chick rearing period. Sparky, Cobra and Spot clearly were more independent in this respect than Phantom and Quark (overall, Quark and Phantom spent more time near and accompanying the parents).

Also similar to what was observed with the West Lagoon family, parental rebuffs of chicks (first observed on 8/22) were an important element of the weaning process. As with the West Lagoon family, after first being observed parental rebuffs quickly became daily occurrences though feedings by the parents nevertheless continued right on through to the end of the chick rearing period (however, by the end segment of the chick rearing period the feedings were being directed toward only a selected few chicks).

Also playing a role in the path of each chick toward attaining independence was the fact that the parents recognized each chick as a distinct individual. That this was so was unequivocal because Sinbad and Remy consistently behaved toward and treated each chick in a manner specific to that particular chick, leaving no doubt that they recognized each chick as an individual (note that their specific behavior toward each chick was not static—it evolved during the course of the chick rearing period as the chick aged). Furthermore, the behavior of each of the parents toward a particular chick could differ significantly. For example, during the period when Cobra was nearing her independence Remy still allowed Cobra to solicit her from up close and to closely accompany her. In contrast, Sinbad would not allow this and was rather intolerant of Cobra; he readily rebuffed her and on a few occasions even chased her away from close proximity to Remy. [Relating to their recognition of the individual chicks, it appeared that the parents were also cognizant regarding the presence of the full complement of their brood. This was evident in the episode on 8/20 when Quark became isolated in the West Lagoon while the rest of the family was in the SC (this was during the period when the chicks were still fairly small and vulnerable and before the chicks had begun going to the West Lagoon by themselves or remaining in one of the main sections of the territory—the SC and West Lagoon—by themselves when neither of the parents was present there). In this incident Sinbad clearly became aware of Quark's absence and went to the West Lagoon with the specific intent of finding and retrieving Quark (see the 1315-1355 entry in the field notes for 8/20).]

For bird species that fly, achieving the ability for flight is one of the major milestones on a chick's road toward becoming independent. The first observed flights taken by the chicks were seen on 9/5 (on this day Quark and Cobra were seen taking short flights just above the surface of the water). These occurred on the 42$^{nd}$ day after the last egg hatched (considering 7/26 as the day of last hatching and counting it as day one). Given that Quark (who had been the smallest of the chicks and probably the last one to hatch) was capable of flight by 9/5 then it is a safe bet that all of the other chicks were also probably capable of at least short flights by that time. Thus, the indications are that all of the chicks had entered into the fledgling stage of their development by this point. (Note again here that my daily observations covered only a small fraction of their lives. Because of the large gaps between my daily observations and the fact that the chicks seldom took flight the odds of my having been able to witness flights by all five of the chicks by that early point were extremely slim. Eventually, I did get to see all of the chicks engage in at least a short flight.) As previously noted, PB grebes are seldom seen in flight. Because of this, the odds are that the chicks were probably capable of at least short flights before I actually got to see any of them fly. For example, on 9/2 I was afforded a look at the wings of several of the chicks. At that time, based on the size and completeness of the flight feathers it appeared as though the wings on the chicks, and especially the two largest (Sparky, who had always been a little ahead of the others, and Spot), were capable of flight. The most notable of the observed chick flights were two very long, beautiful, back to back flights taken by Cobra in which she gained altitude during the flight (see the field notes and comments for 9/16). Four to five days later Cobra left the natal range and became fully independent.

The affinity and grouping behavior among the chicks evolved as they developed. In general, this process involved a gradual loosening of their tendency to stay together as a group. Initially, when the chicks were very young they kept together as a cohesive group. At this earliest stage when the chicks were at their most vulnerable they obviously perceived and appreciated the greater safety and security of remaining in a group (and this certainly made it easier for the parents to look after and protect them). The loosening process started out modestly, this initial phase occurring in the period when the chicks first emerged from the reeds and carrying into the time when they first became mobile—this was a period when the chicks were keeping together as a cohesive group. It began with the chicks sometimes spreading out a bit and with one or another individual occasionally venturing off by itself a short distance into the reeds or to a parent. They soon also began to sometimes break into subgroups.

As the chicks increased their mobility and expanded their ranging this loosening process gradually progressed. However, they tended to regress somewhat during occasions when a parent(s) led and introduced them into new (for the chicks) areas of the territory or areas in which the chicks as yet felt poorly familiar. At such times the youngsters still tended to stay together as a group. For example, in the period when the parents first began taking the chicks into the West Lagoon the contrast in the chicks' grouping behavior when there versus when in the SC was striking (they were very noticeably looser in their grouping behavior and obviously felt more secure when in the familiar confines of the SC). The chicks fairly quickly reached the point where they were commonly split between the parents and all of the chicks were spending some time off by themselves.

The loosening of the grouping behavior steadily progressed in this manner as the chicks spent increasingly more time ranging and foraging independently of the parents. However, while doing the latter they still continued to spend time ranging, foraging and resting in the reeds in company with one or more of their sibs. In general, the chicks remained gregarious and still tended to gravitate toward one another throughout the chick rearing period, though as they aged they didn't do so nearly to the extent that they did when they were young (nor did they do so in as close a manner). This steady loosening of the chicks' grouping behavior was reflected in the fact that the times when all five of the chicks were seen gathered together in proximity became less and less frequent as the chick rearing period progressed. This eventually reached the point where an entire observation period might pass without once seeing all five of the chicks gathered in proximity or even in the same general section of the territory. In addition, in the latter portion of the chick rearing period there were times when a particular chick would be off by itself for most or all of an observation period. Observations of the East Lagoon chicks revealed that, at least for some of the chicks, this affinity persisted into the period after the chicks become fully independent (I regularly saw two of the fully independent East Lagoon first brood chicks keeping company). Alternatively, some chicks largely kept to themselves immediately upon becoming independent. Regardless, based on my regular sightings of paired adults which appears to be the typical adult situation for the species the indications are that (under normal circumstances) eventually each chick will seek a mate to pair with.

It was through the differential administration of rebuffs by the parents (and their behavior in the more directed feedings in the latter half of the chick rearing period) that it became most obvious that the parents recognized each chick as a distinct individual. These parental rebuffs of chicks were mainly delivered in feeding contexts (specifically, in response to food solicitations or after a chick received or snatched a prey item from a parent) and played a major role in the weaning process. As a general observation, this differential behavior by the parents toward the individual chicks was mainly (but not exclusively) related to the particular level that each chick was at in its developmental process. It was basically analogous to the way human parents behave toward and treat an older child differently than they do a younger one (or more generally, the differential manner in which the parents of any species that has dependent offspring of different ages behaves toward and treats their various aged offspring—while the age differences in PB grebe broods is measured in days rather than months or years this seemingly small increment of time can nevertheless be of significance for offspring that develop so quickly). Specifically, Remy and Sinbad treated the more advanced chicks in a manner that fostered their becoming more independent while at the same time catering more to the greater dependence of the less advanced chicks. Stated more generally, the parents applied their weaning behavior on an individual basis to the chicks according to where each chick was at along its developmental progression with their treatment of each chick evolving accordingly as the chick aged. Interestingly however, similar to the manner in which human parents often tend to "baby" the youngest child of their brood, Remy and Sinbad each "babied" and were very much more tolerant with the two (probably) youngest of the brood (i.e., the two who had probably been the last ones to hatch); this was especially notable in the final stage of their rearing.

The first parental rebuffs of chicks were observed on 8/22. From that point they quickly became daily occurrences. Initially the rebuffs were rather intermittent but as the chicks rapidly progressed they became increasingly more frequent. All of the chicks received rebuffs but there was striking variability regarding the frequency and readiness with which Remy and Sinbad doled these out to each chick.

By early September it was already obvious that the two largest chicks, Sparky and Spot, were the targets of parental rebuffs notably more often than the three smaller chicks (though all five of the chicks received rebuffs from the parents). In addition, Quark, Phantom and Cobra were allowed to sit with and solicit a parent without getting rebuffed considerably more often than those two. Despite the rebuffs, all five of the chicks still spent time accompanying and taking food from the parents at that time. This trend of targeting Sparky and Spot notably more often continued right on through their rearing period, with the parents stepping up the frequency of rebuffs and becoming increasing less tolerant of them. Carrying into mid-September the situation eventually reached the point where Sparky and Spot were no longer being allowed to solicit the parents from up close (i.e., no beak to beak solicitations were being allowed) or to sit close next to them. They then had to resort to conducting their solicitations from a short distance away. Furthermore, because the frequency of rebuffs had increased so much they tended to approach and conduct their solicitations in a wary manner, clearly leery of possibly being rebuffed by the parent. Sparky and Spot were still being allowed to take feedings from the parents but at this time the parents were beginning to sometimes be more specific regarding which of the chicks they wanted to be the recipient of their feedings—thus, the parents began to sometimes rebuff these two larger chicks immediately after either had taken a prey item from them. In these cases the parents appeared to be upset after Sparky or Spot had taken the proffered prey item, apparently because the parent had intended to give it to one of the other chicks.

Moving from mid-September into the latter part of the month the parents became increasingly brusque with Sparky and Spot as these chicks approached and reached the point at which they were capable of surviving independently. The parental rebuffs directed at these two became even more frequent; they also became more aggressive and oftentimes more extended and persistent, with some of the rebuffs involving long chases. In addition, the parents began to conduct rebuffs and chases outside of feeding contexts, such as when Sparky and Spot were just sitting at rest or happened to be in the general area. Spot apparently finally left the natal area on 9/27 or 9/28 (Spot was last seen on 9/27). Though independent, Sparky remained in the natal area through the end of September into early October. By the time that Sparky apparently left the natal area on 10/6 or 10/7 (Sparky was last seen on 10/6) Sparky was being chased away essentially every time that the chick was sighted by Sinbad or Remy when they were near enough to do so (at that point Sparky had become fairly adept at avoiding them).

In regard to parental rebuffs and weaning behavior in general, Cobra's situation through her rearing period was basically similar to that of Sparky and Spot but only in respect to how Sinbad behaved toward her. From the time that the rebuffs began, Sinbad's rebuffs of Cobra became increasingly more frequent (though not as frequent as with Sparky and Spot) and he became increasingly less tolerant of her as she progressed toward independence (Cobra nevertheless continued to receive feedings from him until the time she left the natal area). However, this was in stark contrast to Remy's behavior toward and relationship with Cobra. As the chick rearing period progressed Cobra and Remy developed a particular affinity. For some reason, throughout Cobra's rearing period Remy was very lenient toward her. Remy did occasionally rebuff Cobra but generally she nearly always allowed Cobra to solicit her from up close and to closely accompany her whenever the chick sought to do so (of course, for Cobra the basis of this association was to seek and acquire feedings from Remy). Though Cobra was one of the most independent of the brood it nevertheless became common to see her in close association with and accompanying Remy (she also managed to regularly isolate herself with Remy). Even in the latter stage of Cobra's rearing period (when she was spending the great majority of her time ranging and foraging independently) she still spent some time closely accompanying and seeking feedings from Remy. The strong affinity between these two persisted all the way to the end of Cobra's rearing period when Cobra left the natal range and became fully independent (Cobra continued to seek and receive some feedings from Remy until she left). In fact, Cobra was observed closely accompanying Remy on the last day (9/20) that she was seen (i.e., the day before she apparently left the natal range). The contrast between the differing attitudes of the two parents toward Cobra were starkly evident in the several instances in which Sinbad deliberately chased Cobra away from close proximity to Remy. In each of these cases Cobra returned to Remy after Sinbad moved away from the area and continued to accompany her (this being allowed by Remy).

Phantom and Quark started out as the two smallest chicks of the brood and were probably the last ones to hatch (with Quark being the smallest and probably the last to hatch). As the chicks neared their ultimate adult sizes in the latter half of the chick rearing period both surpassed Cobra in size. However, they always remained slightly behind Cobra and their other sibs in their developmental progression. Perhaps being the two smallest and less advanced of the brood was a factor in why the parents were much more lenient toward Quark and Phantom (and Quark even more so than Phantom) over the course of the chick rearing period in regard to rebuffs and in being allowed to closely solicit and accompany them. Quark and Phantom did receive parental rebuff but these generally were more sporadic and they were never specifically targeted as were Sparky and Spot (and sometimes Cobra by Sinbad). There were numerous times when Sinbad or Remy rebuffed one of the other three while simultaneously allowing Quark or Phantom to remain close. There were also occasions when, after handing a prey item to Quark or Phantom and one of the other three was nearby, the parent chased that chick away in order to keep it from possibly stealing the prey item from Quark or Phantom (in a few such instances Phantom was the one who was chased away after the parent had handed Quark a prey item). Overall, Quark and Phantom were somewhat "babied" by the parents (in a similar manner, Cobra was "babied" by Remy but not by Sinbad). These two chicks were the only ones who remained in the inner circle of the family through to the end of the chick rearing period. [Sparky also remained in the natal range until the end but was peripheral to the family by that point. By that time all three of these chicks had been independent (i.e., were capable of surviving on their own) for some time but had remained in the natal range.]

The situations for Quark and Phantom began to narrow through the latter half of the chick rearing period. Quark was the last of the chicks to be allowed to closely solicit and closely accompany both Remy and Sinbad (but eventually only did so with Remy). At that point Phantom was only allowed to do so with Sinbad. There then came a short period when Sinbad ceased to allow this. However, shortly after this he reversed his behavior toward Phantom and again allowed this of Phantom. In the final segment of the chick rearing period two distinct dyads formed within the family: Remy-Quark and Sinbad-Phantom. During this time the two chicks still spent the great majority of their time ranging and foraging independently but continued to seek out, spend some time with, and receive some feedings from the parent of their respective dyads. These dyads dissolved about a day or two before the chick rearing period came to an end at which point the parents ceased to conduct feedings (and the remaining chicks ceased to look to the parents for any feedings) and the parents and chicks ceased to associate.

As with the staggered hatching (and as observed with the West Lagoon chicks) the achieving of independence by the chicks was also staggered (there seems little doubt that these two things are related). Also variable was the fact that in the initial period after reaching independence some of the chicks remained in the natal range and others left it. Regarding the latter, there was still about a month or so left in the breeding/chick rearing season by the time that all of the chicks had achieved independence. Thus, on the surface there seemed no pressing need for the chicks to leave yet and it may even have been somewhat advantageous to remain a bit longer. For one thing, at that late point Remy and Sinbad weren't going to be producing and rearing another brood in this breeding season so there was no need for them to run the chicks out for that reason (though they nevertheless displayed some notable intolerance of the independent chicks, especially Sparky). Also, the SR pair's revised territory (i.e., with the addition of the West Lagoon after having lost the northern half of their territory to the NR pair) had proven itself well capable of supporting all seven of the family members so there was no need for the chicks to leave on that account (and with it being too late for the parents to produce another brood there would be no new mouths drawing food resources from the territory). Furthermore, staying longer would allow the chicks to be a little more developed and gain some more experience at being independent before striking out on their own or in company with another sib(s). It also meant that the chicks would have their parents as a safety net for a little longer (though, of course, at some point the parents would eventually cease doing any feedings). (Note that at least some of the East Lagoon first and second brood chicks continued to reside in and near their own natal area in the period after becoming independent and they were still doing so at the time that this study concluded.)

The dispersal story for each of the chicks is as follows.

Right from the start Sparky was always a little ahead of the other chicks. Thus, it was no surprise that Sparky was the first of the brood to achieve independence. It appeared that Sparky had done so by 9/13. This was on the 50$^{th}$ day from when the last egg hatched (considering 7/26 as the day of the last hatching and counting that as day one). If Sparky had indeed been the first of the brood to hatch then this was on the 54$^{th}$ day after Sparky had hatched (considering 7/22 as the day of the first hatching and counting that as day one). At that point, though still seeking and taking some feedings from the parents it was clear that Sparky was capable of surviving on its own (and was largely doing so): Sparky had already fledged and was spending the majority of the time ranging and foraging independently (in fact, on 9/13 Sparky was even observed to have captured a medium-sized fish—an impressive display of how well developed its skills and abilities were at that point). Despite being capable of flying away and surviving independently, Sparky continued to reside in the natal range. As noted above, there was no pressing need to leave the natal range at that time and this is apparently how Sparky perceived things. Sparky perhaps also perceived that, as the dominant chick of the brood there was some advantage in remaining and continuing to reap the benefits of being able to outcompete its sibs during feedings (and therefore continue to supplement its own foraging with more easily obtained "free" meals from the parents). Even after the parents had greatly stepped-up their rebuffing of Sparky and had begun to prefer that their feedings go to chicks other than Sparky, this chick still managed to regularly get some feedings from them. This was because Sparky became quite adept at hanging around the periphery of a feeding or a foraging parent and sneaking or barging in and taking or snatching the proffered prey item from the parent (or sometimes stealing it from a sib after it had taken the prey item from the parent). Sparky remained in the natal range all the way through the end of the chick rearing period, though in the latter stage Sparky was only a peripheral member of the family. The last day that Sparky was observed to take (snatch) a feeding from a parent was probably 10/2 (the probable prey exchange in this episode occurred out of my view so I couldn't be certain that Sparky had obtained it from a parent; the last day that Sparky was unequivocally observed taking prey from a parent was 9/30). The chick rearing period had definitely concluded as of 10/5. At that time Sparky was still residing in the natal range. Daily observations were continued for five more days with the study concluding on 10/10. Sparky was last sighted on 10/6. It's quite probable that Sparky finally left the natal range at that time but it's uncertain as to whether or not this was the case. For one thing, the four consecutive day period (from 10/7 to the end of the study) during which Sparky wasn't seen wasn't quite long enough to be 100% certain that Sparky had actually left (especially given that Sparky had become a peripheral member of the family and sightings of Sparky had become much less frequent anyway). In addition, by the last days of the study period Sparky's juvenile markings were probably completely gone meaning that I would no longer be able to identify this chick. Thus, I couldn't be certain whether or not any of the unidentified matured chicks that I saw in the natal range during those last days was Sparky.

Spot and Cobra probably achieved independence at least several days after Sparky had done so. They had always been close to each other in their developmental progression and achieved independence at the same time. It is quite likely that they were the second and third of the brood to hatch and that their hatching times were probably close. It seems a virtual certainty that both left the natal range (though not at the same time) and did so not too long after achieving independence. Cobra was the first of the brood to leave the natal range. She achieved independence in mid-September and left shortly after, most likely on 9/20 or 9/21 (she was last seen on 9/20). Thus she apparently left on the 57$^{th}$ or 58$^{th}$ day after the last egg hatched (again, considering 7/26 as the day of last hatching and counting that as day one). She continued to seek and receive some feedings from the parents right up until the day she left. In addition, as noted above she developed a special affinity with Remy and this also persisted to the day that she left (in contrast, by that time Sinbad's intolerance of Cobra had grown considerably and he regularly rebuffed her in the pertinent situations which allowed him the possibility to do so and no longer allowed her to solicit him from up close or to closely accompany him).

Spot remained in the natal range a little longer than did Cobra and apparently left on 9/27 or 9/28 (Spot was last seen on 9/27). This was on the 64$^{th}$ or 65$^{th}$ day after the last egg of the clutch hatched. Spot was the second largest of the brood and even slightly surpassed Sparky in size in the period just before leaving. As discussed above, Spot (along with Sparky) was targeted for parental rebuffs much more often than the other three sibs and the ever increasing intolerance of the parents toward Spot (and Sparky) was much greater. Though Spot continued to seek and take some feedings from the parents right up until leaving the natal range, by that point the parents had become quite intolerant of Spot. They no longer allowed Spot to closely solicit or to closely accompany them, they had begun to oftentimes rebuff/chase Spot immediately after the chick had taken or snatched a prey item from them during a feeding, and they had begun to sometimes chase Spot in situations other than a feeding context (such as when Spot was just sitting at rest or happened to be passing nearby). As with Sparky, Spot had also become adept at hanging around the periphery of a parent who was conducting a feeding or foraging for that purpose and sneaking or barging in and taking or snatching the proffered prey item from the parent.

Phantom and Quark both achieved independence in mid-September, probably at least a few days after Spot and Cobra had done so (with Phantom perhaps doing so slightly before Quark). Despite this, both remained in the natal range through the end of the chick rearing period. In the last segment of the chick rearing period after Spot had left, the brood was essentially down to just these two chicks. Though three chicks (Phantom, Quark and Sparky) still remained in the natal range, Phantom and Quark were the only chicks whom the parents were attending to (Sparky was still residing in the natal range but was living independently and was only a peripheral member of the family at that point). As noted above, in this last segment two distinct dyads formed among the parents and the two remaining "inner circle" chicks. Remy and Quark comprised one of the dyads and Sinbad and Phantom the other. Accordingly, Remy's attentions were mainly focused on Quark and Sinbad's mainly on Phantom. At that point Remy allowed only Quark to solicit her from up close (including beak to beak solicitations) and to closely accompany her while Sinbad allowed this only of Phantom. In addition, each parent was sometimes observed chasing their non-focal chick away from competing with their dyad partner during a feeding situation. Phantom and Quark were both well aware of this situation and readily went to their respective parent while being cautious and somewhat leery about approaching the other parent. Aside from the time that they spent with the parents, Phantom and Quark spent the majority of time ranging and foraging independently. These dyads dissolved a day or two before the chick rearing period unequivocally ended on 10/5 (the last observed parental feedings of chicks occurred on 10/3; note that within the continued five day post-chick rearing daily observation periods both Phantom and Quark were at some point seen being aggressively chased away by their respective former parent dyad partner). It is known for certain that Phantom continued to live independently in the natal range up to 10/8 (he was last sighted on 10/8). Phantom was not seen during the last two days of the study. Thus, for the same reasons noted regarding Sparky's case, it remained uncertain as to whether or not Phantom had left the natal range by the conclusion of the study on 10/10. Quark is definitely known to have remained in the natal range (living independently) right on through to the conclusion of the study.

There is an intriguing possibility regarding Quark's case that came to light during the biweekly follow-up observations that were conducted from the end of the study (10/10) to the end of October. After the first week of these observations, Sinbad and Remy were never again sighted in the SC during this period nor were they even seen going north of the RR bridge. From that point they were seen only in the West Lagoon and the west half of the inter-lagoon channel. The SC then apparently wasn't being defended and was left open for incursion/use by other grebes (of course, with the breeding/chick rearing season having concluded there was no longer any pressing need for stringent defense of the territory, especially considering that there would probably be at least a few times during the upcoming rainy season when the West Lagoon-south river area would temporarily be drastically transformed when rainstorms caused the river to flow strongly and burst through the south end of the lagoon to the ocean).

It was during this period of the SR pair's absence from the SC that a pair of what almost certainly were mature chicks from this breeding season began to regularly reside together in the SC and the section of river to the east of it. The pair consisted of a male and a female. The fact that they were often seen in peaceful proximity to each other and interacted amicably left little doubt that they probably were a mated pair, either newly formed or in the process of forming one. Their regular presence in the SC and surrounding area over this period raises the obvious question of whether they are attempting to stake out a territory here (and if they are still here as the next breeding season approaches will Sinbad and Remy attempt to evict them—providing, of course, that Remy and Sinbad are still here at that time).

What makes this situation particularly intriguing is that the male of this new pair appears as though he could possibly be Quark! Unfortunately, by the time these follow-up observations began Quark and the other SR chicks had fully transitioned into the plumage of a non-breeding adult. In other words, the juvenile markings that were used to identify the individuals were completely gone. Thus, though I suspect that this male may be Quark, in no way can I be certain of this (in fact, my certainty level is no higher than about 60%). In any event, if it was indeed Quark then this would present some interesting insights into the range of possible life history strategies available to young PB grebes as well as into the pair formation process. In addition, it would also raise some interesting possibilities regarding the behavior of parents toward their chicks in the immediate post-chick rearing period (for example, if this chick was indeed Quark did the SR pair deliberately stay away from the SC in order to leave it for Quark's benefit?). Regardless of whether or not that male is Quark, this situation generates some interesting possibilities to explore in future research on this species. [Note: There are two important considerations that are of particular relevance when weighing the possibilities regarding the composition of this pair. The first is that at this point in my familiarity with this species I'm not yet able to distinguish between mature chicks (i.e., those that have transitioned to the plumage and beak characteristics of non-breeding adults) and adults in non-breeding plumage with 100% certainty. Therefore, though the pair noted above appear to me to be mature chicks rather than adults in non-breeding plumage I am not completely certain of this. Second, recall that for the reason previously stated I have not read the literature pertaining to pied-billed grebes and therefore I'm not familiar with the typical life history of individual grebes. Of particular note here is that I don't know whether individuals typically form pair bonds and breed in their first breeding season following hatching or instead go through a non-breeding period of adolescence of more than one year (as is the case with many avian species). Obviously, if it is known from previous observations and studies that the latter is the case then (unless this is an atypical case) this would mean that the individuals of this pair couldn't be mature chicks and this would also exclude the possibility that Quark is one of them.]

To briefly summarize, the dispersal story for the individual chicks as of the conclusion of the study (on 10/10) is as follows. It is almost certain that Cobra and Spot left the natal range. Quark was fully independent but was still residing in the natal range. The cases for Sparky and Phantom are uncertain. Sparky went unseen over the final four days of the study and Phantom went unseen over the final two days. It is thus possible that either or both of them left the natal range. However, given the circumstances noted above it can't be stated with certainty whether or not either of them left (though I feel fairly certain that they did). All that can be stated with certainty regarding these two is that as of 10/6 Sparky was fully independent but still residing in the natal range and that as of 10/8 the same can be stated for Phantom.

Considering possible threats from other animals, the protection and defense of the chicks by the parents in regard to other bird species was discussed in the comments for the 9/4 field notes and the reader is here referred to those (this matter will also be discussed in the following section). As for potential mammalian threats, raccoons and feral cats were definitely seen in the area; foxes, coyotes and opossums are probably also present and perhaps bobcats. None of these were seen during my observations of the family so I never had the opportunity to see how the parents and chicks react to their presence. The only mammals that I observed the family react to with apprehension and alarm were humans (and their contraptions). The family fled to and hid within the reeds on every occasion that a paddleboard or kayak entered and passed through their area. On several occasions they were also driven to take cover in the reeds when a person in the area loudly and repeatedly shouted. In a few cases the family was compelled to flee and hide in the reeds when a person took a dip and swim in the SC. In yet another instance they were caused to hide out in the reeds by a low flying helicopter that was attending to a nearby grass fire. Finally, the chicks typically reacted apprehensively (though usually more mildly so) to the trains that regularly passed over the RR bridge during the course of a day.

Because my observations covered only two breeding seasons and my sample size was so small I can make no definitive statement regarding two of the basic general aspects associated with

PB grebe breeding: the permanence of breeding pairs, and breeding territory fidelity. Still, at the very least the observations were somewhat suggestive.

Do PB grebes form lifelong pair bonds? Three pairs (the West Lagoon, East Lagoon and SR pairs) held territories in my general study area in the 2016 breeding season. Only two of these (the SR and East Lagoon pairs—both remained intact) held territories there in the 2017 season; probably for the reasons previously discussed (see the notes for 6/23 and 7/10) the West Lagoon pair did not stay and nest in the West Lagoon in 2017. Nothing conclusive can be drawn from a paltry sample size of two. Nevertheless, it is noteworthy that both of the pairs who afforded me the opportunity to monitor their composition over the course of the two years were known to have remained together. This suggests that at least some pairs may form lifelong bonds and that this is a credible possibility that could be examined in future long term studies. It should be noted here that I lack any knowledge regarding the average life span of PB grebes. Obviously, this is an important factor regarding the potential for the formation and duration of long term pair bonds (for example, pair bond longevity potential would have to fall within life span limits; also, if mortality was high and average life span relatively short then it would be expected that members of pairs would frequently be lost and surviving pair members would then have to take on a new partner if they wanted to continue to breed—such a situation clearly would impact the potential for forming and maintaining long term pair bonds).

In regard to breeding territory fidelity the hugely extensive breeding range distribution of PB grebes alone intuitively suggests variability in this matter. Two general probable sources of variability in particular became evident to me from my studies here. One such source revolves around whether pairs in the particular breeding area are resident or migratory. PB grebes reside year-round in my general study area. However, in seasonal breeding areas such as those where the waters in the breeding area freeze over in the winter the grebes are forced to migrate away. It seems probable that pairs who are able to reside year-round in or near their breeding territory would be much more likely to be able to hold onto and continue to use that same territory in subsequent breeding seasons than pairs who reside in their breeding areas only seasonally. For example, a pair from a seasonal breeding area might return in the following season with the intent to use the same territory as in the previous season but lose out to a pair that arrived earlier and had already laid claim to it. In contrast, a pair that is residing in their territory year-round has much more control over the situation. Even if the pair is lax in regard to territorial defense and excluding intruders during the nonbreeding season they can still maintain a presence and can attempt to deal with potentially threatening issues that may arise. Then, as the breeding season approaches they are already there to begin the process of reasserting their claim to the territory. In my study area both the SR and East Lagoon pairs displayed fidelity to their respective breeding territories over the two year period (though in the case of the SR pair they retained their core area but were forced to change their overall territory after the NR pair usurped the northern half of their territory).

Note that the breeding pairs in my study area were only seasonally territorial; that is, they were territorial only during the nesting/chick rearing season. As the end of the chick rearing period and breeding season approached and concluded, both of these pairs became increasingly less keen in defending their territories. The frequency of territorial calls diminished and then ceased—Sinbad was last heard calling on 10/12/17 (which was one week after their chick rearing period had concluded) and no other calls were heard from any other pairs in the area thereafter in the aftermath of the breeding season. During casual, brief biweekly checks in the winter (nonbreeding) months neither of the pairs was seen patrolling their respective territories nor were any of them seen excluding other grebes. In fact, I once sighted as many as 11 grebes in the East Lagoon at one time and 5 in the SC at one particular time (also, as noted above the SR pair did not evict that apparent newly formed pair from the SC during this time and were mainly hanging out in the West Lagoon). This territorial "looseness" was seen during the 2016-2017 nonbreeding season but the pairs resumed their keen territorial maintenance when the 2017 breeding season came around (note that this was the only time during my study period that I had the chance to follow the transition from the nonbreeding season to the breeding season).

A second source of variability regarding breeding territory fidelity derives from the relative stability of the particular breeding territory areas. For example, consider the situations of PB grebe pairs who establish breeding territories in permanent lakes and ponds versus those who do so in end-river estuarine areas (such as my study site) which have distinct dry and rainy seasons. Permanent lakes and ponds generally are much more stable from year to year (though water levels may fluctuate) than are estuarine areas such as these which typically experience considerable seasonal changes.

This variability relating to the relative stability of the breeding territory area was clearly evident within my study site. Recall that my study site consisted of the estuarine area at the terminal end of the Ventura River and that the mild climate in this region includes distinct dry and rainy seasons. Reiterating the typical yearly cycle here, during the dry season the flow of the river is so diminished that it typically doesn't break through the sandy berm that forms on the ocean beach (which is built up by the surf due to tidal action) and thus doesn't empty into the ocean. The water then collects (as though behind a dam) and backs up and two lazy lagoons (the East and West Lagoons) form at the river's terminus (a reminder here that the sections north of the RR bridge in my study area that I refer to as the south river and north river sections are actually just a northern extension of the West Lagoon). These lazy lagoons and the habitat that they foster are favored by the PB grebes (especially the dense, reedy habitat that forms along the shorelines of the lagoons). It is during the dry season when these lagoons are well formed and generally stable that they nest and rear their chicks here. Typically, the habitat here undergoes drastic changes (at least periodically) in the rainy season. At that time, periods of sufficient rainfall in the Ventura River watershed will cause the river to flow strongly. On these occasions the river will break through to the ocean in a wide mouth and the West Lagoon (including the south river and north river sections) is transformed from a lazy lagoon into a proper river. This creates some significant changes to the habitat during the relatively short periods that the river flows strongly. One such change is that the water depth is considerably lowered. Also, the area covered by water is greatly reduced as, instead of spreading out as happens when the system is backed up and the lagoon is formed, the faster flowing water keeps to channels. Because of these latter two conditions much of the area that is underwater when the West Lagoon is formed is then exposed and left high and dry—this includes much (and sometimes all) of the vital dense reedy habitat that grows along the shorelines of the lagoon). When this occurs, the habitat of the transformed West Lagoon is no longer suitable for the PB grebe pairs whose territories are situated within the lagoon and they are forced to retreat to nearby suitable habitat. They aren't able to return fulltime until the flow of the river diminishes enough for the West Lagoon to reform—typically, these periods of forced abandonment aren't particularly long, generally lasting on the order of several weeks to a month or so; several such periods may be interspersed during a rainy season (note that when the flow of the river diminishes to a small outlet channel or ceases to get through to the ocean the West Lagoon reforms surprisingly quickly).

The situation for the East Lagoon differs notably from that of the West Lagoon. The West Lagoon is directly in line with and forms the terminal end of the river but the East Lagoon is a side or overflow lagoon. Thus, while the West Lagoon is transformed to a proper river when the river is flowing strongly, the East Lagoon remains a lagoon and remains a suitable habitat. (When the river is flowing strongly this diverts much of the flow that is typically fed into the East Lagoon when the system is backed up. The East Lagoon then becomes much smaller during these periods but it remains a lagoon and sizeable sections of dense, reedy shoreline habitat remain submerged.)

From the above it can be seen that the East Lagoon is relatively more stable from breeding season to breeding season than the West Lagoon. Examining how the three pairs who held territories here in 2016 faired from the 2016 breeding season to the 2017 season provides an example of how the relative stability of the breeding territory area can cause variability in regard to breeding territory fidelity. As previously noted, a large portion of the West Lagoon reed habitat was eroded away so badly during the 2016-2017 rainy season that the West Lagoon was no longer capable of supporting a breeding effort by the West Lagoon pair (see the field note comments for 6/23 and 7/10). It is quite likely for this reason that the West Lagoon pair did not remain and reclaim/re-establish their West Lagoon territory in 2017. By comparison, the East Lagoon was essentially stable. It experienced some periods when it shrank considerably in size during occasions when the river was running strongly but the general habitat wasn't altered and remained intact. This relative stability allowed the East Lagoon pair to remain in their territory and they keenly reasserted their territorial claim for the 2017 breeding season (in fact, they reared three broods during the 2017 season). Unlike the West Lagoon, the "south river" section of the West Lagoon complex (which is somewhat more stable than the West Lagoon) didn't sustain serious habitat damage and the SR pair remained in their territory there for the 2017 breeding season (and reared the brood that was the focus of my 2017 study).

Besides apparently causing the abandonment of the West Lagoon territory by the West Lagoon pair, the changes to the area wrought during the 2016-2017 rainy season and the absence of the West Lagoon pair also affected some changes in the territorial boundaries of the SR pair and in the usage of the area by the SR and East Lagoon pairs. During my 2016 study period the south ribbon of the river was a common (unclaimed) area that was utilized by all three pairs. After the "reset" of the estuarine/lagoon area for the 2017 season this area was claimed by the SR pair as part of their territory. With the absence of the West Lagoon pair the West Lagoon lay unclaimed by any grebe pair through the first part of the 2017 breeding season. It was used as a common area by the East Lagoon pair (especially while rearing their first two broods) and the SR pair until the SR pair later annexed it to their territory after losing the northern half of their territory to the NR pair. The latter, of course, constituted a change in the overall boundaries of the SR pair's territory, though they retained the same core area during both breeding seasons. For the SR pair, the absence of the West Lagoon pair during the 2017 breeding season was quite fortunate because it allowed them a ready option with which to compensate for the loss of the northern half of their territory.

From what was observed in the study area it can minimally be concluded that nesting territory fidelity does at least sometimes occur, at least for some pairs, when the circumstances allow for it (though I couldn't say how commonly it occurs or over how many breeding seasons). Also, that relative season to season habitat stability is one of the enabling factors involved.

***Postscript:*** Out of curiosity I continued to make brief visits to the study site area (usually twice a week) after the SR family study concluded at the end of October, 2017. By the time that I had finished writing this book in August, 2018, some interesting developments had occurred there. The indications are that Sinbad and Remy continued to restrict their ranging to the West Lagoon and western half of the inter-lagoon channel (these are the only places where they were seen on every occasion that they were sighted during this period). However, the pair left the area in early spring and was not seen here again (i.e., they did not stay and reclaim/re-establish their breeding territory and did not breed here in the 2018 breeding season, at least, not as of late August). The latter brings up a point relating to my lack of familiarity with the overall breeding pattern of PB grebes: do pairs typically attempt to breed in each successive breeding season or is it common for them to sometimes take a nonbreeding break between active seasons (and if the latter is so, is this the case with Sinbad and Remy here)?

Interestingly, the new pair which contained the male that possibly was Quark and appeared to be staking a claim to the SR pair's territory (minus the West Lagoon) did end up establishing a territory there (once again, note the important considerations that were discussed above regarding the probable composition of this pair). When the lagoons reset after the rainy season this pair solidified their claim and established a breeding territory in what was the SR pair's original 2017 territory minus the south ribbon of the river—in other words, the section of the river that ran from the RR bridge north to a point about halfway to the Highway 101 overpass bridges, including the SC (of particular note here is that the NR pair left the area long before the start of the 2018 breeding season and did not return—thus they were not present to contest the northern section of the territory that they had usurped from the SR pair which this new pair has claimed as part of its territory). [Recall that the SR pair's original 2017 breeding territory consisted of the SC, the section of the river that ran from the RR bridge north to a point about halfway to the Highway 101 overpass bridges (this, of course, included the section of river east of the SC), and the south ribbon of the river. As was chronicled in the notes, the SR pair subsequently lost the northern half of this territory to the NR pair and annexed the West Lagoon to compensate for that loss.] The new pair made good use of this and produced a brood of five chicks. As of late August four of these chicks were still alive. Reiterating the point noted above, if the male of this pair is indeed Quark then this poses an intriguing question: When Sinbad and Remy left the area prior to the 2018 breeding season was this (as well as the pair restricting their ranging to the West Lagoon after the chick rearing period concluded) a case of the pair deliberately ceding (handing over, passing on) their territory to a chick from their most recent brood? As a further consideration here, it is interesting to note that Quark (who was probably the last of the brood to hatch) was the only one of the brood who was definitely known to still be residing in the natal range when the study concluded.

Another interesting happening here is that a pair of American coots nested within the new pair's territory during this (2018) breeding season. They produced a brood of four chicks and were simultaneously rearing this brood in the territory while the grebe pair was rearing their brood. Parents and/or chicks of the grebe and coot families commonly came within proximity to each other (and sometimes very closely). As is typical of the relations between these two species, in all such observed instances the coots and grebes paid no heed to each other and just went about their own business. As of late August all four of the coot chicks were still alive and appeared to be doing well.

The situation in the West Lagoon was also intriguing. The extra-strong El Nino event in the 2016-2017 rainy season brought only a brief respite from the severe drought that has been plaguing this region. When that rainy season ended, the severe drought conditions resumed. As a consequence the 2017-2018 rainy season once again was a poor one, though there were several periods when sufficient rain fell to cause the river to flow strongly and break through to the ocean (and thus transform the West Lagoon from a lazy lagoon into a proper river during these periods). What is of interest here is that during these periods the river didn't flow nearly as strongly or for as long a total a period as it did during the previous rainy season. Of particular note, it didn't scour and erode away the east shore of the lagoon as happened during the 2016-2017 rainy season. Recall that during the latter rainy season the lush, dense, extensive growth of reeds that had lined the east shore south of the mouth of the inter-lagoon channel was completely obliterated and washed away by the erosive force of the river (in contrast the dense, extensive growth of reeds along the west shore—the inside bank of the river—remained intact). As a consequence, the comparatively small West Lagoon (now minus the vital reed habitat of the east shore) almost certainly no longer contained sufficient resources to support a nesting/ breeding attempt within the lagoon and neither the West Lagoon pair nor any other grebe pair attempted one during that breeding season (fortunately for the South River pair the West Lagoon thus was available for them to annex to their territory after losing the northern half of their original territory to the North River pair). [Note that the results of the West Lagoon pair's 2016 breeding effort suggest that the resources in the West Lagoon were somewhat limited anyway. The fact that two of their six chicks were lost, very likely due to starvation, perhaps indicates that the lagoon is only capable of supporting a brood of four chicks or so even under the best of circumstances.]

By the time that the 2018 breeding season came around the east shore had benefitted from a year of recovery time, thanks to its having been left intact by the relatively less strongly flowing river during the preceding rainy season. During that time the vegetation began regenerating and the east shore was once again covered with dense plant growth, though only some small patches of reeds (cattails and bulrushes) had yet grown back. In addition, the prey base along the east shore and the lagoon in general faced less drastic conditions on its road toward replenishing its abundance. As a result, a new grebe pair established a territory in the West Lagoon a short time after Sinbad and Remy left, apparently having determined that it now contained sufficient resources for a nesting/breeding attempt. This new pair produced a brood of four chicks. Unfortunately, as of late August the brood was down to two chicks. It should be noted that the pair's territory also included the western half (and perhaps all) of the inter-lagoon channel. In fact, from my limited observations of this family it appeared that the pair was using the inter-lagoon channel (for foraging for the chicks) much more heavily than did the West Lagoon pair in 2016. The indications are that, though the resource conditions in the West Lagoon are recovering, it probably won't be able to again stand mainly on its own as a respectable breeding territory until (or unless) the dense growths of reeds along the east shore have been re-established.

Observations even over this short three year period have illustrated how changeable the habitat conditions can be from one breeding season to the next (the changes can range from subtle to dramatic) in a habitat such as this estuarine area which undergoes such distinct seasonal changes.

# *Chapter 6*

## Chick Rearing In the West Lagoon and South River Families: A Comparison

It is interesting to look at some of the similarities and differences in regard to the parenting and the interactions among the chicks that were observed in the West Lagoon and SR families. Among the similarities are basic elements that are common to PB grebe breeding in general in this particular estuarine area such as: forming a monogamous pair bond (which is the basic breeding unit for the species); conducting breeding/chick rearing during the dry season (spring, summer, early fall); the establishment of a well-maintained breeding territory in fairly expansive, quiet waters which are generously ringed along much of their shorelines with continuous dense growths of tall, emergent reeds.

It appears extremely likely that another common aspect of PB grebe breeding in this particular location is that the pairs build their nest within the dense growths of emergent reeds. The only nests that I had the chance to observe during my studies in this specific estuarine area were the two that were used by the SR pair (the one that contained the clutch that they lost and the one in which the brood successfully hatched). Both of these were built within the dense reeds. When I first discovered the West Lagoon brood the chicks were already mobile and were no longer using their nest. However, the fact that the young chicks were using a section of reeds just north of the snags along the mid-west shore as a kind of home base area suggests that their nest may have been located within the reeds there. [Considering the extremely large geographic distribution of PB grebes, the variety of the types of bodies of water that they utilize (such as estuarine areas, ponds, lakes, calm sections of river) and the accompanying variety of vegetation found along the shorelines and open water sections of these various bodies of water throughout their large range it would hardly be surprising to find variability in the general types of locations in which pairs choose to build their nests. In fact, I found this to be the case even just within Ventura. In one of the large ponds at a water treatment facility in town I discovered four PB grebe nests whose locations were in stark contrast to those in my study site. Rather than being well concealed within dense growths of emergent reeds that ringed the shorelines, these four floating nests were located out in the open, well away from shore. The nests were situated within continuous, dense carpets of low, floating surface vegetation a considerable distance out from the shore where they were easily visible (they were essentially out in the open—the nests rose quite conspicuously above the height of the low lying extensive carpet of vegetation). The shorelines of this large pond were well and densely vegetated. However, they lacked the dense, extensive, continuous growths of emergent cattail and bulrush reeds such as those present in my study area. There were some modest, patchy sections of this but most of the shoreline vegetation consisted of lower growing reeds (whose stems grew out from a central mass, mainly in an arced pattern rather than vertical) and closely clumped bushes and trees as well as dense growths of various weeds.]

The indication from my observations is that another common element of the PB grebe reproductive process is that the hatching of a clutch is staggered over a period of days. The SR pair's clutch is the only PB grebe clutch that I have had the opportunity to monitor over its hatching period. Hatching for this clutch was staggered over a period of 5-7 days (with 5 days being most likely). After hatching had concluded it was seen that there was very notable size disparity among the chicks of the young brood—this almost certainly having resulted from the staggered hatching. When I first discovered the West Lagoon brood and began my study of that family the chicks had already hatched and had probably only recently become mobile. However, similar to what I observed with the SR brood in the following breeding season, there was considerable size disparity among these young chicks. Thus, having actually observed staggered hatching in a PB grebe clutch, in retrospect it is probably a safe inference that hatching was staggered over a period of days for the West Lagoon brood also (the East Lagoon broods also had notable size disparity indicating that staggered hatching is probably the norm for PB grebes, though perhaps the situation may be variable for small clutches containing just two or three eggs).

In considering what the adaptive advantage of this staggered hatching process is (i.e., why this system evolved in this species) I lack the relevant information to do anything more than speculate on this matter. One possibility that comes to mind is that perhaps this system evolved as an adaptation for ensuring the survival of the optimum number of chicks within a brood that is being reared in a situation in which there is "uncertainty" as to just how many of the chicks the food resources within the territory is capable of supporting. In the scenario being envisioned here consider a situation in which a brood of 5 chicks is produced within a territory where (unbeknownst to the parents) the food resources are actually only sufficient for the survival of 4 of the chicks and this, only if the lion's share of the feedings go to 4 of the chicks; that the food resource situation is such that if the feedings are distributed more evenly among the 5 chicks the food distribution will be diluted to the point where ultimately only 3 of the chicks will survive through the chick rearing period. In such a scenario natural selection would thus favor a system which created a situation in which this optimum number of chicks will obtain the requisite lion's share of the feedings. One such type of system would be one that produced a very strong dominance hierarchy among the brood in which a significant trickle-down effect of the feedings existed whereby each successive dominant individual as you go down the hierarchy is capable of securing a lion's share of the feedings over its subordinates. As was observed in these studies (and as was especially evident with the West Lagoon brood in which the two smallest, most subordinate chicks were lost ) staggered hatching that occurs over a period of days is capable of producing a strong dominance hierarchy such as this. Once again, this is all merely speculative on my part and this is but one of the possible hypotheses regarding the adaptive nature of this staggered hatching. Ultimately, it will take a focused, long term study to solve this matter.

The two broods contained a nearly similar complement of chicks. The SR brood contained five chicks and the West Lagoon brood had six when first discovered (this most likely was the full complement after hatching but it can't be ruled out that one or more chicks could have been lost in the period before I first saw this brood). While there was considerable size disparity among the young chicks in both of these broods, the situation was notably more marked in the West Lagoon brood. Two of the chicks in that brood were particularly smaller than the others: one was runty and the other nearly so. In contrast, none of the SR brood chicks were runty. Neither of the two small West Lagoon chicks was able to survive. Ultimately, the West Lagoon pair successfully reared four of their six chicks to independence while the SR pair reared all five of their chicks to independence.

Both members in each pair shared equally in the feeding of their respective broods and for both families it was common to see the chicks split between the two parents. The physical manner in which the feedings were conducted was similar for each pair. For both pairs crayfish was the predominant prey item that they fed to their chicks. However, aside from this overall similarity the pairs did differ in regard to what they fed to their chicks. The West Lagoon pair's feedings consisted almost exclusively of crayfish (only four of the many observed feedings involved a prey item other than crayfish). While crayfish were the predominant prey item fed by the SR pair their feedings also consisted of a notable percentage of fish (see the comments at the end of the field notes for 10/10).

Perhaps as a consequence of this difference in the type of prey that was fed, the foraging habits also differed somewhat between the pairs. The West Lagoon pair did their foraging nearly exclusively along the edges of and within the reeds. The SR pair also did most of their foraging along the edges of and within the reeds but they also regularly did some foraging in open water. (It is interesting that as the chicks of the respective broods aged and spent increasingly more time foraging independently they generally mirrored their parents in regard to their foraging habits.) Whether these differences between the pairs in foraging habits and in the type of prey captured and fed to the chicks are related to individual differences and/or preferences or were related to differences in the abundance, relative availability, relative ease of capture, etc. of the various prey I can't say.

During the first third or so of their chick rearing periods the parents of the SR family were more often seen leading their entire brood to various destinations than were the West Lagoon parents. This almost certainly was due to differences in the size and layout of their respective territories. The

SR pair's territory was notably larger and its two main areas (the SC and the West Lagoon) were relatively distant from each other. Also, when a parent was in one of these areas it wouldn't be able to see and monitor any chicks that were in the other main area. In addition, initially (i.e., until the chicks became more independent in their ranging and foraging) the chicks clearly felt much more comfortable and secure in the more familiar confines of the SC (the place where they hatched and first began ranging about) than in the strange and less familiar West Lagoon. Under these circumstances the parents (either singly or together) led the chicks through an introductory period to the West Lagoon during which they typically took the entire brood over with them (and given the initial unease of the chicks regarding the West Lagoon they were more inclined to stay together as a group and ensure that they followed the parent(s) when it left the lagoon). The situation for the West Lagoon family obviously was quite different in this respect because their territory was comprised almost solely of the West Lagoon (it also included the western half of the inter-lagoon channel) where the chicks quickly felt at home (though they initially kept to the southern half of the lagoon). For PB grebes (and especially for chicks) the West Lagoon was also a much more sheltered place during the 2016 breeding season when it was occupied by the West Lagoon family than in the 2017 season when it was utilized by the SR family (see below).

Overall, the chicks from the West Lagoon and SR broods employed the same general strategies for obtaining food from the parents (see the comments at the end of field notes for 8/28). Also similar is that in both broods the chicks often solicited the parents for food and did so in the same manner either from up close (including beak to beak solicitations) or from a short distance away from the parent (in both broods the latter type solicitations became increasingly more frequent after the parents began rebuffing chicks).

Though the West Lagoon was utilized by both families during the respective breeding seasons in which they were observed there was a considerable difference regarding the areas of the lagoon that were used by the chicks (and the parents) of these two broods. The chicks of the West Lagoon brood regularly used the areas along both the east and west shores of the lagoon while the SR chicks basically kept only to the west shore. This difference is undoubtedly related to the difference in the distribution of the shoreline reed habitat that existed in the lagoon between the 2016 and 2017 breeding seasons. As has previously been noted, the chicks generally kept to the areas near and within the dense, continuous growths of emergent reeds (especially when younger). Also as previously noted, in 2016 when the West Lagoon brood was being reared there, both the east and west shores of the lagoon were completely lined with dense, continuous growths of reeds and this distribution was reflected in the ranging done by the chicks. During the 2016-2017 rainy season the shoreline reed habitat along the entire east shore (with the exception of the reeds at the mouth of the inter-lagoon channel) was devastated, having been eroded away by the river during the periods when it was flowing strongly into the ocean. Thus, in 2017 when the SR chicks were being reared there only the west shore was lined with dense, continuous growths of reeds and the chicks mainly ranged along and within these west shore reeds. [As they began to range more widely and independently they also began venturing over to what remained of the east shore reed habitat at the mouth of the inter-lagoon channel, doing so either alone or with a parent and/or other sib(s).]

One of the most conspicuous differences seen between the SR and West Lagoon parents was the way in which they reacted to certain other bird species that ventured into or over their territory during their respective chick rearing periods. The SR pair generally was rather sensitive regarding the presence of other large (or fairly large) birds in the general vicinity of their chicks. On a number of occasions the pair was observed reacting at least mildly (in a defensive manner for the sake of the chicks) to the presence of individuals or groups of birds of various species (see the comments for the 9/4 field notes). Their reactions included: voicing alarm calls to the chicks; leading the chicks to cover or otherwise causing them to do so on their own; chasing away or getting in between the bird(s) that was perceived as a possible threat and their chick(s) and facing it down; attempting to draw attention away from the chicks to itself instead; and "rescuing" the chicks by retrieving and leading them away from the proximity of the other bird(s). Some of the birds whose presence they reacted to (such as western gulls and great-blue herons) perhaps genuinely were potential hazards to the chicks. However, most of them posed no real threat to the youngsters (the latter included ospreys, turkey vultures, double-crested cormorants, American coots, mallards, belted kingfishers and even a Forster's tern). This was in sharp contrast to the West Lagoon pair who were never observed to react to any other bird species—this despite the common presence in the lagoon of species such as the coot, osprey, double-crested cormorant and mallard whom the SR pair had reacted to. It appeared that the West Lagoon pair recognized that these particular birds posed no threat to the chicks and they thus ignored them accordingly. Note that I never saw any western gulls or great-blue herons in the lagoon during my observations of the West Lagoon family in the 2016 breeding season. Therefore I never had the opportunity to see how the West Lagoon parents would have reacted to birds who possibly did pose an actual threat to the chicks. [In regard to the reaction of the SR pair to the birds listed above it should be noted that for some of those species their reaction was variable—that is, they didn't react on every occasion that one of them ventured within general proximity of the chicks. For example, they did so in only a handful of the many occasions when coots were near the chicks and it was only Sinbad who reacted in this manner toward coots.]

The reactions of the chicks in these two families toward these various bird species reflected the attitudes of their parents. Just as their parents did, the chicks of the West Lagoon brood ignored those other birds as they went about their daily business. Similarly, the SR chicks reacted with apprehension and alarm to most, but not all, of the bird species that their parents reacted to (the most notable exception is that the chicks didn't perceive the coots as a threat and basically ignored them, as did the West Lagoon family). However, as the SR chicks grew and matured their attitudes toward some of these other birds changed (for example, as the chicks progressed toward adult size they reached a point where they were no longer apprehensive of western gulls). Though it can't be stated with certainty, the similar reactions (or lack of reaction) of the chicks of each brood to that of their parents suggests that their reactions/attitudes may have, at least in part, been learned from the parents by taking cues from the behavior of the parents toward these various species.

In both the West Lagoon and SR broods the competition between chicks during feedings was intense and a dominance hierarchy developed among the chicks. Within the two broods this competition was basically similar in its general nature and form and in the manner in which it developed as the chicks aged (this competition was described in detail in the previous section and the reader is referred to that discussion). However, though basically similar the competition within the broods did differ in particular details. The most notable such difference was as pointed out in the previous section and involved the relative frequency of specifically directed aggressive actions among the chicks: "*For the SR brood these direct aggressive actions taken by more dominant chicks toward subordinate ones in competitive feeding situations were notably less frequent and played a lesser role compared to what was observed among the West Lagoon chicks during my observations of that family. Since becoming mobile the feeding competition among the SR brood consisted more prominently of spotting and reaching a provisioning (prey bearing) parent first or barging in and muscling competing sibs aside.*"

As observed with both the West Lagoon and SR families parental rebuffs of chicks were an important element of the weaning process. The manner in which the parents conducted these rebuffs was basically similar for both pairs. The general progression of this behavior was also similar: once the rebuffs began (about four weeks after the last chick hatched for the SR pair and perhaps a bit later for the West Lagoon pair) they quickly became daily occurrences and increased in frequency and intensity as the chicks aged. Another similarity is that the parents in both pairs targeted specific chicks in their respective broods notably more frequently (this was particularly so regarding Sparky and Spot in the SR brood and in general, the more advanced chicks) while being notably more lenient toward certain others.

There were some differences observed between the two pairs concerning their rebuffs but these were mainly a matter of degree. In the latter segment of their chick rearing periods the SR pair exhibited a much higher level of intolerance toward certain chicks than was seen with the West Lagoon pair. This was especially noticeable with regard to Sparky—over the final portion of the chick rearing period Sparky was being chased away essentially every time that this chick was sighted by the parents on the occasions when they were near enough to do so. The SR pair also became notably more persistent and aggressive with their rebuffs than the West Lagoon pair (this included an increasing tendency to rebuff certain chicks even immediately after having handed a prey item to them). It appears certain that a major factor (if not the sole factor) behind these latter differences is the fact that the chicks in the SR brood generally stayed in their natal range considerably longer after becoming independent than did the West Lagoon chicks (in other words, these lengthier stays tended to foster increasing intolerance in the parents, at least in regard to certain chicks).

This latter point was one of the more notable differences between these two families. All of the surviving West Lagoon chicks left their natal range shortly after becoming capable of surviving on their own. Cobra was the only chick from the SR brood to do so and possibly as many as three of the chicks were still residing in the natal range at the time that the study was concluded on 10/10 (one chick, Quark, was definitely still residing there). This difference is very likely related to the different times during the breeding season that each brood was produced and reared. The West Lagoon chicks became independent (i.e., were capable of surviving on their own) near the end of the breeding season whereas there was at least a month or so remaining in the breeding season when the SR chicks became independent (note that some of the East Lagoon first and second brood chicks also remained in their natal range long after becoming independent, which in each case was well before the end of the breeding season). The probable factor operating here is it seemed that, for whatever reason, the chicks felt more compelled to disperse away by the conclusion of the breeding season or shortly thereafter.

Because these studies were strictly observational, for many of the differences between the families that are noted above I can make no definitive statements regarding whether they were attributable to individual differences in the parents and chicks of the respective families, differences in the parenting "styles" of the two pairs, differences relating to the time of the breeding season that the broods were produced and reared, or differences relating to various environmental (physical and biological)/habitat factors and conditions.

As a general observation, the coordination and cooperation among the individuals of these two pairs as they reared their chicks was very impressive (and from a qualitative perspective was quite beautiful to watch). Furthermore, the sharing of parental duties displayed by these two pied-billed grebe pairs in aspects such as nest construction/maintenance, incubation of the eggs, babysitting the chicks at the nest, feeding and protection of the brood, and weaning of the chicks was about as equitable as can be seen among the many avian species.

# Epilogue: Some Final Thoughts

During their respective study periods the West Lagoon and South River pairs added a combined total of nine new pied-billed grebes to the world. The extreme dedication and hard work that the parents devoted to the rearing of these chicks was quite remarkable. However, after achieving independence there was nothing more that the parents could do to help them. The chicks are now on their own and I can't help but wonder how they and new crops of chicks in the coming years will fare in this heavily human-impacted world; a world in which the deleterious impacts by our species continue to mount.

Currently, pied-billed grebes are doing well overall (but see below) and enjoy a wide geographic distribution. Their future may seem secure but there are many examples of species that seemed similarly secure and are now imperiled or extinct due to human impact. The obvious lesson here is that no species' situation can be taken for granted. New, unmonitored and unforeseen human-created threats continually appear and can seriously impact species that had once seemed secure. Some of the more recent of these threats such as human-caused climate change are potentially devastating. As with many children who are interested in wildlife and the natural world, I watched numerous documentaries and often read about African wildlife while growing up. At that time I would never have guessed that iconic African species such as the lion, African elephant and the giraffes would be facing uncertain futures during my lifetime. And fifty years ago there are few people who would have predicted that the polar bear, a species that lives in one of the most sparsely human populated areas on the planet, would also be facing an uncertain future as a consequence of human activities (in this case, due to human-caused climate change).

While on the subject of lost or imperiled species whose continued existence seemed secure one can't help but recall the most infamous example of all: the human-caused extinction of the passenger pigeon. This bird was likely the most gregarious avian species of all time. It was often seen to gather in flocks of astounding size numbering in the millions. There was even a report of a flock that was estimated at 2.2 BILLION! Such mega-flocks were known to blacken the sky and pass overhead in a continuous flow over the course of an entire day. The night roosts utilized by these mega-flocks reached staggering proportions. One such roost reported by the famed naturalist John J.

Audubon was estimated to cover an area forty miles in length by over three miles in width—in other words, an area of over 120 square miles! It seemed unimaginable that a species which existed in such incomprehensible abundance could be lost. Yet, humans drove even this impossibly numerous species to extinction and did so essentially within a thirty year period.

The troubling reality is that in this time of relentless human population growth and our ever worsening impact on the environment it is now necessary to maintain vigilance on all species, including those whose futures currently give the impression of being secure. Thus, despite the present comfortable situation of the pied-billed grebe with its extensive geographic distribution one cannot offhand rule out the possibility that sometime in the near future some human-related cause(s) may send the species into a precipitous decline toward extinction. Our species has created a global situation in which one can't just dismiss the possibility that the pied-billed grebe may be thrust into a pattern of decline similar to what has been seen among other species that once had extensive geographic distributions: a pattern in which the overall population became fragmented into smaller, isolated populations that declined to the point where they were no longer biologically viable and, one by one, began to disappear. In fact, an ominous pattern of decline has already appeared in the northeastern U.S. The situation there has progressed to the point where, for example, the pied-billed grebe has been listed as endangered by the states of New Jersey, Connecticut and Massachusetts, as threatened by New Hampshire, and as a species of special concern by Vermont. A major cause of their decline in this region is habitat degradation and destruction. In light of this disturbing regional trend, given the track record of our species one can't help but get that sinking feeling of déjà vu.

Of the many human-created environmental problems, perhaps the one whose full range of potential impacts (including its potential impact on each species on a case by case basis) is most unclear is that of human-caused climate change. While numerous of the currently observed impacts of this crisis are things that had been predicted, there are also impacts which had not been foreseen. The fact is that human-caused climate change is a novel phenomenon. It is something that we have never previously experienced and there is much about its effects and consequences that we simply don't yet know about. It is quite probable that, in addition to the potentially devastating problems that we are currently aware of, there may be some horrific impacts that we never saw coming, things that we have never even considered due to our lack of knowledge.

There is much that needs to be learned (and as soon as possible) about this serious global problem and this is why research on this matter is so crucial. Yet in the United States (an extremely important player in this research) the environmentally unconscious, corporate and fossil fuel friendly Trump regime has significantly squelched governmental research on human-caused climate change and is intent on snuffing it out altogether. Trump (himself an ardent climate change denier) has even gone the despicable route of appointing fellow climate change denying corporate and fossil fuel cronies to important environmentally related cabinet posts. He has completely undermined the Environmental Protection Agency, rendering it impotent in regard to performing the agency's stated purpose of protecting public health by ensuring a safe and healthy environment, largely through regulating pollutants and other environmentally harmful human agents and activities (i.e., essentially to protect us from ourselves). Such callous disregard reflects the general lack of character displayed by our species. We are well aware of the climate change and mass extinction crises that humans are perpetrating and yet we have chosen to do nothing of true significance to address them. Not only aren't we taking truly serious actions to try and halt these crises, we are also attempting to eliminate, circumvent, emasculate and undermine the paltry few measures and laws that might aid in combating these devastating problems that we have unleashed upon the planet. And what is the purpose behind all of this devastation, chaos and death? Certainly not for anything noble—just simple greed, profit seeking and selfish human gratification.

All throughout my studies of these two grebe families the signs of human impact on their world were ever present. The waters in which the pairs and their chicks swam, dove and lived their lives contained an endless supply of human-made trash and debris of all kinds, notably plastic bags and bottles. There was even a mostly submerged shopping cart and some bicycles. Some of this refuse posed a genuine threat of possibly trapping and drowning a diving grebe. Physically, the water itself looked filthy and is known to contain various pollutants. [With the amount of agriculture that is conducted within the watershed area it is hardly surprising that some pesticides and fertilizers wash into the river and are among the pollutants. If bioaccumulation of any of the chemical pollutants occurs then the predatory grebes regularly take in more concentrated doses of these through the fish and crayfish that they eat.] I would never dream of taking a drink from these waters unless they had undergone major purification treatment. Nor would I even consider eating a fish that had been living here. In regard to other forms of human impact the grebes' activities were interrupted periodically throughout each day when they were spooked by the noisy trains that regularly passed across the railroad bridge. On numerous occasions their activities were also interrupted by paddleboarders. A number of oil rigs are in the area offshore and tar balls regularly wash up on the beach at the south end of the West Lagoon. In addition to all of this was the increased threat of predation posed by the numerous feral cats that inhabit the area.

It was sad and disturbing to watch the grebe parents and their chicks swimming and diving amidst the trash and having to deal with all of the other human-related issues that impacted their lives. Rearing chicks is a difficult enough endeavor without the additional human-created impediments being piled onto this crucial task. Unfortunately, this is merely a microcosm of what is happening on a global scale. It's a reflection of the astounding lack of responsibility that humanity has shown for its actions in regard to the environment and the millions of other species that share the Earth with us. Humanity is quite enamored with itself. People around the world love to pontificate about how great, magnificent and benevolent we are; about our exalted preeminence among all of the other lowly life forms. What most people won't acknowledge is that this is an extremely biased, self-serving view espoused by members of a rather characterless species that generally refuses to take a brutally honest look at itself; a species that prefers the cheap gratification of fallacy over hard truths. And there are no outside assessments to challenge our own because no other species on the planet has the capacity to voice its opinion of us. Yet this isn't really necessary. A simple, honest look at the reality of what our species is doing to this planet and its other life reveals an image of humanity that is decidedly less than flattering. The most crucial of these realities are the human-caused mass extinction crisis and human-caused climate change. Both of these crises are ongoing and are becoming increasingly more desperate. They require immediate, significant, decisive actions by us. Instead, our response has been essentially token when weighed against the immense magnitude of these crises. Our callous disregard and lack of significant remedial action has allowed these crises to progress to the point where the best that we can do now is damage limitation. A mounting toll has already been exacted by each crisis. The only question now is how severe humanity will allow the ultimate toll to be?

# Diagram of Study Area

***Note:** This is only a rough diagram, taken from my field notes. It is not drawn to scale nor is it entirely accurate in its shape and proportions. It is meant only to illustrate the overall layout of the study area.

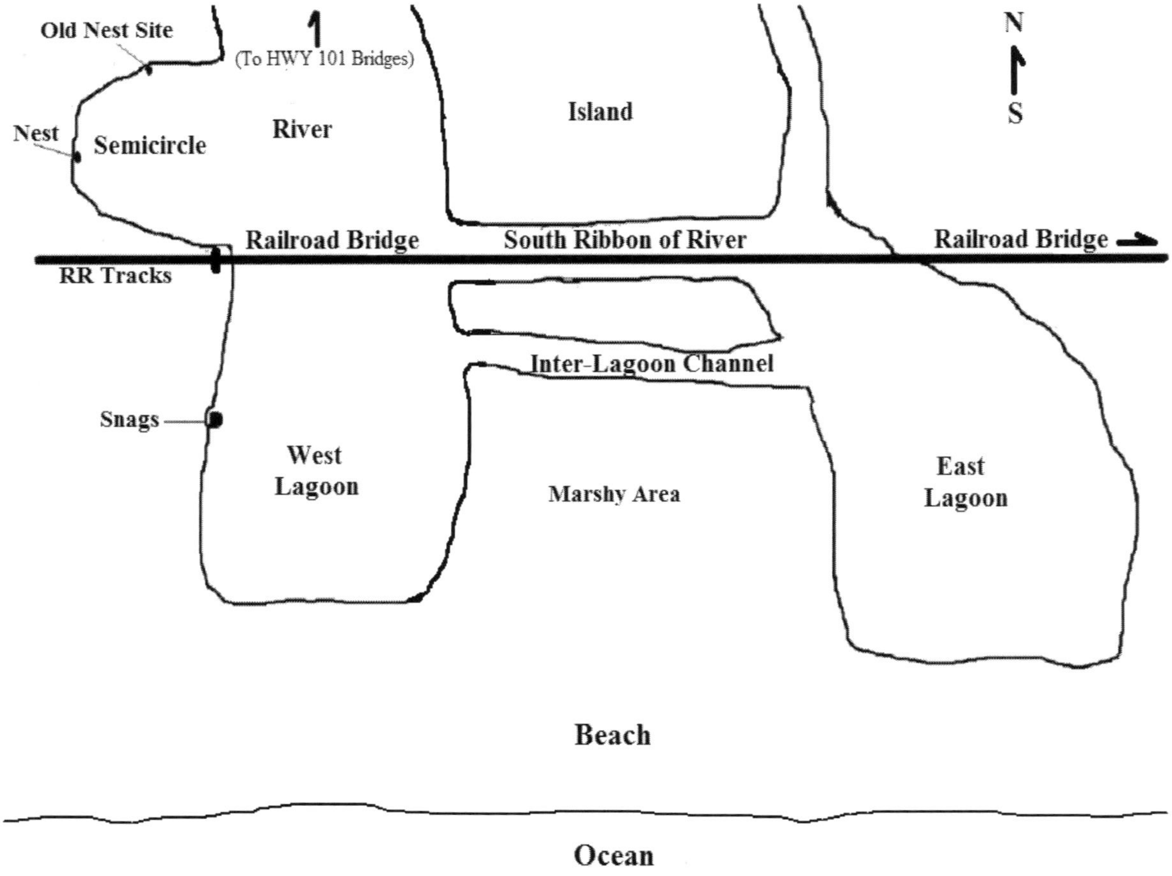

# Timeline of Significant Events for the South River Family

6/9/17 – The pair's nest is discovered.

6/13/17 – It is confirmed that eggs are present in the nest. Daily observations of the pair begin.

6/21/17 – The pair's nest and clutch are lost due to a sudden, dramatic rise in the water level.

6/23/17 – The first day that the pair is observed constructing their new nest.

6/24/17 – The first observed copulation is seen.

6/26/17 – A second copulation is observed.

6/27/17 – A third copulation is observed.

6/28/17 – A fourth copulation is observed.

6/29/17 – The first egg of the second clutch is seen in the nest.

7/2/17 – A fifth copulation is observed (note that this one has occurred after egg laying had begun).

7/8/17 – The first observed territorial encounter between the South River and North River pairs is observed (this occurred at the north end of the South River pair's territory before they later lost this section to the North River pair).

7/22/17 – Hatching begins—one chick is seen.

7/23/17 – The territorial conflict with the North River pair begins. Two savage fights between members of the pairs are seen in the South River pair's territory.

7/26/17 – Hatching concludes (on 7/26 or 7/27, most likely on 7/26).

7/28/17 – A fight is observed between Remy and the North River male in the South River pair's territory.

7/29/17 – The first observed territorial standoff between Sinbad and the North River male occurs (some fighting between them also occurs). This takes place at the new (revised) territorial boundary and provides the first confirmation that the South River pair has lost the entire northern half of their territory to the North River pair.

8/1/17 – The chicks are seen venturing outside of the reeds for the first time.

8/5/17 – The chicks become mobile. A second territorial standoff between Sinbad and the North River male is seen.

8/6/17 – A third territorial standoff between Sinbad and the North River male is seen.

8/7/17 – The last day that the parents were observed spending any time sitting on the nest guarding and babysitting the chicks. This thus concluded the phase during which the parents utilized the nest in their chick production/chick rearing endeavor.

8/8/17 – A fourth territorial standoff between Sinbad and the North River male is seen.

8/9/17 – The chicks are seen in the West Lagoon for the first time.

8/12/17 – A fifth territorial standoff between Sinbad and the North River male is seen (some fighting also occurs between them).

8/15/17 – A sixth territorial standoff between Sinbad and the North River male is seen. The first specifically directed aggressive attack among the chicks during feeding competition is seen.

8/22/17 – The first parental rebuffs of chicks are observed.

8/28/17 – A seventh territorial standoff between Sinbad and the North River male is seen. The territorial border has been pushed slightly farther south by the North River pair.

9/5/17 – The first observed flights by a chick are seen (Quark and Cobra are seen taking brief, short flights).

9/10/17 – An eighth territorial standoff between Sinbad and the North River male is seen (this is the last territorial standoff that was observed).

9/16/17 – Two long flights by Cobra are observed in which she gained altitude during the flights.

9/20/17 or 9/21/17 – Cobra leaves the natal range and becomes fully independent (Cobra was last seen on 9/20).

9/25/17 – Phantom is observed taking two long flights.

9/27/17 or 9/28/17 – Spot leaves the natal range and becomes fully independent (Spot was last seen on 9/27).

10/5/17 – The chick rearing period has concluded.

10/10/17 – The study concludes—daily observations end and biweekly follow-up observations begin.

10/30/17 – Biweekly follow-up observations conclude.

**Note about the author:** Andrew Abate received his master's degree from Arizona State University. His career as a wildlife biologist has included work with some of North America's most endangered species such as the black-footed ferret, the California condor, the southern sea otter, the island fox and Kemp's Ridley turtle.

Printed in Dunstable, United Kingdom

91218259R00279